Vectors of Plant Pathogens

Vectors of Plant Pathogens

EDITED BY

KERRY F. HARRIS
Department of Entomology
Texas A & M University
College Station, Texas

KARL MARAMOROSCH
Waksman Institute of Microbiology
Rutgers University
New Brunswick, New Jersey

ACADEMIC PRESS 1980
A Subsidiary of Harcourt Brace Jovanovich, Publishers
New York London Toronto Sydney San Francisco

ACADEMIC PRESS, INC.
111 Fifth Avenue, New York, New York 10003

United Kingdom Edition published by
ACADEMIC PRESS, INC. (LONDON) LTD.
24/28 Oval Road, London NW1 7DX

Library of Congress Cataloging in Publication Data

Main entry under title:

Vectors of plant pathogens.

Includes index.
1. Insects as carriers of plant disease.
2. Invertebratea as carriers of plant disease.
I. Harris, Kerry F. II. Maramorosch, Karl.
SB931.V28 632'.3 80-17470
ISBN 0-12-326450-2

PRINTED IN THE UNITED STATES OF AMERICA

80 81 82 83 9 8 7 6 5 4 3 2 1

CONTENTS

List of Contributors xi
Preface xiii

Chapter 1. Aphids, Leafhoppers, and Planthoppers **1**
Kerry F. Harris
1.1 Introduction 1
1.2 Aphids 2
1.3 Leafhoppers 9
1.4 Planthoppers 11
1.5 Conclusion 11
1.6 References 11

Chapter 2. Mealybugs **15**
Osmo Roivainen
2.1 Introduction 15
2.2 Transmission Characteristics 19
2.3 References 34

Chapter 3. Whiteflies **39**
V. Muniyappa
3.1 Introduction 39
3.2 Whitefly Vectors 41
3.3 Yellow Mosaic Diseases 45
3.4 Yellow Vein Mosaic Diseases 54
3.5 Leaf Curl Diseases 57
3.6 Mosaic Diseases 65
3.7 Suspected Whitefly Transmitted Diseases 71
3.8 Discussion and Conclusions 72
3.9 Acknowledgments 73
3.10 References 73

Chapter 4. Psyllids **87**

George H. Kaloostian

4.1 Introduction 87
4.2 Pear Decline 88
4.3 Greening Disease of Citrus 89
4.4 Proliferation Disease of Carrots 90
4.5 References 90

Chapter 5. Membracids **93**

John N. Simons

5.1 Introduction 93
5.2 Symptomatology of Pseudo-Curly Top (PCT) Disease 94
5.3 Transmission of PCT Agent 94
5.4 Life History of the Treehopper Vector 94
5.5 Epidemiology 95
5.6 Nature of the PCT Agent 95
5.7 Relationship of PCT to Other Curly Top Diseases 95
5.8 References 95

Chapter 6. Piesmids **97**

Gerhard Proeseler

6.1 Introduction 97
6.2 *Piesma quadratum* (Fieb.) and the Beet Leafcurl Virus 98
6.3 *Piesma quadratum* (Fieb.) and Beet Latent Rosette Disease 110
6.4 *Piesma cinereum* (Say) and Sugarbeet Savoy Virus 111
6.5 References 112

Chapter 7. Beetles **115**

J. P. Fulton, H. A Scott, and Rodrigo Gamez

7.1 Introduction 116
7.2 The Viruses 116
7.3 The Beetles 121
7.4 Beetle Vectoring of Viruses 122
7.5 Association of Virus with Beetles 126
7.6 Speculation on Mechanisms of Transmission 127
7.7 Acknowledgment 128
7.8 References 128

Chapter 8. Bark Beetles, *Ceratocystis ulmi* and Dutch Elm Disease **133**

Francis W. Holmes

8.1 Introduction 134
8.2 Transmission 136
8.3 Elm Bark Beetle Characteristics: Contrasting *Scolytus multistriatus* and *Hylorgopinus rufipes* 138
8.4 Disease Control by Insect Control 141
8.5 Closing Remarks 143
8.6 Acknowledgments 143
8.7 References 143

Chapter 9. Thrips **149**

T. N. Ananthakrishnan

9.1 Introduction 149
9.2 Thrips and Gall Formation 150
9.3 Thrips as Vectors of Bacterial, Fungal, and Viral Pathogens 156
9.4 Acknowledgment 161
9.5 References 161

Chapter 10. Flies **165**

T. A. Zitter and J. H. Tsai

10.1 Introduction 165
10.2 Diptera as Vectors of Disease Agents 167
10.3 Biology of *Liriomyza* Flies 170
10.4 Mode of Transmission and Natural Spread 172
10.5 References 174

Chapter 11. Lethal Yellowing of Coconut Palm: Search for a Vector **177**

James H. Tsai

11.1 Introduction 177
11.2 Patterns of Spread 179
11.3 Symptomatology 179
11.4 Causal Agent 179
11.5 Historical Perspectives in the Search for a Vector(s) of Lethal Yellowing Agent 180
11.6 Vector Research in the United States 181
11.7 Attempts to Mechanically Transmit LY Agent 190
11.8 Determination of Incubation Period in Palm 190

11.9 Attempts to Culture Coconut Embryo and Tissue *in Vitro* 190
11.10 Discussion 191
11.11 Acknowledgments 197
11.12 References 197

Chapter 12. Insect Involvement in the Transmission of Bacterial Pathogens 201

Monty D. Harrison, J. Wayne Brewer, and Laura D. Merrill

12.1 Introduction 201
12.2 Insect Transmission of Bacterial Pathogens 202
12.3 Summary 225
12.4 Acknowledgments 276
12.5 References 276

Chapter 13. Insect Involvement in the Transmission of Fungal Pathogens 293

George N. Agrios

13.1 Introduction 293
13.2 Role of Insects in the Transmission of Fungal Plant Pathogens in the Soil 295
13.3 Role of Insects in the Transmission of Fungi Affecting Stalks, Stems, Trunks, and Branches 297
13.4 Role of Insects in the Transmission of Fungi Causing Foliar Diseases 305
13.5 Role of Insects in the Transmission of Fungal Diseases Affecting Buds and Blossoms 308
13.6 Role of Insects in the Transmission of Fungal Diseases Affecting Fruits or Seeds in the Field 309
13.7 Role of Insects in the Transmission of Fungal Diseases Affecting Fruits or Seeds After Harvest 315
13.8 Concluding Remarks 316
13.9 References 317

Chapter 14. Mites 325

J. T. Slykhuis

14.1 Introduction 326
14.2 Characteristics of the Eriophyids 326

14.3 Vectors of Agents that Cause Diseases of Monocots 328
14.4 Vectors of Agents that Cause Diseases of Dicots 336
14.5 Diseases Probably Caused by Mites without an Infectious Agent 345
14.6 Doubtful or Erroneous Reports of Mite Transmission 347
14.7 Disease Agents Suspected to be Transmitted by Eriophyid Mites 349
14.8 Conclusions 350
14.9 Acknowledgments 351
14.10 References 351

Chapter 15. Fate of Plant Viruses in Mite Vectors and Nonvectors **357**

Y. C. Paliwal

15.1 Introduction 357
15.2 General Characteristics and Anatomical Features of Eriophyid Mites 358
15.3 Behavior of Viruses in Their Eriophyid Vectors 359
15.4 Behavior of Viruses in Nonvector Eriophyids 362
15.5 Viruses in Tetranychid Mites 365
15.6 Mechanisms of Virus Transmission 370
15.7 References 372

Chapter 16. Nematodes **375**

C. E. Taylor

16.1 Introduction 375
16.2 Viruses and Diseases 376
16.3 Nematode Vectors 383
16.4 Virus Transmission Characteristics and Mechanisms 392
16.5 Control 401
16.6 References 404

Chapter 17. Fungi **417**

David S. Teakle

17.1 Introduction 417
17.2 Vectors 418
17.3 Viruses 421
17.4 Virus–Vector Relationships 425

17.5	Suspected Fungus Transmission	429
17.6	Concluding Remarks	433
17.7	Acknowledgments	434
17.8	References	435

Index 439

LIST OF CONTRIBUTORS

Numbers in parentheses indicate the pages on which the authors' contributions begin.

George N. Agrios (293), Department of Plant Pathology, University of Massachusetts, Amherst, Massachusetts 01003

T. N. Ananthakrishnan (149), Zoological Survey of India, 34 Chittaranjan Avenue, Calcutta 700 012, India

J. Wayne Brewer (201), Department of Zoology and Entomology, Colorado State University, Fort Collins, Colorado 80523

Joseph P. Fulton (115), Department of Plant Pathology, University of Arkansas, Fayetteville, Arkansas 72701

R. Rodrigo Gamez (115), Cellular and Molecular Biology Research Center, University of Costa Rica, San José, Costa Rica

Kerry F. Harris (1), Department of Entomology, Texas A&M University, College Station, Texas 77843

Monty D. Harrison (201), Department of Botany and Plant Pathology, Colorado State University, Fort Collins, Colorado 80523

Francis W. Holmes (133), Shade Tree Laboratories, College of Food and Natural Resources, University of Massachusetts, Amherst, Massachusetts 01003

George H. Kaloostian (87), United States Department of Agriculture, Science and Education Administration, Boyden Entomology Laboratory, University of California, Riverside, California 92502

Laura D. Merrill (201), Department of Zoology and Entomology, Colorado State University, Fort Collins, Colorado 80523

V. Muniyappa (39), Department of Plant Pathology, University of Agricultural Sciences, Hebbal, Bangalore 560 024, India

Yogesh C. Paliwal (357), Chemistry and Biology Research Institute, Agriculture Canada, Ottawa, Ontario K1A 0C6, Canada

Gerhard Proeseler (97), Institute for Phytopathology, Academy of Agricultural Sciences, Theodor-Roemer-Weg 4, Aschersleben, German Democratic Republic

Osmo Roivainen (15), National Board of Agriculture, S-551 83 Jönköping, Sweden

Howard A. Scott (115), Department of Plant Pathology, University of Arkansas, Fayetteville, Arkansas

John N. Simons (93), JMS Flower Farms, Inc., 1105 25th Avenue, Vero Beach, Florida 32960

John T. Slykhuis (325), Research Station, Agriculture Canada, Summerland, British Columbia VOH 1ZO, Canada

Charles E. Taylor (375), Scottish Horticultural Research Institute, Invergowrie, Dundee DD2 5DA, Scotland

David S. Teakle (417), Department of Microbiology, University of Queensland, St. Lucia, Queensland, Australia

James H. Tsai (165, 177), Agricultural Research Center, University of Florida, 3205 S. W. 70th Avenue, Fort Lauderdale, Florida 33314

Thomas A. Zitter (165), Department of Plant Pathology, Cornell University, Ithaca, New York 14853

PREFACE

"Vectors of Plant Pathogens " is the third in a four-volume series of books on the general topic of vectors, pathogens and plant diseases. The first two volumes, "Aphids as Virus Vectors" and "Leafhopper Vectors and Plant Disease Agents," are intended to provide up-to-date, coherent syntheses of data pertaining to the interrelationship of plant pathogens with their aphid, leafhopper and plant-hopper vectors, and plant hosts, to suggest promising frontiers for further research, and to call attention to possible practical applications of our present understanding of vector–pathogen–plant interactions in preventing or lessening the threat of vector-borne disease agents to food and fiber crops. The editors' intents in the present volume remain the same; we have merely expanded the scope of the treatise to include numerous vector groups not covered in the first two volumes.

To realize our objectives in the most comprehensive and scholarly manner possible, we called on the talents of twenty-two, world-renowned, research scientists from a total of eight different countries. The opening chapter presents a synoptic overview of the three major homopterous vectors (aphids, leaf-hoppers, and planthoppers), with primary emphasis being placed on our current knowledge of vector–pathogen interactions. But the reader is referred to "Aphids as Virus Vectors" and "Leafhopper Vectors and Plant Disease Agents" for detailed information on the multifarious aspects of disease agent transmission by these insects. Chapters 2–17 report the latest findings regarding plant pathogen transmission by mealybug, whitefly, psyllid, membracid, piesmid, beetle, thirps, fly, mite, nematode, and fungal vectors. Also included among the latter chapters are special individual treatises on insect involvement in the spread of bacterial and fungal pathogens.

"Vectors of Plant Pathogens," like its predecessors, is intended for use by students, teachers, and researchers of vector–pathogen–plant relationships. All major vectors of plant pathogens are included, thus making the book ideally suited as a supplemental text for graduate entomology and plant pathology courses on insect involvement in the transmission and spread of plant disease agents. In addition, each chapter concludes with an extensive list of complete literature citations. Together, these references represent a formidable bibliography of vector-related research, and they make "Vectors of Plant Pathogens" desirable as a core reference and background work for libraries.

The editors thank the authors for their scholarly contributions and the staff of Academic Press for their part in producing and publishing this volume.

Chapter 1

APHIDS, LEAFHOPPERS AND PLANTHOPPERS

Kerry F. Harris

Department of Entomology
Texas A & M University
College Station, Texas

1.1 INTRODUCTION . . . 1
1.2 APHIDS . . . 2
1.2.1 Vectors . . . 2
1.2.2 Pathogens . . . 3
1.2.3 Vector-Pathogen Interactions . . . 3
1.3 LEAFHOPPERS . . . 9
1.3.1 Vectors . . . 9
1.3.2 Pathogens . . . 9
1.3.3 Vector-Pathogen Interactions . . . 9
1.4 PLANTHOPPERS . . . 11
1.5 CONCLUSION . . . 11
1.6 REFERENCES . . . 11

1.1 INTRODUCTION

Most vectors of arthropod-borne pathogens are insects, and most of these insect vectors are in the order Homoptera (superorder Hemiptera). Homopterous vectors include aphids (Aphidoidea: mainly Aphididae), mealybugs (Coccoidea: Pseudococcidae), whiteflies (Aleyrodoidea: Aleyrodidae) and psyllids (Psylloidea: Psyllidae) in the suborder Sternorrhyncha, and leafhoppers (Cicadoidea: Cicadellidae), spittlebugs (Cicadoidea: Cercopidae), membracids (Cicadoidea: Membracidae) and planthoppers (Fulgoroidea: Delphacidae and Cixiidae) in the suborder Auchenorrhyncha (Harris, 1980; Harris and Maramorosch, 1980). In this chapter, only the three major homopterous vectors (aphids, leafhoppers and planthoppers) are discussed, with primary emphasis being placed on current knowledge of vector-pathogen interactions.

ISBN 0-12-326450-2

1.2 APHIDS

1.2.1 Vectors

The biology, feeding behavior and worldwide distribution of aphids make them ideally suited for transmitting plant viruses (Harris and Maramorosch, 1977). The total number of described aphid species is 3,742 (Eastop and Hille Ris Lambers, 1976; Eastop, 1977). Only a few species, ca. 300, have been tested as vectors of any of 300 different viruses in about the same number of plant species. Over 180 of the 300 aphid species tested have been reported as vectors of at least one plant virus. More than half of the aphid species and most economically important virus vectors occur in the Aphidinae. Although this subfamily contains a few polyphagous aphids, most of its members are host specific. Many genera of Aphidinae have Rosaceae as overwintering primary hosts and migrate to herbaceous secondary hosts in the summer.

Eastop (1977) has collated the geographical, climatological and host-plant distributions of aphids and plant viruses. The known plant viruses are distributed fairly equally through the major plant groups, when the number of plant species in each group is taken into account. And the proportion of viruses transmitted by aphids does not appear to be a function of the number of aphids specific to a particular plant group (Eastop, 1977). As might be expected, however, plant families containing species of economic importance, such as Chenopodiaceae, Rosaceae, Leguminosae, Solanaceae, and Gramineae, have been recorded as hosts of most viruses. The distribution of aphid-borne viruses (nonpersistent = np, semipersistent = sp, persistent = p, and unknown mode of transmission = uk) is as follows: Coniferae (1 np); Moraceae (1 uk) and other Hamamelidae (1 np, 1 uk); Caryophyllaceae (2 np), Chenopodiaceae (4 np, 2 sp and 3 p) and other Caryophyllidae (1 np); Malvaceae (1 np, 2 p and 1 uk), Cucurbitaceae (5 np), Cruciferae (2 np, 1 p and 1 uk) and other Dilleniidae (6 np); Rosaceae (4 np, 4 sp, 4 p and 2 uk), Leguminosae (20 np, 5 p and 1 uk), Rutaceae (1 np, 1 sp and 1 p), Umbelliferae (5 np, 1 sp and 5 p) and other Rosidae (2 np and 2 p); Solanaceae (14 np and 8 p), Convolvulaceae (2 np and 2 uk), Compositae (8 np and 2 p) and other Asteridae (3 np and 3 uk); Gramineae (4 np, 2 p and 1 uk), Liliaceae (6 np, 2 p and 1 uk), Iridaceae (4 np and 3 uk), Orchidaceae (1 np) and other Liliatae (4 np and 1 p).

Relationships between plants, viruses and insects have evolved during the last 200,000,000 years. The agricultural practices of man have obviously strongly influenced the distribution of aphid vectors and aphid-borne viruses. For instance, over a third of the 62 known viruses of Solanaceae are aphid-transmitted, and most are transmitted by vectors that did not encounter potatoes until about 400 years ago (Eastop, 1977). With the possible exception of *Macrosiphum euphorbiae*, most virus transmission in North America is by introduced vector species on introduced plants.

1.2.2 Pathogens

Aphids are only known to transmit plant viruses; however, there are a few unconfirmed reports of their transmitting mycoplasmalike organisms (MLOs). Aphid-borne viruses can be classified as noncirculative (nonpersistent and semipersistent) or circulative on the basis of how they are transmitted (see section 1.2.3). Of the 164 known aphid-borne viruses, 109 are noncirculative (101 nonpersistent and 8 semipersistent), 38 are circulative (persistent), and the status of 17 is uncertain.

1.2.2.1 Noncirculative Viruses. The Plant Virus Subcommittee of the International Committee for Taxonomy of Viruses (ICTV) has endorsed eight groups of plant viruses, each of which contains one or more noncirculative, aphid-transmitted members: potyviruses (poty, sigla from *pot*ato *Y*); carlaviruses (carla, sigla from *car*nation *la*tent); caulimoviruses (caulimo, sigla from *cauli*flower *mo*saic); cucumoviruses (cucumo, sigla from *cucum*ber *mo*saic); two monotypic groups, as yet unnamed, based on alfalfa mosaic virus and broad bean wilt virus, respectively; closteroviruses (clostero, from Greek *kloster*, meaning thread), and a monotypic group, also unnamed, based on parsnip yellow fleck virus. For information on the characteristics of viruses in each of these groups the reader is referred particularly to a recent review by Shepherd (1977) and also to ones by Pirone and Harris (1977) and Harris (1978).

1.2.2.2 Circulative Viruses. These viruses encompass at least five taxonomic groups. Three groups have been formally described and endorsed by the ICTV—luteoviruses (luteo, from Latin *luteus*, meaning yellow), rhabdoviruses (rhabdo, from Greek *rhabdos*, meaning rod), and an as yet unnamed monotypic group based on pea enation mosaic virus. Two other groups based on potato leafroll and carrot mottle viruses seem likely to be endorsed by the ICTV. For information on the characteristics of viruses in these groups the reader is referred to reviews by Shepherd (1977) and Harris (1979).

1.2.3 Vector-Pathogen Interactions

1.2.3.1 Nonpersistent, Noncirculative Transmission. Aphids can acquire and transmit these viruses during very brief probes. Acquisition and inoculation threshold periods may be as brief as 5 sec. In general, acquisition and inoculation probes of 15-60 sec are adequate for optimal transmission. As the duration of acquisition probes is increased beyond 1 min, there is a corresponding decrease in the level of transmission. Starving aphids prior to allowing them acquisition probes can increase the level of transmission several times over that obtained using nonstarved controls. The probability that infection will result after single prolonged inoculation probes is about the same as with single brief probes; however, after making such prolonged probes, aphids are rarely viruliferous.

Since a detectable latent period is absent, the acquisition, carry-over and inoculation phases of the transmission cycle can be completed in a minute or two. Vector inoculativity is lost through ecdysis (nontransstadial). As a rule, individual viruliferous aphids cease to be inoculative minutes after being placed on uninfected test plants. Aphids that make only brief probes in test plants retain inoculativity longer than those that make prolonged penetrations. Groups of starved viruliferous aphids remain inoculative longer than nonstarved ones. But with both starved and nonstarved insects, loss of inoculativity follows a somewhat exponential curve.

Analyses of the observable characteristics of nonpersistent, noncirculative transmission and of data obtained from membrane-probing and feeding experiments, from electronic monitoring of aphid probing and feeding, and from experiments involving sequential probing by aphids, first in isotopically labeled and then in nonlabeled leaf disks, permit the following generalizations regarding this type of transmission (Harris, 1977, 1978, 1979). The transmission event is epidermal and intracellular in nature. Sap-sampling (host-selection) behavior plays an important, perhaps essential, role in the transmission process. Sap-sampling behavior is stimulated by subjecting the aphids to preacquisition starvation. Sap-sampling, when occurring on a virus-infected plant, serves to contaminate the fore alimentary canal with virus-laden material ("cell sap" or protoplasm). The transmission cycle is completed when all or a portion of this virus-laden material is egested during subsequent sap-sampling probes in healthy plants: an *ingestion-egestion mechanism* of transmission. This host-selection behavior serves to bring plant material in contact with the epipharyngeal gustatory organ, thus permitting a quantitative and qualitative analysis of the plant's suitability. The volumes of plant sap that are transmitted during host-selection probes can be many times greater than what could be physically accommodated either on the stylets or in the maxillary food canal lumen which has a maximum carrying capacity of ca. 60 μm^3. Subjecting aphids to preacquisition starvation produces corresponding increases both in the numbers of aphids that transmit virus and those that acquire and release the minimal infective dose of plant material.

This unique, host selection behavior of aphids—when coupled with their finely tipped stylets and their habit of inserting only the maxillary tips into cells from which they ingest sap, and then closing these feeding sites with a salivary plug during stylet withdrawal—makes them ideally suited as vectors of noncirculative viruses. Aphids function more like flying syringes than "flying needles." Sap-sampling behavior provides for a continuum of transmission mechanisms ranging from simple stylet-tip contamination (?) to any number of simple to complex variations on the basic ingestion-egestion theme. If stylet-borne transmission does occur, it would presumably be masked by transmission resulting from ingestion-egestion behavior; and, at present, we have no single technique for distinguishing between these two mechanisms. Perhaps an approach that combined the techniques of radioisotope methodology and electronic monitoring would make such a distinction possible (Harris, 1977).

Recently, using enzyme-linked immunosorbent assay (ELISA), Gera *et al.* (1978) were able to detect an aphid-transmissible strain of cucumber mosaic virus associated with single viruliferous aphids. No reaction was obtained with aphids that had been exposed, by probing or feeding, to a non-aphid-transmissible strain of CMV. The researchers (Gera *et al.*, 1978; Gera, 1978) suggest that aphid-transmissibility or nontransmissibility is determined by the ability or inability, respectively, of virus to attach to an aphid's stylet.* However, since whole aphids or heads of aphids were used in the ELISA tests (Gera, 1978), proof that the detectable virus is located on the stylet is lacking. For the many reasons discussed elsewhere (Harris, 1977), I consider it more likely that the virus is associated with the anterior portion of the insect's fore alimentary canal—especially the maxillary food canal, pharynx and cibarial pump regions (Harris and Bath, 1973).

1.2.3.2 Semipersistent and Bimodal, Noncirculative Transmission. Semipersistent viruses resemble circulative ones in that they are not generally acquired during brief acquisition probes; and the probability of transmission increases with increased acquisition-access feeding periods up to 12-24 hr or longer. The latter property suggests that these viruses are somehow able to accumulate in or on and resist being separated from the vector (Harris, 1977). No one has yet been able to demonstrate a latent period for semipersistent transmission, and viruliferous aphids do not retain infectivity through ecdysis (nontransstadial). There is no evidence to suggest that virus is present in the hemocoele of inoculative insects. Also, injection of virus directly into the hemocoele does not render the recipient insect inoculative.

Semipersistent transmission characteristics are compatible with an ingestion-egestion transmission mechanism. Semipersistence and increases in the probability of transmission with increased acquisition feeding periods suggest that virus can accumulate in the vector's anterior alimentary canal and resist being quickly flushed through with virus-free sap from healthy plants. One way virus might persist in the vector is by selective adsorption or attachment to cellular and/or noncellular surfaces in the fore alimentary canal, either directly or indirectly via an intermediate helper agent. The probability of such a mechanism has been confirmed by studies on aphid transmission of semipersistent anthriscus yellows virus which appears to be retained in a matrix material adsorbed to the ridged intima lining the ventral wall of the posterior region of the pharynx where it passes over the tentorial bar (Murant *et al.*, 1976). During the inoculation phase of transmission, release of virus from binding sites might be mediated by pH changes in the milieu of the fore gut, as has been suggested for the transmission of tobra- and nepo-viruses by dorylaimid nematodes. In the latter system, "egestion" represents a backflow of material brought about by the ejection of salivary fluids into and their forward passage through the esophageal lumen (Taylor, 1980). Such pH changes would influence adsorption and release of virus by altering the surface charge density of the virion relative to that of the retention site (like to opposite or vice versa).

Some noncirculative viruses are transmitted both nonpersistently and semi-

*Since the writing of this chapter, preliminary experiments by Gera *et al.* (1979) showed that ELISA detection of CMV is almost entirely associated with the foregut.

persistently by certain aphid vectors. This "biomodal" transmission is predictable in terms of an ingestion-egestion transmission mechanism (Harris, 1977). Such viruses could be acquired from or inoculated to either superficial or deep tissues of the host plant. Additionally, they would be able to accumulate and persist in the fore gut.

1.2.3.3 Innovative Approaches to Control-Oriented Research. Most approaches to control are aimed at eradicating or altering one or more of the primary participants in the transmission process (vector, virus, host plant and man) or at preventing their coming together. Breeding plants for immunity, resistance or tolerance to virus infection has proven effective in a number of cases where a particular crop is threatened exclusively or primarily by a single virus.

Insecticides, including systemics, can sometimes effectively control the spread of circulatively transmitted viruses because longer acquisition and inoculation feedings are required for transmission and because there is a vector latent or incubation period between the acquisition and inoculation phases. This is not the case with noncirculative transmission, however. In nonpersistent, noncirculative transmission, for example, brief probes in the epidermis, the absence of a detectable latent period, relatively slow-acting insecticides, and the habit that alatae newly-arrived in a field have of making a series of alightings, brief probes and take offs before settling on a host plant for prolonged feeding, all combine to form a transmission cycle that is highly resistant, if not immune, to interruption or suppression by insecticides alone (Harris, 1977).

Recent advances in our knowledge of "transmission ecology" may provide the basis for interesting new approaches to controlling the spread of aphid-borne viruses. Ultrastructural studies led Wensler (1974) to conclude that the paired dendrites serving the mandibular tips of aphids are mechanoreceptors that enable them to monitor the movement and position of the stylets. I (Harris, 1977) hypothesized that the dendrites might represent simple contact chemoreceptors ("ingestion receptors") that can stimulate aphids to ingest plant material to the epipharyngeal gustatory organ and allow them to follow a glucoside gradient to the phloem. Whatever is their mode of action, these receptors and others (e.g., chemoreceptors of the pharyngeal gustatory organ and mechanoreceptive pegs at the labial tip) are apparently part of the sensory transduction system that makes possible such behavioral manifestations as anticlinal groove localization, epidermal cell entry, sap-sampling, deep probing, and feeding site localization. In other words, aphid feeding behavior is a function of the physical and chemical stimuli to which aphids' mechano- and chemo-receptors are exposed during probing, and of the condition, receptivity, and sensitivity of such receptors to stimulation.

If sap-sampling is responsible for nonpersistent transmission, then it might be possible to manipulate this behavior in favor of nontransmission by permanently or temporarily impairing the normal functioning of the vector's sensory transduction system, by exposing the system to antiphagostimulatory stimuli or to phagostimulatory ones at the "wrong" time, or by preventing ingestion inducing stimuli from reaching and interacting with the system (Harris, 1977).

Mineral and vegetable oils and milk lipids might inhibit noncirculative transmission by modifying the probing and feeding behavior responsible for transmission. The physico- and electro-chemical (dielectric) properties of oils might insulate the sensory transduction system of the aphid's feeding apparatus from or inhibit its interpretation of sap-sampling and feeding-inducing stimuli (Harris, 1978). Oil on the labium or in anticlinal grooves on the surface of leaves (Simons and Beasley, 1977) could inhibit anticlinal groove localization by altering the functioning of the labial mechanoreceptors. The preprobing times of oil-exposed aphids are significantly longer than those of nonexposed controls (Simons *et al.*, 1977). Whether they be mechano- or chemo-receptors, oil might also inhibit the normal functioning of the dendrites serving the mandibular tips (Harris, 1978).

Vanderveken (1977) and others, thinking in terms of a stylet-borne transmission mechanism, suggest that oil inhibits transmission by impeding virus adsorption to or its elution from the stylets. Presumably, ingested oil would have a similar effect on virus adsorption sites in the fore alimentary canal. Aphids ingest oil from oil-treated leaves (Vanderveken, 1973a), and this oil could insulate the chemoreceptors of the pharngeal gustatory organ from phagostimulatory chemicals. This, in turn, would logically inhibit sap-sampling, feeding site (phloem) localization, and prolonged feeding, thus preventing the acquisition of transmissible titers of virus (Harris, 1978). This latter effect would be particularly important in the case of semipersistent transmission. The need for ingestion of oil would explain why optimal inhibition of beet yellows virus (BYV) transmission occurs after a prolonged period (24 hr) (Vanderveken, 1973b) of contact between aphids and oil-treated leaves (Harris, 1978). Once having reached a sieve element of a BYV-infected plant, aphids require a minimal additional 5 min and 22 sec of fluid ingestion to become inoculative (Chang, 1968). Acquisition-access feeding periods of more than 12 hr and inoculation feeds of at least 6 hr are necessary for optimal transmission of BYV.

The discovery that helper agents are necessary parts of the transmission ecology of certain viruses leads to yet another control possibility (Pirone, 1977; Pirone and Harris, 1977). Studies on the genetic regulation of helper component production in plants could lead to the development of plants that do not produce helper component ("helperless varieties") as a result of infection. Helperless varieties would eliminate the secondary spread of viruses that require helper and would be particularly effective in situations where the primary source of virus inoculum for aphids is within the crop itself, as with seedborne viruses (Pirone, 1977; Pirone and Harris, 1977). This type of approach might complement present programs that concentrate on breeding plants for vector-resistance, and immunity, resistance, or tolerance to virus infection.

The testing of naturally occurring phagostimulatory or antiphagostimulatory phytochemicals for their effect on both sap-sampling behavior and virus transmission would also seem appropriate (Harris, 1977). Also, keeping aphids on plants might prove as effective as keeping them off in programs aimed at controlling virus spread. Some type of phytochemical "masking" or "interference" (with the aphid's sensory transduction system) phenomenon could explain why aphids would continue to probe plants, on which they are unable to feed, until they literally starve to death (Taylor, 1959). In general, if the ingestion-egestion conceptualization of the nonpersistent, noncirculative, transmission mechanism is valid, then any practical approaches to preventing or manipulating sap-sampling behavior could prove valuable in planning integrated control strategies.

1.2.3.4 Circulative Transmission. Circulative aphid-borne viruses can be separated into at least five distinct taxonomic groups. The International Committee on Taxonomy of Viruses (ICTV) has assigned latinized names to two groups, the luteoviruses and rhabdoviruses, and recognizes a third, monotypic group, as yet unnamed, founded on pea enation mosaic virus (PEMV). Two additional groups based on potato leafroll and carrot mottle viruses, respectively, are likely forthcoming.

Those circulative viruses that have been studied in their vectors at the organ, tissue, or cellular level can be conveniently categorized, according to their particle morphologies, as polyhedral viruses and rhabdoviruses (Rhabdoviridae) (Harris, 1979). The polyhedral viruses that have been localized in their aphid vectors are barley yellow dwarf (BYDV), potato leafroll (PLRV), pea enation mosaic, and beet western yellows (BWYV) (Harris, 1979). There are no unequivocal data indicating that any of these viruses multiply in their vectors; and for one, BYDV, there is convincing evidence that it does not. Eskandari *et al.* (1978) recently confirmed the earlier report by Harrison (1958) that PLRV is not transmissible (beyond the lst transfer) from insect to insect using serial injection technique. Transovarial passage is only known to occur in the case of PLRV and its vector *Myzus persicae* (Sulzer) (Miyamoto and Miyamoto, 1966).

Several of the phytorhabdoviruses have known vectors; nine are aphid-borne, six leafhopper-borne, six planthopper-borne, and one, sugarbeet leaf curl virus, is transmitted by a piesmid. All these viruses appear to multiply in their vectors. Those (circulative-propagative) that have been localized in both their plant and aphid vectors are lettuce necrotic yellows (LNYV), sowthistle yellow vein (SYVV), strawberry crinkle, broccoli necrotic yellow, and *Sonchus* yellow net viruses. Transovarial passage has been recorded for SYVV and LNYV. The vector-virus relationships of aphid-borne circulative viruses have recently been reviewed by Harris (1979).

Our understanding of circulative transmission ecology has also advanced significantly during the last few years. Recent studies on two isometric, aphid-borne viruses, barley yellow dwarf virus (Rochow *et al.*, 1975) and pea enation mosaic virus (PEMV) (Harris *et al.*, 1975; Harris, 1977, 1979), indicate that vector-virus and virus-vector specificity phenomena associated with some circulative transmissions are mediated by vector salivary gland-virus coat protein interactions. It appears that reciprocity (nature unknown) between recognition sites on virus coat protein and salivary gland membranes is required for passage of virus through the vector's salivary system. Slight variations in either the coat protein of viruses or virus strains, or in the salivary membrane systems of vector species or biotypes can effect the aphid-transmissibility of the virus by altering the permeability of the salivary glands to virions. In the case of PEMV, an extra protein found in aphid-transmissible but not in non-aphid-transmissible variants has been identified as the "transmissibility factor." (See Harris, 1979, for a detailed discussion of these phenomena.)

1.3 LEAFHOPPERS

1.3.1 Vectors

Leafhoppers (Cicadellidae), with 130 known vector species covering 10 subfamilies and 58 genera, account for more than 80% of all auchenorrhynchous vectors (Harris, 1979). The greatest number of vector genera and species occur in the subfamily Deltocephalinae, which also encompasses the greatest number of nonvector species of leafhoppers. Vector species (numbers in parentheses) occur in the following subfamilies: Agalliinae (13), Macropsinae (3), Gyponinae (2), Coelidiinae (2), Aphrodinae (2), Cicadellinae (28), Deltocephalinae (75) and Typhlocybinae (3). The reader is referred to Nielson (1979) for the latest information on the taxonomy and phylogeny of leafhopper vectors, as well as a complete listing of the various vector species and the disease agents they transmit.

1.3.2 Pathogens

Leafhoppers are known to transmit ca. 77 disease agents (38 viruses, 31 MLOs, 4 spiroplasmas, and 4 RLOs). Suspected or known spiroplasmas include the pathogens responsible for citrus stubborn, corn stunt, aster yellows and citrus little leaf diseases, whereas RLOs include the causative agents of almond leaf scorch, phony peach, Pierce's disease of grapevines (lucerne or alfalfa dwarf) and clover club leaf.

1.3.3 Vector-Pathogen Interactions

Most leafhopper-borne viruses (presumably MLOs and spiroplasmas too) are transmitted in a circulative manner and multiply in their vectors—the circulative, nonpropagative beet curly top virus and the transitory ("semipersistent") noncirculative rice tungro virus group and maize chlorotic dwarf virus being notable exceptions (Harris, 1979). More data are needed regarding the mechanism of RLO transmission by leafhoppers; however, as pointed out below, in at least one case (Pierce's disease agent of grapevines) the transmission appears to be noncirculative via an ingestion-egestion mechanism.

The aforementioned cases of noncirculative virus transmission by leafhoppers might also be explained on the basis of an ingestion-egestion mechanism in combination with an ability of virus to accumulate and persist in the anterior portion of the alimentary canal (Harris, 1977, 1978, 1979). The virus-vector relationships of rice tungro and maize chlorotic dwarf virus are characterized by the absence of a demonstrable latent period, the absence of evidence for virus entering the hemocoele of the vector and exiting via the salivary system, a gradual decline in inoculativity once a viruliferous insect is separated from a virus source, and nontransstadial passage. Since there are no known cases of leafhoppers transmitting virus in a nonpersistent manner, and since "semipersistent" connotes a state intermediate between nonpersistent and persistent, it has been suggested that such noncirculative

TABLE I. List of Planthopper Vectors and Associated Plant Diseases[a]

Vector	Disease
Dicranotropis hamata (Boheman)	Oat sterile dwarf (=*Arrhenatherum* blue dwarf and *Lolium* enation)
	Cereal tillering disease
Javesella discolor (Boheman)	Oat sterile dwarf
J. dubia (Kirsch.)	European wheat striate mosaic
	Oat sterile dwarf
J. obscurella (Boheman)	European wheat striate mosaic
	Oat sterile dwarf
J. pellucida (Fabricius)	European wheat striate mosaic
	Maize rough dwarf
	Oat sterile dwarf
Laodelphax striatellus (Fallén)	Barley yellow striate mosaic
	Cereal tillering disease
	Maize rough dwarf
	Northern cereal mosaic
	Oat pseudorosette (*zakuklivanie ovsa*)
	Rice black-streaked dwarf
	Rice stripe
Megadelphax sordidulus (Stål)	*Phleum* green stripe
Muellerianella fairmairei (Perris)	Northern cereal mosaic
Nilaparvata lugens Stål	Rice grassy stunt
Peregrinus maidis Ashmead	Maize mosaic
	Maize rough dwarf
Perkinsiella saccharicida Kirkaldy	Sugarcane Fiji disease
P. vastatrix Breddin	Sugarcane Fiji disease
Sogatella furcifera (Horváth)	Pangola stunt
S. kolophon (Kirkaldy)	*Digitaria* striate mosaic
S. vibix (Haupt)	Maize rough dwarf
Sogatodes cubanus (Crawford)	Rice hoja blanca
S. orizicola (Muir)	Rice hoja blanca
Tarophagus proserpina (Kirkaldy)	Bobone disease
Terthron albovittatus (Matsumura)	Northern cereal mosaic
	Rice stripe
Ukanodes albifascia (Matsumura)	Northern cereal mosaic
	Rice black-streaked dwarf
	Rice stripe
U. sapporona (Matsumura)	Northern cereal mosaic
	Rice black-streaked dwarf
	Rice stripe

[a]Disease agents are known or presumed viruses, except for rice grassy stunt (MLO).

leafhopper-borne viruses be labeled transitory rather than semipersistent (Ling and Tiongco, 1979).

Ingestion-egestion behavior could also be involved in the transmission by leafhoppers of disease agents other than viruses (Harris, 1977, 1979; Harris *et al.*, 1980). For example, the transmission characteristics of Pierce's disease agent (PDA) of grapevines suggest that the vector-pathogen (bacterium) relationship is a noncirculative one. Retention of the xylem-restricted pathogen at adsorption sites in the fore alimentary canal and a simple inoculation mechanism, such as egestion, seem most compatible with the characteristics of a brief or nonexistent latent period, prolonged retention of inoculativity by vectors, a broad vector range (low specificity), and persistent retention of inoculativity by adults (Harris, 1979).* The PDA is transmitted by 24 species of leafhoppers and 4 species of spittle bugs. In the fore gut of adults, PDA could thrive and possibly multiply (persistence) while being bathed in medium (xylem fluid) in which it is accustomed to multiplying.

The vector-virus relationships of viruses that are transmitted by leafhoppers in a circulative manner have been recently reviewed in great detail (Harris, 1979; Banttari and Zeyen, 1979; Maramorosch and Harris, 1979; Shikata, 1979a, 1979b).

1.4 PLANTHOPPERS

As vectors, planthoppers (Delphacidae) have received far less attention from researchers than have aphids and leafhoppers. However, ca. 21 vector species (13 genera) have been recorded; and, as a group, these vectors are responsible for transmitting at least 1 MLO (rice grassy stunt) and ca. 17 viruses or suspected viruses (Table I).

1.5 CONCLUSION

The reader is referred to "*Aphids as Virus Vectors*" (Harris and Maramorosch, 1977) and "*Leafhopper Vectors and Plant Disease Agents*" (Maramorosch and Harris, 1979) for more comprehensive treatises on aphid, leafhopper and planthopper vectors.

1.6 REFERENCES

Banttari, E. E., and Zeyen, R. J. (1979). Interactions of mycoplasmalike organisms and viruses in dually infected leafhoppers, planthoppers and plants. *In* "Leafhopper Vectors and Plant Disease Agents" (K. Maramorosch and K. F. Harris, eds.), pp. 327-347. Academic Press, New York and London.

Chang, V. C-S. (1968). Intraspecific variation in the ability of the green peach aphid, *Myzus persicae* (Sulz.) to transmit sugar beet yellows virus. Ph. D. thesis, University of California, Davis. 123 pp.

Eastop, V. F. (1977). Worldwide importance of aphids as virus vectors. *In* "Aphids as Virus Vectors" (K. F. Harris and K. Maramorosch, eds.), pp. 3-47. Academic Press, New York and London.

Eastop, V. F., and Hille Ris Lambers, D. (1976). "Survey of the World's Aphids." Published by Dr. W. Junk, The Hague. 586 pp.

Eskandari, F., Sylvester, E. S., and Richardson, J. (1978). Evidence for lack of propagation of

*This hypothesis was recently confirmed by data indicating that vectors do not retain inoculativity through ecdysis (nontransstadial) (Purcell and Finlay, 1979).

potato leaf roll virus in its aphid vector, *Myzus persicae. Phytopathology* 69: 45-47.
Gera, A. (1978). Personal communication.
Gera, A., Loebenstein, G., and Raccah, B. (1978). Detection of cucumber mosaic virus in viruliferous aphids by enzyme-linked immunosorbent assay. *Virology* 86: 542-545.
Gera, A., Loebenstein, G., and Raccah, B. (1979). Protein coats of two strains of cucumber mosaic virus affect transmission by *Aphis gossypii. Phytopathology* 69: 396-399.
Harris, K. F. (1977). An ingestion-egestion hypothesis of noncirculative virus transmission. *In* "Aphids as Virus Vectors" (K. F. Harris and K. Maramorosch, eds.), pp. 165-220. Academic Press, New York and London.
Harris, K. F. (1978). Aphid-borne viruses: ecological and environmental aspects. *In* "Viruses and Environment" (E. Kurstak and K. Maramorosch, eds.), pp. 311-337. Academic Press, New York and London.
Harris, K. F. (1979). Leafhoppers and aphids as biological vectors: vector-virus relationships. *In* "Leafhopper Vectors and Plant Disease Agents" (K. Maramorosch and K. F. Harris, eds.), pp. 217-308. Academic Press, New York and London.
Harris, K. F. (1980). Transmission concepts: horizontal transmission of plant viruses. *In* "Comparative Aspects of Animal and Plant Pathogen Vectors" (J. J. McKelvey, Jr., K. Maramorosch, and B. Eldridge, eds.). Praeger Scientific, New York (in press).
Harris, K. F., and Bath, J. E. (1973). Regurgitation by *Myzus persicae* during membrane feeding: its likely function in transmission of nonpersistent plant viruses. *Ann. Entomol. Soc. Am.* 66: 793-796.
Harris, K. F., and Maramorosch, K., eds. (1977). "Aphids as Virus Vectors." Academic Press, New York and London. 559 pp.
Harris, K. F., and Maramorosch, K., eds. (1980). Vectors of Plant Pathogens. Academic Press, New York and London (in press).
Harris, K. F., Bath, J. E., Thottappilly, G., and Hooper, G. R. (1975). Fate of pea enation mosaic virus in PEMV-injected pea aphids. *Virology* 65: 148-162.
Harris, K. F., Treur, B., Tsai, J., and Toler, R. (1980). Observations on the ingestion-egestion behavior of a leafhopper vector during membrane feeding. In manuscript.
Harrison, B. D. (1958). Studies on the behavior of potato leaf roll virus and other viruses in the body of their aphid vector *Myzus persicae* (Sulz.). *Virology* 6: 265-277.
Ling, K. C., and Tiongco, E. R. (1979). Transmission of rice tungro virus at various temperatures: a transitory virus-vector interaction. *In* "Leafhopper Vectors and Plant Disease Agents" (K. Maramorosch and K. F. Harris, eds.), pp. 349-366. Academic Press, New York and London.
Maramorosch, K., and Harris, K. F., eds. (1979). "Leafhopper Vectors and Plant Disease Agents." Academic Press, New York and London. 654 pp.
Miyamoto, S., and Miyamoto, Y. (1966). Notes on aphid-transmission of potato leafroll virus. *Sci. Rep. Hyogo Univ. Agric.* 7: 51-66.
Murant, A. F., Roberts, I. M., and Elnager, S. (1976). Association of virus-like particles with the foregut of the aphid *Cavariella aegopodii* transmitting the semi-persistent viruses anthriscus yellows and parsnip yellow fleck. *J. Gen. Virol.* 31: 47-57.
Nielson, M. W. (1979). Taxonomic relationships of leafhopper vectors of plant pathogens. *In* "Leafhopper Vectors and Plant Disease Agents" (K. Maramorosch and K. F. Harris, eds.), pp. 3-27. Academic Press, New York and London.
Pirone, T. P. (1977). Accessory factors in nonpersistent virus transmission. *In* "Aphids as Virus Vectors" (K. F. Harris and K. Maramorosch, eds.) pp. 221-235. Academic Press, New York and London.
Pirone, T. P., and Harris, K. F. (1977). Nonpersistent transmission of plant viruses by aphids. *Ann. Rev. Phytopathol.* 15: 55-73.
Purcell, A. H., and Finlay, A. (1979). Evidence for noncirculative transmission of Pierce's disease bacterium by sharpshooter leafhoppers. *phytopathology* 69: 393-395.
Rochow, W. F., Foxe, M. J., and Muller, I. (1975). A mechanism of vector specificity for circulative aphid-transmitted plant viruses. *Ann. N. Y. Acad. Sci.* 266: 293-301.

Shepherd, R. J. (1977). Intrinsic properties and taxonomy of aphid-borne viruses. *In* "Aphids as Virus Vectors" (K. F. Harris and K. Maramorosch, eds.), pp. 121-136. Academic Press, New York and London.

Shikata, E. (1979a). Cytopathological changes in leafhopper vectors of plant viruses. *In* "Leafhopper Vectors and Plant Disease Agents" (K. Maramorosch and K. F. Harris, eds.), pp. 309-325. Academic Press, New York and London.

Shikata, E. (1979b). Rice viruses and MLOs, and leafhopper vectors. *In* "Leafhopper Vectors and Plant Disease Agents" (K. Maramorosch and K. F. Harris, eds.), pp. 515-527. Academic Press, New York and London.

Simons, J. N., and Beasley, C. A. (1977). Visualization of oil on leaf surfaces–a technique. *J. Econ. Entomol.* 70: 307-308.

Simons, J. N., McLean, D. L., and Kinsey, M. G. (1977). Effects of mineral oil on probing behavior and transmission of stylet-borne viruses by *Myzus persicae. J. Econ. Entomol.* 70: 309-315.

Taylor, L. R. (1959). Abortive feeding behavior in a black bean aphid of the *Aphid fabae* group. *Entomol. Exp. Appl.* 2: 143-153.

Taylor, C. E. (1980). Nematodes. *In* "Vectors of Plant Pathogens" (K. F. Harris and K. Maramorosch, eds.). Chapter 16. Academic Press, New York and London.

Vanderveken, J. 1973a). Recherche du méchanisme de l'inhibition de la transmission aphidienne des phytovirus par des substances huileuses. *Parasitica* 29: 1-15.

Vanderveken, J. (1973b). Considérations sur les modalités de la transmission aphidienne d'un virus semi-persistent. *Parasitica* 29: 155-162.

Vanderveken, J. J. (1977). Oils and other inhibitors of nonpersistent virus transmission. *In* "Aphids as Virus Vectors" (K. F. Harris and K. Maramorosch, eds.), pp. 435-454. Academic Press, New York and London.

Wensler, R. J. D. (1974). Sensory innervation monitoring movement and position in the mandibular stylets of the aphid, *Brevicoryne brassicae. J. Morphol.* 143: 349-364.

Chapter 2

MEALYBUGS

Osmo Roivainen

National Board of Agriculture
S-551 83 Jönköping
Sweden

2.1 INTRODUCTION. 15
2.1.1 Extent of Virus Transmission by Mealybugs. 16
2.1.2 The Virus-Mealybug-Cocoa Complex 17
2.2 TRANSMISSION CHARACTERISTICS 19
2.2.1 Preliminary Fasting and Preliminary Feeding 19
2.2.2 Availability of Virus in Source Plants. 22
2.2.3 Acquisition of Virus by Vectors. 23
2.2.4 Persistence of Virus in Vectors 24
2.2.5 Inoculation of Virus by Vectors. 26
2.2.6 Transmission Specificity and Efficiency of Vectors. 29
2.2.7 Effect of Physical Factors on Virus Transmission. 32
2.2.8 Relationship between Virus Uptake, Persistence and Infectivity of Vectors. 32
2.2.9 Type of Transmission. 33
2.3 REFERENCES 34

2.1 INTRODUCTION

Mealybugs or Pseudococcidae, the false scale insects as they lack the scale, can be found almost everywhere amongst terrestrial plant life. They are scarce in the cold climates but represented by many species and high population densities in the subtropics and the tropics. They are remarkably like other homopterous insects furnished with the necessary implements required for transmission of plant viruses, however, together with the mite and the nematode vectors of plant disease agents, the mealybugs transmit virus particles without having wings. To rely on legs or unpredictable wind currents as means of virus transmission is hazardous, and

ISBN 0-12-326450-2

special conditions are required to maintain a continuous virus-vector-plant relationship. The transmission of virus particles by mealybugs from a cocoa plant to another with records on some alternative virus host plants is the only well documented case of a firmly established relationship between viruses, mealybugs and plants. The relationship manifests itself in a disease of the cocoa tree which "is one of the most important factors limiting the production of cocoa and is one of the most economically important plant diseases in the world" (Thresh, 1958a). The unfortunate lack of detailed studies on other possible virus-mealybug-plant relationships, and the extensive research carried out with the virus-mealybug-cocoa complex together with the serious economic losses, make it natural that a review on mealybugs as virus vectors cannot be much else than a product well flavored with cocoa.

This review attempts to evaluate and summarize the accumulated knowledge on mealybugs as vectors of plant viruses. The aim has been not to favor any of the three participants in the transmission process. Various aspects of the subject have been reviewed earlier (Dale, 1962; Entwistle, 1972; Roivainen, 1973, 1976; Thorold, 1975; Thresh, 1958a, 1958b; Thresh and Tinsley, 1959).

2.1.1 Extent of Virus Transmission by Mealybugs

Among the different families of *Coccoidea* only the *Pseudococcidae* have been reported to transmit plant viruses, although the general ecology and feeding behavior is remarkably alike in all the families. Virus transmission by other homopterous insects, notably by aphids, is thus the nearest parallel of virus transmission by mealybugs.

Transmission of tobacco mosaic virus (TMV) by *Planococcus citri* was reported by Olitsky (1925). The same vector was found capable of transmitting TMV from a tobacco plant infected with both TMV and Storey's leaf curl (Newton, 1953). In a reappraisal of TMV transmission, *P. citri* did not transmit the virus from source plants infected with TMV alone (Orlob, 1963). Nontransmission of TMV by *Pseudococcus maritimus* has been reported also (Brcak, 1959). The current thinking of TMV transmission includes clawing by aphids as a means of transmission (Bradley and Harris, 1972; Harris and Bradley 1973a, 1973b; Harris, 1977). Furthermore, purified, poly-L-ornithine treated TMV can be transmitted by aphids, and vector transmission of some other viruses can be assisted by a helper component (Harris, 1977, 1979; Pirone, 1977, Pirone and Harris, 1977). Clawing of mealybugs and/or the presence of Storey's leaf curl as a helper may be considered as attributes of transmission. Therefore, it appears premature to exclude mealybugs altogether as possible vectors of TMV in some cases.

Other sporadic observations on mealybugs as virus vectors include *P. maritimus* which was recorded as a vector of an unspecified mosaic disease (Elmer, 1922, 1925). An unidentified mealybug species was reported vector of a bean mosaic disease (Fajardo, 1930). Hughes and Lister (1953) noted that lime dieback was transmitted by *Ferrisia virgata*.

Mealybug wilt of pineapple was long considered to be caused by wilt inducing toxic secretions of *Dysmicoccus brevipes* and *D. neobrevipes* introduced into the plant while feeding. In a reappraisal Carter (1963) concluded, however, that a transmissible latent factor is also involved which "is presumed to be a latent virus."

Most recently Gollifer *et al.* (1977) reported that the dasheen mosaic virus (DMV) is transmitted by *P. citri* and *Pseudococcus longispinus*.

Some of these reports have not been confirmed or contradictory evidence is available. In other cases there is not enough evidence to sustain virus entity, or the information reported is insufficient to help the case study of viruses, mealybugs and cocoa towards a more generalized picture.

2.1.2 The Virus-Mealybug-Cocoa Complex

The evidence of mealybug transmission of plant viruses is founded on the extensive research with the cocoa swollen shoot disease. The disease was first reported by Steven (1936) in Ghana, and Posnette (1940) showed through successful graft transmission that the disease was caused by a virus. A detailed description of the virus followed (Posnette, 1947).

The search for vectors started in about 1940 (Cotterell, 1943; Posnette, 1941), and mealybugs were confirmed to be vectors of the virus (Box, 1945; Posnette and Strickland, 1948). Subsequently, mealybugs were reported vectors of cocoa viruses elsewhere, notably in Ivory Coast, Nigeria, Sierra Leone, Sri Lanka, Togo and Trinidad. Thorold (1975) gives a detailed account of these discoveries together with suspected virus diseases of cocoa. The most recent addition to the collection of cocoa viruses is from Sumatra (Kenten and Woods, 1976).

2.1.2.1 The Viruses. Cocoa virus isolates collected in different localities differ usually in symptoms and may also differ in host range, virulence and characteristics of mealybug transmission. The relationship between the different virus isolates is not quite clear. Thresh and Tinsley (1959) classified the West African mealybug transmitted isolates into cocoa swollen shoot virus (CSSV) and cocoa mottle leaf virus (CMLV). Kenten and Legg (1967, 1971) showed, however, that the two viruses have serological and other similarities and should be referred to as CSSV. The status of cocoa Trinidad virus (CTV) is not clear even if the dissimilarities between CSSV and CTV seem to point out that these are different viruses (Baker and Dale, 1947; Kirkpatrick, 1950; Posnette, 1944). Less is known about the Ceylon cocoa virus (CCV) which also is transmitted by mealybugs (Carter, 1956). For the sake of convenience CCV, CMLV, CSSV and CTV are treated as distinct viruses in this review.

Two other cocoa viruses are known, the cocoa yellow mosaic virus (CYMV) and the cocoa necrosis virus (CNV), but they are not transmitted by mealybugs (Attafuah and Brunt, 1960; Blencowe *et al.*, 1963; Owusu, 1971a; Posnette, 1950).

2.1.2.2 The Vectors. Several valid and incompletely identified species of mealybugs are known to transmit one or more of the four cocoa viruses (Table I). Most of the information on the characteristics of transmission comes from the numerous contributions dealing with *Planococcoides njalensis*. Some other species, notably

TABLE I. Vectors of Cocoa Viruses

Vector species	Virus	Reference
Delococcus tafoensis (Strickland)	CSSV	Anon., 1958
Dysmicoccus brevipes (Cockerell)	CMLV	Posnette, 1950
	CSSV	Posnette, 1950
	CTV	Kirkpatrick, 1950
D. sp. near *brevipes* (Cockerell)	CTV	Kirkpatrick, 1950
Ferrisia virgata (Cockerell)	CSSV	Posnette and Strickland, 1948
	CTV	Kirkpatrick, 1950
Maconellicoccus ugandae (Laing)	CSSV	Posnette, 1950
Paracoccus sp. near *proteae* (Hall)	CSSV	Lister, 1953
Paraputo anomalus (Newstead)	CSSV	Posnette, 1950
Planococcoides njalensis (Laing)	CMLV	Posnette, 1950
	CSSV	Posnette and Strickland, 1948
Planococcus sp. near *celtis* (Strickland)	CMLV	Posnette, 1950
	CSSV	Posnette, 1950
P. citri (Risso)	CMLV	Posnette, 1950
	CSSV	Posnette, 1950
	CTV	Kirkpatrick, 1950
	CCV	Carter, 1956
P. kenyae (Le Pelley)	CMLV	Dale, 1957
	CSSV	Dale, 1957
P. lilacinus (Cockerell)	CCV	Carter, 1956
Pseudococcus comstocki Kuwana	CTV	Kirkpatrick, 1953a
P. concavocerarii James	CMLV	Attafuah and Brunt, 1960
	CSSV	Posnette, 1950
P. sp. near *fragilis* Brain	CMLV	Attafuah and Brunt, 1960
	CSSV	Posnette, 1950
P. hargreavesi Laing	CMLV	Posnette, 1950
	CSSV	Posnette, 1950
P. longispinus (Targioni Tozzetti)	CSSV	Posnette, 1950
P. sp. near *masakensis* James	CSSV	Posnette, 1950
Tylococcus westwoodi Strickland	CSSV	Lister and Thresh, 1954

D. brevipes, F. virgata and *P. citri*, have been studied also but to a lesser extent, and little is known about the transmission characteristics of the other vectors. *Delococcus tafoensis* can be said to be the least investigated. It appears in table I as a vector because it was preliminarily reported vector in a mimeographed report (Anon.,1958).

It has not always been generally agreed that only the mealybugs are vectors of cocoa viruses. During the search for vectors all other insects, particularly aphids and psyllids, which were at first recorded vectors of cocoa viruses (Cotterell, 1943), failed to transmit virus in repeated tests (Box, 1945; Lister, 1953; Posnette and Strickland, 1948).

Vectors other than mealybugs came into focus again as Martini (1961, 1962) reported successful transmission of two isolates of CSSV by *Aphis gossypii* in tests where 250 aphids were confined on 20 mm diameter areas on symptom bearing flush leaves for an acquisition access time (AAT) of 18 hr, and then transferred on cocoa half bean test plants at a rate of 50 aphids per half bean. Virus infection was recorded in 5-12% of the test plants. Longworth (1964a) repeated these tests with the same technique but no transmissions were recorded in 376 tests. Recently, Eguagie (1970) used about 28,050 individuals of *A. gossypii* with the same technique but 18-36 hr AAT and inoculation access time (IAT) of 12-30 hr. Virus symptoms developed in one out of the 1,038 test plants. The single transmission was attributed to mealybug contamination of test seedlings in the insectary. In front of all the evidence available it appears justified to conclude that *A. gossypii* is not a vector of cocoa viruses and that only some mealybugs are the vectors of these viruses.

2.1.2.3 The Virus Host Plant. The experimental host range of CMLV and CSSV is limited to about 30 species in the families *Bombacaceae, Sterculiaceae, Tiliaceae* and *Malvaceae* (Posnette *et al.*, 1950; Tinsley and Wharton, 1958). About 20 of these species are indigenous or well-established in West Africa. Besides cocoa, natural infection by CMLV or CSSV has been found in *Adansonia digitata* L., *Ceiba pentandra* (L.) Gaertn., *Cola chlamydantha* K. Schum., *C. gigantea* A. Chev. var. *glabrescens* Brenan and Keay, and *Sterculia tragacantha* Lindl. (Attafuah and Tinsley, 1958; Legg and Agbodjan, 1969, Posnette *et al.*, 1950). Virus isolates from *C. chlamydantha* are related to CSSV and those from *A. digitata* are related to CMLV (Attafuah and Tinsley, 1958; Legg and Bonney, 1967). Alternative host plants are not known for cocoa viruses other than CMLV and CSSV in West Africa.

2.2 TRANSMISSION CHARACTERISTICS

The process of virus transmission by mealybugs can be divided into the following consecutive stages or periods.

1. Period before acquisition feeding, during which the vectors may either starve or feed. The term used is preliminary fasting or preliminary feeding.
2. The time the vectors spend on virus infected source plant is called acquisition access time (AAT).
3. The period between the end of AAT and the next feed is termed post acquisition fasting or postacquisition feeding to indicate fasting or feeding after AAT but before the actual test feeding on healthy plants.
4. The time the vectors stay on virus indicator or test plants is called inoculation access time (IAT).

When the feeding of mealybugs is discussed in this review, the term "feeding" indicates access to feed rather than actual feeding, or that the duration of the actual feeding is not known. The term infection rate is used to indicate the proportion of test plants that show symptoms of virosis after IAT. The relationships between viruses, mealybugs and cocoa are described in the following and, as much as possible, in the same sequence as the different four stages occur.

2.2.1 Preliminary Fasting and Preliminary Feeding

It can be expected that mealybugs settle down to feed more rapidly if they have starved for some time before access to food is given. It is likewise possible that for some time immediately after fasting, the starved mealybugs actually feed more rapidly than mealybugs which are transferred from one food source to another without delay. It is possible also that virus transmission would be affected by increased settling of mealybugs to feed, and faster uptake of food and virus.

According to Kirkpatrick (1950, 1953a) preliminary fasting does not increase the ability of *P. citri* to transmit CTV, but after fasting for 6-24 hr, the mealybugs settle down to feed more readily. The results of Posnette and Robertson (1950) with CSSV suggest, however, that preliminary fasting increases also the infectivity of *P. njalensis* per se. The settling and infection rates rose with the increasing length of AAT and more rapidly with mealybugs that had starved for 12 hr than with unstarved mealybugs. The infection rate was largely independent of the settling rate when the starved and the unstarved mealybugs were compared, e.g. 48% of the starved mealybugs settled down to feed at 10 hr AAT and gave an infection rate of 30%, while 47% of the unstarved mealybugs settled at 12 hr AAT gave an infection

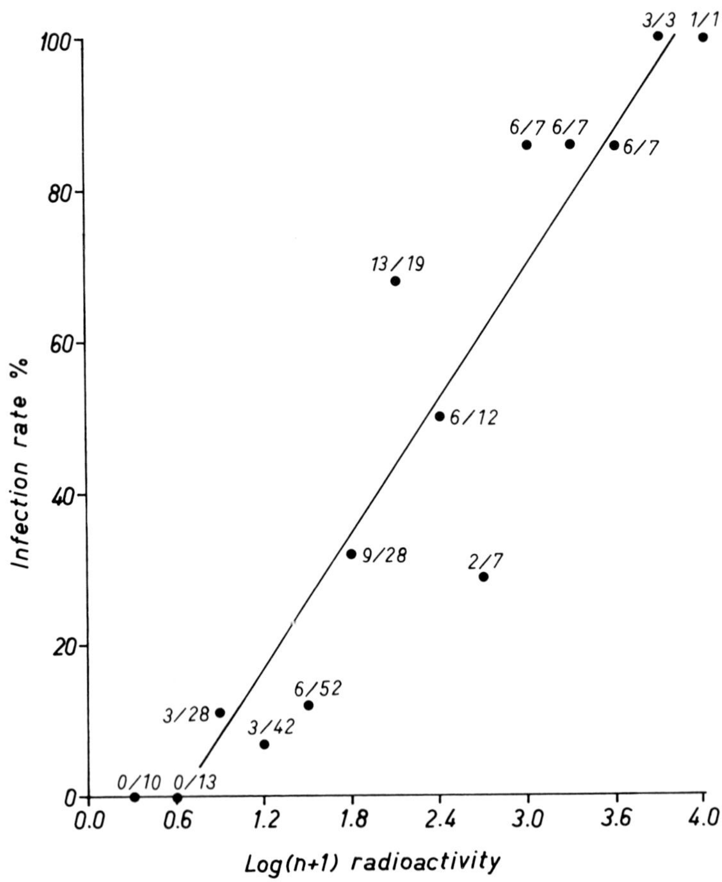

FIG. 1. Relationship between the amount of ^{32}P uptake by *P. njalensis* and infection rate. $Y = -19.11 + 29.65x$; $r = +0.956^{***}$ (from Roivainen, 1976).

rate of only 18%. Thus the infection rate increased through preliminary fasting but only part of the increase could be related to increased settling.

Dale (1954a, 1955a) tested the effect of preliminary fasting on the infection rate with CSSV and *P. njalensis*. The vectors were starved for 0-48 hr and then given an AAT of 16 hr. No consistent differences were found between the infectivity of the unstarved mealybugs and those starved for various period. Martini (1959a, 1961) used an AAT of 2 hr and various fasting periods of 15-120 min with CSSV and *P. njalensis* and F. *virgata* as vectors. Equal infection rates were recorded for both starved and unstarved mealybugs.

Roivainen (1976) used virus source plants and parafilm covered sucrose solutions both labelled with ^{32}P to obtain more direct evidence on feeding and virus transmission of *P. njalensis*. Radioactivity in the mealybugs, after access had been given to the labelled sources, was taken as evidence of settling and feeding, while the amount of radioactivity in the mealybugs was assumed to reflect the amount of food uptake. A positive relationship was found to exist between the food uptake of mealybugs and the infection rate (Fig. 1). Preliminary fasting of 24-48 hr increased the settling rate of mealybugs during the first 8 hr and probably up to 16 hr of feeding. This difference between the starved and the unstarved mealybugs evened out with longer feeds. Preliminary fasting also increased the rate of food uptake of the mealybugs (Fig. 2). With short acquisition feeds of 6-18 hr the starved mealybugs were more efficient vectors of CSSV than the unstarved ones.

These findings on the effects of preliminary fasting are in accordance with the results reported by Posnette and Robertson (1950), but not with those of Dale (1954a, 1955a) and Martini (1959a, 1961). However, the experimental details are not given in the four reports and, therefore, the right evaluation is impossible.

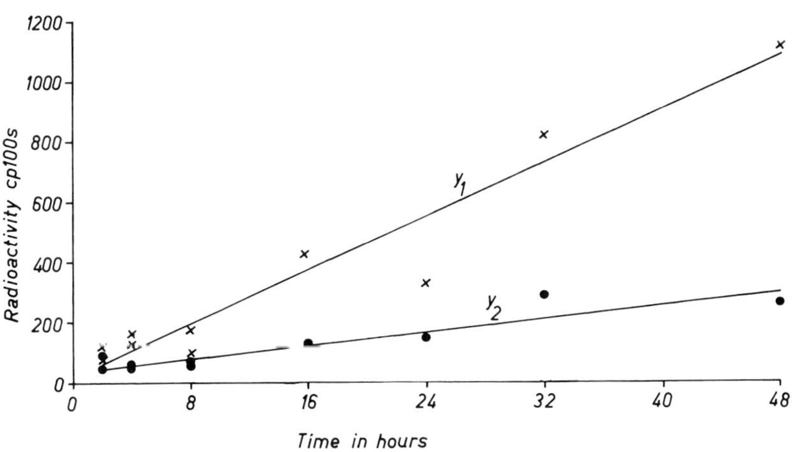

FIG. 2. Uptake of radioactivity by starved and unstarved nymphs of *P. njalensis* fed on ^{32}P-active 10% sucrose solution via a parafilm membrane. Starved nymphs: $y_1 = 19.60 + 22.03x$; $r = +0.965$***. Unstarved nymphs: $y_2 = 36.70 + 5.36x$; $r = + 0.928$*** (from Roivainen, 1976).

Based on statistically significant evidence (Posnette and Robertson, 1950; Roivainen, 1976) it is concluded that preliminary fasting increases the settling rate and amount of food uptake of *P. njalensis*, and also that these two factors contribute to the increase of infectivity of this mealybug species. Both increased settling and food uptake make it more probable that the vectors acquire virus.

2.2.2 Availability of Virus in Source Plants

The chance of mealybugs becoming infective depends on the availability of virus in the plant tissues on which the mealybugs are feeding. In general mealybugs can pick up virus from any part of the plant (Posnette and Strickland, 1948; Anon., 1950, 1963; Dale, 1954b). The only exception to this general rule is that virus is not available in the seed (Posnette, 1947).

Recently infected plants in the acute phase of infection and with severe symptoms of virus are good sources of virus while availability decreases with the age of infection when the infection turns into the chronic phase (Dale, 1958; Igwegbe, 1966a; Posnette, 1947; Posnette and Robertson, 1950). There is evidence that virus is not always fully systemic in the source plant (Posnette and Robertson, 1950).

Availability of virus does not need to be indicated by the presence of visible symptoms of disease, even if virus is more readily available in infected trees with symptoms than in trees without them (Thresh, 1958c). It is a well-established fact that cocoa viruses can "hide themselves" in the host plants, which may not show any symptoms at all or they become symptomless when the infection turns into the chronic phase. The erratic relationship between the visible symptoms and the availability of virus for mealybugs has been the main reason to hamper the control of virus diseases in cocoa through eradication of infected trees. The relationship between symptoms and availability together with the anomalies involved is well exemplified by Owusu (1972) who reported "that two plants yielded virus without ever producing disease symptoms while two plants which showed symptoms throughout most of the period of the experiment never yielded virus."

According to Posnette and Todd (1955) a virulent isolate of CSSV was more easily transmitted by *P. njalensis* than a mild isolate. From newly infected seedlings, the two isolates were transmitted almost equally at first, but the mild isolate became much less readily transmitted as the symptoms became less conspicuous.

It is generally known that tolerance to virus exists among certain types of cocoa. A tolerant plant produces mild and often transient symptoms of infection. There is evidence that virus is more easily available or more readily transmitted by mealybugs from virus sensitive than virus tolerant cocoa, however, the difference in availability is relatively small (Blencowe, 1962; Dale, 1958; Igwegbe, 1966b; Legg and Bonney, 1968; Longworth, 1964b). At most, the difference of availability of virus between sensitive and tolerant cocoa types appears to be about the same as the difference between the probability of infection while using one and two mealybugs per test plant during the IAT (Owusu, 1969, 1970, 1971b, 1973).

2.2.3 Acquisition of Virus by Vectors

During acquisition feed virus is taken up by the vectors which may subsequently transmit the virus into other plants during the next feed. Several factors may influence virus acquisition, length of the feeding being one of the most obvious. Also viruses and virus isolates may differ in their characteristics of acquisition, and there can be differences between vector species.

With different cocoa viruses the infection rate generally rises with the increasing length of AAT. Posnette and Strickland (1948) used 1A isolate of CSSV with *F. virgata* and *P. njalensis* to test the effect of AAT of 4-48 hr on the rate of virus transmission. The infection rate increased with the increasing length of the AAT and the highest rate of transmission was obtained with an AAT of 48 hr. Posnette and Robertson (1950) tested both starved and unstarved populations of *P. njalensis* with 1M isolate of CSSV. The length of AAT was 2-18 hr and an overall increase of infection rates was recorded with the increasing length of the AAT.

Dale (1954a, 1954c) used starved *P. njalensis* and CSSV isolate 1A with the AAT at 2-64 hours. The infection rate rose with an AAT of up to 16-32 hr but apparently decreased with the long AAT of 64 hr. In further studies with starved individuals of the same species and virus, the length of AAT was 1-6 days and about equal infection rates of 80-90% were obtained in all the treatments (Dale, 1954a). Dale (1958) also reported infection rates of 43-55% with the length of AAT at 1-3 days.

After a short AAT of one hour *P. njalensis* did not transmit the 1A isolate of CSSV (Dale, 1958; Roivainen, 1976), and only few transmissions occurred after an AAT of 90 minutes (Dale, 1957). With CTV and *P. citri* as the vector, the minimum length of the AAT required for transmission was 33 min, while an AAT longer than one hour did not substantially increase the infection rate (Kirkpatrick, 1950).

Roivainen (1976) studied the effect of the length of the acquisition feed on the infectivity of *F. virgata* and *P. njalensis*. Both CSSV and CMLV were used with an AAT of 6-120 hr. The optimum length of AAT for maximum rate of transmission with starved mealybugs was 48-72 hr, with a shorter or longer AAT the infection rate was lower.

There is enough evidence to conclude that with CSSV isolate 1A and *P. njalensis* as the vector, the infectivity of mealybugs increases with the length of the AAT and maximum infectivity is reached with an AAT of about two to three days. The settling rate of mealybugs causes a delay in the actual commencement of feeding and thus the optimum length of the acquisition feed proper must be shorter (Posnette and Robertson, 1950; Roivainen, 1976). With a longer AAT of four days or more the infectivity decreases with mealybugs that have gone through a preliminary fasting long enough to cause considerable increase of settling and feeding in the beginning of the AAT. These phonomena are discussed further in Section 2.2.8. The contradictory evidence reported by Dale (1954a), that equal and high infection rates were obtained with an AAT of 1-6 days, can be explained by the excessively high level of infection rate which obscured differences between the treatments.

Though there is less information on other cocoa viruses or mealybug species, no controversial evidence has been reported to the general statement given here for CSSV isolate 1A and *P. njalensis*, except the findings of Kirkpatrick (1950) which seem to indicate that the effect of length of the AAT on virus transmission by mealybugs may differ between the West African viruses and the CTV.

Liquid leaf extracts of CSSV in agar were used as sources of virus by Okusanya (1969, 1970). Virus transmission was obtained with *P. njalensis* while feeding via parafilm on the virus source during the AAT. However, none of the infected test plants showed the typical CSSV symptoms and it was suggested that a new virus or a new isolate of CSSV is involved. According to Roivainen (1976) CSSV was acquired from infected seedlings by *P. njalensis* while feeding via parafilm which, therefore, is not a barrier for transmission. Uptake of liquid leaf extract in virus preserving solution via parafilm membrane by the mealybugs was demonstrated, however, the subsequent transmission of virus into test plants was rare.

It appears to be too early to decide upon the right status of these results before further evidence from confirmatory tests is available.

2.2.4 Persistence of Virus in Vectors

After cessation of the acquisition feed, infective mealybugs may carry over virus into other feeding sites. There are factors that can influence the fate of virus during the period between the acquisition feed and the next feed. Time is the most obvious among these factors, but postacquisition fasting or postacquisition feeding and also moulting of vectors may influence the resulting retention of virus or loss of infectivity.

2.2.4.1 Persistence in Fasting or Feeding Vectors. Posnette and Robertson (1950) tested the persistence of CSSV 1M isolate after AAT in starving *P. njalensis*. The infectivity declined gradually up to 20 hr of fasting and then more rapidly, and the mealybugs infected one test plant after 34 hr of fasting. With *P. citri* as the vector there was no decline in the infection rate up to 18 hr of fasting, but after this the infectivity declined and the virus did not persist longer than 36 hr in the fasting vectors. Lister (1953) found that the 1A isolate of CSSV persisted in starving adults of *P. njalensis* up to 49 hr and in first instar nymphs up to 23 hr. In Trinidad the CTV persisted in *P. citri* up to 23 hr and in *D. brevipes* up to 8 hr after leaving the infected plant (Kirkpatrick, 1950). Martini (1961) reported that mealybugs bred on infected plants retained their infectivity unchanged up to 17 hr if starved on glass. After 24 hr the infectivity declined to about 50% and was lost entirely after 30-40 hr. However, the infectivity of vectors was lost after 17 hr on glass when a short AAT of 4 hr was given. If postacquisition fasting was done on filter paper, the infectivity was lost more rapidly than on glass.

Carter (1961) kept *P. njalensis* after AAT on agar for various periods. Infectivity of the mealybugs persisted up to 24 hr but it remained unknown whether the mealybugs actually fed or voluntarily starved on the agar. Longworth and Entwistle (1965) allowed infective mealybugs to probe into agar. Subsequently, as the stylets were withdrawn they were passed through tannic acid solution; no transmissions were obtained. In further studies infective adults of *F. virgata* were confined in

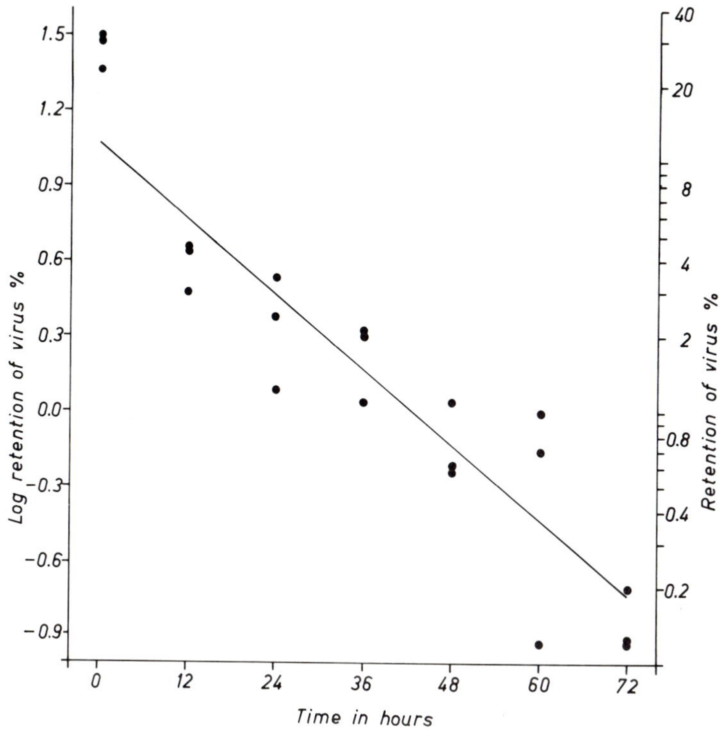

FIG. 3. Persistence of infectivity in starving nymphs of *P. njalensis* after acquisition access time. y = 1.07 – 0.0249x ; r = –0.867***. T ½ persistence = 12.1 hr (from Roivainen, 1976).

special agar cells and the stylets were observed as they probed the agar. After probing continuously in agar for up to three hours the mealybugs were still infective.

Roivainen (1976) reported that the infectivity of mealybugs declined gradually during the postacquisition fasting. The decline was best described as exponential with time (Fig. 3). In terms of a maximum the virus did not persist longer than 96 hr in the starving mealybugs. Persistence of CMLV in *P. njalensis* and CSSV in *F. virgata* were closely similar to persistence of CSSV in *P. njalensis*. These results agree largely with those reported by other authors, however, the infectivity did not remain unchanged up to 17-18 hr of postacquisition fasting as reported by Posnette and Robertson (1950) and Martini (1961), but rather started to decline earlier.

When the postacquisition feeding was carried out on healthy ^{32}P-active plants, uptake of food by mealybugs was recorded and the food uptake did not reduce the infectivity of the mealybugs (Roivainen, 1976). Thus, in line with the evidence on mealybug probing in agar (Carter, 1961; Longworth and Entwistle, 1965), it is doubtful whether postacquisition feeding on healthy plants has any effect different from postacquisiton fasting on virus persistence as such, yet postacquisition feeding may cause a delay in the settling to feed on test plants and thus reduce the infection rate.

2.2.4.2 Persistence Through Moulting. The critical test of whether a virus is transmitted after a moult is considered to give conclusive evidence on the mechanism of virus transmission. If virus is transmitted after a moult the mechanism is circulative, if not, the virus is stylet-borne, for during the moult the stylets and other sites where the virus is carried are cast with the exuvium.

Martini (1959b) used nymphs of *F. virgata* and *P. njalensis* taken from infected plants during moulting, but no transmissions were obtained with these insects. Longworth and Entwistle (1965) tested some 100 moulted nymphs of *F. virgata* but none was infective. However, transmission rates with the unmoulted control insects were so low as to make comparison impossible.

Successful transmission of CSSV isolate 1A by moulted nymphs of *P. njalensis* was reported by Roivainen (1971). Young nymphs were given an AAT of 120 hr on infected seedlings and after this they were transferred into moulting cages for 20 hr. Moulting was indicated by the presence of exuvial skins in the cages. In seven tests the moulted insects gave a total infection rate of 35/259 and the unmoulted control insects 45/254. Roivainen (1976) carried out further studies with CMLV, but only one and six test plants out of 65 became infected by the moulted and unmoulted vectors, respectively.

In accordance with discussion by Longworth and Entwistle (1965) and Roivainen (1971, 1976), the experimentation with moulted mealybugs was limited by several factors, e.g., considerable loss of infectivity with time during the postacquisition moulting period, generally low and erratic rate of moulting, and excessive handling of mealybugs. Obviously a large material is required to obtain consistent results. It is not known whether settling and feeding behavior of moulted and unmoulted mealybugs on test beans differs to such an extent as to make direct comparison of infection rates unjustified.

It is suggested that CSSV 1A isolate is transmitted after a moult and thus the virus is circulative in the vector *P. njalensis*. More data on the transmission mechanism of cocoa viruses in their vectors would be highly desirable.

2.2.5 Inoculation of Virus by Vectors

The final stage, the transmission of virus by mealybugs into a healthy plant, takes place when an infective mealybug settles down to feed on the test plant at the same time inoculating the virus. Several factors may contribute to the final observed infection rate, e.g. length of the IAT, length of latent period of virus in the test plants, number of infective mealybugs on the test plants and degree of resistance to virus of the test plants.

Virus infection is manifested in symptoms of virus in the test plants after inoculation feed. With 1A isolate of CSSV, accumulated experience has shown that Amelonado-type cocoa seedlings, infected as beans, show symptoms of virus on the first or second leaf flushes, or 17-25 days and 40-50 days after planting the beans, respectively. Symptomless seedlings can be considered healthy (Posnette and Strickland, 1948). Such "healthy" seedlings can be used to calculate the infection rate, but they cannot be considered as healthy for any critical experiment because occasionally symptoms develop during later flushes or do not develop at all as discussed in Section 2.2.2. It is generally accepted that symptom development is largely the same in other cocoa types and also if infected by other viruses transmitted by mealybugs.

2.2.5.1 Length of Inoculation Feed. Posnette and Strickland (1948) used *P. njalensis* and isolate 1A of CSSV while testing 3-192 hr IAT on the infection rate. No clear relationship was found between the infection rate and the length of the IAT. In more detailed studies with shorter feeds, the 1M isolate of CSSV was transmitted during an IAT of 20 min and the infection rate increased with the length of the IAT up to 50 min (Posnette and Robertson, 1950). Dale (1954b, 1954d, 1955a) reported that *P. njalensis* could often infect a test plant during an IAT of 15 min and the infection rate rose rapidly with increasing length of the IAT up to one hour and more slowly up to four hours. When the mealybugs settled rapidly, maximum infection rate was obtained with an IAT of two hours.

Martini (1959b) reported that *P. njalensis* transmits most frequently during the 45-75 min of IAT, and only few transmissions occur during the first 30 min or after 150 min on the test plants. *F. virgata* transmitted the CSSV isolate 1A during an IAT of six hours (Posnette and Strickland, 1948). According to Kirkpatrick (1950) the CTV was transmitted equally during an IAT of 3-7 and 18-24 hr. Transmissions were obtained also with a short IAT of 90-100 min.

The available data indicate that the inoculation access threshold is about 15 min. Probability of infection rises rapidly during the first hours of the IAT up to about three hours and there is usually no further increase after longer inoculation feeds. According to Roivainen (1976) the settling rate of mealybugs on cocoa seedlings or parafilm covered sucrose solutions reaches the maximum in 8-24 hr, and though mealybugs may settle down to feed on cocoa half beans faster (Posnette and Robertson, 1950), it is possible, in view of the short IAT required for maximum transmission, that mealybugs transmit virus into cocoa beans also during short probes preceding actual feeding on the beans. It is of some significance to note that, even if only a short IAT of some hours is required to inoculate the virus, only in few occasions a single vector has infected two test plants in a series (Posnette and Strickland, 1948; Posnette and Robertson, 1950; Roivainen, 1976). The persistence of virus in mealybugs as discussed in Section 2.2.4.1 would seem to indicate that the potential to inoculate several plants exists.

The possibility that mealybug transmitted cocoa viruses undergo a latent period in the vector before inoculation can occur was discussed by Dale (1955a). Although virus has been acquired and transmitted by *P. njalensis* in 7 hr, composed of a 4 hr AAT and a 3 hr IAT (Posnette and Strickland, 1948), or within 5 hr (Dale, 1955a), much longer periods are required on source plants to achieve maximum rates of transmission. Roivainen (1976) tested a short AAT of one hour with postacquisition fasting of 0-4 hr and IAT of one hour but no transmissions occurred. There is no direct evidence to conclude on the presence or absence of a latent period of virus in the mealybugs.

2.2.5.2 Multiple Inoculation. The infection rate generally increases with the number of vectors used during the IAT on the test plants, as demonstrated with the viruses CSSV and CTV and the vectors *F. virgata, P. citri* and *P. njalensis* (Adegbola, 1965; Dale, 1955b; Kirkpatrick, 1950; Posnette and Strickland, 1948; Posnette and Robertson, 1950; Roivainen, 1976). The increase of infection rate may be obscured by variation in and between tests as experienced by Kenten and Legg (1970) and Roivainen (1976). The infection rate increases with rising mealybug numbers according to the laws of probability (Posnette and Robertson, 1950; Roivainen, 1976) (Fig. 4). Therefore, there is no mass action effect of the virus involved, or in other words, individual mealybugs do not inject subminimal doses of virus which can accumulate to produce infection when groups of these insects are used.

Among the numerous reports where the number of mealybugs on test plants and the resulting infection rate are discussed, the probability of transmission with single mealybugs shows wide variation. With the accumulated knowledge on optimum conditions for transmission a probability of about 0.2-0.3 can be maintained with single vectors, and with more elaboration of the experimental technique a probability of 0.5 or more can be achieved (Dale, 1958; Roivainen, 1976).

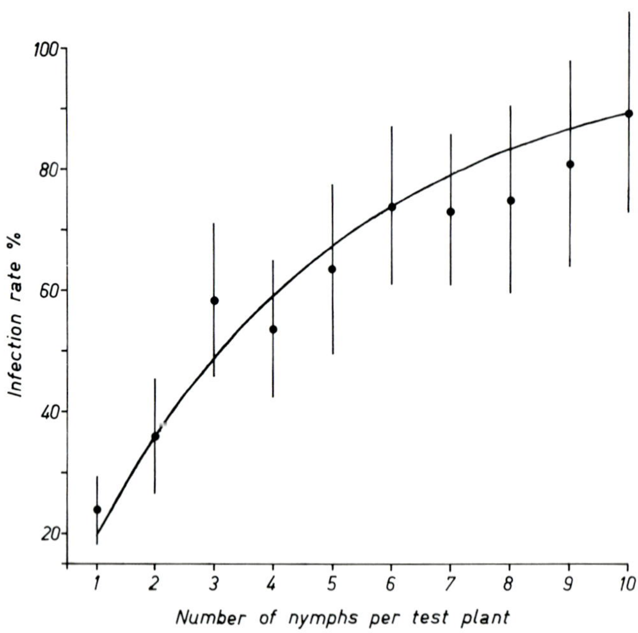

FIG. 4. Effect of number of nymphs of *P. njalensis* during inoculation access time on virus transmission. The curve describes the probability of transmission if one nymph per test plant gives an infection rate of 20%.

2.2.5.3 Inoculation of Virus Resistant/Tolerant Plants. Posnette and Todd (1951) showed that some Upper Amazon cocoa types are more resistant to infection and more tolerant of infection with CSSV than the Amelonado-type. This observation is now generally accepted. Mealybugs have been used as vectors of virus to show that some cocoa types are more difficult to infect than others, e.g. Blencowe and Attafuah (1959) reported that five infective mealybugs per bean were required to infect 50% of the susceptible Amelonado beans, while nine mealybugs were needed to attain the same level of infection with the more resistant Amazon beans. Since the discovery of mechanical transmission of CSSV by Brunt and Kenten (1960), suitable techniques were developed to test resistance and tolerance with the manual transmission of virus. It is obvious that the mealybugs received less attention during the development of the manual technique. Indeed, Kenten and Legg (1970) reported that the manual inoculation technique not only improves the sensitivity of the estimation of resistance but also considerably simplifies such studies, because the mealybug method is so difficult to standardize and is so unreliable that it is impracticable to use mealybugs for large scale tolerance and resistance studies.

During the recent years the mealybug technique has improved with indications that the technique is suitable for the selection of cocoa progenies resistant to virus (Legg and Lockwood, 1977). Comparisons between different cocoa types with the mealybug inoculation technique show that one mealybug per test plant may cause an infection rate of about 40% in susceptible cocoa, while five mealybugs are required per test plant to cause about the same infection rate in resistant cocoa (Kenten and Lockwood, 1977; Roivainen, 1976).

2.2.6 Transmission Efficiency and Specificity of Vectors

2.2.6.1 Differences Between Vector Species. Results obtained with *D. brevipes, D.* sp. near *brevipes, F. virgata* and *P. comstocki* transmitting the CTV in Trinidad suggest that there is little if any difference in the transmission efficiency between the species (Kirkpatrick, 1950, 1953b). According to Dale (1955a) *F. virgata* was much inferior to *P. njalensis* as a vector of CSSV. However, *F. virgata* seemed unsuited to the method used because of the large amount of wax produced. The insects became entangled within the paper cone attached to the seedlings and few settled down to feed. Consequently a fair comparison was not obtained between the species.

Longworth (1964c) reported infection rates of 50% from stems and 6% from leaves with *P. njalensis* but 4% from stems and 48% from leaves with *F. virgata.* This difference of CSSV transmission was explained by Entwistle and Longworth (1963) who reported that in cocoa stems 164 out of 345 stylets of *P. njalensis* were seen to end in the phloem, whereas the stylets of *F. virgata* were seen in the phloem of stems less often. The stylets of a non-vector, *P. madeiriensis*, did not end in the phloem. The authors suggested that mealybugs acquire virus from the phloem and must deposit virus in the phloem to infect a plant. Thus the less efficient transmission by *F. virgata* from the stem can be explained by the frequency its stylets reach the phloem. The low infection rate obtained with *P. njalensis* from leaves may indicate feeding preference, for Longworth (1964c) reported that *P. njalensis* prefers leaf axils, crevices between veins and pulvinus and scars and crevices on the stem. According to Igwegbe (1966a) there was little difference between the two species when leaves were used as the source, but from stems, petioles and whole seedlings *P. njalensis* was a more efficient vector. Roivainen (1976) noted that *P. njalensis* transmitted CSSV more efficiently when confined in feeding cages on stems than on leaves during the AAT. Also, *P. njalensis* was a more efficient vector than *F. virgata* when both species were kept in feeding cages on stems during the AAT.

Adegbola (1965) tested the relative efficiency of *P. njalensis* and *F. virgata* with CSSV infected source plants and either seedlings or beans as test plants. *P. njalensis* transmitted more efficiently to the beans, while *F. virgata* transmitted more efficiently to the seedlings. Adegbola (1971) did not find any difference between the transmission efficiency of the two vectors with CSSV, however, differences caused by operator skills were obvious and *P. njalensis* was claimed to be more difficult to handle than *F. virgata*.

It is clear that under identical conditions, *P. njalensis* and *F. virgata* transmit CSSV with different efficiency. The reason for this is apparently the preference for different feeding sites and the frequency the phloem is reached by the feeding insects during AAT or IAT.

Some mealybug species such as *P. njalensis, P. hargreavesi* and *P. citri* transmit a wide range of cocoa virus isolates. Indeed, *P. njalensis* has transmitted virtually every West African isolate of cocoa viruses known to have mealybug vectors. In some mealybug species, however, the ability to transmit seems to be restricted to certain viruses or isolates (Affafuah and Brunt, 1960; Legg and Bonney, 1967; Posnette, 1950). Thus, *F. virgata* is a more or less efficient vector of a number of isolates, but has consistently failed to transmit different isolates of CMLV. *P. longispinus* has failed to transmit all other isolates except CMLV 1 C and CSSV 1 M. Both *F. virgata* and *P. longispinus* seem to be unable to transmit virus from cocoa infected with CMLV isolates collected from naturally infected *A. digitata* (Dale, 1958; Legg and Bonney, 1967). The two striking controversies are that *F. virgata* may or may not transmit CSSV 1M (Attafuah and Brunt, 1960; Posnette, 1950) and, that *F. virgata* and *P. longispinus* may or may not transmit the Gandorhun isolate of CSSV (Attafuah and Brunt, 1960; Blencowe, 1961).

2.2.6.2 Differences Within Vector Species. The failure reported by Posnette and Robertson (1950) to obtain transmissions in certain experiments suggested that strains of mealybugs unable to transmit might occur within the species. In tests that followed, 10 progenies of *P. njalensis* collected from widely separated points transmitted CSSV 1M with no striking differences in the infection rates. Lister (1953) and Roivainen (1976) also compared different progenies and populations of *P. njalensis* but the results did not provide any evidence that there are non-transmitting progenies of the species or differences in efficiency of transmission between populations.

Posnette and Robertson (1950) reported that non-transmitting strains of *P. citri* do occur, however. Evidence on this was obtained with CSSV 1A, but some strains were lost through parasitism before the tests could be concluded. Experiments with CMLV 1C showed that a progeny of *P. citri* consistently failed to transmit under conditions suitable for transmission by mixed progenies of five females.

The effect of vector age on virus transmission has been a subject of considerable experimentation and discussion. Posnette and Strickland (1948) suggested that crawlers (= first and second instar nymphs) of *P. njalensis* may be more efficient vectors of CSSV 1A than older nymphs or adults. Ten to 14 crawlers infected 31%, 5-8 older nymphs 12 % and 4-5 adults 13% of the test plants.

Dale (1955a, 1955b) used the same virus and vector with three insects on test plants. The infection rate was 68% with crawlers, 74% with older nymphs and 75% with young adults. With single insects on test beans, adults infected 57% and nymphs 54% of the test plants (Dale, 1958). If unskilled assistants carried out the transfer of vectors and two adults were compared with six nymphs, the infection rates were 75 and 95%, respectively. Similarly, five adults infected 26% and 15 nymphs 61% of the test plants. These tests indicated that unskilled handling did not favor the more robust adults (Dale, 1957).

Posnette and Robertson (1950) compared adults and nymphs of *P. citri* by using five insects per test plant. The adults infected 47% of the test plants and the nymphs 29%. With the virus CTV and *P. citri* as the vector, Kirkpatrick (1950) obtained transmission rates of 33% with the first instar, 22% with the second, 24% with the third and 17% with the adults. The difference was thought to be due to insect behavior as usually more nymphs than adults settled and apparently fed on the test beans. No transmissions were obtained with parturient females or second instar males.

Five adults, four third instar nymphs and seven crawlers of *F. virgata* on test plants were compared by Posnette and Strickland (1948). The infection rates were 1/20 for adults, 2/20 for the third instar nymphs and 4/16 for the crawlers.

Nymphs of *P. njalensis* removed at birth from females, colonizing CSSV 1A infected seedlings, did not transmit virus (Dale, 1958). It is generally accepted that passage of virus into nymphs in the female reproductive organs does not occur.

At the Cocoa Research Institute of Ghana, nymphs for routine screening tests are from field collected females of *P. njalensis* which are cleaned from debris and adhering nymphs and then kept in reproduction cages. Every second day for six days nymphs are collected from the cages and the virus source plants infested. Thus nymphs after the usual 48 hr AAT vary in age from two to four days when used for the IAT. Kenten (1976) investigated the effect of age of the nymphs on transmission efficiency with batches of nymphs aged 2-3 and 3-4 days after AAT. The 3-4 days old nymphs gave an infection rate of 52% and were seemingly more efficient transmitters than those aged 2-3 days which gave an infection rate of 38%. The three successive batches of nymphs which were taken from the fasting adults in the reproduction cages were also compared. As the adults aged their nymphs were less efficient as vectors.

It is quite obvious that there are no consistent differences in the efficiency of vectors of varying age, apart from old adult females, or males that may not feed during later developmental stages. As the amount of food uptake by vectors from virus source plants is positively related to the infectivity of vectors (Roivainen, 1976), it is possible that reported efficiency differences were caused by differences in feeding behavior.

2.2.7 Effect of Physical Factors on Virus Transmission

Asomaning and Kwaka (1968) and Asomaning and Lockard (1964a, 1964b) studied the effects of low and high intensity solar radiation on the development of CSSV 1A in cocoa seedlings. Development of virus symptoms in stems and roots was faster and more prominent in seedlings exposed to radiation of low intensity. In general, the seedlings showed the usual leaf symptoms of CSSV 1A infection, however, the low intensity seedlings produced the "fern-leaf" pattern while the high intensity seedlings produced the "speckled" pattern. Longworth (1965) performed similar studies with CSSV Egbeda isolate. Bean infected cocoa seedlings were kept in darkness and in full sunlight. Germination and growth of the beans planted in full sunlight was poor and symptom recognition was confused by symptoms of scorch and mineral deficiencies. There was an indication, however, that placing the plants in the dark immediately after infection and storing them in the dark for several weeks may increase transmission rates. It appears from these studies that the intensity of solar radiation influences development of that part of CSSV infection which is visible in symptoms, but further studies are needed to show if radiation can change the proportion of actually infected plants.

Different combinations of light and dark during preliminary fasting and acquisition feed did not influence the infection rate when the inoculation feed was carried out in the dark. However, the amount of food uptake by mealybugs was higher and subsequent transmission of virus more frequent in 29-36°C. than in 19-27° C. (Roivainen, 1976).

2.2.8 Relationship Between Virus Uptake, Persistence and Infectivity of Vectors

It has been discussed earlier in this paper that the amount of food uptake by mealybugs from virus infected source plants is positively related to the infectivity status (Fig. 1). There is evidence also that the infectivity of vectors increases with the length of acquisition feed (Section 2.2.3), and that virus persistence in vectors is best described as exponential with time (Fig. 3). Roivainen (1976) demonstrated that the amount of food uptake by vectors increases with time in a linear manner in a large population of vectors. Assuming that the amount of virus uptake is also linear with time and the loss of infectivity exponential, a mathematical expectation of the infectivity increase of vectors was calculated (Fig. 5). The experimental infection rate means were in accordance with the mathematical expectation up to 72 hr AAT. However, with longer acquisition feeds there was a significant decrease in the infection rate which was considered due to the decreased amount of food uptake by pre-

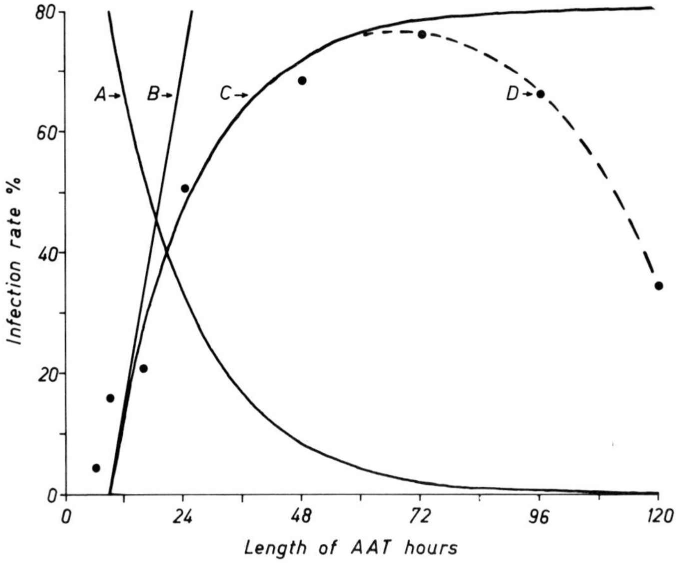

FIG. 5. Relationship between linear uptake of food (virus) by mealybugs and exponential loss of infectivity (virus) in mealybugs. A case with 50% settling rate at 9 hr AAT. Infection rate maximum 80%. A: exponential loss of infectivity with time (I) = $80e^{-0.0578(t-9)}$. B: linear uptake of food with time (F) = $-41.62 + 4.62t$. C: increase of infectivity during AAT (I) = $80[1 - e^{-0.0578(t-9)}]$. D: suggested decrease of infection rate with long AAT due to disappearance of increased feeding caused by preliminary fasting (from Roivainen, 1976).

starved vectors at later stages of prolonged feeding, or in other words, the disappearance of increased amount of food uptake caused by preliminary fasting.

2.2.9 Type of Transmission

Twelve characteristics of stylet-borne (non-persistent) and circulative (persistent) viruses transmitted by homopterous insects have been selected to analyze properties of cocoa viruses with their mealybug vectors (Table II). The accumulated data partly presented in this paper has been used in the analysis. However, where evaluation of data has been more difficult or definition of a characteristic is not exact, the judgment may reflect the opinion of the present author.

Although the mealybug transmitted cocoa viruses have some anomalous properties, on the whole they have more characteristics of circulative viruses than stylet-borne viruses. In view of this, these viruses are best grouped with persistent circulative viruses than non-persistent stylet-borne viruses.

TABLE II. Properties of Cocoa Viruses Transmitted by Mealybugs According to Characteristics of Stylet-borne and Circulative Viruses[a]

Stylet-borne viruses	Circulative viruses		Property of cocoa viruses
Short transmission cycle (seconds)	Long transmission cycle (hours)	C	Minimum cycle 5-7 hr, longer more typical
No latent period in vector	Latent period in vector	NK	
Pre-AAT fasting increases vector efficiency	Pre-AAT fasting does not increase vector efficiency	S	Increases settling, food uptake and vector efficiency
Short AAT (minutes) more efficient than long	Long AAT (hours) more efficient than short	C	Minimum about 2 hr, 48-72 hr most efficient
Vector capable of infecting one or few plants	Vector capable of infecting many plants	S	Two plants
Vector specificity not marked	Vector specificity marked	C	Only mealybug vectors, specificity of *F. virgata* and *P. longispinus*
Virus non-persistent in vector	Virus persistent in vector	C	Maximum persistence 72-96 hr, 50% persistence about 12 hr
Virus not retained in vector through a moult	Virus retained in vector through a moult	C	
Virus generally transmitted by mechanical means	Virus not generally transmitted by mechanical means	C	Mechanical transmission occurs after chemical processing
Virus affects mainly epidermal tissue	Virus affects mainly conductive tissue	C	
Virus not recoverable from haemolymph of vector	Virus recoverable from haemolymph of vector	NK	
Vector not infective after inoculation of virus into haemocoele	Vector infective after inoculation of virus into haemocoele	NK	

[a]S = property of cocoa viruses is characteristic of stylet-borne viruses; C = property of cocoa viruses is charactertistic of circulative viruses; NK = property of cocoa viruses not known.

2.3 REFERENCES

Adegbola, M. O. K. (1965). The relative efficiency of *Planococcoides* (=*Pseudococcus*) *njalensis* (Laing) and *Ferrisiana virgata* (Ck11) in the transmission of CSSV. *Ann. Rep. Cocoa Res. Inst. Nigeria 1963-64*, pp. 43-44.

Adegbola, M. O. K. (1971). The efficiency of mealybug handlers as a factor in laboratory transmission of cocoa swollen shoot virus (CSSV). *Niger. Ent. Mag.* 2: 75-82.

Anon. (1950). Virus-host relationships. A newly-discovered locus of infection in the plant. *Ann. Rep. W. Afr. Cacao Res. Inst. 1948-49*, pp. 13-14.

Anon. (1958). *Quart. Rep. W. Afr. Cocoa Res. Inst.* 52: 5-6.

Anon. (1963). *Half-Yearly Rep. Cocoa Res. Inst. Ghana* 2: 13.

Asomaning, E. J. A., and Kwakwa, R. S. (1968). Effects of degrees of shading on the growth of virus-infected seedlings. *Ann. Rep. Cocoa Res. Inst. Ghana 1965-66*, pp. 114-125.

Asomaning, E. J. A., and Lockard, R. G. (1964a). The effects of solar radiation on the growth of cocoa infected with New Juaben virus. *Ann. Rep. Cocoa Res. Inst. Ghana 1962-63*, p. 88.

Asomaning, E. J. A., and Lockard, R. G. (1964b). Studies on the physiology of cocoa (*Theobroma cacao* L.). I. Suppression of swollen-shoot virus symptoms by light. *Ann. Appl. Biol.* 54: 193-198.

Attafuah, A., and Brunt, A. A. (1960). Classification of isolates. *Ann. Rep. W. Afr. Cocoa Res. Inst. 1958-59*, pp. 16-17.

Attafuah, A., and Tinsley, T. W. (1958). Virus diseases of *Adansonia digitata* L. (*Bombacaceae*) and their relation to cocoa in Ghana. *Ann. Appl. Biol.* 46: 20-22.

Baker, R. E. D., and Dale, W. T. (1947). Notes on a virus disease of cacao. *Ann. Appl. Biol.* 34: 60-65.

Blencowe, J. W. (1961). A virus disease of cocoa in Sierra Leone, *Ann. Rep. W. Afr. Cocoa Res. Inst. 1959-60*, p. 20.

Blencowe, J. W. (1962), Future plans for cocoa breeding in Ghana. *Rep. Cocoa Conf. London 1961*, pp. 141-145.

Blencowe, J. W., and Attafuah, A. (1959). Resistance and tolerance. *Ann. Rep. W. Afr. Cocoa Res. Inst. 1957-58*, pp. 24-25.

Blencowe, J. W., Brunt, A. A., Kenten, R. H., and Lovi, N. K. (1963). A new virus disease of cocoa in Sierra Leone. *Trop. Agric.*, Trin. 40: 233-236.

Box, H. M. (1945). Insect transmission of the 'swollen-shoot' virus in West Africa. *Nature*, London 155: 608-609.

Bradley, R. H. E., and Harris, K. F. (1972). Aphids can inoculate plants with tobacco mosaic virus by clawing. *Virology* 50: 615-618.

Brcak, J. (1959). Otnoshenie nasekomih k virusu tabachnoy mozaiki. *Vop. Virusol.* 8: 171-176.

Brunt, A. A., and Kenten, R. H. (1960). Mechanical transmission of cocoa swollen-shoot virus. *Virology* 12: 328-330.

Carter, W. (1956). Notes on some mealybugs (*Coccidae*) of economic importance in Ceylon. *Pl. Prot. Bull. F. A. O.* 4: 49-52.

Carter, W. (1961). Report to the International Cooperation Administration and USOM/Ghana. 31pp. (available at Cocoa Res. Inst. Ghana).

Carter, W. (1963). Mealybug wilt of pine-apple; a reappraisal. *Ann. N. Y. Acad. Sci.* 105: 741-764.

Cotterell, G. S. (1943). Swollen-shoot virus. *Rep. Centr. Cocoa Res. Sta. Tafo 1938-42*, pp. 51-55.

Dale, W. T. (1954a). *Quart. Rep. W. Afr. Cacao Res. Inst.* 36: 5.

Dale, W. T. (1954b). *Quart. Rep. W. Afr. Cacao Res. Inst.* 33: 7.

Dale, W. T. (1954c). *Quart. Rep. W. Afr. Cacao Res. Inst.* 35: 9.

Dale, W. T. (1954d). *Quart. Rep. W. Afr. Cacao Res. Inst.* 34: 6.

Dale, W. T. (1955a). Optimum conditions for transmission. *Ann. Rep. W. Afr. Cocoa Res. Inst. 1954-55*, pp. 33-35.

Dale, W. T. (1955b). *Quart. Rep. W. Afr. Cacao Res. Inst.* 37: 5-6.

Dale, W. T. (1957). Insect transmission studies. *Ann. Rep. W. Afr. Cocoa Res. Inst. 1955-56*, pp. 30-31.

Dale, W. T. (1958). Insect-transmission studies. *Ann. Rep. W. Afr. Cocoa Res. Inst. 1956-57*, pp. 22-24.

Dale, W. T. (1962). Diseases and pests of cocoa. A. Virus diseases. *In* "Agriculture and Land Use in Ghana" (J. B. Wills, ed.), pp. 286-316. Oxford University Press, London.

Eguagie, W. E. (1970). Non-transmittability of Offa-Igbo isolate of cocoa swollen shoot virus

(CSSV) by *Aphis gossypii* Glover *(Homoptera: Aphididae)*. *Niger. Ent. Mag.* 2: 49-55.

Elmer, C. H. (1922). Mosaic cross-inoculation and insect transmission studies. *Science* 56: 370.

Elmer, C. H. (1925). Transmissibility and pathological effects of the mosaic disease. *Res. Bull. Iowa Agric. Exp. Sta.* 82: 39.

Entwistle, P. F. (1972). Pests of cocoa. Longman Group Ltd., London.

Entwistle, P. F., and Longworth, J. F. (1963). The relationship between cacao viruses and their vectors: the feeding behaviour of three mealybug (*Homoptera: Pseudococcidae*) species. *Ann. Appl. Biol.* 52: 387-391.

Fajardo, T. C. (1930). Studies of the mosaic disease of the bean (*Phaseolus vulgaris*). *Phytopathology* 20: 469.

Gollifer, D. E., Jackson, G. V. H., Dabek, A. J., Plumb, R. T., and May, Y. Y. (1977). The occurrence and transmission of viruses of edible aroids in the Solomon Islands and the southwest Pacific. *Pest Artic. & News Summ. (PANS)* 23: 171-177.

Harris, K. F. (1977). An ingestion-egestion hypothesis of noncirculative virus transmission. *In* "Aphids as Virus Vectors" (K. F. Harris and K. Maramorosch, eds.), pp. 165-220. Academic Press, New York.

Harris, K. F. (1979). Aphid-borne plant viruses: ecological and environmental aspects. *In* "Viruses and Environment" (E. Kurstak and K. Maramorosch, eds.), pp. 311-338. Academic Press, New York.

Harris, K. F., and Bradley, R. H. E. (1973a). Importance of leaf hairs in the transmission of tobacco mosaic virus by aphids. *Virology* 52: 295-300.

Harris, K. F., and Bradley, R. H. E. (1973b). Tobacco mosaic virus: can aphids inoculate it into plants with their mouthparts? *Phytopathology* 63: 1343-1345.

Hughes, W. A., and Lister, C. A. (1953). Lime dieback in the Gold Coast, a virus disease of the lime, *Citrus aurantifolia* (Christerum) Swingle. *J. Hort. Sci.* 28: 131.

Igwegbe, E. C. K. (1966a). The preferences shown by *Planococcoides njalensis* (Laing) and *Ferrisiana virgata* (Ck11) for different feeding sites and its effect on their efficiency as vectors. *Ann. Rep. Cocoa Res. Inst. Nigeria 1964-65*, pp. 60-62.

Igwebe, E. C. K. (1966b). Availability of C.S.S.V. in different cocoa clones to *Planococcoides njalensis* (Laing). *Ann. Rep. Cocoa Res. Inst. Nigeria 1964-65*, pp. 62-65.

Kenten, R. H. (1976). Factors affecting the transmission of virus by mealybugs. *Ann. Rep. Cocoa Res. Inst. Ghana 1973-74*, pp. 45-50.

Kenten, R. H. and Legg, J. T. (1967). Some properties of cocoa mottle leaf virus. *J. Gen. Virol.* 1: 465-470.

Kenten, R. H., and Legg, J. T. (1970). Methods of assessing the tolerance and resistance of different types of cocoa to cocoa swollen-shoot virus. *Ann. Appl. Biol.* 65: 419-424.

Kenten, R. H., and Legg, J. T. (1971). Serological relationships of some viruses from cocoa (*Theobroma cacao* L.) in Ghana. *Ann. Appl. Biol.* 67: 195-200.

Kenten, R. H., and Lockwood, G. (1977). Studies on the possibility of increasing resistance to cocoa swollen-shoot virus by breeding. *Ann. Appl. Biol.* 85: 71-78.

Kenten, R. H., and Woods, R. D. (1976). A virus of the cocoa swollen shoot group infecting cocoa in North Sumatra. *Pest Artic. & News Summ. (PANS)* 22:488-490.

Kirkpatrick, T. W. (1950). Insect transmission of cacao virus disease in Trinidad. *Bull. Ent. Res.* 41: 99-117.

Kirkpatrick, T. W. (1953a). Insect pests of cacao and insect vectors of cacao virus disease. *Rep. Cacao Res. Imp. Coll. Trop. Agric. Trin. 1945-51*, pp. 122-125.

Kirkpatrick, T. W. (1953b). Notes on minor insect pests of cacao in Trinidad. *Rep. Cacao Res. Imp. Coll. Trop. Agric. Trin. 1952*, pp. 62-71.

Legg, J. T., and Agbodjan, F. X. (1969). Swollen shoot disease. *Ann. Rep. Cocoa Res. Inst. Ghana 1967-68*, pp. 23, 25.

Legg, J. T., and Bonney, J. K. (1967). The host range and vector species of viruses from *Cola chlamydantha* K. Schum., *Adansonia digitata* L. and *Theobroma cacao* L. *Ann. Appl. Biol.* 60: 399-403.

Legg, J. T., and Bonney, J. K. (1968). Availability of CSSV in sensitive and tolerant varieties of cocoa. *Ann. Rep. Cocoa Res. Inst. Ghana 1966-67*, pp. 27-28.

Legg, J. T., and Lockwood, G. (1977). Evaluation and use of a screening method to aid selection of cocoa (*Theobroma cacao*) with field resistance to cocoa swollen-shoot virus in Ghana. *Ann. Appl. Biol.* 86: 241-248.

Lister, R. M. (1953). Search for new vectors. *Ann. Rep. W. Afr. Cacao Res. Inst. 1952-53*, pp. 9-10.

Lister, R. M. and Thresh, J. M. (1954). Mealybug vectors. *Ann. Rep. W. Afr. Cacao Res. Inst. 1953-54*, p. 13.

Longworth, J. F. (1964a). *Half-Yearly Rep. W. Afr. Cocoa Res. Inst. Nigeria* 2: 20-21.

Longworth, J. F. (1964b). Availability of CSSV in different cocoa types. *Ann. Rep. Cocoa Res. Inst. Nigeria 1962-63*, pp. 31-33.

Longworth, J. F. (1964c). The preference shown by *P. njalensis* and *F. virgata* for different feeding sites, and its effect on their efficiency as vectors. *Ann. Rep. Cocoa Res. Inst. Nigeria 1962-63*, pp. 29-30.

Longworth, J. F. (1965). The effect of light on transmission of cocoa swollen shoot virus by insects and mechanical methods. *Ann. Rep. Cocoa Res. Inst. Nigeria 1963-64*, p. 46.

Longworth, J. F., and Entwistle, P. F. (1965). Virus/vector relationships. *Ann. Rep. Cocoa Res. Inst. Nigeria 1963-64*, pp. 40-43.

Martini, C. K. H. (1959a). *Quart. Rep. W. Afr. Cocoa Res. Inst.* 56: 16.

Martini, C. K. H. (1959b). *Quart. Rep. W. Afr. Cocoa Res. Inst.* 55: 12-13.

Martini, C. K. H. (1961). Transmission of Nigerian cocoa viruses. *Ann. Rep. W. Afr. Cocoa Res. Inst. 1959-60*, pp. 68-69.

Martini, C. K. H. (1962). Vectors of Nigerian cocoa viruses. *Ann. Rep. W. Afr. Cocoa Res. Inst. 1960-61*, pp. 75-76.

Newton, W. (1953). Transmission of tobacco mosaic by citrus mealybug. *Pl. Prot. Bull. F. A. O.* 2: 40.

Okusanya, B. A. M. (1969). Cocoa swollen shoot virus. *Ann. Rep. Cocoa Res. Inst. Nigeria 1967-68*, pp. 88-92.

Okusanya, B. A. M. (1970). Transmission of CSSV. *Ann. Rep. Cocoa Res. Inst. Nigeria 1968-69*, 55-56.

Olitsky, P. K. (1925). The transfer of tobacco and tomato mosaic diseases by *Pseudococcus citri. Science* 62: 442.

Orlob, G. B. (1963). Reappraisal of transmission of tobacco mosaic virus by insects. *Phytopathology* 53: 822-830.

Owusu, G. K. (1969). The availability of swollen shoot virus to mealybug vectors in tolerant and senstitive cocoa. *Ann. Rep. Cocoa Res. Inst. Ghana 1967-68*, pp. 32-34.

Owusu, G. K. (1970). Availability of cocoa swollen shoot virus to mealybugs from tolerant and sensitive cocoa. *Ann. Rep. Cocoa Res. Inst. Ghana 1968-69*, pp. 31-33.

Owusu, G. K. (1971a). Cocoa necrosis virus in Ghana. *Trop. Agric.*, Trin. 48: 133-139.

Owusu, G. K. (1971b). Availability of cocoa swollen shoot virus to mealybugs from tolerant and sensitive cocoa. *Proc. 3rd Intern. Cocoa Res. Conf. Accra 1969*, pp. 512-517.

Owusu, G. K. (1972). Acquisition of swollen shoot virus by mealybugs from cocoa plants during the period of latent infection. *Ann. Rep. Cocoa Res. Inst. Ghana 1969-70*, pp. 60-61.

Owusu, G. K. (1973). Availability of cocoa swollen shoot virus to mealybugs from tolerant and sensitive cocoa. *Ann. Rep. Cocoa Res. Inst. Ghana 1970-71*, p. 63.

Pirone, T. P. (1977). Accessory factors in nonpersistent virus transmission. *In* "Aphids as Virus Vectors" (K. F. Harris and K. Maramorosch, eds.), pp. 221-235. Academic Press, New York.

Pirone, T. P. and Harris, K. F. (1977). Nonpersistent transmission of plant viruses by aphids. *Ann Rev. Phytopathol.* 15: 55-73.

Posnette, A. F. (1940). Transmission of swollen-shoot. *Trop. Agric.*, Trin. 17: 98.

Posnette, A. F. (1941). Swollen-shoot virus disease of cacao. *Trop. Agric.*, Trin. 18: 87-89.

Posnette, A. F. (1944). Virus diseases of cacao in Trinidad. *Trop. Agric.*, Trin. 21: 105-106.

Posnette, A. F. (1947). Virus diseases of cacao in West Africa. I. Cacao viruses 1A, 1B, 1C and 1D. *Ann. Appl. Biol.* 34: 388-402.

Posnette, A. F. (1950). Virus diseases of cacao in West Africa. VII. Virus transmission by different vector species. *Ann. Appl. Biol.* 37: 378-384.

Posnette, A. F., and Robertson, N. F. (1950). Virus diseases of cacao in West Africa. VI. Vector investigations. *Ann. Appl. Biol.* 37: 363-377.

Posnette, A. F., and Strickland, A. H. (1948). Virus diseases of cacao in West Africa. III. Technique of insect transmission. *Ann. Appl. Biol.* 35: 53-63.

Posnette, A. F., and Todd, J. McA. (1951). Virus diseases of cacao in West Africa. VIII. The search for virus resistant cacao. *Ann. Appl. Biol.* 38: 785-800.

Posnette, A. F., and Todd, J. McA. (1955). Virus diseases of cacao in West Africa. IX. Strain variation and interference in virus 1A. *Ann. Appl. Biol.* 43: 433-453.

Posnette, A. F., Robertson, N. F., and Todd, J. McA. (1950). Virus diseases of cacao in West Africa. V. Alternative host plants. *Ann. Appl. Biol.* 37: 229-240.

Roivainen, O. (1971). Circulative transmission of cocoa swollen shoot virus by the mealybug *Planococcoides njalensis* (Laing) (*Homoptera: Pseudococcidae*). *Proc. 3rd Intern. Cocoa Res. Conf. Accra 1969*, pp. 518-521.

Roivainen, O. (1973). Viruses and Coccoidea. *In* "Viruses and Invertebrates" (A. J. Gibbs, ed.), pp. 455-462. North Holland Publishing Company, Amsterdam and London.

Roivainen, O. (1976). Transmission of cocoa viruses by mealybugs (*Homoptera: Pseudococcidae*). *J. Scient. Agric. Soc Finl.* 48: 203-304.

Steven, W. F. (1936). A new disease of cocoa in the Gold Coast. *Gold Coast Farm.* 5: 122. (reprinted *Trop. Agric.*, Trin. 14: 84).

Thorold, C. A. (1975). Diseases of cocoa. Clarendon Press, Oxford.

Thresh, J. M. (1958a). The control of cacao swollen shoot disease in West Africa. *Tech. Bull. W. Afr. Cocoa Res. Inst.* 4:1-36.

Thresh, J. M. (1958b). The spread of virus disease in cacao. *Tech. Bull. W. Afr. Cocoa Res. Inst.* 5: 1-36.

Thresh, J. M. (1958c). The availability of cacao swollen-shoot virus to mealybugs feeding on infected trees. *Ann. Rep. W. Afr. Cocoa Res. Inst. 1956-57*, pp. 78-81.

Thresh, J. M. and Tinsley, T. W. (1959). The viruses of cacao. *Tech. Bull. W. Afr. Cocoa Res. Inst.* 7: 1-32.

Tinsley, T. W. and Wharton, A. L. (1958). Studies on the host ranges of viruses from *Theobroma cacao* L. *Ann. Appl. Biol.* 46: 1-6.

Chapter 3

WHITEFLIES

V. Muniyappa

Department of Plant Pathology
University of Agricultural Sciences
Hebbal, Bangalore
India

3.1	INTRODUCTION	39
3.2	WHITEFLY VECTORS	41
	3.2.1 Species of Whiteflies Involved in Transmission	41
	3.2.2 Races of *Bemisia tabaci*	45
3.3	YELLOW MOSAIC DISEASES	45
3.4	YELLOW VEIN MOSAIC DISEASES	54
3.5	LEAF CURL DISEASES	57
3.6	MOSAIC DISEASES	65
3.7	SUSPECTED WHITEFLY TRANSMITTED DISEASES	71
3.8	DISCUSSION AND CONCLUSIONS	72
3.9	ACKNOWLEDGMENTS	73
3.10	REFERENCES	73

3.1 INTRODUCTION

Whitefly transmitted diseases mainly occur in tropical countries (Bird and Maramorosch, 1975; Bird and Sanchez, 1971; Nene, 1972; Pal and Tandon, 1937; Thung, 1932) but are also known to occur in sub-tropical and temperate countries (Fig. 1) (Costa, 1965; Duffus, 1965; Hildebrand, 1959; Kirkpatrick, 1931; Loebenstein and Harpaz, 1960). Whitefly transmitted diseases have become important on legume crops, tomato, cotton and chilli in various parts of the world (Bird and Maramorosch, 1975; Cohen and Harpaz, 1964; Gamez, 1971; Nene, 1972; Pierre, 1975; Varma, 1962). Most evidence indicates that they are caused by viruses. Isometric, rod-shaped and dimers or siamese twin particles have been associated with infectivity (Galvez and Castano, 1976; Goodman *et al.,* 1977; Hollings *et al.,* 1976; Maramorosch, 1975; Sharp and Wolf, 1949).

ISBN 0-12-326450-2

FIG. 1. Distribution of whitefly transmitted diseases in various parts of the world. The circles indicate the prevalence of disease, and numbers indicate the places. 1. California, 2. Arizona, 3. S. Texas, 4. Missouri, 5. Georgia, 6. Maryland, 7. Washington, 8. Florida (N. America), 9. Cuba, 10. West Indies, 11. Puerto Rico (Caribbean countries), 12. Venezuela, 13. Columbia, 14. Ecuador, 15. Sao Paulo, 16. Brazil (S. America), 17. Rhodesia, 18. Congo, 19. Nigeria, 20. Ghana, 21. Chad, 22. Sudan, 23. Egypt (Africa), 24. England, 25. Belgium, 26. Germany, 27. Rumania, 28. Bulgaria (Europe), 29. Israel, 30. W. Pakistan, India (31. Delhi, 32. Uttar Pradesh, 33. Poona, 34. Hyderabad, 35. Karnataka, 36. Kerala, 37. Madras), 38. Ceylon, 39. Sumatra, 40. Java, 41. Philippines, 42. Japan (Asia).

Three species of whiteflies, *Bemicia tabaci*, *Trialeurodes vaporariorum* and *T. abutilonia*, are known to transmit plant viruses. There are about 1156 known whitefly species worldwide. Many species may be acting as virus vectors (Mound and Halsey, 1978).

Many reviews have been published on whitefly transmitted diseases (Bird and Maramorosch, 1978; Costa, 1969, 1976; Orlando and Silberschmidt, 1946; Varma, 1962). More than 70 diseases have been reported on cultivated and weed plants. Most of the reports are based on symptomatology and transmission. The nature of the causal agent is known in only a few instances. Some of these diseases might have been caused by the same or related strains of the same agent. Costa (1976) grouped whitefly transmitted dieseases into (a) mosaic, (b) leaf curl, and (c) yellowing types. The aim of the present review is to describe individual diseases occurring naturally in various cultivated plants and weed hosts, so that it might serve as a ready reference for those involved in research on whitefly transmitted diseases. Based on the symptoms, the various diseases have been grouped into (a) yellow mosaic (Figs. 2 & 3), (b) yellow vein mosaic (Fig. 4), (c) leaf curl (Fig. 5), and (d) mosaic and other types of diseases (Fig. 6).

3.2 WHITEFLY VECTORS

The whiteflies are small piercing and sucking insects belonging to the family Aleyrodidae in the order Homoptera. Since whiteflies are not always white and are not flies, the name is a misnomer. They suck plant juices through their slender stylets which in the resting position lie protected in a groove in the rostrum. Eggs are laid on the lower surface of leaves. There are four nymphal instars. The nymphal instars of whiteflies are firmly attached to the host plant except for a short period during the first stadium. Whiteflies penetrate plant tissues intercellularly and feed in the phloem (Capoor, 1949; Avidoz, 1956; Varma, 1962; Costa, 1969).

3.2.1 Species of Whiteflies Involved in Transmission

Bemisia tabaci Gen. (Fig. 7) is responsible for transmitting a very large number of viruses in many parts of the world (Capoor, 1959; Karam Singh, 1961; Varma, 1962; Costa, 1969). Heinze (1959) reported the following species as vectors of plant viruses: *Bemisia fascialis* Jacq., *B. goldingi* Corb., *B. inconspicua* Russel, *B. manihotis* Frappa, *B. nigeriensis* Corb., *B. rhodesiaensis* Corb., *B. tabaci* Gen., *B. tuberculata* Bondar, *B. voyssierei* Frappa, *Trialeurodes abutilonea* Hold., *T. vaporariorum* Westw., *T. natalensis* Corb., *Aleurotrachelus socialis* Bondar, *A. floccosus* Mask. Russel (1957) concluded that *B. inconspicua, B. costa-limai, B. signata, B. bahiana, B. longispina, B. gossypiperda* var *mosaicivectura, B. goldingi, B. rhodesiaensis, B. hibisci* and *B. nigeriensis* were synonyms of *B. tabaci*.

Tobacco leaf curl was transmitted by *B. tuberculata* and *A. socialis* in Venezuela

FIG. 2. Yellow mosaic symptoms on the leaves of *Phaseolus lunatus* caused by yellow mosaic virus.

FIG. 3. Yellow mosaic symptoms on the leaves of *Dolichos lab-lab* caused by yellow mosaic virus.

FIG. 4. Yellow vein mosaic symptoms on the leaves of *Abelmoschus esculentus* caused by yellow vein mosaic virus.

FIG. 5. Leaves of tomato showing typical leaf curl symptoms caused by leaf curl virus.

FIG. 6. *Jasminum samba* leaves showing chlorotic ring spot mosaic caused by chlorotic ring spot virus.

FIG. 7. *Bemisia tabaci* adults feeding on leaf of *Lycopersicon esculentum*.

(Wolf *et al.*, 1949) and *T. natalensis* in Transval (McClean, 1940), where the whitefly vector *B. tabaci* has not been recorded. Sweet potato yellow dwarf was transmitted by *T. abutilonea* in Maryland (Hildebrand, 1960). The ability of *T. natalensis* to transmit tobacco leaf curl is doubtful, but *B. gossypiperda* is involved in its transmission (Costa, 1969). Russel (1957) stated that *B. gossypiperda* is a synonym of *B. tabaci*. The ability of *B. tuberculata* and *A. socialis* to vector plant viruses is also questionable (Wolfe *et al.*, 1949). *Trialeurodes vaporariorum* was found to be the vector of beet pseudo yellows virus in California (Duffus, 1965), and *T. abutilonea* reportedly transmits sweet potato yellow dwarf virus in Maryland and sweet potato mosaic in Georgia (Hildebrand, 1959, 1960).

B. tabaci, T. abutilonia and *T. vaporariorum* definitely are known vectors of plant viruses (Duffus, 1965; Hildebrand, 1959; Costa, 1969; Bird and Maramorosch, 1978).

3.2.2 Races of Bemisia Tabaci

Bird (1957) and Bird and Sanchez (1971) noticed that *B. tabaci* reared on *Jatropha gossypifolia* would not feed or breed on *Sida carpinifolia*. Colonies of the species from *Sida* would not feed or breed on *J. gossypifolia*. The races have been called ecological biotypes by Flores and Silberschmidt (1958).

Pruthi and Samuel (1942), Costa and Bennett (1950), Flores and Silberschmidt (1958) and Nene (1972) reared *B. tabaci* on several different species of plants.

Mound and Halsey (1978) listed 1156 species of Aleyrodidae from around the world. At present, very few species are known vectors. But considering the total number of species and the few that have been tested as vectors, it seems likely that many more species are involved in transmission.

3.3 YELLOW MOSAIC DISEASES

Yellow mosaic of Acalypha indica. Chenulu and Phatak (1965) reported on yellow mosaic diesease of *Acalypha indica* from India. It is characterized by mottling, pronounced chlorosis and, in severe cases, size reduction and puckering of leaves. The host range appears to be confined to *Acalypha* spp. The disease was transmitted by *Bemisia tabaci*.

Yellow mosaic of hollyhock. Yellow mosaic disease of hollyhock (*Althaea rosea*) was reported in England in 1957 (Smith, 1957). Singh and Misra (1971) reported yellow mosaic on hollyhock from India. The disease is characterized by typical mosaic symptoms, usually consisting of pronounced yellow and green mottle. Sometimes leaves show yellowing of the veins. Often the bands are mixed up to form yellow pockets and may cover a large part of the leaf lamina (Singh and Misra, 1971). The virus was transmitted by *Bemisia tabaci* to *Malva sylvestris*, *Physalis peruviana* and *Nicotiana tabacum* (Singh and Misra, 1971).

Yellow mosaic of pigeon pea. The incidence of yellow mosaic disease of pigeon pea (*Cajanus cajan*) from India was first reported by Williams *et al.* in 1968 and, subsequently, by Nene in 1972. The disease first appears in the form of yellow diffused spots (not limited by veins and veinlets) scattered on leaf lamina. Leaflets show broad yellow patches alternating with green color. Often the entire lamina turns yellow in color (Nene, 1972).

Nene (1972) reported that pigeon pea is a host for mung bean yellow mosaic virus and that the yellow mosaic which is seen in nature on pigeon pea is caused by mung bean yellow mosaic virus. Muniyappa and Reddy (1976) transmitted horsegram yellow mosaic virus from horsegram to *Cajanus cajan* and vice versa. Newton and Peiris (1953) reported a similar yellow mosaic from Ceylon which was transmissible by sap to *Psophocarpus tetragonolobus*. Nene (1972), unable to obtain mechanical transmission with his virus, mentioned that the pigeon pea yellow mosaic in Ceylon is caused by a different virus. Maramorosch *et al.* (1974) reported that the mosaic of *Rhynchosia minima* and the yellow mosaic of *Cajanus cajan* are one and the same disease. No information was obtained concerning the nature of the whitefly-borne agent when thin sections of infected material were examined by electron microscopy.

Yellow mosaic of jute. The disease was first noticed at Poona, India in 1962 by Varma. Jute plants (*Corchorus trilocularis*) infected by yellow mosaic virus are slightly dwarfed and tend to flower early. The infected plants show bright yellow patches on leaves, with slight reduction of leaf lamina. The severe strain is characterized by yellowing and stunting accompanied by the absence of flowers or seed, and the mild strain produces yellowing flecking (Varma *et al.*, 1966; Bisht and Mathur, 1964).

The yellow mosaic strains were successfully transmitted by *Bemisia tabaci* to *Corchorus capsularis*, *C. olitorius* and wild *Corchorus* sp., and by grafting to *Hibiscus sabdoriffa* and *H. cannabinus* (Varma *et al.*, 1966; Bisht and Mathur, 1964).

Yellow mosaic of horsegram. Williams *et al.* (1968) observed the yellow mosaic symptoms on horsegram (*Dolichos biflorus*) in South India. The disease is transmitted by whiteflies, *Bemisia tabaci* (Muniyappa *et al.*, 1975). Symptoms appear 7 to 12 days after inoculation. Faint yellow discolorations appear on young leaves in the beginning and, as the disease progresses, the leaves show mosaic mottling. These mottles are irregular and greenish yellow in color intermixed with normal green patches. Later on, these mottles become enlarged and turn bright yellow. Finally the infected leaves have a bleached appearance. Severely infected plants are stunted with leaves reduced in size (Williams *et al.*, 1968; Muniyappa *et al.*, 1975). Muniyappa *et al.* (1976a, 1977) transmitted the disease to eleven hosts: *Cajanus cajan*, *Centrosema* sp., *Glycine max*, *Phaseolus aconitifolius.*, *P. atropurpureus*, *P. aureus*, *P. lathyroides*, *P. lunatus*, *P. mungo*, *P. vulgaris*, and *Teramnus uncinatus*. When 100 horsegram cultures were screened for yellow mosaic resistance, 68 cultures showed severe infection, 22 moderate infection, and 10 mild infection (Muniyappa *et al.*, 1976b, 1979).

Single insects, *B. tabaci*, are able to transmit the virus (Muniyappa and Reddy, 1976). The minimum acquisition feeding period was found to be 30 min, after which another 6 hr is required for insects to become viruliferous. The minimum inoculation feeding period was found to be 10 min. The whiteflies remain infective for up to 12 days.

The incidence of yellow mosaic disease of horsegram–up to 100 per cent in summer and early *kharif*, but much lower in late *kharif* and *rabi* seasons–is correlated with the vector (*B. tabaci*) population in the field. Disease incidence and vector density were highest from February to early June (Muniyappa *et al.*, 1978).

Dolichos biflorus has been reported to be the host of yellow mosaic virus of *P. lunatus* (Capoor and Varma, 1948) and of mung bean yellow mosaic virus (Nariani, 1960). The symptoms and host range of horsegram yellow mosaic are comparable with yellow mosaic of mung bean and *P. lunatus* (Capoor and Varma, 1948; Nariani, 1960). Cross transmission of virus from horsegram to leguminous plants and from leguminous hosts to horsegram indicate that the disease found on *Cajanus cajan, Centrosema* sp., *Dolichos biflorus, Cylcine max, P. aureus, P. latheroides, P. lunatus, P. vulgaris, P. aconitifolius, P. mungo, P. atropurpureus* and *Teramnus uncinatus* is caused by the same virus or by strains of the same virus (Muniyappa and Reddy, 1976).

Yellow mosaic of Dolichos lab-lab. This disease was first reported by Capoor and Barma in 1950 from Poona (India). Detailed investigations on the disease were later published by Ramakrishnan *et al.* (1972).

Capoor and Varma (1950a) described the symptoms as faintly discolored patches that become bright yellow and eventually spread till only a few spots and stripes of green tissue remain. Ramakrishnan *et al.* (1972) observed three patterns of symptoms: (i) yellow mosaic with pitting, (ii) yellow mosaic with alternating yellow and green patches, and (iii) green mosaic with alternating light and green areas. These different types of symptoms were found to be manifestations of the same *Dolichos* yellow mosaic virus.

Dolichos yellow mosaic was reportedly transmitted to *Phaseolus aureus, P. mungo, Dolichos biflorus*, and *Vigna sinensis* (Samuel *et al.*, 1972; Subramanian *et al.*, 1971). But the author was unable to transmit to *D. biflorus, P. aureus* and *Vigna sinensis*. The virus can be transmitted only by the whitefly, *Bemisia tabaci*. The incubation period was found to be 14 to 20 days (Capoor and Varma, 1950a). Ramakrishnan *et al.* (1972) reported that the whitefly, *B. tabaci* was an inefficient vector, transmitting to only 3 per cent of the plants tested.

Chinnadurai and Nair (1971) analyzed diseased and healthy leaves for chlorophyll, chlorophyllase, chloroplastic protein and cytoplasmic protein contents and found that the chloroplastic protein is utilized for virus protein synthesis, thus resulting in a depleted chlorophyll level.

Yellow mosaic of Euphorbia prunifolia. In Sao Paulo, Brazil, the common weed *Euphorbia prunifolia* is affected at all seasons by yellow mosaic virus (Costa and

Bennett, 1950). The disease is characterized by light green, bright yellow, golden or whitish mottling on a green background, either along the veins or in interveinal areas (Costa and Bennett, 1950; Bird *et al.*, 1975).

The vector is *B. tabaci* race *sidae*. Both the acquisition and inoculation feeding periods were found to be 30 - 60 min. The minimum incubation period of the virus in the vector is 4-5 hr. Vectors can retain inoculativity for at least 20 days. The female whitefly transmits the virus almost twice as efficiently as the male (Costa and Bennett, 1950; Bird *et al.*, 1975). Virus is transmissible to a low percentage of plants by juice inoculation, but not through seeds of diseased plants (Bird *et al.*, 1975). It was transmitted by whiteflies to *D. stramonium*, *P. vulgaris*, *N. tabacum* and *G. max*, and by mechanical means (31% infection) from *E. prunifolia* to *D. stramonium* (Bird *et al.*, 1975).

The virus appears to be closely related to the one reported from Brazil (Costa and Bennett, 1950) but unrelated to other rugaceous entities studied in Puerto Rico. It does not infect *Rhynchosia minima*. The virus has been partially purified by differential and density gradient centrifugation, and its spherical particles visualized via electron microscopy (Bird *et al.*, 1975).

Yellow mosaic of soybean. Reifman and Polivanova (1969) observed the yellow mosaic on soybean (*Glycine max*) in the Soviet Far East. The percentage of plants infected by yellow mosaic virus increased with age up to 49 days, and reduction in yields varied from 21 to 61 per cent (Singh *et al.*, 1971). The virus is transmissible both by grafting and *Bemisia tabaci* (Singh *et al.*, 1971; Suteri, 1974; Suteri and Srivastava, 1975). Its host range is restricted to legumes. Aldicarb applied twice to the soil was highly effective in reducing virus infection (Singh *et al.*, 1971; Rataul and Singh, 1974). The yellow mosaic increased the protein content of seeds. However, seeds from plants of susceptible varieties that have been exposed to infection contain less oil than those from healthy plants (Suteri, 1975). The virus resembles mung bean yellow mosaic virus in symptomatology, host range and transmissibility (Suteri, 1974).

Yellow mosaic of Jacquemontia tamnifolia. A weed, *Jacquemontia tamnifolia*, was infected with yellow mosaic virus in Puerto Rico. Foliage of affected plants that are exposed to full sunlight develops a golden yellow mosaic, whereas plants that fail to receive enough sunlight usually show rather pale yellow mosaic symptoms. The virus infects and produces leaf curl symptoms in *Jacquemontia pentantha, J. tamnifolia, Phaseolus aborigineus, P. acutifolius, P. coccineus, P. vulgaris, Vigna aconitifolius, V. angularis, Nicotiana tabacum* (Bird *et al.*, 1975).

The virus, which is transmitted by *Bemisia tabaci* race *sidae*, causes leaf distortion on beans and *D. stramonium* but fails to protect tobacco from invasion by *Rhynchoia* mosaic virus. The agent produces striking symptoms on tobacco (Bird *et al.*, 1975).

Yellow mosaic of Jatropha gossypifolia. Bird (1957) reported on a yellow mosaic disease of *Jatropha gossypifolia* Linn. from Puerto Rico. Capoor (1939) transmitted through grafts a similar disease affecting *Jatropha* at Poona. The disease produces bright yellow symptoms on leaf laminae. Seedling (but not mature) tobacco plants were inoculated in the greenhouse by whiteflies from infected *J. gossypifolia* and developed symptoms closely resembling those of tobacco leaf curl virus (Bird, 1957). The virus is transmissible by grafting and single *B. tabaci* (Bird, 1957). The minimum acquisition feeding period of 2 hr and a minimum inoculation feeding period of 10 min are required for transmission.

The virus infects *Croton lobatus, Jatropha gossypifolia, J. multifida, J. podagrica, Jacquemontia tamnifolia, Nicotiana tabacum, Phaseolus vulgaris, Glycine max* (Bird *et al*., 1975). It could not be transmitted to some of the wellknown hosts of tobacco leaf curl virus like *Datura stramonium* and tomato (Bird, 1957). The *Jatropha* virus was thought to be similar to the agent studied on *Euphorbia prunifolia* in Brazil (Costa and Bennett, 1950), but the result of host range studies indicated that the two diseases were not related (Bird *et al*., 1975).

Yellow mosaic of tomato. Debrot *et al*. (1963) and Malaguti (1963) reported on the tomato (*Lycopersicon esculentum*) yellow mosaic disease from Venezuela, and Varma *et al*. (1975) noted the yellow mosaic disease of tomato from India. The disease is characterized by yellow mosaic, reduction in size and twisting of leaves. Typically all leaves are wrinkled and speckled with sharp-edged, irregular yellow spots, varying from very small to large, covering the whole surface. Dwarfing is usual and, sometimes, apical leaves are thin, folded inwards and twisted to one side (Debrot *et al*., 1963; Varma *et al*., 1975). The virus is acquired by whiteflies, *B. tabaci*, in 30 min, with a minimum inoculation feeding time of 15 min. A single whitefly can transmit the disease. The latent period of the virus is 21-24 hr. It is acquired and retained by the vector throughout life (Debrot *et al*., 1963; Varma *et al*., 1975). Lastra and Vzcategui (1975) transmitted the virus mechanically. The virus resembles yellow leaf curl virus of tomato reported from Israel (Cohen and Nitzany, 1966).

The yellow mosaic or golden yellow mosaic virus has been partially purified from infected *Nicotiana glutinosa*. Infectivity is associated with twin spheroidal particles, each measuring 12-13 nm in diameter (Matyis *et al*., 1975).

Yellow top virus of tomato. Cohen *et al*. (1961) reported tomato (*Lycopersicon esculentum*) yellow top virus from Israel. Summer and winter tomatoes in the Jordan valley were seriously attacked by tomato yellow top virus. The virus was transmitted to *Datura stramonium*, samsun tobacco and potato by *B. tabaci*. The vector acquired the virus after several hours of feeding on diseased plants; the latent period in the insect was 24 hr; the inoculation feeding time was 1 hr; insects retained infectivity for over 10 days and could acquire virus from inoculated tomato before symptom expression. The disease could be controlled by spraying with endrin every 4-5 days (Cohen *et al*., 1963).

Yellow mosaic of Merremia quinquefolia. Mosaic disease of *Merremia quinquefolia* has been known to occur in Puerto Rico for a number of years (Bird and

Sanchez, 1971). Affected *Merremia* plants show bright yellow blotches intermixed with green color. The virus is transmitted by *B. tabaci* race *sidae* to a number of plant species (Bird *et al.*, 1975). The virus infects species within the Convolvulaceae, Leguminosae, Malvaceae and Solanaceae, producing brilliant yellow mosaic symptoms. The species infected were *Calonyction aculeatum, Ipomoea angustifolia, I. crassucaulis, I. dissecta, I. tiliaceae, I. triloba, P. tuberosa, Merremia aegyptia, M. quinquefolia, Pachyrrhizus erosus, Phaseolus aborigineus, P. acutifolius, P. coccineus, P. lunatus, P. vulgaris, Vigna umbellata, Malachra capitata, Datura stramonium, Lycopersicon esculentum,* and *Nicotiana tabacum* (Bird *et al.*, 1975).

Yellow mosaic of mung bean. Nariani (1960) first described a yellow mosaic disease of mung bean (*Phaseolus aureus*) and reported that the causative agent is transmissible to urd bean by *Bemisia tabaci*. Ahmed and Harwood (1973) reported yellow mosaic from West Pakistan. Mung bean yellow mosaic virus transmitted by whitefly is very widespread, causes serious losses, and affects several pulses, including soybean (Nene, 1972, 1973b; Nair and Nene, 1974b; Williams *et al.*, 1968).

First symptoms of the disease on mung bean appear on young leaves in the form of mild scattered yellow specks or spots. The next tri foliate leaf emerging from the growing apex shows irregular alternating yellow and green patches. Leaf size is generally not much affected; but, sometimes, the green areas are slightly raised, and the leaves do show slight puckering and reduction in size (Nariani, 1960; Nene, 1972). Nene (1972) described two types of yellow mosaic symptoms on urd bean—a yellow (chlorotic) mottle and a necrotic mottle—and suggested that these may represent two different diseases caused either by different strains of the same virus or by different viruses. Nair (1971) established that the yellow mottle and necrotic mottle are two types of symptoms produced by mung bean yellow mosaic virus in different cultivars.

The whitefly, *Bemisia tabaci*, can transmit the virus after a 15 min feeding on a source plant. The incubation period in the insect is ca. 4 hr. The inoculation feeding period is ca. 15 min. Single viruliferous insects can transmit the virus. Most efficient female adults of a population can retain infectivity for 10 days. Adult females proved to be more than three times as efficient as males in transmitting the virus (Nene, 1972; Nair and Nene, 1973a, 1973b; Rathi and Nene, 1974). The adults preferred healthy green leaves to diseased yellow ones for oviposition. Transovarial transmission does not appear to occur. The vector is found in nature all the year round and can be carried long distances on wind currents (Nene, 1972; Nair and Nene, 1974a).

Using *B. tabaci*, Nariani (1960) transmitted the virus to *Phaseolus acutifolius, P. aconitifolius, P. lathyroides, P. aureus* and *P. mungo, Glycine max,* and *Dolichos biflorus*. Nene *et al.* (1971) transmitted the virus to *Brachiaria ramosa, Eclipta alba, Xanthium strumarium* and *Cajanus cajan*. Rathi (1972) reported *Cosmos bipinnatus* and *P. vulgaris* as hosts for mung bean yellow mosaic. Muniyappa and Reddy (1976) transmitted horsegram yellow mosaic virus to mung bean. Two selections of urd bean (CUT-9 and D-6-7) and two selections of mung bean (P 364-68 and P 366-68) have been found to be resistant to yellow mosaic (Nene *et al.*, 1972; Dey and Singh, 1973). Muniyappa *et al.* (1976c) failed to identify the resistance source in mung bean varieties.

The insecticides baythion, malathion and parathion give quicker kill of whitefly adult populations when followed by ambithion. Mineral oil is best and can give 100% kill in 30 min, compared to 1 hr by baythion, malathion and parathion (Nene, 1972, 1973a).

A fungus, *Paecilomyces farinosus*, was discovered on the whitefly (Nene, 1972). The possibility of using this fungus for biological control of whitefly populations has not been tested.

Yellow mosaic of Phaseolus longepedunculatus. Flores and Silberschmidt (1966) reported on a mosaic disease of *P. longepedunculatus* from S. Paulo, Brazil. The virus, which induces clearcut, angular, chlorotic spots on leaflets, is transmitted by *Bemisia tabaci*. It is tentatively regarded as a strain of infectious chlorosis of Malavaceae (*Abutilon* mosaic virus) (Flores and Silberschmidt, 1966) distinct from that described on *Leonurus sibiricus* on which it produces yellow mosaic symptoms (Flores and Silberschmidt, 1962). *P. longepedunculatus* is a host for bean golden yellow mosaic (Flores and Silberschmidt, 1966). It appears that the yellow mosaic disease of *P. longepedunculatus* is caused by bean golden yellow mosaic virus.

Yellow mosaic of lima bean. Capoor and Varma (1948) reported on the yellow mosaic disease of bean, *Phaseolus lunatus*, from Poona (India). *Phaseolus lunatus* yellow mosaic is perhaps the most important bean virus occurring on the island of Puerto Rico (Bird *et al.*, 1972). The leaves of infected plants develop scattered, slightly discolored patches which gradually turn bright yellow. Infected plants are not dwarfed and continue to grow normally, but pod production is greatly reduced (Capoor and Varma, 1948). Severe golden mosaic and yellow vein banding symptoms were observed at Ibadan (Williams, 1975).

Bemisia tabaci is the vector (Capoor and Varma, 1948; Williams, 1975). *B. tabaci* requires 1 hr to acquire virus and 15 min to inoculate it. The incubation period in the vector is 8 hr. Vectors retain infectivity for 15 to 20 days. And single insects can transmit the virus.

The virus infects *Macroptilium lathyroides, Phaseolus aborigineus, P. coccineus, P. lunatus, P. limensis, P. vulgaris, P. aureus, Dolichos biflorus, Canavalia ensiformis* and *Vigna radiata* (Capoor and Varma, 1948; Bird *et al.*, 1975). Of the *P. vulgaris* varieties tested, 'Asgrow Stringless Greenpod' and 'Davis Whitewax' were found to be resistant (Capoor and Varma, 1948).

Yellow mosaic of *Phaseolus longepedunculatus* from Brazil (Flores and Silberschmidt, 1966), *Phaseolus lunatus* yellow mosaic from India (Capoor and Varma, 1948) and Puerto Rico (Bird *et al.*, 1975), *Sida carpinifolia* mosaic from Puerto Rico (Bird *et al.*, 1975), *Phaseolus vulgaris* yellow mosaic from Brazil (Costa, 1965), *Dolichos biflorus* yellow mosaic (Muniyappa *et al.*, 1975), and mung bean yellow mosaic (Nariani, 1960) may be caused by the same virus. Muniyappa *et al.*

(1976a) transmitted yellow mosaic virus of *D. biflorus* to 11 leguminous hosts, and yellow mosaic from *P. aureus, P. lathyroides, P. lunatus, P. mungo, P. vulgaris* and *Glycine max* to *D. biflorus*, thus suggesting that the yellow mosaic diseases of all these hosts are caused by the same virus.

Yellow mosaic of French bean (bean golden mosaic virus). The disease was first recorded in Brazil in 1961 (Costa, 1965). It is present in practically all important bean growing areas of that country (Costa, 1975). The disease is also severe in coastal Pacific plains of America (Gamez, 1971). It produces a bright yellow (golden) mosaic on the leaves of French bean, *P. vulgaris*, and other leguminous plants. The first symptoms may be noticed as downward rolling of young leaves which later show the golden mosaic symptoms. Infected plants show definite stunting and reduction in size of leaves and yield (Costa, 1975).

The vector, *Bemisia tabaci*, acquires virus after feeding for 3 hr, inoculates it to plants after a test feed of 3 hr, and retains it for up to 21 days (Costa, 1965; Gamez, 1971). Bean cultivars 'Topcrop' and 'Stringless Green Refugee' that are inoculated mechanically develop typical golden mosaic symptoms (Meiners *et al*., 1975). The virus infects *Canavalia ensiformis, Cyamopsis tetragonolobus, Desmodium aculeatum, Blycine hispida, Nicandra physaloides, P. longependunculatus, Solanum esculentum, Teramnus uncinatus* (Flores and Silberschmidt, 1966). *Sida rhombifolia* and *S. micrantha* show resistance to golden mosaic virus, and *Malva parviflora* is a host plant of bean golden mosaic virus (Costa, 1965; Flores and Silberschmidt, 1966).

Bean golden mosaic virus from Brazil was described by Costa (1965) as perhaps related to the virus that causes yellow mosaic in lima beans (Capoor and Varma, 1948) and yellow mosaic in horsegram (Muniyappa and Reddy, 1976) in India. Yellow mosaic of *Dolichos lab-lab* seems to be different from bean golden mosaic virus (Ramakrishnan *et al*., 1972). Muniyappa and Reddy (1976) and Muniyappa *et al*. (1976a) indicated that yellow mosaics occurring in nature on *Phaseolus aureus, P. lunatus, P. vulgaris, Glycine max, P. lathyroides, P. mungo* and *Dolichos biflorus* are probably caused by the same virus. *Dolichos lab-lab* yellow mosaic seems to be entirely different from these diseases.

In ultrathin sections of bean leaves, isometric particles measuring 20-25 nm in diameter were observed in the sieve tube lumina of golden mosaic-infected plants but not in healthy controls (Kitajima and Costa, 1974). Galvez and Castano (1976) were able to secure highly infectious partially purified extracts that contained numerous "doublet" particles measuring about 32 x 19 nm (single particles: 15-20 nm). Goodman *et al*. (1977) found similar particles in purified extracts from 'Top Crop' plants inoculated with a Puerto Rican isolate of bean golden yellow mosaic virus (BGYMV). The DNA of BGYMV sediments in sucrose density gradients and in an analytical ultracentrifuge as a single component. It is the first plant pathogenic virus described with a single-stranded DNA genome (Goodman, 1977).

Yellow mosaic of Rhynchosia minima. Bird (1962) reported on this disease of *R. minima* in Puerto Rico. The virus is transmitted from diseased to healthy *R.*

minima by *B. tabaci*, but it is not mechanically transmissible. It produces an early diffused yellowing that later becomes very bright. In beans the *Rhynchosia* virus causes leaf malformation, yellowing, witches' broom and stunting (Bird *et al.*, 1975). The virus infects *Salvia splendens, Cajanus cajan, Canovalia ensiformis, C. marithima, Crotalaria juncea, Glycine max, Macroptilium lathyroides, Pachyrrhizus erosus, Phaseolus aborigineus, P. acutifolius, P. coccineus, P. lunatus, P. trichocarpus, P. vulgaris, Rhynchosia reticulata, Vigna aconitifolia, V. angularis, Abelmoschus esculentus, Gossypium hirsutum, Malachra capitata, Oxalis barrelieri, Nicotiana acuminata, N. alata, N. bonariensis, N. clevelandii, N. glauca, N. glutinosa, N. knightiana, N. maritima, N. paniculata,* and *N. tabacum* (Bird, 1962; Bird *et al.*, 1975). The *R. minima* mosaic virus is similar to bean double yellow mosaic virus in India (Bird, 1962).

Yellow mosaic of rose. This disease of rose (*Rosa* sp.) is reported from India (Bhargava and Joshi, 1962) and is characterized by the appearance of yellow colored bands adjacent to veins. In severe cases these bands spread and cause clearing of the veins and veinlets. As the disease progresses, veins and veinlets become more yellow and exhibit a yellow net pattern. The virus is transmissible by *B. tabaci* and by budding.

Yellow mosaic of Sida carpinifolia. Bird (1958) reported on a virus from Puerto Rico causing yellow blotches on leaves of *Sida carpinifolia*. *B. tabaci* acquired the virus in 15 min and took a minimum of 20 min to inoculate it to plants. The virus persisted in the vector for 7 days (Bird, 1958; Varma, 1962).

The virus infects plants in two families, Malvaceae and Solanaceae: *Althaea rosea, Bastardia viscosa, Corchorus aestuans, Hibiscus brasilensis, Sida acuminata, S. aggregata, S. carpinifolia, S. cordifolia, S. glabra, S. glomerata, S. humilis, S. procumbens, S. rhombifolia,* and *S. urens* in Malvaceae, and *Nicotiana tabacum* in Solanaceae (Bird *et al.*, 1975). It is considered related to *Sida* viruses (infectious chlorosis of Malvaceae) studied by Orlando and Silberschmidt (1946) and by Costa (1954) in Brazil. Attempts made in Puerto Rico to transmit the *Sida* mosaic virus to *P. vulgaris* and other legumes have not been successful. Thus it seems that this virus is not closely related to the Brazilian virus. It produces leaf curl symptoms on tobacco, suggesting that it may be related to tobacco leaf curl virus (Bird *et al.*, 1975).

Yellow mosaic of Cowpea. Nene (1972) reported on the yellow mosaic disease of cowpea, *Vigna sinensis*, from Pantnagar, India. The disease has also been reported from Pakistan (Ahmed, 1978). Sharma and Varma (1976) observed yellow flecks on cowpea. The first symptoms of the disease appear in the form of irregular, bright yellow dots scattered all over the young trifoliates. These yellow spots later enlarge, covering the entire lamina with yellow patches alternating with green color. Sometimes, necrotic areas also develop in the center of such yellow spots. The virus, transmitted by *B. tabaci* (Ahmed, 1978), does not seem related to mung bean yellow mosaic virus. Also, Muniyappa and Reddy (1976) were unable to transmit yellow mosaic of horsegram to cowpea.

3.4 YELLOW VEIN MOSAIC VIRUS DISEASES

Yellow vein mosaic of bhendi. Yellow vein mosaic of bhendi (*Abelmoschus esculentus* Syn. *Hibiscus esculentus*) was first mentioned by Kulkarni (1924) in a report on virus diseases from Bombay, India. It has been reported by Jha and Mishra (1955) and Varma and Mukherjee (1955) that the disease was there all the time in other States also. Fernando and Uduravan (1942) reported yellow vein banding disease of Bhendi from Ceylon. Uppal *et al.* (1940) first established the viral etiology of the disease. Yellow vein mosaic is extremely severe in the Matale district of Ceylon where it reduces yield by 50% (Newton and Peiris, 1953). *Ochrovena hibiscae* and *Hibiscus* virus 1 are names that have been proposed for the virus (Capoor and Varma, 1950b).

The first perceptible symptom is clearing of the small veins, and then of the larger ones. In severely diseased plants, the young leaves develop a generalized chlorosis rather than actual mosaic patterns. All growth produced subsequent to infection is stunted, the leaves being undersized and the petioles abnormally short. Flowering is sparse and few fruits are formed. Thickening of the veins on the lower leaf surface is a feature of the disorder (Uppal *et al.*, 1940).

The virus is transmitted by *Bemisia tabaci* and by grafting (Uppal *et al.*, 1940; Capoor and Varma, 1950b; Varma, 1952, 1955a, 1955b). *B. tabaci* acquires the virus in 30 min and, sometimes, in 15 min. The minimum incubation period in the vector is 7 hr. Infective flies retain their infectivity until death. The virus cannot pass from the fly to its offspring through the egg. Females are slightly more efficient than males in transmitting the disease. Ten whiteflies per test plant were required to obtain 100% infection (Varma, 1952). And, *Bemisia tabaci* could carry three different viruses, yellow vein mosaic of pumpkin, yellow vein mosaic of Behndi and yellow mosaic of lima beans simultaneously for 6 days (Varma, 1955b).

The yellow vein mosaic is transmissible to *Hibiscus abelmoschus, H. moscheutas, H. tetraphyllus, H. cannabinus, H. sabdariffa, Althaea rosea, Malvastrum tricuspidatum, Croton sparsiflorus, Ageratum conyzoides,* and *Zinnia elegans.* The species *A. monihot* var. *pungens, A. crinitus, H. vitifolius,* and *H. panduraeformis* are immune (Capoor and Varma, 1950b; Jha and Mishra, 1955; Nariani and Seth, 1958).

Chlorophyll a and b are reduced and chlorophyllase enzyme activity increased due to virus infection. Carotene and xanthophyll contents are reduced, while the respiration of infected tissue is increased (8.33%) and photosynthesis decreased. Infected plants show lower contents of total sugars and higher N, crude fiber and carbohydrates than healthy plants; the C:N ratio is lower in infected plants (Mandahar and Singh, 1971, 1972; Ramiah *et al.*, 1972; Potty and Wilson, 1974).

Equal amounts of bacteria and actinomycetes were found in phyllospheres of both infected and healthy *A. esculentus* plants at flowering. *Aspergillus candidus, A. tamarii* and *Alternaria* sp. were found on infected leaves, and *Pythium* sp. and *Phytophthora* sp. on healthy plants (Obliswami *et al.*, 1974; Potty and Wilson, 1975).

Capoor (1953) and Capoor and Varma (1950b) recommend eradicating *H. tetraphyllus*, spraying the crop once every 3 weeks with fish oil to control the whitefly population, and keeping the fields clear of weeds. The virus spread in *H. esculentus* was restricted by 4 sprays each of parathion (0.02%), oxydemeton-methyl (0.02%) or dimethioate (0.05%) at 10-day intervals from germination, or one application of phorate 10-G (15 kg/ha) at sowing (Sastry and Singh, 1973a, 1973b).

Yellow vein mosaic of Ageratum conyzoides. Gadd and Loos (1941) reported on the yellow vein mosaic disease of *A. conyzoides* from Ceylon. The yellow veins of top leaves of infected plants appear as yellow bands narrowing and growing less yellow towards the leaf base. On the undersurface, the vein banding is less conspicuous, but the veins are yellow and stand out prominently owing to the interveinal areas being concave. Infected plants are stunted (Gadd and Loos, 1942).

The virus is transmitted by grafting and by *Bemisia tabaci*. Single whiteflies are able to transmit the virus. After acquiring the virus, vectors require 8 hr to transmit the disease (Gadd and Loos, 1941; Varma, 1962; Nair and Wilson, 1970).

The virus infects *Browallia elata* and *Vernonia* sp. On the former host a typical curl results, whereas on *Vernonia* typical yellow vein mosaic symptoms occur (Varma, 1962). The virus responsible for vein banding is considered to be tobacco leaf curl virus (Gadd and Loos, 1941).

Yellow vein mosaic of Blumea neilgherrensis. Wilson and Potty (1972) reported on yellow vein mosaic disease of *B. neilgherrensis* from Kerala, India. The disease is characterized by pronounced vein clearing in the leaves. In severely affected leaves, the yellowing diffuses into surrounding interveinal areas, thus leading to mottling. Veins on the lower side are sometimes slightly thickened. The virus is transmitted by *Bemisia tabaci*. Cross inoculation with the virus causing yellow vein mosaic on *Hibiscus esculentus* and *Ageratum conyzoides* in the vicinity failed.

Yellow vein mosaic of Croton sparsiflorus. *Croton sparsiflorus*, a common weed throughout India, was found infected with yellow vein mosaic virus. In the beginning, slight yellow clearing is seen along the veins, and this yellowing becomes very bright as the disease progresses. The virus is transmitted by grafting and by *Bemisia tabaci* (Varma, 1962; Nair and Wilson, 1969, 1970).

Yellow vein mosaic of pumpkin. This disease of pumpkin, *Cucurbita pepo*, was first noticed at Poona, India (Varma, 1962). Infected plants show a conspicuous yellow vein mosaic in leaves, with no reduction of leaf lamina. The disease is transmitted by *Bemisia tabaci* (Varma, 1955b). The vector needs 15 min to acquire the virus and the minimum incubation period in the vector is 9 hr. The virus is transmitted by its whitefly vector to vegetable marrow, *C. moschata, Cucumis sativus, Citrullus fistula* and *Luffa acutangulata* (Varma, 1962). This mosaic is similar to the mosaic described from Israel in symptom expression and transmission (Cohen and Nitazany, 1960), but it differs in that it is mechanically transmissible (Varma, 1962).

Yellow vein mosaic of cucurbits. Cohen and Nitazany (1960) and Nitazany *et al.* (1964) reported on a yellow vein mosaic disease of *Cucumis sativus* from Israel. It produces conspicuous vein clearing, chlorosis and, finally, a general necrosis in cucumber plants. The virus is transmitted by *Bemisia tabaci*, and it is the first cucurbit virus known to be transmitted mechanically (Cohen and Nitzany, 1960; Harpaz and Cohen, 1965). The virus does not persist in its vector for more than 4-6 hr. Vector efficiency is low, and at least 30-35 whiteflies per plant are needed to infect 80% of test cucumbers, though females were 3 times as efficient as males. Fasting before acquisition did not influence vector infectivity, but fasting afterwards delayed loss of it.

Yellow vein mosaic of Legendra mollisima. A virus disease occurring in *Lagendra mollisima* was recorded by Varma (1962) in Poona, India. The virus produces typical yellow vein symptoms on leaf laminae. The virus is transmissible by grafting and by *Bemisia tabaci*. The virus could not be transmitted to tobacco, tomato, bhendi, *Ageratum conyzoides* and *Malvastrum coromandelianum*, thus suggesting that the virus is different from other yellow vein mosaic diseases.

Yellow vein mosaic of Leucas. Capoor observed typical yellow vein mosaic disease symptoms in two species of *Leucas*, *L. cilliata* and *L. stelligera*, in nature at Poona, India. Transmission was obtained through grafting and by *Bemisia tabaci* (Varma, 1962). The author also observed the yellow vein mosaic symptoms in *Leucas aspera* at Bangalore. Its relation to other whitefly transmitted diseases is not known.

Yellow vein mosaic of Malvastrum coromandelianum. Rao and Varma (1961) observed yellow vein mosaic symptoms in *Malvastrum coromandelianum* at Poona, India. The disease is not sap transmissible but is easily transmissible through grafting and *Bemisia tabaci*. The insect can acquire the virus in 30 min. The incubation period in the vector is 4 hr and vector inoculativity persists for 16 days (Rao and Varma, 1964). The virus infects malvaceous hosts, safflower, tobacco and petunia. Because of its inability to cause typical leaf curl infection in *Zinnia* and tobacco and also because it does not cross protect tobacco from leaf curl virus, it is considered distinct. It seems to be an *Abutilon* mosaic virus strain similar to that found in America and other countries (Schuster, 1964), though in India it does not infect *A. indicum* (Rao and Varma, 1961; Varma, 1962; Rao and Varma, 1964).

Yellow net vein virus of Mulberry. Raychaudhuri *et al.* (1961) reported on the yellow net vein virus of mulberry, *Morus indica*, from India. The virus produces typical yellow netting in leaf laminae. It is transmissible only by inarch grafting, and no vector is known, though a species of whitefly (*Bemisia* sp.) has been suggested as a possible vector.

Yellow net virus of tobacco. Dhingra and Nariani (1961) reported yellow net virus in tobacco, *Nicotiana tabacum* variety White Burley, from Delhi. The disease causes a yellow network in veins and veinlets. Transmission is by grafting and *Bemisia tabaci*. The virus is not sap transmissible. The virus infects several *Nicotiana* sp., *Beta vulgaris*, aster and tomato. Cross protection tests revealed that the virus is distinct from tobacco leaf curl virus.

Yellow vein mosaic of Rosa bourbiana. Sastry (1966) reported yellow vein mosaic from India. The veins and veinlets become yellow in color, and leaves exhibit a yellow net appearance. The plants flower normally except for slight reduction in flower size. The virus is transmitted by grafting, budding and *B. tabaci*.

Yellow vein mosaic of Salvia splendens. Varma (1974a) first reported the yellow vein mosaic on *Salvia splendens* from India. It produces vein clearing, vein necrosis, dwarfing and reduced flower production. It was easily transmitted by grafting and *Bemisia tabaci*, but not by sap inoculation. Details about the disease are not known.

Yellow net disease of Zinnia elegans. A disease of *Zinnia elegans* that produces vein yellowing, rugosity of leaves, and stunting of the plant was observed in 1975 in Lucknow, India (Srivastava *et al.*, 1977). The disease could not be transmitted by mechanical means, but it is easily transmissible by *Bemisia tabaci*. *B. tabaci* acquired the causal agent in an hour. The latent period of the causal agent in the vector was 15-18 hr. Inoculation feeding period was found to be 1 hr. Once the agent was acquired by whiteflies, it persisted in them for their life span. The disease was transmitted to *Datura stramonium, Gomphrena globosa, Althaea rosea* and *Ageratum conyzoides*.

3.5 LEAF CURL VIRUS DISEASES

Leaf curl of chilli. Leaf curl of chilli, *Capsicum annum*, has been reported by several investigators (Husain, 1932; Pal and Tandon, 1937; Park and Fernando, 1938; Mishra *et al.*, 1963; Dhanraj and Seth, 1968). Johnpulle (1939) and Cherian (1936) considered it to be caused by mites or thrips *Scirtothrips dorsalis*. Mishra *et al.* (1963) established the viral etiology of the disease.

The most characteristic symptom of the leaf curl is the abaxial curling of the leaf blade. The interveinal areas of the leaf are sometimes distorted, and laminae are partially suppressed, especially near the petiole, resulting in the formation of narrow strap shaped leaves (Park and Fernando, 1938). Curling of the leaves accompanied by puckering and blistering of interveinal areas and thickening and swelling of the veins is observed in infected plants. In advanced stages axillary buds are stimulated to produce clusters of leaves that are reduced in size. The whole plant assumes a bushy appearance with stunted growth (Mishra *et al.*, 1963).

The virus is transmissible by wedge grafting and *Bemisia tabaci* (Park and Fernando, 1938; Mishra *et al.*, 1963; Dhanraj and Seth, 1968).

The virus is transmitted by *B. tabaci* to *Nicotiana tabacum* var. Harrison's Special,

N. tabacum var. White Burley, *N. glutinosa, Lycopersicon esculentum, Petunia hybrida, Capsicum annum, C. frutescens, C. microcarpum, C. sinensis, C. pubescens, C. pendulum*, and *Crotalaria juncea* (Mishra *et al.*, 1963). *Capsicum* varieties puri red and puri orange were found to be resistant to the disease (Mishra *et al.*, 1963; Dhanraj *et al.*, 1968).

Infection is caused by two strains of tobacco leaf curl virus. One strain does not produce enations in solanaceous hosts; whereas, the other, a more severe strain, does (Dhanraj and Seth, 1968).

Leaf curl of papaya. The leaf curl of papaya, *Carica papaya*, was first recorded by Thomas and Krishnaswami (1939) from Madras, India. The disease is characterized by severe curling, crinkling and distortion of the leaves accompanied by vein-clearing and reduction in leaf size. Leaves become leathery and brittle and the interveinal areas are raised on the upper surface due to hypertrophy which produces rugocity. Rolling of leaves downward and inward in the form of an inverted cup and thickening of veins is also observed. The petioles are twisted in a zig zag manner. Infected plants fail to flower or bear fruit. In advanced stages of the disease, defoliation takes place and growth of the plant is arrested (Thomas and Krishnaswami, 1939; Nariani, 1956). Nariani (1956) reported that papaya leaf curl virus is not mechanically transmitted, but it is easily transmitted by grafting and *Bemisia tabaci*.

Leaf curl virus of papaya infects tomato, tobacco, zinnia and hollyhock. Leaf curl virus of papaya is transmitted to tobacco and tobacco leaf curl to papaya, indicating that papaya leaf curl is caused by tobacco leaf curl virus (*Nicotiana* virus 10) (Nariani, 1960).

Leaf curl of geranium. A virus disease affecting geranium, *Geranium pelargonium*, was reported by McWhorter (1957) in the Pacific Northwest, USA. Diseased plants exhibit hyaline spots on young leaves which may turn more opaque or yellow later on and also develop small necrotic areas. The young leaves are wrinkled and deformed. The disease occurs in England (Pethybridge and Smith, 1932), in Belgium (Verplancke, 1932), in Canada (Berkeley, 1938) and Washington (Jones, 1940). The insect vector has been shown to be *Bemisia* sp.

Crinkle mosaic of soybean. Crinkle mosaic of soybean, *Glycine max*, was observed in Sao Paulo, Brazil (Costa, 1975). Crinkle mosaic may become important in areas where the presence of *Abutilon* mosaic virus in Malvaceae is abundant and there is a high density of the vector (*Bemisia tabaci*). The leaves of infected plants are smaller and show mosaic and crinkle symptoms. Crinkling of soybean is caused by *Abutilon* mosaic virus, as in the case of bean dwarf mosaic virus.

Leaf curl of cotton. Cotton (*Gossypium hirsutum*) leaf curl disease is very widely distributed in Sudan and Nigeria (Jones and Mason, 1926; Golding, 1930; Kirkpatrick, 1930a, 1931). The viral etiology of the disease was demonstrated by Golding (1930) in Nigeria and by Kirkpatrick in Sudan (1930a, 1931).

Symptoms first appear on very young leaves. Leaves that are fully grown at the time of infection usually do not develop symptoms, but partially grown ones may develop thickening of the lower surfaces of smaller veins. Such thickening starts at a number of points which gradually tend to join up until all the veins are affected. If a diseased leaf is viewed from beneath against the light, it will be seen that these thickened veins are a darker green than the rest of the leaf and, consequently, opaque instead of translucent. This appears to be the best spot characteristic for the disease, especially when it is necessary to decide if the disease is present in mild form on a young leaf or an apicalyx. New leaves produced after infection are small, exceedingly crinkled and curled at the edges, either upward or downward. Oval cup-like foliar growths are produced on the greatly thickened lower sides of primary veins. The primary stem of the plant often tends to grow taller than normal, the internodes being elongated and irregularly curved; but, sometimes, the whole plant is stunted in growth. All parts of a badly affected plant are very brittle and readily broken (Kirkpatrick, 1930a, 1930b; Tarr, 1957b; Mustafa, 1961; Nour and Nour, 1964; Couteax *et al.*, 1968; Elnur and AbuSalih, 1970; Husain, 1930). It has also been called cotton leaf crinkle disease. Tarr (1957b) named it loggerhead. Two forms of leaf curl, type 1 (LC1) and type 2 (LC2) are described on varieties Mono and Hyti in Togo. LC1 is transmitted by the insect vector, *Bemisia gossypiperda*, and LC2 possibly by jassids (Couteaux *et al.*, 1968).

Golding (1930) in Nigeria and Kirkpatrick (1930a) in Sudan first transmitted the disease by means of unidentified species of whiteflies. Golding (1930) also implicated a jassid as a vector. Kirkpatrick (1931) reported that the whitefly *Bemisia gossypiperda* is a very efficient and, in all probability, the only vector of the disease. Once infected the whitefly remains capable of transmitting the disease at least 7 days and probably throughout its life. The virus is picked up by the adult insect in just over 3 hr, but maximum infectivity is attained only after a minimum stay of 4.5 hr on the source plant. The disease is transmitted to healthy plants in 30 min. The virus is not transmitted through the egg of the whitefly. The incubation period of the virus in the cotton plant varied from 8 to 34 days. Successful transmission was obtained through *B. gossypiperda* (Cowland, 1934; Tarr, 1951; Couteaux *et al.*, 1968; Yassin and Elnur, 1970).

The virus is transmissible to *Gossypium hirsutum, G. peruvianum, G. barbadense, Hibiscus esculentus, H. cannabinus, H. sabdariffa, Althaea rosea, Malvaviscus arboreus, Corchorus facicularis, Phyllanthus niruri, Petunia* spp., and possible susceptible species include *Sida cordifolia, S. urens, Urena lobata, Abutilon* sp. (Tarr, 1951, 1957b; Nour and Nour, 1964; Elnur and AbuSalih, 1970; Kirkpatrick, 1931; Cowland, 1934). Response of 671 strains of cotton to leaf curl virus showed that *Gossypium arboreum* and *G. herbaceum* are immune, *G. hirsutum* and its varieties *punctatum* and *marie-galante* are highly resistant, and *G. barbadense* contains a large proportion of susceptible strains (Hutchinson *et al.*, 1950; Lagiere, 1968; Siddiq, 1968). The levels of total phenols, leuco anthocyanins and flavenols are higher in resistant than in susceptible strains (Mustafa, 1961).

Cotton leaf curl virus and its whitefly vector are very abundant on sakel cotton in North Gezira and Sudan. The disease is controlled by removal of sources of virus and vector control (Massey, 1934; Andrews, 1936; Boughey, 1947; Tar, 1951, 1957a, 1957b).

Leaf crumple of cotton. Dickson *et al.* (1954) reported the leaf crumple of cotton, *Gossypium hirsutum*, from Coachella and from southeastern California in the U.S.A. The primary symptom is hypertrophy of the interveinal leaf tissue, accompanied in severe cases by shortening of the veins. It may be restricted to a few spots, causing elevations of 1 to 10 mm in diameter. Vein clearing and interveinal chlorosis are sometimes observed in the foliage. A severe strain of cotton leaf crumple virus caused mosaic, puckering and downward cupping of the leaves and a symmetrical ruffing of the petals in the varieties Acla 4-41, Delta pine, Glandless and R 15 of *G. hirsutum* inoculated by grafting (Dickson *et al.*, 1954; Erwin and Meyer, 1961). The symptoms of leaf crumple are distinct from those of leaf curl in Africa and of crazytop in Arizona (Dickson *et al.*, 1954). Texas cotton vein clearing virus symptomatology includes vein clearing, leaf crumple and severe stunting resembling those of the California leaf crumple virus (Halliwell and Rosberg, 1964).

Substantial reductions in yield (ranging from 16 to 81%) have been recorded in cotton due to cotton leaf crumple virus infection (Allen *et al.*, 1960; Vanschaik *et al.*, 1962).

Transmission was obtained by grafting and the whiteflies *Trialeurodes abutilonea* and *Bemisia inconspicua* (Dickson *et al*., 1954). Laird and Dickson (1959) reported that the leaf crumple agent is transmitted by *Bemisia tabaci (B. inconspicua)* but not by *Trialeurodes abutilonea*. The acquisition threshold was 4-8 hr and inoculation feeding 1-2 hr. The virus was retained by the adult for 5 days and was not transmitted transovarially. The virus overwinters in stub or abandoned cotton (Dickson and Laird, 1960). Amorphous, spherical inclusion bodies measuring 2-10 μm in diameter were observed in nuclei of cells of infected plants (Tsao, 1963).

Leaf curl of Hibiscus rosasinensis. The leaf curl of *H. rosasinensis* is prevalent in many parts of India (Varma, 1962). The disease is characterized by curling of affected leaves to varying degrees and the formation of enations on their undersurfaces. The virus is transmitted by grafting and *Bemisia tabaci*. It is transmissible through grafting to *Althaea rosea* and *Abelmoschus moschatus* (Vasudeva *et al.*, 1953) and through *B. tabaci* to *A. esculentus, A. tuberculatus* and *H. manihot* (Varma, 1962). The virus is different from that which causes yellow vein in bhendi, and leaf curl in tobacco (Varma, 1962).

Leaf curl of balsam. Varma and Singh (1973a) observed leaf curl disease on balsam, *Impatients balsamina*, at the I.A.R.I., Delhi. Infected plants showed thickening of the lower surface of the veinlets and leaf distortion. The virus is transmissible by grafting and *B. tabaci*.

Leaf curl of Jatropha curcas. Nair and Wilson (1970) reported on the leaf curl disease of *J. curcas* from Kerala, India. Irregular chlorotic patches on young leaves and curling, crinkling and distortion of the older leaves were noted in infected plants which were generally stunted. The virus is experimentally transmitted by inarch grafting and *Bemisia tabaci*.

Leaf curl of Lupin hartwegii. Infected *Lupin hartwegii* plants in Delhi are dwarfed, have curled leaves, and produce few or no flowers. The virus is transmitted by grafting and *Bemisia tabaci*. Its host range is restricted to *Lupin hartwegii* (Varma, 1974c).

Leaf curl of tomato. Leaf curl disease of tomato, *Lycopersicon esculentum*, occurring for several years in India, caused great damage to the crop (Vasudeva and Sam Raj, 1948). The disease is widespread in many countries causing yield losses ranging from 10 to 95 per cent (Nour-Eldin *et al.*, 1969; Yassin and Abu Salih, 1972; Sastry and Singh, 1973). Symptomatology includes a pronounced dwarfing and puckering of the leaves, vein clearing and excessive branching. Dark green vein enations occur on the undersurfaces of leaflets. In case of late infections the plant bears a few fruits; but in earlier ones the symptoms are much more pronounced, and the few flowers that are formed bear reduced floral structures that never attain normal size. Such flowers seldom open and soon drop off. Flowers that develop after infection are invariably sterile. Diseased plants usually develop purple patches, especially on the older leaves (Vasudeva and Sam Raj, 1948; Nariani, 1968; Butter and Rataul, 1973).

The virus is transmitted by grafting and *Bemisia tabaci* (Vasudeva and Sam Raj, 1948; Yassin and Abusalih, 1972). Its host range includes tobacco (varieties White Burley, German Samsun, and Harrison's special), the potato variety Craigs Defiance, *Datura stramonium, Nicotiana sylvestris, N. glutinosa, Lycopersicon peruvianum, L. pimpinellifolium, L. hirsutum, L. pississi, L. glandulosum, Capsicum annum, Hibiscus esculentus* and *Acanthospermum hispidum* (Vasudeva and Sam Raj, 1948; Nariani and Vasudeva, 1963; Nariani, 1968; Yassin and Abusalih, 1972; Mariappan and Narayanaswamy, 1972).

In infected leaves the content of soluble amino acids is higher, whereas that of total protein and non-protein nitrogen, as well as reducing sugars and total sugars is lower. Infected leaves contain more histidine, lysine, aspargine, aspartic acid, glycine and serine (Lodh *et al.*, 1971).

Singh and Lal (1964) reported 3 strains of tomato leaf curl virus. A "typical strain" that induced characteristic symptoms was designated as *Nicotiana* virus 10A, a second, more severe strain as *Nicotiana* virus 10B, and a third that induced upward rolling of leaf margins and islands of golden yellow color scattered in the green tissue ("golden yellow mottle strain") as *Nicotiana* virus 10C. Tomato leaf curl virus from Philippines (Retuerma *et al.*, 1971), not identical with either tobacco leaf curl or tomato yellow leaf curl virus was transmitted by grafting and a whitefly vector. The leaf curl disease of tomato is caused by tobacco leaf curl virus (Vasudeva and Sam Raj, 1948; Lal and Singh, 1961; Singh and Lal, 1964; Yassin and Abusalih, 1972).

Stunting and leaf curl symptoms were suppressed by spraying (150 ppm) and by dipping seedling roots in gibberellic acid (Lal and Singh, 1961; Mukherjee and

Raychaudhuri, 1966). Better control of the disease was obtained when antibiotic DPB was sprayed at 75, 100, 150 and 200 ppm (Thirumalachar *et al*., 1973). Tomato yield was increased by spraying with BSOE, EWOS or E-9267, each at 2% (Butter and Rataul, 1973). Virus incidence was reduced greatly by regular spraying with insecticides (Nour-Eldin *et al*., 1969; Yassin and Abusali, 1972; Singh *et al*., 1974; Yassin, 1975).

In trials of tomato hybrids at Ludhiana, it was noted that plants with severe leaf curl virus symptoms had higher root knot incidence than plants with less severe symptoms (Mayee *et al*., 1974).

Yellow leaf curl of tomato. Tomato (*Lycopersicon esculentum*) yellow leaf curl virus is predominant in Israel where it causes heavy damage to the crop (Cohen and Harpaz, 1964). The virus causes chlorosis and curling of the leaves. It also causes thickening of leaves and dwarfing of the plants (Cohen and Harpaz, 1964; Cohen and Nitzany, 1966).

The virus is transmitted by *B. tabaci*. The vector acquires the virus in 15-30 min and inoculates it in 15-30 min. The incubation period in the vector is 21 hr. The virus persisted in the vector for up to 20 days and could be acquired by insects in the larval stage, but it is not transmitted to progeny (Cohen and Nitzany, 1966). The vector is, however, unable to re-acquire virus until it ceases to transmit from the previous acquisition. Cohen and Harpaz (1964) suggested the term periodic acquisition for this unique type of virus-vector relationship.

Cohen and Marco (1970) reported that the anti tobacco mosaic virus factor occurred periodically in homogenates of whiteflies after 1, 2 and 3-day acquisition feedings on tomato yellow leaf curl virus infected plants, but not after 4 or 7-day ones. The capacity of whiteflies to acquire and transmit tomato yellow leaf curl virus successfully was reduced when the insect was membrane fed on a solution of an antiviral factor from virus infected plants (Cohen, 1969; Cohen *et al*., 1974a).

The following species are susceptible to the virus: *Datura stramonium; Lycopersicum esculentum; L. hirsutum; L. peruvianum; L. pimpinellifolium, Lens esculenta, Malva nicaensis, N. glutinosa, N. tabacum* and *Phaseolus vulgaris* var. Bullgarit (Cohen and Nitzany, 1966). By inoculations of progenies from crosses between a tomato yellow leaf curl homozygous resistant line of *L. pimpinellifolium* and the susceptible tomato cultivar, Pearson showed incomplete dominance of resistance over susceptibility, thus indicating a monogenic control of resistance (Pilowsky and Cohen, 1974). The virus resembles yellow mosaic of tomato from Venezuela (Malaguti, 1963). The spread of the disease in the field was delayed due to straw mulching at sowing, spraying every 4 days with methidathion (Melamed-Madjar *et al*., 1970; Cohen *et al*., 1974b).

Enation leaf curl of tomato. The disease is characterized by curling, twisting and rolling of the leaves and dark green vein enations on the undersurface of the leaflets. The virus, transmitted by grafting and *Bemisia tabaci* (Nariani, 1968), infects *Capsicum annum*, *Datura stramonium* and *N. glutinosa*. It is considered to be a strain of tobacco leaf curl virus and is designated *Nicotiana* virus 10A or tomato enation leaf curl virus (Nariani, 1968).

Leaf curl of Malvaviscus arboreus. The disease is characterized by faint leaf mottling and chlorosis in early stages of disease development; later, vein thickening becomes pronounced. Severely affected leaves frequently curl upwards. The virus is transmitted by wedge-grafting and *Bemisia tabaci*. The virus infects *Hibiscus rosa-sinensis*, *Abelmoschus esculentus*, *A. tuberculatus* and *A. manihot* and is thought to be the same as leaf curl virus of *H. rosasinens* is (Mukherjee and Raychaudhuri, 1964).

Leaf curl of tobacco. The tobacco (*Nicotiana tabacum*) leaf curl virus is distributed in Belgian Congo, Brazil, Japan, Nigeria, India, Philippines, Rumania, S. Rhodesia, Nayasaland, Venezuela, Java, Sumatra, and the East Indies (Jochems, 1928; Thung, 1932; Hopkins, 1932; Hornby, 1933; Mayne and Ghesquiere, 1934; West, 1936; Pal, 1937; Pal and Tandon, 1937; Deighton, 1940; Wolf *et al.*, 1949; Martinez, 1964; Hill, 1969; Kiriyama and Nishimura, 1969). Tobacco leaf curl virus has been referred to by such names as cabbaging, frenching, kroepoek, krulziekte and Gila. Thung (1932) described three types of symptoms: (1) the common kroepoek in which leaves have edges curled in places towards the dorsal side and show thickenings and outgrowths (enations) of veins, (2) curl disease, characterized by curling of the whole leaf edge towards the dorsal side, with enations of the veins and the distance between the latter being much reduced, and (3) transparent kroepoek distinguished by curling of the leaves towards the ventral side, clearing of the veins, and the absence of enations. Pruthi (1945) observed five types of symptoms (which are illustrated but not described) and are designated as A, B, C, D and X. Each type is transmitted by the *B. tabaci*.

The virus is transmitted by *Bemisia tabaci* (Storey, 1931; Hopkins, 1932; Thung, 1932; Chang, 1973). Later, the tobacco leaf curl was transmitted by *Trialeurodes natalensis* in Transvaal (Storey, 1932, 1935; West, 1936; Pruthi and Samuel, 1937; McClean, 1940). The vector of this virus has been identified as *Bemisia tuberculata* and *Aleurotrachelus socialis* in Venezuela (Wolf *et al.*, 1949). Mastsumoto and Tateoka (1940) reported the vector as *B. gossypiperda*. The host range is fairly wide and Wolf *et al.* (1949) recorded a list of 63 susceptible species from 14 families. The tobacco leaf curl virus was transmitted to *Acanthospermum hispidum, Ageratum conyzoides, Asystasis coromandeliana, Althaea* sp., *Brassica chinensis, B. juncea, Capsicum annum, Carica papaya, Datura stramonium, Euphorbia geniculata, E. hirta, Helichrysum monstrosum, Hibiscus esculentus, Lycopersicon esculentum, L. pimpinellifolium, Launia asplenifolia, Nicandra physoloides, Nicotiana glutinosa, N. tabacum, Physalis peruviana, Rhaphanus sativus, Schizanthus* sp., *Scoparia dulcis, Sesamum indicum, Sida carpinifolia, S. cordifolia, S. rhombifolia, S. veronicaefolia, Solanum melongena, S. nigrum, Stachytarpheta jamaicensis, Synedrella nodiflora, Vigna sinensis* var. *sesquipedalis, Withania somnifera, Zinnia elegans, Vernonia cinerea, Nicotiana glauca, N. rustica, Malvastrum tricuspidatum, Petunia hybrida, Polanisia viscosa* (Pruthi and Samuel, 1939, 1941, 1942; Shepherd, 1940; McClean, 1940; Nariani and Pathanian, 1953; Garga, 1949; Wolf *et al.*, 1949; Phatak Raychaudhuri, 1967; Hill, 1968; Olivares *et al.*, 1972; Mathur, 1932, 1933; Nariani, 1956; Mishra *et al.*, 1963; Nariani, 1968; Flores, 1961).

Electron micrographs of concentrates of the material reveal large numbers of soft roughly spherical bodies, with a mean diameter of 39 nm which become flattened on the supporting film in the course of preparation (Sharp and Wolf, 1951). Sharp and Wolf (1949) reported that purified preparations were found to contain rod like particles up to 1,144 nm in length and about 15 nm in width. Spherical, 30-nm particles were seen in electron microscope preparations. Serological tests confirmed the presence of tobacco leaf curl virus (TLCV) specific antibodies in anti-TLCV serum obtained from rabbits that had been injected with purified virus (Kiriyama, 1972). Wolf *et al.* (1949) stated that leaf curl probably represents a complex of at least three distinct viruses, namely tobacco leaf curl, cotton leaf curl and cassava leaf curl (cassava mosaic).

Incidence of the disease in tobacco depends on the whitefly population, which reaches a maximum in the autumn, falls in the winter and mounts again in March (Pruthi and Samuel, 1942; Hill, 1968). The importance of *Schizanthus* as a host of the virus depends on its role as a source of infection to other economic crops (Garga, 1949). The size of whitefly populations is positively correlated with increases in environmental temperature which favor the vector (Munshi and Choudhary, 1964). *Withania somnifera*, a common weed, may serve as a potential source of infection (Pathak and Raychaudhuri, 1967).

For control, emphasis is placed on the necessity of clearing up old tobacco lands during the winter in order to deprive the insect vectors (whiteflies). By this means, the perpetuation of the disease may be largely prevented (Moore, 1933). Practical control of the disease was secured in 1933 by the timely removal (in the middle of June, July and August) of *Azeratum conyzoides*, *Synedrella nodiflora* and *Vernonia cinerea* which serve as sources of infection. Practical control of tobacco leaf curl in southern Nigeria may be effected, in seasons of normal rainfall distribution, by late planting to coincide with the gradual decline in the whitefly vector population from the end of July onwards, supplemented by subsidiary sanitary measures, such as the destruction of all crop remains at the end of each season, of all volunteer plants before planting of the new crop and of any plants developing leaf curl symptoms shortly after transplanting (West, 1940). Control would also be strengthened by removal of *Acanthospermum hispidum, Malvastrum tricuspidatum* and *Sida carpinifolia* from the field (Shepherd, 1940). Three applications of 100 ppm giberellic acid suppresses the symptoms (Nariani, 1963).

Leaf curl of cape-gooseberry. This disease of *Ribes grossularia* is characterized by curling, puckering and smalling of the leaves and the presence of dark green enations on the veins underneath. The veins and veinlets are very prominent and thickened. In severe cases, leaf margins are curled upwards, forming cup like structures, but in mild cases curling is only slight. The virus is transmitted by grafting and *B.*

tabaci. It is transmissible to *Nicotiana tabacum* Cv. Harrison's Special (Nariani and Pathanian, 1953).

Leaf curl virus of soapwort. Varma (1974c) recorded stunted growth and twisted, curled leaves on soapwart, *Saponaria vaccaria*, in India. The virus is transmitted by grafting and *Bemisia tabaci*.

Leaf curl of sesamum. Symptoms of leaf curl virus in sesamum, *Sesamum orientale*, were observed in India by Sahambi (1958). The disease is characterized by very leathery and thickened veins on the underside of the leaves. Severely affected plants remain stunted and bear few fruits. The virus is transmitted by *Bemisia tabaci*.

Leaf curl of potato. This desease of potato, *Solanum tuberosum*, was reported from Venezuela (Wolf *et al.*, 1949). The disease is characterized by reduced leaf size, curling of leaf margins and rough surface. The virus is transmitted by *Bemisia tuberculata* and *Alerotrachelus socialis*.

Leaf curl virus of Zinnia elegans. Leaf curl of *Zinnia elegans* is very prevalent at Dehra Dun, India (Mathur, 1933). Characteristic effects of the disease are thickening of the lower surface of the veinlets and curling of the leaf blades. The diseased plants often fail to attain a height of more than one foot, and flowers are dwarfed and partially sterile. Leaf curl is most prominent during the period from July to September (Mathur, 1933; Varma and Singh, 1973b).

Kerling (1933) described two types of symptoms on *Zinnia elegans*. The "common kroepoek" is characterized by a disturbance of sugar transport throughout the leaf, an increase of the primary phloem in the veins, enlargement of the pericycle through cell division, and formation of new woody vessels surrounded by cambium. The "transparent" type of kroepoek produces in the leaf veins typical swellings of the ends of the xylem vessels and sieve tubes, and enlargement of the pericycle and cortical parenchyma cells. The sieve tubes are curved, and the cell walls of the primary phloem as well as those of the pericycle are irregularly swollen. The two types of kroepoek are transmissible by the same insect, *Bemisia tabaci*.

3.6 MOSAIC AND OTHER TYPES OF VIRUS DISEASES

A virus disease of Anthurium andraeanum. Herold (1967) reported a virus disease on *Anthurium andraeanum* from Venezuela. The virus is transmitted by *Bemisia tabaci* and grafting. It causes severe mosaic and leaf deformation.

Pseudo yellows virus of beet. Plants infected by beet (*Beta vulgaris*) pseudo yellows virus show stunting, interveinal yellowing and chlorotic spotting (Duffus, 1975). The greenhouse whitefly, *Trialeurodes vaporariorum* transmits yellowing virus to sugarbeet and some weed plants. A single whitefly (*T. vaporariorum*) is able to transmit the disease. Insects may become viruliferous after a 1-hr acquisition-access feeding period. Viruliferous whiteflies are capable of transmitting virus within 1-hr inoculation feedings. The latent period in the vector is less than 6 hr. The virus is retained by the vector for a maximum of 4 days (Duffus, 1975).

The beet pseudo yellows virus (BPYV) infects the following hosts: *Gomphrena globosa, Spergula arvensis, Beta macrocarpa, B. vulgaris, Chenopodium album, C. amaranticolor, C. capitatum, C. murale, Spinacea oleracea, Callistephus chinensis, Cichorium endiva, Lactuca sativa, L. serriola, Senecio vulgaris, Sonchus oleraceus, Taraxacum officinale, Zinnia elegans, Capsella bursa-pastoris, Cucumis sativus, C. melo, Cucurbita moschata, Erodium cicutarium, Gevanium dissectum, Linum grandiflorum, L. usitatissimum, Malva parviflora, Claytonia perfoliata, Acquilegia* sp., *Nicotiana clevandii, N. glutinosa, N. tabacum, Physalis ixocarpa, P. wrightii, Solanum dulcamara, Conium maculatum, Daucus carota,* and *Urtica californica* (Duffus, 1975).

Symptoms induced by BPYV on common host plants are quite distinct from symptoms of the other whitefly transmitted viruses which are transmitted by *B. tabaci* and seem to be much more like those induced by the aphid transmitted yellowing viruses. Excluding the fact that BPYV is transmitted by *T. vaporariorum*, the virus appears to be more closely related to the aphid transmitted viruses of the yellows group than to the typical whitefly transmitted viruses (Hildebrand, 1960; Duffus, 1975).

Mosaic virus of Abutilon (infectious chlorosis of Malvaceae). Baur (1906) observed attractive yellow and green patches on *Abutilon* leaves. The disease has been reported from Trinidad (Owen, 1946), Brazil (Orlando and Silberschmidt, 1946; Costa, 1954, 1955), Florida (Costa and Bennett, 1953; Kunkel, 1930). Buchwald (1948) reported the history of infectious variegation in *Abutilon* spp. *Abutilon* mosaic is widespread among indigenous species of Sida and other Malvaceae in Sao Paulo and Brazil (Costa, 1955). The disease has been called *Abutilon* infectious chlorosis virus, *Abutilon* infectious variegation virus, infectious chlorosis of Malvaceae and *Abutilon* mosaic virus (Silberschmidt and Tommasi, 1955).

The attractive bright yellow and green variegation in various species of *Abutilon* is well known and is the chief reason for the propagation of the species as an ornamental plant. Hertzsch (1927) described two distinct types of infectious chlorosis in Malvaceae, which are termed A and B. The former, exemplified in *Abutilon striatum thompsonii*, is characterized by yellow spotting of leaves and yellow color of veins, and the latter, found in *A. darwini tesselatum*, by pale green spots and stripes on the foliage, with veins being mostly green. He also mentioned that A type of infectious chlorosis of Malvaceae is stated to have first appeared in a single plant of *A. striatum* imported into England from the West Indies in 1868. The variegated strain propagated from this plant became known as *A. striatum thompsonii*. The origin of B type is unknown, but it is thought to have probably appeared spontaneously in *A. dorwini*.

The virus is transmitted through the seeds of *Mucuna pruriens* (Crandell, 1954). Abutilon mosaic virus from *Sida* spp. is mechanically transmissible to *Malva parvi-*

flora and *M. rotundifolia* (Costa and Carvalho, 1960a), but mosaic from the leaves of *Abutilon striatum* var. *thompsonii* is not. Keur (1933) and Costa (1955) failed to transmit the virus mechanically.

Orlando and Silberschmidt (1946) first reported the transmission of *Abutilon* variegation mosaic by whiteflies, *Bemisia tabaci*. The period elapsing between the infestation of the plants by inoculative *B. tabaci* and the initial appearance of disease symptoms ranged from 5 to 28 days in *S. rhombifolia* and from 12 to 45 days in *A. striatum* var. *spurium*. A single whitefly is able to transmit the virus, females being more efficient than males (Orlando and Silberschmidt, 1946). Costa and Carvalho (1960b) reported that individual insects were able to carry *Euphorbia* mosaic virus and *Abutilon* mosaic virus simultaneously and that *B. tabaci* can transmit concomitantly *Leonurus* mosaic virus (strain of infectious variegation virus) and infectious chlorosis of Malvaceae (Flores and Silberschmidt, 1963).

Sida rhombifolia, S. micrantha, G. hirsutum infected by the vector were highly resistant to mechanical transmission. *S. rhombifolia* and *Malva parviflora* were good sources of inoculum and *Nicandra physaloides*, tomato, *D. stramonium* and some infected legumes generally gave an inoculum of low activity. The virus remained active in infective sap for 24 hr at room temperature and for 48 hr when treated with buffer and sodium sulphite. A 1:25 dilution of sap produced infection. Most activity was lost in an inoculum heated at 50°C for 10 min but was not completely lost at 55°C (Costa and Carvalho, 1960a). *Datura stramonium* is a local lesion test plant for *Euphorbia* mosaic virus while *Malva parviflora* and *Nicotiana glutinosa* proved to be good test plants for *Abutilon* mosaic virus. The physical properties of both viruses *in vitro* were remarkably similar; although they may be related, the evidence indicates that they are not identical (Costa and Carvalho, 1960b).

The host range of *Abutilon* mosaic virus (transmitted either through mechanical means, budding, grafting or the whitefly *Bemisia tabaci*) includes *Abutilon tesselatum, A. indicum, A. megapotamicum, A. sellovianum, A. striatum* var. *spurium, A rotundifolia, A. thompsonii, A. mulleri, A. striatum,* var. *thompsonii, A. mauritianum, Althaea officinalis, A. venosum, A. rosea, Arachis hypogaea, Anada wrightii, Castanea* sp., *Cyamopsis tetragonaloba, Cyphomandra betacea, Datura stramonium, Euonyraus* sp., *Euphorbia heterophylla, E. hirta, E. prunifolia, Fraxinus* sp., *Glycine max, Gossypium hirsutum, Hibiscus cannabinus, H. esculentus, Jasminum* sp., *Lavatera arborea, L. cachemiriana, L. cretica, L. arborea, Lycopersicon esculentum, Malva borealis, Malachra adeifolia, M. crispa, Mucuna pruriens, Malvastrum coromandelianum, Malope trifida, Malva parviflora, M. hispanica, Nicandra physoloides, Nicotiana tobacum, N. glutinosa, Phenzx sonneratii, Phaseolus vulgaris, rhynchosia minima, Sida glomerata, S. linifolia, S. urens, Sida acuta* var. *carpinifolia, S. cordifolia, S. carpinifolia, S. micrantha, S. napaea, S. rhombifolia, Solanum tuberosum, Stizolobium* sp., *Triumfetta lappula*, lentil, and white lupin (Blakeslee, 1921; Hertzsch, 1927; Kunkel, 1930; Silberschmidt, 1943, 1948; Crandell, 1954; Costa, 1954, 1955; Silberschmidt and Tommasi, 1956; Silberschmidt *et al.*, 1957; Flores and Silberschmidt, 1958, 1962, 1966, 1967; Klebakin, 1931; Hollings, 1959; Flores *et al.*, 1960, 1967; Costa and Carvalho, 1960a, 1960b; Owen, 1946; Costa and Carvalho, 1962).

The portions of *Abutilon* leaves suffering from infectious chlorosis are characterized by a much lower catalase activity, a higher proportion of amino nitrogen, and less chlorophyll, xanthophyll, carotene, and tryptophane than the normal green areas (Davis, 1929; Euler *et al.*, 1930).

Spheroidal, 80-nm particles were detected in ultrathin sections of *Abutilon spriatum* plants affected by the *Abutilon* virus transmitted by *B. tabaci*. Isometric particles, 20-25 nm in diameter, were observed in the sieve-tube elements of *Sida* spp. affected by *Abutilon* mosaic virus (Kitajima and Costa, 1974).

Pseudo mosaic of Eupatorium. Van DerLaan (1940) from Sumatra reported that the pseudo mosaic of *Eupatorium odoratum* was transmitted by the whitefly *Bemisia tabaci*. The virus infects *Ageratum conyzoides*, *Nicotiana glutinosa* and *Synedrella nodiflora*, all known to act as reservoirs of "gila" infection.

Mosaic of cotton. The mosaic of cotton (*Gossypium hirsutum*) was first noticed in 1968, and it spread alarmingly on variety BJA 592 during 1969 in Chad. The disease was transmitted by three species of whiteflies: *Bemisia tabaci*, *Trialeurodes lubia* and *T. ricini* (Lourens *et al.*, 1972; Bink, 1973). The disease may be the same as that caused by cotton leaf mottle virus described in Sudan in 1959 (Nour, 1959).

Mosaic disease of sweet potato. A mosaic of sweet potato, *Ipomoea batatas*, causing mild or severe chlorosis of leaf veins and designated as sweet potato virus B, has been described by Sheffield (1957) from East Africa. The symptoms on sweet potato are stunting, with shortening of the internodes, reduction in leaf size and general chlorosis. Leaves may be leathery in consistency. In the varieties with much dissected leaves, segments may be strap-shaped. Yellow mottling is more obvious in the broad-leaved than in the narrow-leaved varieties.

The virus is transmitted by *Bemisia tabaci*. The whitefly can acquire the virus in 5 min but is an inefficient vector (Sheffield, 1958; Girardeau and Ratcliffe, 1960). The virus infects *Gomphrena globosa, Datura ferox, Lycopersicon esculentum, Nicotiana glutinosa, N. tabacum, Petunia* sp, and *Physalis peruviana* (Sheffield, 1957, 1958).

It appears to be different from all other sweet potato viruses and from all known viruses transmitted by whiteflies (*B. tabaci*). All attempts to transmit virus B to members of the Chenopodiaceae, Compositae, Euphorbiaceae and Malvaceae have failed (Sheffield, 1958). The virus is different from the virus reported by Eusign (1919) in the United States and by Newton and Periris (1953) from Ceylon. Mechanical and insect transmission needs confirmation.

Yellow dwarf virus of sweet potato. Observations of sweet potato, *Ipomoea batatas*, plantings at Beltsville, Md., in 1955 and 1956 revealed relatively low percentages of natural spread of feathery mottle virus. It was subsequently discovered that the whitefly *Trialeurodes abutilonea* is the natural vector of sweet potato yellow dwarf (feathery mottle) in Maryland (Girardeau, 1958; Hildebrand, 1958, 1959, 1961).

Vein clearing virus of sweet potato. Loebenstein and Harpaz (1960) reported sweet potato vein clearing virus from Israel and Clerk (1960) reported it from Ghana. The symptoms on sweet potato are pronounced clearing of the veins and slight reduction in leaf size. It causes minute chlorotic specks between veins mostly on veinlets. Subsequent vein clearing is followed by a brilliant mosaic, and plants become stunted and chlorotic (Loebenstein and Harpaz, 1960).

The virus is transmitted by grafting and *Bemisia tabaci*. The virus has been transmitted to *Ipomoea purpurea*, *I. mexicana* and *I. bonanon* (Loebenstein and Harpaz, 1960). Three strains are distinguishable: a chlorotic streak strain, milder and masked at 25°C; a medium chlorosis strain, producing faint chlorosis along the midrib with protrusions along the main veins; and an angular spot strain, characterized by irregular spots often made up of 3-5 cleared veinlets (Loebenstein and Harpaz, 1960).

Vein clearing in Ghana shows some similarities to virus B of East Africa (Sheffield, 1957) and vein clearing in Israel (Loebenstein and Harpaz, 1960), though the latter is not sap transmissible to other hosts as is virus B.

Chlorotic ring spot of jasmine. Wilson (1972) first recorded chlorotic ring spot virus on jasmine (*Jasminum samba*) from Kerala, India. Hertzsch (1927) recorded a graft transmissible chlorosis of *Jasminum* sp. from Germany. The infected plants show typical mosaic symptoms consisting of yellow chlorotic spots or areas of varying size and shape, intermingled with the green color of the leaves. Frequently yellow colored rings with green centers appear on some of the leaves along with the mosaic pattern. The virus is transmitted by inarch grafting and *Bemisia tabaci*.

Electron microscopy of crude extracts from yellow ring spot mosaic infected *Jasminum samba* plants revealed filamentous virus particles measuring 700-750 nm in length (Benigno *et al.*, 1975).

Mosaic disease of cassava. Mosaic disease of cassava, *Manihot utilissima*, has been reported from East Africa as 'Kräuselkrankheit' (Warburg, 1894), Belgian Congo (Lefevre, 1935; Kufferath and Ghesquiere, 1932), Nigeria (Golding, 1930) and Brazil (Silberschmidt and Campos, 1944). Muller (1931) reported the disease from Java and believed it to be the same as the Kräuselkrankheit reported in East Africa and the mosaic of cassava found in West Africa.

The main symptom of cassava mosaic is the presence on the upper surface of the leaves of yellowish white spots and streaks, 2 to 3 mm broad and, on the under surface, depressed necrosis immediately beneath the spots and streaks. The incidence of crinkling and distortion appears to be greater in older than in younger leaves (Jolly, 1931; Silberschmidt, 1938; Pascalet, 1932). Storey and Nichols (1938) divided symptoms into two well-defined groups: (1) severe chlorosis with usually large, more or less uniformly distributed, but sometimes localized, yellow or nearly white chlorotic areas, and (2) often quite symptomless, characterized by slight chlorosis with small, either generally distributed or localized chlorotic areas, only slightly paler than normal.

Staner (1931) reported that the mosaic of cassava is probably due to a bacterium.

The disease was transmitted by *Bemisia gossypiperda* var. *mosaicivecta* and *B. nigeriensis* (Kufferath and Ghesquiere, 1932; Lefevre, 1935; Golding, 1935, 1936a, 1936b; Storey and Nichols, 1938; Chant, 1958; Steyaert, 1946). Lefevre (1935) successfully transmitted the disease mechanically, but Storey and Nichols (1938) attempted to transmit the virus by needle scratch, puncture, by rubbing young leaflets, and by hypodermic injections and obtained negative results. Whiteflies need to feed for 4 hr on young infected leaves to acquire the virus and, after acquisition, an additional 4 hr is needed for them to become inoculative. Adults remain infective for more than 48 hr after ending their infection feed. Single whiteflies are able to transmit the disease (Chant, 1958).

The cassava mosaic virus infects *Manihot utilissima*, *M. aipi*, *M. palmata*, and *M. glaziovii*. The virus seems to have been found infecting only the Euphorbiaceae (Kufferath and Ghesquiere, 1932; Golding, 1936a). Varieties Calabar II and Bankye Sareso, clones 4723A/26, 26106/27, and 4763/16, were found to be resistant (Dade, 1931; Doughty *et al.*, 1955 and Jennings, 1957, 1960). Bitter types of cassava appear to be more susceptible than sweet ones (Lefevre, 1935; Golding, 1936b). Cross inoculation with whiteflies gave no evidence that cassava mosaic and cotton leaf curl viruses are related (Chant, 1958).

By encouraging the natives to use seeds instead of cuttings, the disease can be controlled to some extent (Jolly, 1931). The virus was completely inhibited in green stems but was able to survive in a thick woody cutting after hot water treatment (Chant, 1959).

Dwarf mosaic of bean. Bean (*Phaseolus vulgaris*) dwarf mosaic disease results from the infection of the plants by the common strains of the *Abutilon* mosaic virus from malvaceous species in Brazil (Costa, 1975). Infected plants are generally dwarfed and tend to become bunchy or rosetted. The yellow spots in the older leaves are few, and spotting is accompanied by a rolling downward of the leaflets. Infected plants usually do not produce pods; but, if these are present, they may be malformed (Costa, 1975). The virus is transmitted by *B. tabaci*.

Isometric particles measuring 20-25 nm in diameter were observed in ultrathin sections of infected leaves but not in leaves of healthy controls (Kitajima and Costa, 1974). Bean dwarf mosaic is caused by *Abutilon* mosaic virus (Costa, 1975).

Necrosis of potato. Vasudeva and Azad (1948) reported on potato (*Solanum tuberosum*) necrosis from India which is transmitted by *Bemisia gossypiperda*. In the greenhouse, potato plants (Darjeeling Red Round) raised from virus free seed tubers developed dwarfing and curling of the leaves and necrosis of the growing point which progressed downward, ultimately killing many of the plants. *Bemisia gossypiperda* was found to be the vector of the disease transmitting it to tomato and tobacco after having fed for 24 hr on infected source plants. The symptoms on differential hosts show that potato necrosis is caused by a mild strain of tobacco leaf curl virus (Vasudeva and Azad, 1948).

Mosaic virus of Wissadula amplissima. Schuster (1964) described a mosaic virus of *Wissadula amplissima* in southern Texas that is transmitted by *Bemisia tabaci*.

The virus infects *Gossypium hirsutum, Hibiscus cannabinus, Althaea rosea*, and *Malvastrum coromandelianum*. The virus appears closely related to *Abutilon* mosaic (infectious variegation) virus of Brazil (Silberschmidt, 1943; Costa, 1955), but it produces a milder disease in cotton and is not transmissible to *H. esculentus* and *Sida rhombifolia*.

3.7 SUSPECTED WHITEFLY TRANSMITTED VIRUS DISEASES

Yellow mosaic of Acanthospermum hispidum. The author observed yellow mosaic symptoms on *Acanthospermum hispidum* during 1976 at Bangalore, India. Light yellow patches appear on leaf laminae. In some leaves only a few yellow spots appear, and in others the whole lamina is covered with yellowing.

Yellow net virus of Odontonema nitidum. Ramakrishnan and Subramanian (1961) reported the occurrence of yellow netting on *Odontonema nitidum*. Three varieties of this ornamental shrub are usually found in Southern Indian gardens. One variety has variegated leaves with bright yellow netted veins. The leaves of another variety are dark purplish on the upper surface and purple colored on the lower surface, and a third variety has reddish green leaves with wavy margins and blotched with green. Grafting twigs from the yellow netted variety on purple leaved variety produced conspicuous yellow netting in about 4 weeks. Many yellow net viruses transmitted by *B. tabaci* are similar to yellow net virus of *Odontonema* in symptoms.

Yellow vein mosaic of Passiflora foetida. The disease is characterized by clearing of veins and veinlets and has been reported from India. The virus is transmitted by grafting, but not by sap inoculation (Wilson and Satyarajan, 1970).

Yellow vein mosaic of potato. Vein yellowing of potato, *Solanum tuberosum*, is very common in the southern part of Colombia and seems to have its chief range of distribution in Ecuador (Alba, 1950; Silberschmidt, 1954). The chief symptom of the disease is the bright yellow coloration of the leaf veins. When newly developed, these symptoms are extremely striking. The virus is not sap transmissible and is transmitted by grafting. Possibility of whitefly transmission is indicated (Smith, 1957; Varma, 1962).

Leaf curl of Thevetia nerrifolia. This disease is reported to occur in India and is characterized by curling of the leaf tips downwards and sideways, smalling of the leaves, and waviness of the leaf margin followed by mild chlorosis near the midrib and veins (Garga, 1953).

Leaf curl of Vinca rosea periwinkle. This disease of periwinkle, *Vinca rosea*, is characterized by development of crumpling, curling and mosaic symptoms on the leaves, especially those at the apex of a branch. Discoloration of the leaves and fading of flowers also occur in later stages of infection. The virus is graft transmissible (Bisht and Singh, 1964).

3.8 DISCUSSION AND CONCLUSIONS

Whiteflies have become important in the spread of plant viruses. Three species of whiteflies are involved in transmission. *Trialeurodes abutilonea* transmits sweet potato yellow dwarf virus and sweet potato mosaic, and *T. vaporariorum* transmits beet pseudo yellows virus. All other diseases are transmitted by *Bemisia tabaci*. Thus far, no instance of vectors other than aleyrodids have been encountered.

Most of these diseases are present in tropical areas such as Puerto Rico, Venezuela, Columbia, Ecuador, Egypt, Ghana, Nigeria, Chad, Sudan, India, Ceylon (Bird and Maramorosch, 1975; Nene, 1972; Varma, 1962) and very few diseases occur in temperate areas, such as Washington, Missouri, California, Florida, Maryland, Georgia and S. Texas (Duffus, 1965; Hildebrand, 1959; Kirkpatrick, 1931 and Loebenstein and Harpaz, 1960). Figure 1 reveals the global distribution of whitefly transmitted diseases. Whitefly transmitted diseases have not been reported from many countries. In temperate countries the whitefly population is very low, this might be the reason why few such diseases occur. Even the diseases that have been reported in temperate areas are not very prevalent. Though the whitefly transmitted diseases are not seed-borne, their distribution is seen to include different parts of the globe. The distribution might have taken place through infected plants of ornamental value, as in the case of variegated *Abutilon*.

Costa (1976) divided whitefly transmitted diseases into three groups: (a) diseases of the mosaic type, (b) diseases of leaf curl type, and (c) diseases of yellowing type. The author divided into four groups based on symptomatology. They are (a) yellow mosaic diseases, (b) yellow vein mosaic diseases, (c) leaf curl diseases, and (d) mosaic diseases. In nature one disease might occur on various hosts which have been described separately. In the group yellow mosaic diseases many are related or might be strains of the same virus. Serological relationships of these viruses have to be established. Very recently an antiserum has been produced for bean golden yellow mosaic virus by Guillermo Galvez (Bird and Maramorosch, 1978). The causal agents for many of these diseases are not known with certainty; but most evidence indicates that they are viruses. Isometric, rod-shaped, and dimers or Siamese twin particles have been associated with infectivity (Galvez and Castano, 1976; Goodman *et al.*, 1977; Hollings *et al.*, 1976 and Sharp and Wolf, 1949). Golden yellow mosaic virus has been described as having a single stranded DNA genome, a first for phytopathogenic viruses.

Most whitefly transmitted diseases are not mechanically transmissible. More detailed investigations would be possible, if they were. Bean golden yellow mosaic virus is mechanically inoculable to *Phaseolus lunatus, P. aborigineus*, and *P. coccineus*, and it has been studied in greater detail.

The control of diseases transmitted by whiteflies through the application of insecticides is being investigated in many places but success is limited. Breeding for disease resistance is an obvious method of control, but success has been attained in only a few cases. Various cultural, biological and chemical methods of control have to be evolved to suit to local needs. These diseases are very much conditioned by the season, especially because of the effect of season on vector numbers and activity. In South India tomato crops started in October are practically free of the leaf curl disease because of low whitefly populations, but crops after February are severely affected by this disease because the vector multiplies abundantly in summer. Tomato plants have been immunized on a large scale with mild strains of tobacco mosaic virus in Europe. Finding mild strains of whitefly transmitted viruses and using them for control would be very economical and practical. Also, the possibility of biological control of the vector *Bemisia tabaci* has to be explored.

Since the nature of a large number of these diseases is still not known, they represent an attractive challenge to research workers, especially to those who work in regions where whitefly transmitted diseases are abundant.

3.9 ACKNOWLEDGMENTS

I am grateful to the late Dr. K. Ramakrishnan for encouraging me to write a review on whitefly transmitted diseases. I thank Dr. Kerry F. Harris and Dr. Karl Maramorosch for critically editing the manuscript.

3.10 REFERENCES

Ahmed, M. (1978). Whitefly (*Bemisia tabaci*) transmission of a yellow mosaic disease of cowpea, *Vigna unguiculata. Pl. Dis. Reptr.* 62: 224-226.

Ahmed, M., and Harwood, R. F. (1973). Studies on a whitefly-transmitted yellow mosaic of urd bean (*Phaseolus mungo*). *Pl. Dis. Reptr.* 57: 800-802.

Alba, V. R. (1950). Viropatogenos, longerencia Latino–Americana de especialists en papa Bogota, 52-58.

Allen, R. M., Tucker, H., and Nelson, R. A. (1960). Leaf crumple disease of cotton in Arizona. *Pl. Dis. Reptr.* 44: 246-250.

Andrews, F. W.. (1936). The effect of leaf curl disease on the yield of the cotton plant. *Emp. Cott. Grow. Rev.* 13: 287-293.

Avidoz, Z. (1956). Binomics of the tobacco whitefly *Bemisia tabaci. Gen. Rec. agric. Res. Sta.* Rehovot, Israel. 7: 25-41.

Baur, E. (1906). Über die infektiöse Chlorose der Malvaceas. *S. B. Preuss Akad. Wiss.* 1: 11-29.

Benigno, D. A., Fauali-Hedayat, M. A., and Retuerma, M. L. (1975). Sampagnita yellow ring spot mosaic. *Philippine Phytopathology* 11: 91-92.

Berkeley, G. H. (1938). Leaf curl of geranium. *Canad. Hort.* 938: 1-4.

Bhargava, K. S., and Joshi, R. D. (1962). Yellow mosaic a virus disease of rose in Gorakhpur. *Sci. & Cult.* 28: 184-185.

Bink, F. A. (1973). A new contribution to the study of cotton mosaic in chad. I. Symptoms, transmission by *Bemisia tabaci* Genn., II. Observations on *B. tabaci*, III. other virus diseases on cotton and related plants. *Cotton et Fibres Tropicales* 28: 365-378.

Bird, J. (1957). A whitefly transmitted mosaic of *Jatropha gossypifolia. Tech. Pap. agric. Exp. Sta.* 22: 35.

Bird, J. (1958). *Univ. Puerto Rico Agric. Exp. Sta.* Tech. Pap. 26: 1-23.

Bird, J. (1962). A whitefly-transmitted mosaic of *Rhychosia minima* and its relation to tobacco leaf curl and other virus diseases of plants in Puerto Rico. *Phytopathology* 52: 285-288.

Bird, J., and Maramorosch, K. (1975). Tropical diseases of legumes. Academic Press, N. Y., San Francisco, London, 171 pp.

Bird, J., and Maramorosch, K. (1978). Viruses and diseases associated with whiteflies. *Adv. Virus Res.* 22: 55-109.

Bird, J., and Sánchez, J. (1971). Whitefly-transmitted viruses in Puerto Rico. *J. Agric. Univ. P. Rico.* 55: 461-467.

Bird, J., Perez, J. E., Alconero, R., Vakili, N. C., and Meléndez, P. L. (1972). A whitefly transmitted golden yellow mosaic virus of *Phaseolus lunatus* in Puerto Rico. *J. Agric. Univ. P. Rico.* 56: 64-74.

Bird, J., Sánchez, J., and Lopez-Rosa, J. H. (1970). Whitefly-transmitted viruses in Puerto Rico. *Phytopathology* 60: 1539-1543.

Bird, J. Sánchez, J. Rodríguez, R. L., and Juliá, F. J. (1975). Rugaceous (whitefly-transmitted) viruses in Puerto Rico. *In* "Tropical Diseases of Legumes" (J. Bird and K. Maramorosch, Eds.), pp. 3-25. Academic Press, New York.

Bisht, N. S., and Mathur, R. S. (1964). Occurrence of two strains of Jute mosaic virus in U. P. *Curr. Sci.* 33: 434-435.

Bisht, N. S., and Singh, D. V. (1964). Leaf curl of *Vinca rosea* and *Vinca rosea* var. *alba*. *Proc. Nat. Acad. Sci. Ind. B.* 34: 289-290.

Blakeslee, A. F. (1921). A graft infectious disease of *Datura* resembling a vegetative mutation. *J. Genet.* 11: 17-36.

Boughey, A. S. (1947). The causes of variation in the incidence of cotton leaf curl in the Sudan Gezira. *Mycol. Pap. Imp. Mycol. Inst.* 22: 9.

Buchwald, N. F. (1948). A note on leaf-breaking in Abutilon species. Hereditary and virus variegation. *Gartnertidende* 64: 313-317.

Butter, N. S., and Rataul, H. S. (1973). Control of tomato leaf curl virus (TLCV) in tomatoes by controlling the vector whitefly *Bemisia tabaci* Genn. by mineral oil sprays. *Curr. Sci.* 42: 864-865.

Capoor, S. P. (1939). Mosaic of *Jatropha curcas*. *Progr. Rep. Virus diseases scheme,* Bombay, 1939-40, 18 pp.

Capoor, S. P. (1949). Feeding methods of the whitefly. *Curr. Sci.* 18: 81-83.

Capoor, S. P. (1953). Bhendi mosaic and its control in Poona. *Indian Fmg.* 2: 14-16.

Capoor, S. P., and Varma, P. M. (1948). Yellow mosaic of *Phaseolus lunatus* Linn. *Curr. Sci.* 17: 152-153.

Capoor, S. P., and Varma, P. M. (1950a). A new virus disease of *Dolichos lab-lab*. *Curr. Sci.* 19: 248-249.

Capoor, S. P., and Varma, P. M. (1950b). Yellow vein mosaic of *Hibiscus esculentus* L. *Indian J. agric. Sci.* 20: 217-230.

Chang, Y. C. (1973). Study on methods of transmitting tobacco leaf curl virus by *Bemisia tabaci* acquiring the virus through parafilm membrane. *Taiwan Agric. Q.* 9: 153-160.

Chant, S. R. (1958). Studies on the transmission of cassava mosaic virus by *Bemisia* spp. (Aleyrodidae). *Ann. appl. Biol.* 46: 210-215.

Chant, S. R. (1959). A note on the inactivation of mosaic virus in Cassava. (*Manihot utilissima* Pohl) by heat treatment. *Emp. J. exp. Agric.* 27: 55-58.

Chenulu, V. V., and Phatak, H. C. (1965). Yellow mosaic of *Acalypha indica* L., a new whitefly transmitted disease from India. *Curr. Sci.* 34: 321-322.

Cherian, M. C. (1936). The leaf curl disease of chillies. *Dept. Agric.*, Madra, Leafl. 73.

Chinnadurai, G., and Nair, M. C. (1971). Effect of Dolichos yellow mosaic on chlorophyll and leaf proteins. *Curr. Sci.* 40: 18-19.

Clerk, G. C. (1960). A vein-clearing virus of sweet potato in Ghana. *Pl. Dis. Reptr.* 44: 931-933.

Cohen, S. (1969). *In vivo* effects in whiteflies of a possible antiviral factor. *Virology* 37: 448-454.

Cohen, S., and Harpaz, I. (1964). Periodic, rather than continual acquisition of a new tomato virus by its vector, the tobacco whitefly (*Bemisia tabaci* Genn.) *Ent. exp. and appl.* 7: 155-166.

Cohen, S., and Marco, S. (1970). Periodic occurrence of an anti-TMV factor in the body of white-

flies carrying tomato yellow leaf curl virus. *Virology* 40: 363-368.
Cohen, S., and Nitzany, F. E. (1960). A whitefly-transmitted virus of cucurbits in Israel. *Phytopath. medit.* 1: 44-46.
Cohen, S., and Nitzany, F. E. (1966). Transmission and host range of the tomato yellow leaf curl virus. *Phytopathology* 56: 1127-1131.
Cohen, S., Eisenberg, M., and Sela, I. (1974a). Antiviral effect of a factor from TMV–infected plants on the inoculactivity of an insect vector. *Phytoparasitica* 2: 13-18.
Cohen, S., Melamed-Madjar, V., and Hameiri, J. (1974b). Prevention of the spread of tomato yellow leaf curl virus transmitted by *Bemisia tabaci* Genn. (Homoptera, Aleyrodiadae) in Israel. *Bull. Entomol. Res.* 64: 193-197.
Cohen, S., Nitzany, F. E., and Harpaz, I. (1963). Tests for the control of the yellowing top virus on tomatoes. *Hassadeh* 43: 576-578.
Cohen, H., Nitzany, F. A., and Vilda, T. (1961). Tomato yellow top virus in Israel. *Hassadeh* 42: 139-140.
Costa, A. S. (1954). Identidade Entre O Mosaico Comum do algoderio ea clorose Infeciosa dos Malvaceas. *Bragantia* 13: 23-27.
Costa, A. S. (1955). Studies on Abutilon mosaic in Brazil. *Phytopath. Z.* 24: 97-112.
Costa, A. S. (1965). Three whitefly-transmitted virus diseases of beans in Sao Paulo, Brazil. *Pl. Prot. Bull. FAO.* 13: 121-130.
Costa, A. S. (1969). Whiteflies as virus vectors. *In* "Viruses, Vectors, and Vegetations" (K. Maramorosch, Ed.), pp. 95-119. Wiley (Interscience), New York.
Costa, A. S. (1975). Increase in the populational density of *Bemisia tabaci*, a threat of widespread virus infection of legume crops in Brazil. *In* "Tropical Disease of Legumes" (J. Bird and K. Maramorosch, Eds.), pp. 27-49. Academic Press, New York.
Costa, A. S. (1976). Whitefly-transmitted diseases. *A. Rev. Phytopath.* 14: 429-449.
Costa, A. S., and Bennett, C. W. (1950). Whitefly-transmitted mosaic of *Euphorbia prunifolia. Phytopathology* 40: 266-283.
Costa, A. S. and Bennett, C. W. (1953). A probable vector of *Abutilon* mosaic on species of *Sida* in Florida. *Pl. Dis. Reptr.* 37: 92-93.
Costa, A. S., and Carvalho, A. M. B. (1960a). Mechanical transmission of the Abutilon mosaic virus. *Phytopath. Z.* 37: 259-272.
Costa, A. S., and Carvalho, A. M. B. (1960b). Comparative studies between *Abutilon* and *Euphorbia* mosaic virus. *Phytopath. Z.* 38: 129-152.
Costa, A. S., and Carvalho, A. M. B. (1962). Moletias de virus do Algodoeiro. *Brangantia* 21: 45-62.
Couteaux, L., Lefort, P. L., and Kaukuvi, E. (1968). Some observations on cotton leaf curl in *G. barbadense* at the Anie Mono Station. *Coton Fibr. trop.* 23: 506-507.
Cowland, T. W. (1934). Gezira Entomological section, G. A. R. S. Final report on experimental work 1932-33. *A. Rept. Gezira Agric. Res.* 1933, 107-125.
Crandell, B. S. ((1954). Additions to the host and geographic range of Abutilon mosaic. *Pl. Dis. Reptr.* 38: 574.
Dade, H. A. (1931). Cassava mosaic. *Gold Cst. Dept. Agric. Yb. Bull.* 23: 245-247.
Davis, R. E. (1929). Some chemical and physiological studies on the nature and transmission of infectious chlorosis in variegated plants. *A. Missouri Bot. Gard.* 16: 145-213.
Debrot, C. E., Herold, F., and Dao, F. (1963). Preliminary note on yellowish mosaic of tomato in Venezuela. *Agron. trop.* Maracay 13: 33-41.
Deighton, F. C. (1940). Tobacco leaf curl in Sierra Leone. *Afr. agric. Conf.* 1: 7-8.
Dey, S. K., and Singh, D. R. (1973). Resistance test of mung (*Phaseolus aureus* L.) against yellow mosaic in Uttar Pradesh. *Indian J. Farm Sci.* 1: 113.
Dhanraj, K. S., and Seth, M. L. (1968). Enations in *Capsicum annum* (Chilli) caused by a new strain of leaf curl virus. *Indian J. Hort.* 25: 70-71.
Dhanraj, K. S., Seth, M. L., and Bansal, H. C. (1968). Reactions of certain chilli mutants and varieties to leaf curl virus. *Indian Phytopath.* 21: 342-343.

Dhingra, K. L., and Nariani, T. K. (1961). Yellow net virus disease of tobacco plant. *Indian J. Microbiol.* 1: 94-99.

Dickson, R. C., and Laird, E. F. (1960). Disease of cotton. *Calif. Agric.* 14: 14.

Dickson, R. C., Johnson, M. McD., and Laird, E. F. (1954). Leaf crumple, a virus disease of cotton. *Phytopathology* 44: 479-480.

Doughty, L. R., Jennings, D. L., and Gourlay, D. W. (1955). Cassava breeding. *Rep. E. Afr. Agric. Res. Org.* 36-39.

Duffus, J. E. (1965). Beet pseudo-yellows virus, transmitted by the Greenhouse whitefly (*Trialeurodes vaporariorum*). *Phytopathology* 55: 450-453.

Duffus, J. E. (1975). A new type of whitefly-transmitted disease. A link to the aphid-transmitted viruses. *In* "Tropical Diseases Legumes" (J. Bird and K. Maramorosch, Eds.), pp.79-88. Academic Press, New York.

El Nur, E., and Abu Salih, H. S. (1970). Cotton leaf curl disease. *Pans* 16: 121-131.

Erwin, D. C., and Meyer, R. (1961). Symptomatology of the leaf crumple disease in several species and varieties of *Gossypium* and variations of the causal virus. *Phytopathology* 51: 472-477.

Euler, H. V., Hertzsch, W., Myrback, S., Runehjelm, D., and Forssberg, A. (1930). Chemical changes in infectious chlorosis in leaves of Abutilon. *Arkiv. Kemi Mineral. Geol.* 10B; 6.

Eusign, M. R. (1919). Sweet potato mosaic. *Calif. Agric.* 9: 180-181.

Fernando, H. F., and Uduravan, S. B. (1942). The nature of mosaic disease of Bandaka (*Hibiscus esculentus* L.). *Trop. Agriculturist* 98: 16-24.

Flores, A. B. (1961). Varieties of tobacco infected with leaf curl disease and transmission of its virus to healthy tobacco seedlings. *Araneta J. Agric.* 8: 100-109.

Flores, E., and Silberschmidt, K. (1958). Relations between insect and host plant in transmission experiments with infectious chlorosis of Malvaceae. *Annls Acad. bras. Sci.* 30: 535-560.

Flores, E., and Silberschmidt, K. (1962). Observations on a mosaic disease of *Leonurus sibiricus* occurring spontaneously in Sao Paulo. *Phytopath. Z.* 43: 221-233.

Flores, E., and Silberschmidt, K. (1963). Ability of single whiteflies to transmit concomitantly a strain of infectious chlorosis of Malvaceae and of *Leonurus* mosaic virus. *Phytopathology* 53: 238.

Flores, E., and Silberschmidt, K. (1966). Studies on a new virus disease of *Phaseolus longepedunculatus*. *Annls. Acad. bras. Sci.* 38: 327-334.

Flores, E., and Silberschmidt. K. (1967). Contribution to the problem of insect and mechanical transmission of infectious chlorosis of Malvaceae and the disease displayed by *Abutilon thompsonii*. *Phytopath. Z.* 60: 181-195.

Flores, E., Moraes, W. B. C., and Meneghini, M. (1967). An inhibitory activity of leaf extracts of *Abutilon striatum* on the infectivity of *Abutilon* mosaic virus and tobacco mosaic virus. *Arq. Inst. biol.*, S. Paulo 34: 83-89.

Flores, E. Silberschmidt, K., and Kramer, M. (1960). Studies on infectious chlorosis of Malvaceae in field tomatoes. *Biologica* 26: 65-69.

Gadd, C. H., and Loos, C. A. (1941). A virus disease of *Ageratum conyzoides* and tobacco. *Trop. Agriculturist* 96: 255 261.

Galvez, G. E., and Castaño, M. (1976). Purification of the whitefly-transmitted bean golden mosaic virus. *Turrialba* 26: 205-207.

Gámez, R. (1971). The virus of bean in central America. I. Transmission by whitefly (*B. tabaci*) and host plants of the golden mosaic virus. *Turrialba* 21: 22-27.

Garga, R. P. (1949). *Schizanthus* sp. a host of tobacco leaf curl virus. *Curr. Sci.* 18: 260-261.

Garga, R. P. (1953). Preliminary observations on virus diseases of *Thevetia nerrifolia* Juss. *Curr. Sci.* 32: 243-244.

Girardeau, J. H. (1958). The sweet potato whitefly, *Bemisia inconspicua*, as a vector of sweet potato mosaic in South Georgia. *Pl. Dis. Reptr.* 42: 819.

Girardeau, J. H., and Ratcliffe, T. J. (1960). The vector-virus relationship of the sweet potato whitefly and a mosaic of sweet potatoes in South Georgia. *Pl. Dis. Reptr.* 44: 48-50.

Golding, F. D. (1930). A vector of leaf curl of cotton in Southern Nigeria. *Emp. Cott. Grow. Rev.* 7: 120-126.

Golding, F. D. (1935). A probable vector of cassava mosaic in Southern Nigeria. *Trop. Agriculture,* Trin. 12: 215.

Golding, F. D. (1936a). Cassava mosaic in Southern Nigeria. *Eleventh Bull. agric. Dep.* Nigeria 1-10.

Golding, F. D. (1936b). *Bemisia nigeriensis* Corb., a vector of cassava mosaic in Southern Nigeria. *Trop. Agriculture*, Trin. 12: 182-186.

Goodman, R. M. (1977). Single-stranded DNA genome in a whitefly-transmitted plant virus. *Virology* 83: 171-179.

Goodman, R. M., Bird, J., and Tongmeearkom, P. (1977). An unusual virus-like particle associated with golden yellow mosaic of beans. *Phytopathology* 67: 37-42.

Halliwell, R. S., and Rosbery, D. W. (1964). Texas cotton vein-clearing virus compared with California cotton leaf crumple virus. *Phytopathology* 54: 621-627.

Harpaz, I., and Cohen, S. (1965). Semipersistant relationship between cucumber vein-yellowing virus (CVYV) and its vector, the tobacco whitefly (*Bemisia tabaci*). *Phytopath. Z.* 54: 240-248.

Heinze, K. (1959). "Phytopathogene Viren and ihre Überträger, Dücker and Humblot, Berlin. 290 pp.

Herold F. (1967). Investigation of a virus disease of *Anthorium andraenum. Phytopathology* 57: 8 (abstr.).

Hertzsch, W. (1927). Contributions to infectious chlorosis. *Zeitschr. für. Bot.* 20: 2-3.

Hildebrand, E. M. (1958). Two syndromes caused by sweet potato viruses. *Science* 128: 203-204.

Hildebrand, E. M. (1959). A whitefly, *Trialeurodes abutibrea*, an insect vector of sweet potato feathery mottle in Maryland. *Pl. Dis. Reptr.* 43: 712-714.

Hildebrand, E. M. (1960). The feathery mottle virus complex of sweet potato. *Phytopathology* 50: 751-756.

Hildebrand, E. M. (1961). Relations between whitefly and sweet potato tissue in transmission of yellow dwarf virus. *Science* 133: 282-284.

Hill, B. G. (1968). Occurrence of *Bemisia tabaci* in the field and its relation to the leaf curl disease of tobacco. *S. Afr. J. agric. Sci.* 11: 583-594.

Hill, B. G. (1969). Leaf curl–a virus disease of tobacco. *Fmg. S. Afr.* 45: 6-7.

Hollings, M. (1959). Host range studies with fifty-two plant viruses. *Ann. appl. Biol.* 47: 98-108.

Hollings, M., Stone, A. A., Olwen, M., and Bock, K. R. (1976). Purification and properties of sweet potato mild mottle, a whitefly-borne virus from sweet potato (*Ipomoea batatas*) in East Africa. *Ann. appl. Biol.* 82: 511-528.

Hopkins, J. C. F. (1932). Leaf curl of tobacco in Southern Rhodesia. *Rhodesia Agric. J.* 29: 680-686.

Hornby, A. J. W. (1933). Report of the Asst. Director and Agricultural Chemist. *A. Rept. Dept. Agric.* Nyasaland, 1932, 36-47.

Husain, A. M. (1930). Leaf curl in cotton. *Nature* 126: 958.

Husain, A. M. (1932). Leaf curl in cotton and other plants. *Nature* 130: 312.

Hutchinson, J. B., Knight, R. L., and Pearson, E. O. (1950). Response of cotton to leaf curl disease. *J. Genet.* 50: 100-111.

Jennings, D. L. (1957). Further studies in breeding cassava for virus resistance. *E. Afr. agric. J.* 22: 213-219.

Jennings, D. L. (1960). Observations on virus diseases of cassava in resistant and susceptible varieties. I. Mosaic disease. *Emp. J. exp. Agric.* 28: 23-24.

Jha, A., and Mishra, J. N. (1955). Yellow-vein mosaic of Bhendi (*Hibiscus esculentus* L.) in Bihar. *Proc. Bihar Acad. agric. Sci.* 4: 129-130.

Jochems, S. C. J. (1928). Eeen nienwe viruszieckte van Deli-Tabak de Rotterdam B-ziekte. *Bull. Deli Proefst.* Medan 26: 5-26.

Johnpulle, A. L. (1939). Chilli leaf curl experiments. *Trop. Agriculturist* 92: 28.

Jolly, R. L. (1931). The effects of cassava mosaic. *Rev. de Bot. Appliquée et d'Agric. Trop.* 11: 99-104.

Jones, L. K. (1940). Leaf curl and mosaic of geraniums. *Bull. Wash. Agric. Exp. Sta.* 390: 3-19.

Jones, G. H., and Mason, T. G. (1926). On two obscure diseases of cotton. *Ann. Bot.* 40: 759-772.

Karm Singh. (1931). A contribution towards our knowledge of Aleyrodidae (whiteflies) in India. *Bull. agric. Res. Inst. Pusa Ent. Ser.* 12: 1.

Kerling, L. C. P. (1933). The anatomy of the 'Kroepoek diseased' leaf of *Nicotiana tabacum* and of *Zinnia elegans. Phytopathology* 23: 175-190.

Keur, J. Y. (1933). Seed transmission of the virus causing variegation of *Abutilon. Phytopathology* 23: 20.

Kiriyama, K. (1972). Inoculation, purification and serology of tobacco leaf curl virus. *Ann. Phytopath. Soc. Japan* 38: 323-332.

Kiriyama, K., and Nishimura, N. (1969). On the occurrence of virus diseases in the tobacco field in Northeastern Japan. *Bull. Morioka Tob. Exp. Sta.* 4: 85-97.

Kirkpatrick, T. W. (1930a). Leaf curl in cotton. *Nature* 125: 672.

Kirkpatrick, T. W. (1930b). Preliminary note on leaf crinkle of cotton in the Gezira area, Sudan. *Bull. Entomol. Res.* 21: 127-137.

Kirkpatrick, T. W. (1931). Further studies on leaf curl of cotton in the Sudan. *Bull. Entomol. Res.* 22: 323-363.

Kitajima, E. W., and Costa, A. S. (1974). VII. Reuniao Anual Soc. Bras. Fitop. (Mimeographed summary).

Klebahn, H. (1931). Continuation of the experimental investigations on alloiophylly and virus diseases. *Phytopath. Z.* 4: 1-36.

Kufferath, H., and Ghesquiere, J. (1932). Cassava mosaic. *C. r. Soc. Biol.* 109: 1146-1148.

Kulkarni, G. S. (1924). Mosaic and other related diseases of crops in Bombay Presidency. *Poona agric. Coll. Mag.* 6: 6-12.

Kunkel, L. O. (1930). Transmission of *Sida* mosaic by grafting. *Phytopathology* 20: 129-130.

Lagiere, R. (1968). Notes on leaf curl and anthracnose of cotton in Togoland. *Coton Fibr. Trop.* 23: 394-395.

Laird, E. F., and Dickson, R. C. (1959). Insect transmission of the leaf crumple virus of cotton. *Phytopathology* 49: 324-327.

Lal, S. B., and Singh, R. P. (1961). A note on stimulation of growth of leaf curl-infected tomato plants with gibberellic acid. *Sci. & Cult.* 27: 547-548.

Laloraya, M. M., and Jee, G. (1955). Effect of tobacco leaf curl and tobacco mosaic virus on the amino-acid and amide content of *Nicotiana* sp. *Nature* 175: 907-908.

Lastra, J. R., and Vzcategui, R. C. DE. (1975). Viruses affecting tomatoes in Venezuela. *Phytopath. Z.* 84: 253-258.

Lefevre, P. (1935). Some considerations on Cassava mosaic. *Bull. agric. Congo Belge* 26: 442-447.

Lodh, S. B., Ravel, P., and Singh, S. (1971). Studies on nitrogen metabolism in tomato (*Lycopersicon esculentum* Mill.) infected with tomato leaf curl virus. *Indian J. exp. Biol.* 9: 281-283.

Loebenstein, G., and Harpaz, I. (1960). Virus diseases of sweet potatoes in Israel. *Phytopathology* 50: 100-104.

Lourens, J. H., Laon, P. A. V., and Brader, L. (1972). Contribution to the study of a cotton mosaic in Chad: distribution in a field: Common Aleurodidae; transmission trials from cotton to cotton by Aleurodidae. *Coton Fibr. Trop.* 27: 225-230.

Malaguti, G. B. (1963). Outbreaks and new records. Plant disease situation in Venezuela during 1962. *FAO Pl. Prot. Bull.* 11: 43-45.

Mandahar, C. L., and Singh, J. S. (1971). Destruction of chlorophylls a & b in virus-infected leaves. *Sci. & Cult.* 37: 485-487.

Mandahar, C. L., and Singh, J. S. (1972). Effect of bhindi yellow-vein mosaic virus on its host. *Acta Phytopath. Acad. sin. hung.* 7: 187-191.

Maramorosch, K. (1975). Etiology of whitefly-borne diseases. *In* "Tropical Diseases of Legumes"

(J. Bird and K. Maramorosch, Eds.), pp. 71-77. Academic Press, New York.

Maramorosch, K., Hirumi, H., Kimura, M., Bird, J., and Vakili, N. G. (1974). Diseases of pigeonpea in the Caribbean area: an electron microscopy study. *FAO Pl. Prot. Bull.* 22: 32-36.

Mariappan, V., and Narayanaswamy, P. (1972). *Acanthospermum hispidum* DC, a new host of tomato leaf curl virus. *Madras agric. J.* 59: 355-357.

Martinez, A. L. (1964). Leaf curl of tobacco in the Philippines and its transmission by whiteflies (*Bemisia* sp.). *Philipp. J. Agric.* 26: 27-31.

Matsumoto, T., and Tateoka, R. (1940). Virus diseases of tobacco in Formosa. *Trans. nat. Hist. Soc. Formosa* 30: 197-198, 31-33.

Massey, R. E. (1934). Section of Botany and Plant Pathology, G. A. R. S. Final Report on experimental work in 1932-33. *A. Rept. Gezira agric. Res. Serv.* 126-146.

Mathur, R. N. (1932). Leaf curl of cotton in garden zinnias in N. India. *Nature* 129: 797.

Mathur, R. N. (1933). Leaf curl in *Zinnia elegans* at Dehra Dun. *Indian J. agric. Sci.* 3: 89-96.

Matyis, J. C., Silva, D. M., Oliveira, A. R., and Costa, A. S. (1975). Purification and morphology of tomato golden mosaic virus. *Summa Phytopathologica* 1: 267-274.

Mayee, C. D., Mahajan, R., and Kanwar, J. S. (1974). Relationship of leaf curl virus and root knot nematode incidence in some tomato hybrids. *Nematologia Mediterranea* 2: 181-182.

Mayne, H., and Ghesquiere, J. (1934). Hemioteris nuisibles vegetaux du Congo Belge. *Ann. Gambl.* 40: 41.

McClean, A. P. D. (1940). Some leaf curl diseases in South Africa (i) Leaf curl disease of tobacco, (ii) A new petunia strain of leaf curl and a note on the occurrence of a leaf curl disease of hollyhock. *Sci. Bull. Dep. Agric. S. Afr.* 225: 72.

McWhorter, F. P. (1957). Virus diseases of geranium in Pacific Northwest. *Pl. Dis. Reptr.* 4: 83-88.

Meiners, J. P., Lawson, R. H., Smith, E. F., and Díaz, A. J. (1975). Mechanical transmission of whitefly (*Benisia tabaci*)-borne disease agents of beans in El Salvador. *In* "Tropical Diseases of Legumes" (J. Bird and K. Maramorosch, Eds.), pp. 61-69. Academic Press, New York.

Melamed-Madjar, V., Cohen, S., and Yunis, H. (1970). Control of tobacco whitefly. *Hassadeh* 50: 1033-1035.

Mishra, M. D., Raychaudhuri, S. P., and Ashrafi, J. (1963). Virus causing leaf curl of chilli (*Capsicum annuum* L.). *Indian J. Microbiol.* 3: 73-76.

Moore, E. S. (1933). Important to tobacco growers. The crinkly dwarf menace of tobacco. *Fmg. S. Afr.* 8: 276.

Mound, L. A. and Halsey, S. H. (1978). Whitefly of the world, British Museum (Natural History) 1978. John Wiley & Sons, New York. pp. 340.

Mukherjee, A. K., and Raychaudhuri, S. P. (1964). Leaf curl of *Malvaviscus arboreus* Cav. *Indian J. Hort.* 21: 176-177.

Mukherjee, A. K., and Raychaudhuri, S. P. (1966). A note on the effect of some therapeutics on leaf-curl disease of tomato. *Indian Phytopath.* 19: 399-401.

Muller, H. R. A. (1931). Mosaic disease of cassava. *Inst. Voor Planteziekten Bull.* 24: 17.

Muniyappa, V., and Reddy, H. R. (1976). Studies on the yellow mosaic disease of horsegram (*Dolichos biflorus* Linn.). I. Virus vector relationships. *Mysore J. agric. Sci.* 10: 605-610.

Muniyappa, V., Chandrasekhariah, S. C., and Shivashankar, G. (1979). Horsegram cultures tolerant to yellow mosaic. *Indian J. Genet. Pl. Breed.* 38: 148.

Muniyappa, V., Reddy, H. R., and Mustak Ali, T. M. (1978). Studies on the yellow mosaic disease of horsegram (*Dolichos biflorus*) IV. Epidemiology of the disease. *Mysore J. agric. Sci.* 12: 277-279.

Muniyappa, V., Reddy, H. R., and Shivashankar, G. (1975). Yellow mosaic disease of *Dolichos biflorus* Linn. (Horsegram). *Curr. Res.* 4: 176.

Muniyappa, V., Reddy, H. R., and Shivashankar, G. (1976a). Studies on the yellow mosaic disease of horsegram (*Dolichos biflorus*). II. Host range studies. *Mysore J. agric. Sci.* 10: 611-614.

Muniyappa, V., Reddy, H. R. and Shivashankar, G. (1976b). Studies on the yellow mosaic disease of horsegram (*Macrotyloma uniflorus* Syn. *Dolichos biflorus*). III. Evaluation of germplasm for the disease. *Curr. Res.* 5: 52-53.

Muniyappa, V., Setty, B. A. S., and Shivashankar, G. (1977). Reaction of *Phaseolus* species to horsegram yellow mosaic. *Indian J. Genet. Pl. Breed.* 37: 488-490.

Muniyappa, V., Reddy, H. R., Shivashankar, G., and Sreekantaradhya. (1976c). Varietal response of greengram (*Phaseolus aureus* Roxb.) to the greengram yellow mosaic disease. *Curr. Res.* 5: 100-101.

Munshi, Z., and Chowdhry, A. H. (1964). Some correlation studies between temperature and leaf curl incidence in cigarette tobacco. *Pakist. J. Sci.* 16: 48-52.

Mustafa, E. A. (1961). Leaf curl virus and the flavonoid content of resistant and susceptible strains of cotton. *Nature* 191: 415.

Nair, N. G. (1971). Studies on the yellow mosaic of urd bean caused by mung bean yellow mosaic virus. Ph. D. thesis. UPAU, Pantnagar, India pp. 1-66.

Nair, N. G., and Nene, Y. L. (1973a). Studies on the yellow mosaic of urd bean (*Phaseolus mungo* L.) caused by mung bean yellow mosaic virus. II. Virus-vector relationships. *Indian J. Fm. Sci.* 1: 62-70.

Nair, N. G., and Nene, Y. L. (1973b). Studies on the yellow mosaic of urd bean (*Phaseolus mungo* L.) caused by mung bean yellow mosaic virus. I. Transmission studies. *Indian J. Fm. Sci.* 1: 109-110.

Nair, N. G., and Nene, Y. L. (1974a). Studies on the yellow mosaic of urd bean (*Phaseolus mungo* L.) caused by mung bean yellow mosaic virus. III. Factors influencing transmission and symptom expression. *Indian J. Fm. Sci.* 2: 42-47.

Nair, N. G., and Nene, Y. L. (1974b). Studies on the yellow mosaic of urd bean (*Phaseolus mungo* L.) caused by mung bean yellow mosaic virus. IV. Nature and extent of losses due to infection at various stages of growth. *Indian J. Fm. Sci.* 2: 48-50.

Nair, R. R., and Wilson, K. I. (1969). Studies on some whitefly-transmitted plant virus diseases from Kerala. *Agric. Res. J. Kerala* 7: 123-126.

Nair, R. R., and Wilson, K. I. (1970). Leaf curl of *Jatropha curcas* L. in Kerala. *Sci. & Cult.* 36: 569.

Nariani, T. K. (1956). Leaf curl of papaya. *Indian Phytopath.* 9: 151-155.

Nariani, T. K. (1960). Yellow mosaic of mung (*Phaseolus aurens* L.). *Indian Phytopath.* 13: 24-29.

Nariani, T. K. (1963). Effect of gibberellic acid on tobacco leaf curl affected plants. *Indian Phytopath.* 16: 101-102.

Nariani, T. K. (1968). Enation leaf curl of tomato. *Pl. Dis. Reptr.* 52: 595-596.

Nariani, T. K., and Pathanian, P. S. (1953). *Physalis peruviana* L. a new host of tobacco leaf curl virus. *Indian Phytopath.* 6: 143-145.

Nariani, T. K., and Seth, M. L. (1958). Reaction of *Abelmoschus* and *Hibiscus* species to yellow vein mosaic virus. *Indian Phytopath.* 11: 137-143.

Nariani, T. K., and Vasudeva, R. S. (1963). Reaction of *Lycopersicon* species to tomato leaf curl virus. *Indian Phytopath.* 16: 238-239.

Nene, Y. L. (1972). A survey of viral diseases of pulse crops in UP. *G. B. Pant Univ. Agric. Tech.* 191 pp.

Nene, Y. L. (1973a). Control of *Bemisia tabaci* Genn., a vector of several plant viruses. *Indian J. agric. Sci.* 43: 433-436.

Nene, Y. L. (1973b). Viral diseases of some warm weather pulse crops in India. *Pl. Dis. Reptr.* 57: 463-467.

Nene, Y. L., Naresh, J. S., and Nair, N. G. (1971). Additional hosts of mung bean yellow mosaic virus. *Indian Phytopath.* 24: 415-417.

Nene, Y. L., Srivastava, S. K., and Naresh, J. S. (1972). Evaluation of blackgram (*Phaseolus mungo*) and greengram (*P. aurens*) varieties and germplasms for resistance to yellow mosaic virus of greengram. *Indian J. agric. Sci.* 42: 251-254.

Newton, W., and Peiris, J. W. L. (1953). Virus diseases of plants in Ceylon. *FAO Pl. Prot. Bull.* 2: 17-21.

Nitzany, F. E., Geisenberg, H., and Koch, B. (1964). Tests for the protection of cucumbers from a whitefly-borne virus. *Phytopathology* 54: 1059-1061.

Nour, M. A. (1959). Cotton leaf mottle: a new virus disease of cotton. *Emp. Cott. Grow. Rev.* 36: 32-34.

Nour, M. A., and Nour, J. (1964). Identification, transmission and host range of leaf curl viruses infecting cotton in the Sudan. *Emp. Cott. Grow. Rev.* 41: 27-37.

Nour-Eldin, F., Mazyad, H., and Hassan, M. S. (1969). Tomato leaf curl virus disease. *Agric. Res. Rev.* Cairo 47: 49-54.

Oblisami, G., Kandasamy, D., and Bhaskaran, R. (1974). Phyllosphere microflora of bhendi plants as influenced by yellow vein mosaic virus infection. *Z. Pflkrankh. Pflpath. PflSchutz* 81: 522-527.

Olivares, F. M. Jr., Curz, A. A., Dela, and Salisi, L. B. (1972). Progress report on whitefly transmission of tobacco leaf curl virus to vegetable and weeds. *Philippine Phytopath.* 8: 21-29.

Orlando, A., and Silberschmidt, K. (1946). Studies on the natural dissemination of the virus of infectious chlorosis of the Malvaceae (Abutilon virus 1. Baur) and its relation with the insect *Bemisia tabaci* (Homoptera-Aleyrodidae). *Arq. Inst. biol. S. Paulo* 17: 1-36.

Owen, H. (1946). Mosaic disease of Malvaceae in Trinidad. *Trop. Agric.*, Trin. 23: 157-162.

Pal, B. P. (1937). Leaf curl of toabacco in North India. *Proc. 24th Indian Sci. Congr. B* 372.

Pal, B. P., and Tandon, R. K. (1937). Types of tobacco leaf curl in northern India. *Indian J. agric. Sci.* 5: 263.

Park, M., and Fernando, M. (1938). The nature of chilli leaf curl. *Trop. Agriculturist* 91: 263-265.

Pascalet, M. (1932). Mosaic of leprosy of cassava. *Agron. colon.* 21: 117-131.

Pettybridge, G. H., and Smith, K. M. (1932). A suspected virus disease of zonal pelargoniums. *Gdnr. Chron.* 92: 378-379.

Pathak, H. C., and Raychaudhuri, S. P. (1967). *Withania somnifera*, an additional host of tobacco leaf curl virus. *Sci. & Cult.* 33: 234-235.

Pierre, R. E. (1975). Observations on the golden mosaic of bean (*Phaseolus vulgaris* L.) in Jamaica. *In* "Tropical Disease of Legumes," (J. Bird and K. Maramorosch, Eds.), pp. 55-59. Academic Press, New York.

Pilowsky, M., and Cohen, S. (1974). Inheritance of resistance to tomato yellow leaf curl virus in tomatoes. *Phytopathology* 64: 632-635.

Potty, V. P., and Wilson, K. I. (1974). Studies on the physiology of virus disease of bhendi plant. *Agril. Res. J. Kerala* 11: 65-68.

Potty, V. P. and Wilson, K. I. (1975). Effect of yellow vein mosaic virus infection on the phyllosphere microflora of bhendi (*Abelmoschus esculentus*) plants. *Agril. Res. J. Kerala* 12: 168-171.

Pruthi, H. S. (1945). Leaf curl disease of tobacco in India. *Indian Fmg.* 5: 220-223.

Pruthi, H. S., and Samuel, C. K. (1937). Entomological investigations on the leaf curl disease of tobacco in N. Bihar. I. Transmission experiments with some suspected insect vectors. II. An alternative host of the virus and the insect transmitter. *Indian J. agric. Sci.* 7: 659-670.

Pruthi, H. S., and Samuel, C. K. (1939). Entomological investigations on the leaf curl disease of tobacco in Northern India. IV. Transmission of the disease by whitefly (*Bemisia gossypiperda*) to tobacco, sunn-hemp and a new alternate host of the curl virus. *Indian J. agric. Sci.* 9: 223-275.

Pruthi, H. S., and Samuel, C. K. (1941). Entomological investigations on the leaf curl disease of tobacco in Northern India. IV. transmission of the disease by whitefly (*Bemisia gossypiperda*) from some new alternate hosts. *Indian J. agric. Sci.* 11: 387-409.

Pruthi, H. S., and Samuel, C. K. (1942). Entomological investigations on the leaf curl disease of tobacco in Northern India. V. Biology and population of whitefly vector (*Bemisia tabaci* Gen.) in relation to the incidence of the disease. *Indian J. agric. Sci.* 12: 35-37.

Ramakrishnan, K., and Subramanian, C. P. (1961). A virus disease of *Odontonema nitidum*. *Curr. Sci.* 30: 431-432.

Ramakrishnan, K., Kandaswamy, T. K., Subramanian, K. S., Janarthanan, R., Mariappan, V., Samuel, G. S., and Navaneethan, G. (1972). Investigations on virus diseases of pulse crops in

Tamil Nadu. *Final Tech. Rep.* Tamil Nadu Agric. Univ. Coimbatore, India.

Ramaiah, M., Vidhyasekaran, P., and Kandaswamy, T. K. (1972). Changes in photosynthetic pigments of bhendi infected by yellow vein mosaic disease. *Madras agric. J.* 59: 402-404.

Rao, D. G., and Varma, P. M. (1961). Investigations on yellow vein mosaic of *Malvastrum coromandalianum*, a whitefly-transmitted virus in India. *Proc. 48th Indian Sci. Cong.* Roorkee 499-500.

Rao, D. G., and Varma, P. M. (1964). Studies on yellow vein mosaic of *Malvastrum coromondelianum* in India. *An. Acad. bras Cienc.* 36: 207-215.

Rataul, H. S., and Singh, L. (1974). Control of soybean yellow mosaic virus in soybean, *Glycine max*, by controlling the vector, whitefly *Bemisia tabaci* Genn. in Punjab. *J. Res. Pun. Agric. Univ.* 11: 73-76.

Rathi, Y. P. S. (1972). Mung bean yellow mosaic virus: Host range and relationship with the vector, *Bemisia tabaci* Genn. Ph. D. thesis, G. B. Pant University of Agril. Technology, Pantnagar, India pp. 1-93.

Rathi, Y. P. S., and Nene, Y. L. (1974). Sex of *Bemisia tabaci* Genn. in relation to the transmission of mung bean yellow mosaic virus. *Acta Botanica Indica* 2: 74-76.

Raychaudhuri, S. P., Chatterjee, S. N., and Dhar, H. K. (1961). Preliminary note on the occurrence of the yellow net vein disease of mulberry. *Indian Phytopath.* 14: 94-95.

Reifman, V. G., and Polivanova, T. A. (1969). Virus diseases of soybean in the Soviet Far East. *Trudy biol. Pochv. Inst. Dalnevost. Fil. Sib. otd. ANSSR* 83-104.

Reuterma, M. L., Pableo, G. O., and Price, W. C. (1971). Preliminary study of the transmission of Philippine tomato leaf curl virus by *Bemisia tabaci. Philippine Phytopath.* 7: 29-34.

Russel, L. M. (1957). Synonyms of *Bemisia tabaci* (Homoptera-Aleyrodidae). *Bull. Brooklyn ent. Soc.* 52: 122-124.

Sahambi, H. S. (1958). Proceedings of the Mycological Research Workers Conference, Simla, 181-184.

Samuel, G. S., Janarthanan, R., Kandaswamy, T. K., and Subramanian, K. S. (1972). New hosts for yellow mosaic virus of *Dolichos lab-lab. Curr. Sci.* 41: 686.

Sastry, K. S. M. (1966). Yellow vein mosaic of *Rosa bourbiana. Indian Phytopath.* 19: 316.

Sastry, K. S. M., and Nariani, T. K. (1962). Effect of host plant nutrition on growth and susceptibility of tobacco plants to infection with tobacco leaf curl virus. *Indian J. agric. Sci.* 32: 288-293.

Sastry, K. S. M., and Singh, S. J. (1973). Assessment of losses in tomato leaf curl virus. *Indian J. Myco. Pl. Path.* 3: 50-54.

Sastry, K. S. M., and Singh, S. J. (1973a). Restriction of yellow vein mosaic virus spread in okra through the control of vector, whitefly, *Bemisia tabaci. Indian J. Myco. Pl. Path.* 3: 76-80.

Sastry, K. S. M., and Singh, S. J. (1973b). Field evaluation of insecticides for the control of whitefly (*Bemisia tabaci*) in relation to the incidence of yellow vein mosaic of okra (*Abelmoschus esculentus*), *Indian Phytopath*. 26: 129-138.

Schuster, M. F. (1964). A whitefly-transmitted mosaic virus of *Wissadula amplissima. Pl. Dis. Reptr.* 48: 902-905.

Sharma, S. R., and Varma, A. (1976). Cowpea yellow fleck a whitefly transmitted disease of cowpea. *Indian Phytopath*. 29: 421-423.

Sharp, D. G., and Wolf, F. A. (1949). The virus of tobacco leaf curl. *Phytopathology* 39: 225-230.

Sharp, D. G., and Wolf, F. A. (1951). The virus of tobacco leaf curl. II. *Phytopathology* 41: 94-98.

Sheffield, F. M. L. (1957). Virus diseases of sweet potato in East Africa. I. Identification of the viruses and their insect vectors. *Phytopathology* 47: 582-590.

Sheffield, F. M. L. (1958). Virus diseases of sweet potato in East Africa. II. Transmission to alternative hosts. *Phytopathology* 48: 1-6.

Shepherd, E. F. S. (1940). Tobacco leaf curl. *Pro. Third W. Afr. agric. Conf.* 1: 87-89.

Siddiq, M. A. (1968). Genetics of resistance to cotton leaf curl in sakel cotton. *J. agric. Sci.* Camb. 70: 99-103.

Silberschmidt, K. (1938). Cassava mosaic. *Biologica* 4: 177-181.

Silberschmidt, K. (1943). Studies on the experimental transmission of infectious chlorosis of the Malvaceae. *Arq. Inst. biol.* S. Paulo 14: 105-156.

Silberschmidt, K. M. (1948). Infectious chlorosis of *Phenax sonnerati. Phytopathology* 38: 395-398.

Silberschmidt, K. (1954). Potato viruses in the Americas. *Phytopathology* 44: 415-420.

Silberschmidt, K., and Campas, A. R. (1944). Estudos relations a doenca, "Super-brotamento" in Envassouramento da mandioca. *Arquit. Inst. Biol.* Sao Paulo 15: 1-26.

Silberschmidt, K., and Tommasi, L. R. (1955). Observacees E Estudos sobre Especies de plantas suscetiveis a clorose Infecciosa das Malvaceas. *Ann. Acad. bras. Sci.* 27: 195-214.

Silberschmidt, K., and Tommasi, L. R. (1956). A solanaceous host of the virus of infectious chlorosis of Malvaceae. *Ann. appl. Biol.* 44: 161-165.

Silberschmidt, K., Flores, E., and Tommasi, L. R. (1957). Further studies on the experimental transmission of infectious chlorosis of Malvaceae. *Phytopath. Z.* 30: 387-414.

Singh, D. V., and Lal, S. P. (1964). Occurrence of viruses of tomato and their probable strains at Kanpur. *Proc. natn. Acad. Sci.* India 34: 179-187.

Singh, B. P. and Misra, A. K. (1971). Occurrence of hollyhock yellow mosaic virus in India. *Indian Phytopath.* 24: 213-214.

Singh, H., Sandhu, G. S., and Mavi, G. S. (1971). Control of yellow mosaic virus in soybean, *Glycine max* (L.) Merrill by the use of granular insecticides. *Indian J. Entom.* 33: 272-278.

Singh, S. J., Sastry, K. S. M., and Sastry, K. S. (1974). Effect of oil spray on the control of tomato leaf curl virus in the field. *Indian J. agric. Sci.* 43: 669-672.

Smith, K. M. (1957). Hollyhock mosaic virus. *In* "Plant Virus Diseases" (K. M. Smith, Ed.). Churchill Ltd. 641 pp.

Srivastava, K. M., Singh, B. P., Dwadash Shreni, V. C., and Srivastava, B. N. (1977). *Zinnia* yellow net disease–transmission, host range, and agent–vector relationship. *Pl. Dis. Reptr.* 61: 550-554.

Staner, P. (1931). Mosaic of cassava leaves. *Bull. agric. Congo Belge* 22: 75-80.

Steyaert, R. L. (1946). Plant Protection in the Belgian Congo. *Scient. Mon.*, N. Y. 58: 268-280.

Storey, H. H. (1931). A new virus disease of the tobacco plant. *Nature* 128: 187-188.

Storey, H. H. (1932). Leaf curl of tobacco in Southern Rhodesia. *Rhodesia agric. J.* 29: 186-192.

Storey, H. H. (1935). Virus diseases of East African plants. II. Leaf curl disease of tobacco. *E. Afr. agric. J.* 1: 148-153.

Storey, H. H., and Nichols, R. F. W. (1938). Studies of the mosaic of cassava. *Ann. appl. Biol.* 25: 790-806.

Subramanian, K. S., Samuel, G. S. Janarthanan, R., and Kandaswamy, T. K. (1971). A note on the varietal resistance studies of *Dolichos lab-lab* L. to yellow mosaic disease. *South Indian Hort.* 19: 96-97.

Suteri, B. D. (1974). Occurrence of soybean yellow mosaic in Uttar Pradesh. *Curr. Sci.* 43: 689-690.

Suteri, B. D. (1975). Reduction in oil content of yellow mosaic-infected soybean seeds. *Curr. Sci.* 44: 287.

Suteri, B. D., and Srivastava, K. S. B. (1975). Protein content in healthy and yellow mosaic-infected soybean seeds. *Curr. Sci.* 44: 205-206.

Tarr, S. A. J. (1951). Leaf curl disease of cotton, *Commonw. Mycol. Inst.* Kew 1951, 55.

Tarr, S. A. J. (1957a). The host range of the cotton leaf curl virus in Sudan. *Emp. Cott. Grow. Rev.* 34: 258-262.

Tarr, S. A. J. (1957b). Recent observations on diseases of cotton in the Sudan Gezira. *FAO Pl. Prot. Bull.* 5: 85-88.

Thirumalachar, M. J., Govindu, H. C., and Yaraguntaiah, R. C. (1973). Chemotherapeutic control of tomato leaf curl virus disease. *Indian Phytopath.* 26:330-332.

Thomas, K. M. and Krishnaswami, C. S. (1939). Leaf crinkle–a transmissible disease of papaya. *Curr. Sci.* 8: 316.

Thung, T. H. (1932). The curl and crinkle diseases of tobacco and the causes of their dissemination. *Proefstat. Vorstenlandsche Tabak Meded* 72: 1-54.

Tsao, P. W. (1963). Intranuclear inclusion bodies in the leaves of cotton plants infected with leaf crumple virus. *Phytopathology* 53: 243-244.

Uppal, B. N., Varma, P. M., and Capoor, S. P. (1940). Yellow mosaic of Bhendi. *Curr. Sci.* 9: 227-228.

Van Der Laan, P. A. (1940). Motschilduis en *Eupatorium* als Oorzaken Van Pseudo-mosaic (whitefly and *Eupatorium* as causes of pseudomosaic) *Vlugschr. Deli-Proefst*, Medan 67: 4.

Vaschaik, P. H., Erwin, D. C., and Garber, M. J. (1962). Effects of time of symptom expression of the leaf crumple virus on yield and quality of fiber of cotton. *Crop. Sci.* 2: 275-277.

Varma, P. M. (1952). Studies on the relationship of the Bhendi yellow-vein mosaic virus and its vector, the whitefly (*Bemisia tabaci*). *Indian J. agric. Sci.* 22: 75-91.

Varma, P. M. (1955a). Persistance of yellow-vein virus of *Abelmoschus esculentus* (L.) Moench in its vector *Bemisia tabaci* (Gen.). *Indian J. agric. Sci.* 25: 293-302.

Varma, P. M. (1955b). Ability of the whitefly to carry more than one virus simultaneously. *Curr. Sci.* 24: 317-318.

Varma, P. M. (1962). Transmission of plant viruses by whiteflies. Proc. Symp. on Plant and Animal Viruses held at Cuttack on Dec. 31, 1961 to Jan. 1, 1962. *Bull. Nat. Inst. Sci.*, India 24: 14.

Varma, V. S. (1974a). Salvia yellow-vein mosaic virus. *Gartenbauwissenschaft* 39: 565-566.

Varma, V. S. (1974b). Lupin leaf curl virus. *Gartenbauwissenschaft* 39: 55-56.

Varma, V. S. (1974c). Soapwort leaf curl virus. *Gartenbauwissenschaft* 39: 567-568.

Varma, P. S., and Mukherjee, S. K. (1955). Studies on the varietal classification of virus resistance in Lady's Finger (*Abelmoschus esculentus* Moench)., *Proc. 42nd Indian Sci. Congr.* Baroda 371-372.

Varma, P. M., Rao, D. G., and Capoor, S. P. (1966). Yellow mosaic of *Corchorus trilocularis. Sci. & Cult.* 32: 466.

Varma, V. S., and Singh, S. (1973a). Balsam leaf curl disease. *Hort. Res.* 13: 55-56.

Varma, V. S. and Singh, S. (1973b). Zinnia leaf curl virus. *Gartenbauwissenschaft* 38: 159-162.

Varma, H. N., Srivastava, K. M., and Mathur, A. K. (1975). A whitefly-transmitted yellow mosaic virus disease of tomato from India. *Pl. Dis. Reptr.* 59: 494-498.

Vasudeva, R. S., and Azad, R. N. (1948). Potato necrosis. *Curr. Sci.* 17: 216-217.

Vasudeva, R. S., and SamRaj, J. (1948). A leaf curl disease of tomato. *Phytopathology* 38: 364-369.

Vasudeva, R. S., Varma, P. M., and Capoor, S. P. (1953). Some important whitefly (*Bemisia tabaci* Gen.)-transmitted virus in India. *Atti. VI Int. Congr. Microbiol.* Roma, 5: 520-521.

Verplanke, G. (1932). Une maladie virus filtrant du pelargonium zonale. *Bull. Clin. Sci. Acad. Roy. de Belgique Ser.* 8: 269-281.

Warburg, O. (1894). Die Kultrupflanzen Usambara. *Mitt. dtsch. Schutzgeb.* 1: 131.

West, J. (1936). Leaf curl of tobacco in Southern Nigeria. *Trop. Agric.* Trin. 13: 242-244.

West, J. (1940). The control of leaf curl of tobacco in Southern Nigeria. *Pap. Third W. Afr. agric. Conf.* 1938, 1: 205-206.

Williams, R. J. (1975). A whitefly-transmitted mosaic of lima bean in Nigeria. *Trop. Grain Legume Bull.* 1: 11.

Williams, F. J., Greweal, J. S., and Amin, K. S. (1968). Serious and new diseases of pulse crops in India in 1966. *Pl. Dis. Reptr.* 52: 300-304.

Wilson, K. I. (1972). Chlorotic ring spot of Jasmine. *Indian Phytopath.* 25: 157-158.

Wilson, K. I., and Potty, V. P. (1972). Yellow vein mosaic of *Blumea neilgherrensis. Agril. Res. J.* Kerala 10: 68.

Wilson, K. I., and Satyarajan, P. K. (1970). Yellow vein mosaic of *Passiflora foetida. Sci. & Cult.* 14: 224-228.

Wolf, F.A., Whitecomb, W. H., and Mooney, W. C. (1949). Leaf curl of tobacco in Venezuela. *J. Elisha Mitchell Sci. Soc.* 65: 38-47.

Yassin, A. M. (1975). Epidemics and chemical control of leaf curl virus disease of tomato in the Sudan. *Exp. Agric.* 11: 161-165.

Yassin, A. M. and Abusalih, H. S. (1972). Leaf curl of tomato. *Tech. Bull. agric. Res. Corporation*, Sudan 3. 33 pp.

Yassin, A. M., and Elnur, E. (1970). Transmission of cotton leaf curl virus by single insects of *Bemisia tabaci*. *Pl. Dis. Reptr.* 54: 528-531.

Chapter 4

PSYLLIDS

George H. Kaloostian

USDA, SEA, Boyden Entomology Laboratory
University of California
Riverside, California

4.1 INTRODUCTION 87
4.2 PEAR DECLINE 88
4.3 GREENING DISEASE OF CITRUS 89
4.4 PROLIFERATION DISEASE OF CARROTS 90
4.5 REFERENCES 90

4.1 INTRODUCTION

Psyllids (jumping plant lice) belong to the family Psyllidae, which consists of 180 genera and about 1500 species. Psyllids are highly host specific during the reproductive cycle, but they readily disperse from their breeding hosts during diapause and feed on and may be collected from a wide range of plants on which they cannot reproduce (Kaloostian, 1970). Because the nymphs of some species are highly toxigenic, they cause a number of systemic plant diseases that have been mistaken for virus or virus-like diseases. A classical example is "psyllid yellows" of the potato which was considered a virus disease for over 20 years until Richards (1931) showed that the disease was caused by a toxin injected into potato plants by the nymphs of *Paratrioza cockerelli* (Sulc). Once the nymphs were removed, the plants recovered. On the other hand, a psyllid-transmitted pathogen that actually causes pear decline disease was reported as a toxin (Lindner *et al.*, 1962) before the relationship of the pathogen and the vector, *Psylla pyricola* Foerster, was demonstrated (Jensen *et al.*, 1964).

Of the approximately 1500 species of psyllids, only 6 have been incriminated as vectors of 3 plant disease agents: *P. pyricola, P. pyri* (L.), and *P. pyrisuga* Foerster (pear decline); *Trioza erytrea* (Del Guercio) and *Diaphornia citri* Kuwana (citrus

ISBN 0-12-326450-2

greening disease); and *Trioza nigricornis* Foerster (proliferation disease of carrots). These vector-plant disease relationships are discussed in this report.

4.2 PEAR DECLINE

Pear decline was first reported as a disease of pear in British Columbia, Canada (McLarty, 1948). Soon after this appearance it was found in Washington (Woodbridge *et al.,* 1957) and California (Nichols *et al.,* 1960). According to Refatti (Refatti, 1964), a disease condition of pears in Italy known as "Moria" is the same as the pear decline disease in the Pacific Coast of America. The disease has also been reported from Czechoslovakia and Germany (Blattny' and Vova, 1974) and from Yugoslavia (Grbic, 1974). In addition, Grbic reported *P. pyri* and *P. pyrisuga* as vectors of pear decline in Yugoslavia.

Although the pear psylla was first found in the United States in 1832 (Smith, 1941), the insect was not incriminated as a vector of a disease organism until 1964 (Jensen *et al.,* 1964). In fact, the pear psylla was apparently not a vector of a plant pathogen when it first became established in the Northwest in 1939. Before it became a vector, it did relatively little damage to pears unless heavy infestations developed; and the damage to fruit and "psylla shock" caused by the toxin could be prevented by ordinary orchard pest management. The relationship between the pear decline disease agent and pear psylla toxin on pear trees were described by Kaloostain *et al.* (1968).

Discovery of mycoplasmalike organisms as disease agents in plants in 1967 by the Japanese (Doi *et al.,* 1967) suggested that the pear decline pathogen, formerly thought to be a virus, might be a mycoplasmalike organism. Such an organism was found in trees with pear decline disease (Schneider and Hibino, 1970) and in the bodies of infectious pear psylla (Hibino *et al.,* 1971). Later, Kaloostian *et al.* (1971) used pear psylla to transmit mycoplasmalike bodies from trees with pear decline disease to Madagascar periwinkle, *Vinca rosea* L. = *Catharanthus roseus* (L.).

Since 1962 all pear psylla collected from such northern California localities as El Dorado, Sacramento, Contra Costa, and Santa Clara counties were infectious and transmitted the disease organism readily to Bartlett pear trees on Oriental rootstock, *Pyrus ussuriensis* Maxim (Kaloostian *et al.,* 1968). Non-infectious pear psylla collected from trees free of pear decline in San Diego County became infectious 21 days after a 1- to 2-day acquisition feeding period on pear decline inoculum. The latent period for the disease in greenhouse test seedlings was about 60 days; symptoms appeared in field plot trees the year following inoculation. Groups of 10 or 20 infectious adult pear psylla were sufficient to transmit the pear decline mycoplasma. The insects were infectious for life.

Symptoms of pear decline Bartlett trees were complex and varied according to rootstock and variety. Bartlett trees on Oriental rootstock were characterized by an upward roll of the leaves and premature red foliation; some trees wilted and died (quick decline), but others made little growth and remained in a slow decline condition. Bartlett pear trees on French rootstock, *Pyrus communis* L., developed leaf curl (Kaloostian, 1968): the tips of leaves on diseased trees curled abaxially and downward, sometimes 360 degrees or more; also, they developed a deep, reddish-purple color that usually intensified with the advent of cold weather. In general the symptoms on trees of *P. ussuriensis* or *P. communis* included sieve tube necrosis below the graft union. However, own-rooted Bartlett trees developed leaf curl, and a leaf-drop symptom was associated with own-rooted Magness pear trees (Kaloostian, 1968). A brown leaf-vein symptom was observed only in diseased greenhouse-grown pear trees (Tsai *et al.,* 1966).

4.3 GREENING DISEASE OF CITRUS

Greening disease of citrus in South Africa is also known as leaf mottle in the Philippines, citrus decline or die-back in India, Likubin in Taiwan, yellow shoot in China, and vein-phloem degeneration in Indonesia. It is caused by a mycoplasma-like pathogen morphologically different from a true mycoplasma. The outer membrane of the greening disease organism is twice as thick (200 Å vs. 100 Å) as the unit membrane that envelopes a mycoplasmalike agent (Garnier *et al.,* 1975). Citrus trees affected with greening disease are characterized by a general decline, leaf mottle, yellowing of midribs and lateral veins, complete yellowing of terminal leaves, reduced leaf size, and stunted plants with lopsided fruits (Capoor and Viswanath, 1967; Martinez and Wallace, 1967).

The greening disease agent is transmitted by 2 species of psyllids, *Trioza erytreae* and *Diaphorina citri* (McClean and Oberholzer, 1965; Tirtawidjaja, Martinez and Wallace, 1967). Wherever these insects are found, greening disease, or a disease similar to greening disease, is also found. The two vectors are distributed in widely separated geographical areas except that both species are present on the islands of Mauritius and Réunion. *Trioza erytreae* occurs throughout Africa south of the Sahara including Madagascar, Mauritius and Réunion; *D. citri* is found in India, Nepal, Taiwan, China, Thailand, Hong Kong, Philippines, Réunion, Mauritius, and Brazil (Catling, 1970).

Transmission of the greening disease pathogen by psyllids is more effective when adults are collected from diseased trees and caged on healthy test plants (Martinez and Wallace, 1967). Single adult insects of both species may transmit, and nymphs of *D. citri* transmit occasionally, but no transovarial transmissions have ever been observed in either species (McLean, 1974). The insects acquire the organism after an acquisition period of 24 hr. and are inoculative 3-10 weeks after a latent period of 8-12 days. Vector efficiency is optimal in the cooler spring months as is the receptivity of the plants to infection. Symptoms appear in psylla-inoculated plants in about 60 days (Martinez and Wallace, 1967).

4.4 PROLIFERATION DISEASE OF CARROTS

The psyllid, *Trioza nigricornis,* has recently been reported as a vector of a mycoplasmalike organism that causes proliferation and yellows in carrots (Leclant *et al.,* 1974). High populations of the vector are present throughout the year on infected carrots in the field. It is also reported (Giannotti *et al.*, 1974) that a double infection was observed by electron microscopy: mycoplasma and rickettsia-like microorganisms were found both in plants with proliferation symptoms, and in the psyllid vector. The etiology of this disease needs clarification.

4.5 REFERENCES

Blattny', C., and Vana, V. L. (1974). Pear decline accompanied with mycoplasmalike organisms from Czechoslovakia. *Biol. Plant.* 10(6): 474-475.

Capoor, S. P., Viswanath, D. G. and S. M. (1967). *Diaphorina citri* Kuwan, a vector of the greening disease of citrus in India. *Indian J. Agric. Sci.* 37: 572-576.

Catling, H. D. (1970). Distribution of the psyllid vectors of citrus greening disease, with notes on biology and bionomics of *Diaphorina citri. FAO Plant Prot. Bull.* 18: 8-15.

Doi, Y., Teranaka, N., Yoba, K., and Asuyama, H. (1967). Mycoplasma–or PLT group-like microorganisms–found in the phloem elements of plants infected with mulberry dwarf, potato witches' broom, aster yellows, or witches' broom. *Ann. Phytopathol. Soc. Japan.* 33: 259-266.

Garnier, M., Latrille, J., and Bove, J. M. (1975). *Spiroplasma citri* and the organism associated with Likubin: comparison of their envelopes. *7th Conf. Int. Org. Citrus Virol.* Abstracts, page 18.

Giannotti, J. C. Louis, C., Leclant, F., Marchoux, G., and Vago, C. (1974). Infection due to mycoplasma and to micro-organisms with rickettsia pattern in a plant affected by proliferation and in the psyllid vector of the disease. *C. R. Acad. Sci.*, Ser. D., 278.4: 469-471.

Grbic, V. (1974). Some injurious psylla species in pear plantations in Vojvodina. Poljoprivredni Fakultet, Novi Sad, Yogaslavia. *Zast. Bilja,* 25, 128/129: 121-131.

Hibino, H., Kaloostian, G. H., and Schneider, H. (1971). Mycoplasmalike bodies in the pear psylla vector of pear declin. *Virology* 43: 34-40.

Hibino, H. and Schneider, H. (1970). Mycoplasmalike bodies in sieve tubes of pear trees affected with pear declin. *Phytopathology* 60: 499-501.

Jensen, D. D., Griggs, W. H., Gonzales, C. Q., and Schneider, H. (1964). Pear decline virus transmission by pear psylla. *Phytopathology* 54: 1346-1351.

Kaloostian, G. H. (1968a). A leaf drop symptom associated with own-rooted Magness pear trees inoculated with pear decline virus by the pear psylla, *Psylla pyricola. Plant Dis. Rep.* 52. 363-365.

Kaloostian, G. H. (1968b). Pear leaf curl virus transmitted by pear psylla. *Plant Dis. Rep.* 52: 924-925.

Kaloostian, G. H. (1970). Transitory hosts of the pear psylla. *J. Econ. Entomol.* 63(1): 1039-1041.

Kaloostian, G. H., Oldfield, G. N., and Jones, L. S. (1968). Effect of pear decline virus and pear psylla toxin on pear trees. *Phytopathology* 58(9): 1236-1238.

Kaloostian, G. H., Hibino, H., and Schneider, H. (1971). Mycoplasmalike bodies in periwinkle; their cytology and transmission by pear psylla from pear trees affected with pear decline. *Phytopathology* 61: 1177-1179.

LeClant, F. Marchoux, G., and Giannotti, J. (1974). Evidence of the role of the psylla *Trioza nigricornis* as a vector in the transmission of a proliferation disease of Daucus carota. *C. R. Acad. Sci. Ser.* 278.1: 57-59.

Lindner, R. C., Burts, E. C., and Benson, N. R. (1962). A decline condition in pears induced by pear psylla. *Plant Dis. Rep.* 46: 59-60.

Martinez, A. L., and Wallace, J. M. (1967). Citrus leaf-mottle-yellows disease in the Philippines and transmission of the causal virus by a psyllid, *Diaphorina citri. Plant Dis. Rep.* 51: 692-695.

McClean, A. P. D., and Oberholzer, P. C. J. (1965). Citrus psylla, a vector of greening disease of sweet orange. *S. Afr. J. Agric. Sci.* 8: 297-298.

McClean, A. P. D. (1974). Efficiency of citrus psylla, *Trioza erytreae* (Del G.) as a vector of greening disease of citrus. Phytopathylactica. *Plant Prot. Res. Inst., Pretoria, South Africa.* 6: 45-54.

McLarty, H. R. (1948). Killing of pear trees. *Ann. Rep. Can. Plant Dis. Sur.* 28-77.

Nichols, C. W., Schneider, H. O'Reilly, H. J., Shalla, T. A. and Griggs, W. H. (1960). Pear decline in California. *Calif. Dep. Agric. Bull.* 49: 186-192.

Refatti, E. (1964). La Moria del Pero in Italia. Estratto dal "Notiziario sulle Mallatie delle Piante". N. 68 (N. S. 47). Instituto di Pathologia Vegetale dell' Universita Milano.

Richards, B. L. (1931). Further studies with psyllid yellows of the potato. *Phytopathology* 21: 103.

Smith, L. G. (1941). Pear psylla in Washington. Ext. Bull. No. 255. State College of Washington.

Tirtawidjaja, S., Hadiwidjaja, T., and Laskeen, A. M. (1967). Citrus vein-phloem degeneration virus, a possible cause of citrus chlorosis in Java. *Proc. Am. Soc. Hort. Sci.* 86: 235-243.

Tsao, P. W., Schneider, H., and Kaloostian, G. H. (1966). A brown leaf-vein symptom associated with greenhouse-grown pear plants infected with pear decline virus. *Plant Dis. Rep.* 50(4): April 1966.

Woodbridge, C. G., Blodgett, E. C., and Diener, T. O. (1957). Pear decline in the Pacific Northwest. *Plant Dis. Rep.* 41: 569-572.

Chapter 5

MEMBRACIDS

John N. Simons
JMS Flower Farms, Inc.
Vero Beach, Florida

5.1 INTRODUCTION 93
5.2 SYMPTOMATOLOGY OF PSEUDO-CURLY TOP (PCT) DISEASE 94
5.3 TRANSMISSION OF PCT AGENT 94
5.4 LIFE HISTORY OF THE TREEHOPPER VECTOR 94
5.5 EPIDEMIOLOGY 95
5.6 NATURE OF THE PCT AGENT 95
5.7 RELATIONSHIP OF PCT TO OTHER CURLY TOP DISEASES 95
5.8 REFERENCES 95

5.1 INTRODUCTION

Pseudo-curly top disease (PCT) was first recognized in Florida in 1944. Giddings *et al.* (1951) reported that specimens of infected tomato had been sent from Bradenton, Florida to Riverside, California for diagnosis. The symptoms in these plants were reported to be indistinguishable from those associated with curly top infection. Stoner (1950) reported a similar disease from the East Coast of Florida again in tomatoes. Both authors reported successful graft transmission from tomato to tomato and failure to transmit the disease by rubbing.

During the mid-1950's the author observed a similar disease in tomato throughout south Florida and a serious effort was made to locate and identify an insect vector. This culminated in the discovery that a treehopper (Membracidae) was capable of vectoring the causal agent of PCT (Simons and Coe, 1958). At this writing this is still the only known instance of a treehopper acting as a vector of a virus-type disease.

ISBN 0-12-326450-2

5.2 SYMPTOMATOLOGY OF PSEUDO-CURLY TOP DISEASE

Simons (1962) tested 24 plant species (most of which had been reported to be suscepts of other curly top viruses) for susceptibility to PCT. Species found to be susceptible included tomato (*Lycopersicon esculentum* Mill.), *Datura stramonium* L., *Nicotiana glutinosa* L., *Solanum gracile* Link, *Stellaria medea* (L.) and an *Ambrosia* sp. Symptoms were generally very similar to those reported for other curly top diseases and included stunting, leaf rolling, epinasty and suppression of apical dominance. Sugar beet (*Beta vulgaris* L.) was not a host and bean (*Phaseolus vulgaris* L.) was not infected experimentally although there is considerable suspicion that pole bean is a host since symptoms typical of curly top have been frequently observed in the Dade County area of south Florida.

5.3 TRANSMISSION OF PCT AGENT

The causal agent of PCT is known to be transmitted only by a single species of membracid, *Micrutalis malleifera* Fowler (Simons and Coe, 1958). The vector relationships have been studied by Simons (1962a). He found an acquisition threshold period of 1 hr with a rapid increase in transmission efficiency as the length of acquisition feeding was increased to 24 hr. The inoculation threshold period for insects reared on diseased plants was 2 hr or less with transmission efficiency increasing rapidly through 6 hr of feeding. There was evidence that a 24-48 hr latent period existed for insects given a 24-hr acquisition feeding period. Transmission continued throughout the life of the insect (up to 5 weeks) and inoculativity was higher for insects given 24-hr access to virus than for insects given either 1- or 4-hr access feeding periods.

5.4 LIFE HISTORY OF THE TREEHOPPER VECTOR

The life history of *M. malleifera* has been studied by Simons (1962b). The insect is a warm weather species which is abundant during the summer in south Florida. Populations drop off rapidly during September and the species is difficult to find during the winter months even though host plants are abundant. The only known plant hosts are in the family Solanaceae and include nightshade, ground cherry (Physalis spp.) and eggplant (Solanum melongena L.). Tomato is not a host for the insect.

Under laboratory conditions the species requires about 38 days to complete a generation on eggplant. Females deposit about 50 eggs in stems and leaves over a 3-week period.

The species is very placid in disposition and adults can be collected rather easily by direct aspiration from plants.

5.5 EPIDEMIOLOGY

The occurrence of PCT is highly seasonal with infected crop plants being found only in August and September. Diseased weeds are very common during the summer months and rare during the winter months. Disease incidence in tomato is generally restricted to the margins of fields. Incidence of PCT is usually less than 10% but on occasion outbreaks exceeding 25% have been seen.

5.6 NATURE OF THE PCT AGENT

It appears likely that the agent causing PCT is a virus but this has not been proven. Richardson at the University of California at Berkeley (personal communication) has examined both plant tissue and excised salivary glands from infective insects using electron microscopy and failed to find either virus or mycoplasma. The author has treated infected plants with tetracycline but failed to obtain remission of symptoms.

5.7 RELATIONSHIP OF PCT TO OTHER CURLY TOP DISEASES

Diseases that are characterized as curly top have been described from Puerto Rico (Adsuar, 1955), North America (Severin, 1928) and South America (Bennett *et al.*, 1946; Bennett and Costa, 1949). North American curly top has been reported from the mid-East (Bennett and Tanrisever, 1957; Gibson, 1967). These diseases differ in host range and in vector species but all are transmitted by leafhoppers. Where common host plant species exist the symptoms of the different curly top diseases are usually very similar. The vector-virus relationships of these diseases are all similar and closely resemble those reported for PCT.

PCT may be caused by an agent unrelated to those that cause the other curly top diseases but there is a considerable body of circumstantial evidence to indicate otherwise. Certainly if PCT had been found to be transmitted by a leafhopper the disease would have been considered to be closely related to other described curly top diseases.

5.8 REFERENCES

Adsuar, J. (1955). A disease of tomato in Puerto Rico resembling the Brazilian curly top of tomatoes. *J. Agric. Univ. Puerto Rico* 39: 113-4.

Bennett, C. W., Carsner, E., Coons, G. H. and Brandes, E. W. (1946). The Argentine curly top of sugar beet. *J. Agric. Res.* (US) 72: 19-47.

Bennett, C. W., and Costa, A. S. (1949). The Brazilian curly top of tomato and tobacco resembling North American and Argentine curly top of sugar beet. *J. Agric. Res.* (US) 78: 675-93.

Bennett, C. W., and Tanrisever, A. (1957). Sugar beet curly top disease in Turkey. *Plant Dis. Rptr.* 41: 721-5.

Gibson, K. E. (1967). Possible incidence of curly top in Iran—a new record. *Plant Dis. Rptr.* 51: 976-7.

Severin, H. H. P. (1928). Transmission of tomato yellows, or curly top of the sugar beet, by *Eutettix tenellus* (Baker). *Hilgardia* 3: 251-74.

Simons, J. N. (1962a). The pseudo-curly top disease in south Florida. *J. Econ. Entom.* 55: 358-63.

Simons, J. N. (1962b). Life-history and behavioral studies on *Micrutaleis maellifera,* a vector of pseudo-curly top virus. *J. Econ. Entom.* 55: 363-5.

Simons, J. N., and Coe, D. M. (1958). Transmission of pseudo-curly top virus in Florida by a treehopper. *Virology* 6: 43-8.

Stoner, W. N. (1950). Viruses affecting vegetable crops in the Everglades area. *Florida Expt. Sta. Ann. Rpt.*, p. 206.

Chapter 6

PIESMIDS

Gerhard Proeseler
Institute for Phytopathology, Aschersleben
Academy of Agricultural Sciences
German Democratic Republic

6.1 INTRODUCTION 97
6.2 *PIESMA QUADRATUM* (FIEB.) AND THE BEET LEAFCURL VIRUS 98
 6.2.1 The Vector 98
 6.2.2 Beet Leafcurl Virus 101
 6.2.3 Virus-Vector Interactions 104
 6.2.4 Vector and Virus Under Field Conditions 108
 6.2.5 Vector Control 109
6.3 *PIESMA QUADRATUM* (FIEB.) AND BEET LATENT ROSETTE DISEASE . 110
6.4 *PIESMA CINEREUM* (SAY) AND SUGARBEET SAVOY VIRUS 111
6.5 REFERENCES 112

6.1 INTRODUCTION

Although the order Heteroptera comprises approximately 40,000 species, only two members of the genus *Piesma* have been recognized up to now as vectors of plant-pathogenic viruses and rickettsia-like agents. In North America, *Piesma cinereum* (Say) transmits the beet savoy virus. In Europe *P. quadratum* (Fieb.) is known as the vector of beet leafcurl virus and more recently also as a vector of a rickettsia-like organism that induces sugarbeet latent rosette disease. One can find in the literature both Latin names, *P. quadratum* and *P. quadrata*. Since the gender of the Greek word *Piesma* is neuter *P. quadratum* is the correct name (Proeseler, 1967).

Economic losses caused by the beet leafcurl disease in certain years resulted in extensive studies of the causative agent and its vector. Numerous publications have appeared concerning the bionomics of *P. quadratum*, interactions between the virus and the vector, the morphology of the virus, as well as possible means of vector control. Comparatively few publications are available in regard to the

ISBN 0-12-326450-2

North American beet savoy virus and its vector *P. cinereum*. Despite certain similarities, it is not yet possible to make a comprehensive comparison between the viruses and vectors of the European and American diseases.

Within the biotope of *P. quadratum*, one frequently encounters the related species *P. capitatum* (Wolff) and *P. maculatum* (Laporte). Neither of these, however, can act as vector of the beet leafcurl virus. In this respect, the high degree of specificity that is known in other instances and vector groups seems to apply also to the Heteroptera. The nature of this specificity is not yet understood. The present status of piesmid vectors of disease agents is discussed in the following pages.

6.2 *PIESMA QUADRATUM* (FIEB.) AND THE BEET LEAFCURL VIRUS

6.2.1 The Vector

6.2.1.1 Morphology of Adults. The basic color of adults (Fig. 1A) is variable. Immediately following the last molt, the adults appear green or greyish green. After a few weeks they become darker grey or greyish brown. The eyes are red. The antennae consist of four parts. The terminal one is spindly, and it is covered by numerous tactile hairs (Fig. 1B). The ventrally-placed labium is 4-segmented and extends to the prosternum where it terminates between the coxae of the first pair of legs (Fig. 1C). The prothorax is the most pronounced of the three thoracic regions (Fig. 1D). There are three characteristic wedges on the pronotum of *P. quadratum*. The edges of the pronotum are straight and lack tails. The notum of the mesothorax is triangular and terminates posteriorly in a yellow-white point. When the insect is at rest, the hemelytra cover the hindwings and the abdomen. The pronotum and, except for their membranous apices, the hemelytra are thicked and covered with dark spots (Fig. 1E). Ventrally one can distinguish males and females on the basis of the morphology of the last segments of the abdomen. Males have a rounded knob protruding from the terminal segment. The 6th and 7th segments of the abdomen of females have a deep median groove that divides the segments into halves. The groove covers the sex organs (Fig. 1F). Adults are 3.0–3.5 mm long, and, generally, males are smaller than females (Wille, 1929).

6.2.1.2 The Egg. Eggs are yellow and elongated. Their average size is 0.64 mm by 0.27 mm, and they are rounded at one end. The opposite end is obliquely flattened and covered on its circular surface with small protrusions. The egg shell or chorion is longitudinally deeply ribbed (Wille, 1929).

6.2.1.3 Nymphal Instars. *Piesma quadratum* develop pautometabolously, as do other Heteroptera. The first instar is light yellow when it hatches from the egg. The later instars are green, except for the eyes that are red, as in adults. Successive instars resemble more and more the adult form. There are from four to six nymphal instars. For every eight individuals, one will have 4 instars, six will have 5 instars,

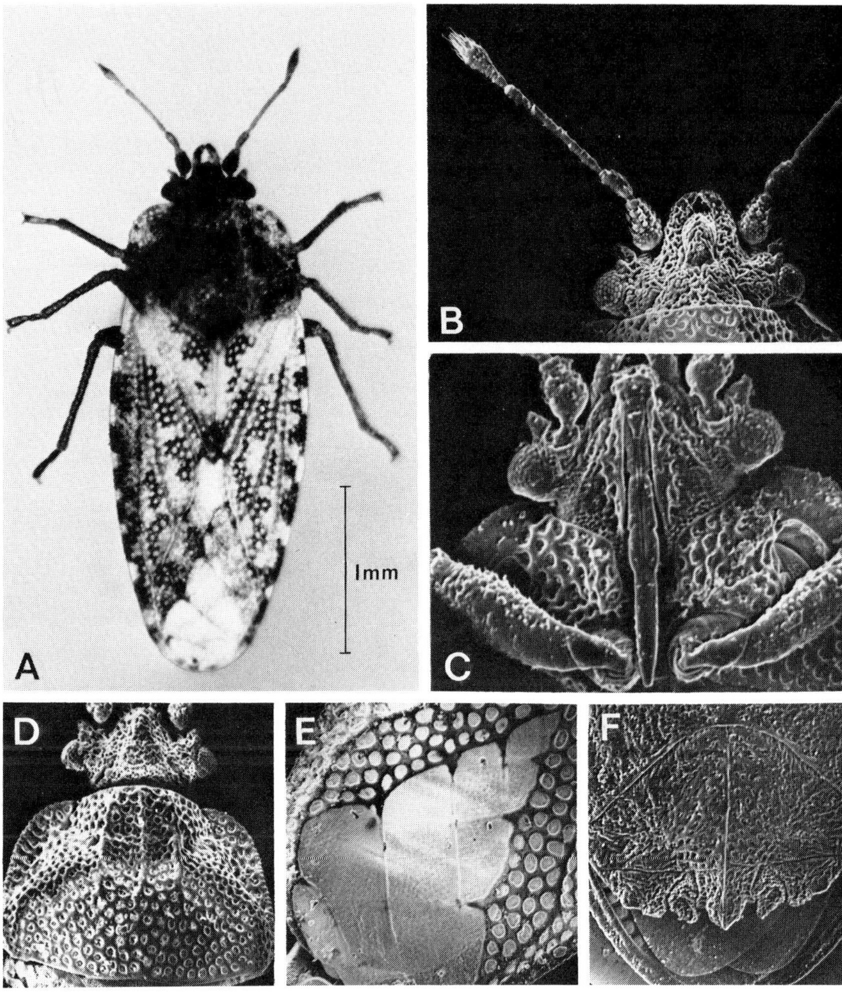

FIG. 1. *Piesma quadratum* (Fieb.). A. adult male, B. head with antennae, C. ventral view with labium, D. pronotum, E. overlapping, membraneous apices of hemelytra, F. caudal segments of female, ventral view. (Original illustration by Proeseler, Institute for Phytopathology, Aschersleben.)

and one will have 6 instars. The same average numbers of nymphal instars are found among males and females (Proeseler, 1978a).

6.2.1.4 Development in Nature. *Piesma quadratum* overwinters in the adult stage only. In late summer or early fall, the adults fly or crawl into dry, protected places, usually patches of grass or twigs at the base of trees and bushes. One can sometimes find the bugs in cracks of tree bark or wood. The insects always select southern or western exposure. Overwintering is always in the proximity of infested beet acreage, particularly on trees at the border of such fields. Where there are larger forest stands, the bugs overwinter only at the edge of the forest. Immobilized, hibernating *P. quadratum* can be revived by raising the temperature. On warm days the bugs leave their hiding places, crawl on tree stems and sun themselves. The cold-induced paralysis can be interrupted experimentally and the bugs then kept at higher temperature and under extended day-length illumination. Under these artificial conditions the bugs resume feeding and subsequently begin to proliferate. With the start of spring and until the end of April the bugs leave their winter places and begin to search for food, especially weeds belonging to the Chenopodiaceae. They move to beets as soon as they appear in the field. The invasion starts at the edges of a field and moves towards the inner areas with increases in the bug population.

A few days after the first meal the bugs begin to copulate. After 5–6 days, usually in the beginning of May, the first eggs are deposited. Ovipositing is irregular, on either the upper or the lower leaf surface. A single female deposits an average of 150 to 160 eggs by the end of July. Sometimes ovipositing is discontinuous. The embryonic development under natural conditions depends on the temperature; it can last from 2 to 3 weeks. Therefore, the first instars appear at the end of May or the beginning of June. Nymphs prefer the undersides of leaves so as to be protected from adverse weather conditions. The developmental cycle is completed by the beginning of July and, usually, only one generation a year is produced. Occasionally, under favorable conditions, a second, summer generation can appear and complete its development to adults by the end of September (Wille, 1929).

6.2.1.5 Rearing. Various types of cages can be used to breed *P. quadratum* in the laboratory. It is important to provide adequate ventilation to prevent excessive humidity and water condensation. The bugs try to hibernate and refuse to feed late in summer or in the beginning of autumn. The urge to hibernate can be overcome by additional illumination and prolonged daytime. In order to proliferate, however, the bugs require not only additional hours of light, but also the proper light spectrum (Krczal and Völk, 1956; Stuben, 1958).

The developmental cycle can be shortened under laboratory conditions. At 25°C eggs hatch one week after ovipositing (Proeseler, 1966a). Under controlled, constant temperature of 20±1°C and 18 hr of illumination at 1500 lux and a relative humidity ranging from 50-70%, postembryonic development is completed in 30 days. When there are 6 instars, completion requires 33 days. Under these conditions, the period from hatching to the final molting is at least 21 days in duration and never exceeds 49 days. The first and last stadia are longer in duration than the middle ones (Table 1). There are no differences insofar as development of males and females is concerned, but the total development of males is an average 1.25 days shorter (Proeseler, 1978a). It is comparatively easy to establish a virus-free colony as there is no transovarial passage of virus from mothers to offspring. The careful removal of eggs from host plants to virus-free plants results in the hatching of nymphs that are then used for the establishment of a stock colony.

TABLE I. Duration (in Days) of *Piesma quadratum* (Fieb.) Instars

Type of Development	Instars I	II	III	IV	V	VI	Total Development Time Median	Minimum	Maximum
A. Four-Instar									
males	7.4	7.4	5.6	7.9			28.0	21	32
females	8.9	7.9	6.4	8.6			31.8	23	38
total	8.2	7.5	6.0	8.3			30.0	21	38
B. Five-Instar									
males	5.8	5.0	5.2	5.6	7.4		29.0	21	49
females	6.7	5.1	5.2	5.2	7.8		30.0	21	45
total[a]	6.3	5.0	5.2	5.4	7.6		29.5	21	49
	(10.1)	(7.3)	(7.6)	(7.7)	(7.6)		(40.2)	(23)	(61)
C. Six-Instar									
males	7.2	4.7	3.9	5.1	5.3	6.3	32.5	27	35
females	7.3	4.3	4.1	5.2	4.4	8.1	33.4	28	38
total	7.2	4.5	4.0	5.2	4.9	7.1	32.9	27	38

[a]The total values in parentheses were established by Wille (1929) at 18-20°C (Proeseler, 1978a).

6.2.2 Beet Leafcurl Virus

6.2.2.1 Symptomatology of Diseased Plants. When infective *P. quadratum* are confined to beets (*Beta* sp.) at 20°C, the first symptoms of leafcurl appear after an average of 42 days. The incubation period varies in individual plants from 20 to 63 days (Proeseler, 1966a). Under field conditions, depending on the age of plants and weather conditions, this period can be even longer. The first symptom is veinclearing in the youngest leaves. Soon afterwards these young leaves manifest vein deformation and growth inhibition symptoms (Fig. 2). In most instances the leaves begin to curve towards the budding point, forming closed braids that resemble the head of lettuce. Older, symptomless leaves wilt and dry up. The inner leaves remain small and turn yellow. Since new leaves are constantly being formed, the bushy plant may continue to grow upwards. The underground growth of the beet ceases prematurely.

6.2.2.2 Damaging Effects of the Virus. The earlier the infection starts, the greater the resulting damage caused by the beet leafcurl virus. The photosynthetic ability of leaves becomes disturbed so that the yield of both leaves and roots decreases. Losses can be as high as 75%, and the sugar content often drops from 19.4 to 15.2%. This affects the technological value of the crop. In certain years, leafcurl disease outbreaks have been very severe and losses considerable. For instance, in 1957 the disease caused losses in the Bydgoszcz and Poznan districts of Poland estimated at 285 million zloty (approximately $12,000,000). In 1961, for all of Poland, 248 thousand hectares were affected, and the sugar yield was reduced by 4180 tons (Kotik-Królicka and Szulc, 1961).

FIG. 2. Beet leafcurl disease in a beet plant. (Illustration courtesy of the Archives of the Institute for Phytopathology, Aschersleben.)

6.2.2.3 Host Plants. The main host plants of the beet leafcurl virus are the following species: sugar beet (*Beta vulgaris* L. spp. *vulgaris* var. *altissima* Doell.) and the fodder beet (*B. vulgaris* L. spp. *vulgaris* var. *alba* D.C.). Other members of the same genus include *B. vulgaris* L. spp. *vulgaris* var. *conditiva* Alef., *B. vulgaris* L. spp. *vulgaris* var. *vulgaris, B. vulgaris* spp. *maritima* (L.) Arcang., *B. trigyna* W. et K., *B. macrocarpa* Guss., and *B. procumbens* Chr. Sm. Other host plants of the virus are: *Atriplex calotheca* Fr., *A. halimus* L., *A. hastata* L., *A. hortensis* L., *A. nitens* Schk., *A. patula* L., *A. tatarica* L., *Chenopodium album* L., *C. amaranticolor* Coste and Reyn., *C. anthelminticum* (L.) A. Gray, *C. capitatum* (L.) Aschers., *C. ficifolium* Sm., *C. foliosum* Aschers., *C. giganteum* Don, *C. multifictum* L., *C. murale* L., *C. opulifolium* Schrad., *C. quinoa* Willd., *C. serotinum* Torn., *C. virgatum* Thunb., *Spinacia oleracea* L., *S. tetranda* Stev., *S. turkestanica* Iljin and *Tetragonia tetragonoides* (Pall.) O. Ktze. With the exception of the sugarbeet and the fodder beet, all other species play a minor role in the epidemiology of the virus, because the disease agent overwinters in its vector and persists in the bugs (Wille, 1920; Proeseler 1966a; Schmutterer and Ehrhardt, 1966b).

6.2.2.4 Physical Properties. Since it has not yet been possible to transmit the beet leafcurl virus mechanically to plants, physical properties of the virus could only be ascertained by injecting virus-free *P. quadratum* with sap extracts from diseased plants. In this manner the following data were obtained: dilution endpoint of 10^{-4} to 10^{-5}; thermal inactivation point of 54 to 58°C, and longevities in vitro at 25, 5, and –20°C of <24 hr, <7 days and >28 days, respectively. In frozen beet leaves the virus remained infectious for more than 84 days. In vitro infectivity was retained at a pH range from pH 5.0 to 9.5 (Proeseler, 1966b).

6.2.2.5 Morphology of Virus Particles and Detection in Plants. The particles of beet leafcurl virus can be observed by electron microscopy techniques using Lee's (1968) partial purification and concentration, or by preparing ultrathin sections. The particles are bacilliform (Fig. 3), with an average length of 225 nm from diseased leaves, and 350 nm from beet roots. The rhabdo-type particles are 80 nm in diameter, with an electron dense core and an outer shell with small protrusions. In addition to the bacilliform particles one encounters also bullet-shaped virions. Large concentrations are found in side-by-side aggregates in phloem parenchyma of leaves as well as parenchyma cells of roots. They are found predominantly in the perinuclear space between the double membrane of the nucleus. The particles are separated from the cytoplasm by the outer nuclear membrane that penetrates deeply and irregularly into the cytoplasm. The normal distance of 10-20 nm that separates the two membranes can be extended by 2.5 μm. Virus particles are absent from the nuclei as well as from other cell organelles, such as mitochondria and chloroplasts. Small accumulations as well as isolated virus particles have been localized in the cytoplasm (Eisbein, 1973, 1976).

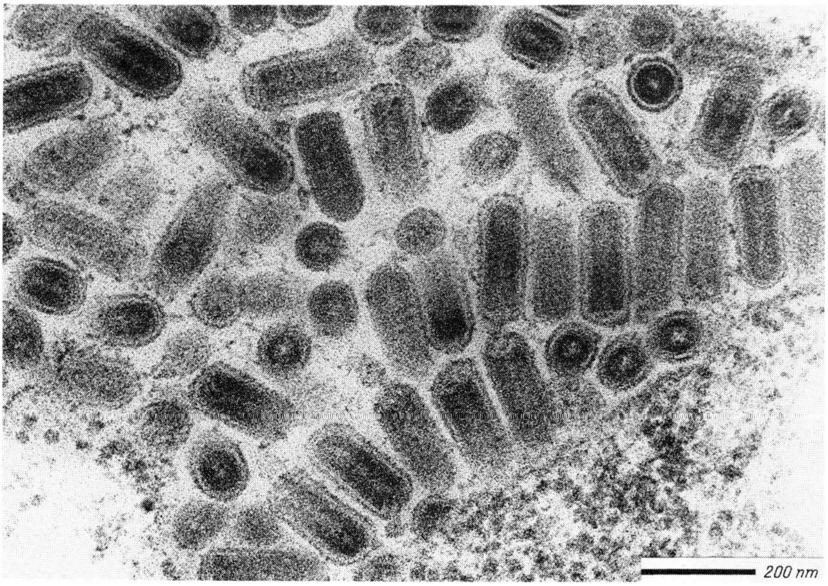

FIG. 3. Bacilliform particles of the beet leafcurl virus. (Original electron micrograph by Eisbein, Institute for Phytopathology, Aschersleben.)

6.2.3 Virus-Vector Interactions

6.2.3.1 Virus Acquisition and Circulation, and Plant Inoculation. The piercing and sucking stylet of *P. quadratum* penetrates plant tissues exclusively intracellularly. Individual parenchyma cells are sucked out so that brown or grey spots become visible (Fig. 4). The salivary sheath usually terminates in the phloem (Ehrhardt and Schmutterer, 1965). Since the beet leafcurl virus is localized in the phloem, short probings are inadequate for virus acquisition. Following a 3-day fasting period, a minimum acquisiton-access feeding period of 30 min is required to render the bugs inoculative. When the acquisition feeding is prolonged to 1, 2, 4, or 8 hr, or to 1 and 3 days, the consecutive transmission rates increase from 1.7 to 4.2%. If the acquisition feeding is extended to 7 days, the transmission rate reaches 25.8% (Table II). The incubation period in the vector, after a 1-hr acquisition-access feeding period, requires a minimum of 14 days before transmission to plants can occur. If bugs are

FIG. 4. Spots resulting from feeding injuries made by *P. quadratum* on a beet leaf. (Illustration courtesy of the Archives of the Institute for Phytopathology, Aschersleben.)

TABLE II. Acquisition-Access Feeding Periods and Incubation Periods of Beet Leafcurl Virus in the Vector *P. quadratum*

Length of Acquisition-Access Feeding	Feeding Periods (days) Following Acquisition							
	1-4	4-7	7-11	11-14	14-18	18-21	21-28	28-35
10 min	0[a]	0	0	0	0	0	0	0
30 min	0	0	0	0	0	0	1	1
1 hr	0	0	0	0	1	1	1	1
2 hr	0	0	0	0	0	0	2	2
4 hr	0	0	0	0	0	0	0	1
8 hr	0	0	1	1	1	1	3	3
1 days	0	0	0	0	1	1	2	4
3 days	0	0	0	0	0	1	3	6
7 days	0	0	6	6	10	11	16	13

[a]Number of infected plants out of a total of 30 exposed ones (after Proeseler, 1978c).

allowed an acquisition feeding of 8 hr, they are able to infect plants after 7-11 days. Longer acquisition feeding does not shorten the incubation period any further. Thus the average incubation period can be calculated as 21 to 28 days (Proeseler, 1978c). When virus-containing inocula are used for the injection of stock bugs, the incubation period in recipient insects increases with increasing dilution of the inoculum. For example, injections of 10^{-2}, 10^{-3}, or 10^{-4} to 10^{-5} dilutions of crude-sap inoculum result in incubation periods of 16, 21, or 29-32 days, respectively. The shortest incubation period, following the injection of bugs, has been reported as 5 days (Proeseler, 1966b). Temperature also affects the duration of the incubation period, and at 10°C a considerable increase occurs, compared to 20 or 30°C. Viruliferous bugs require a minimum of 40 min to transmit the virus to a healthy plant, and prolonged inoculation feeding periods increase the transmission rate (Lassack, 1955a). These observations support the view that the virus requires introduction into phloem cells.

6.2.3.2 Transmission Efficiency of Nymphs. Sometimes there are differences in the transmission efficiencies of immature versus imago stages of vectors such as nematodes, mites and insects. It was also believed up to now that there is a similar difference insofar as the efficiency of virus transmission by *P. quadratum* is concerned. It was generally assumed that the barrier created by the intestinal wall inhibits virus acquisition by nymphal instars. Virus transmission, in turn, was believed possible only after the last molting (Völk and Krczal, 1957). When these assumptions were made it was not yet realized that the incubation period requires an average of 3 to 4 weeks. Therefore, quite often the last molt occurs before the completion of the incubation period. No definite tests have been carried out to ascertain the relationship, if any, between virus transmission and molting, although such tests could now be performed easily.

While the following tests were not aimed at the establishment of the above relationship, a series of experiments were performed to induce nymphs to transmit virus before the final molting. The tests were performed using bug injection or needle puncturing techniques. It was found that nymphs that acquired the virus from diseased plants were rendered infective by abdominal punctures, using glass needles. Moreover, such nymphs transmitted the virus to healthy plants before becoming adults. The same results were obtained when virus-free nymphs were syringe-inoculated with virus-containing plant sap. The incubation period in injected bugs was comparatively short, ranging from 8-15 days. Table III presents the results of a test in which equal volumes of two phosphate-buffered dilutions of plant sap, 1:1 and 1:100, were injected into nymphs and adults. The bugs were then individually tested and transferred four times, at weekly intervals, to young beet seedlings. The injected nymphs became adults only after 4 weeks. At a dilution of 1:1, ten of 10 larvae and 8 of 10 adults were rendered infective. The mean incubation periods in the plants (inoculated nymph versus adult) was 38 and 46 days, respectively, in the first week's series. In the later series an incubation period of only 26 days was encountered. It was concluded that the inoculated bugs of the first-week series released comparatively small amounts of virus.

At the higher dilution, 1:100, seven of 10 nymphs and 6 of 10 adults proved to be infective. The incubation period was longer in both nymphs and adults, while the total number of infected plants was lower (Table III). Consequently, the transmission efficiency of injected nymphs was better than or equal to that of adults, and in no way less efficient (Proeseler, 1966b; Schmutterer and Ehrhardt, 1966a). The injection method also proved that the virus occurred in nymphal hemolymph (Schmutterer and Erhardt, 1963; Proeseler, 1966b). The hypothesis that the intestinal wall constitutes a barrier has not been disproved, however. It was assumed that the injection procedure caused a wounding of the intestinal tract; and, further, specially designed tests were needed to disprove the earlier contention. This was done by confining 1st or 2nd instar nymphs to diseased beets. Afterwards, the insects were confined in groups of 5 on healthy plants. Transfers were carried out every 7 days, and later at 2, or even 1-day intervals. The most efficient group of nymphs transmitted the virus on the 20th, 21st, 22nd, 23rd, 24th, 25th and 28th day, al-

TABLE III. Transmitting Efficiency of Nymphs and Adults of *P. quadratum* (Fieb.) Following Mechanical Inoculation with 1:1 (A) and 1:100 (B) Dilutions of Beet Leafcurl Virus-Laden Plant Sap[a]

	Weeks After Inoculation				Total Number of Infected Plants
	1st	2nd	3rd	4th	
A. Nymphs	4[b]	9	9	7	29[c]
Adults	6	5	5	3	19
B. Nymphs	0	3	5	3	11
Adults	0	0	4	4	8

[a]After Proeseler, 1966b.
[b]Number of infective *P. quadratum*.
[c]Number of infected plants out of a total of 40 exposed ones.

though it was not until the 30th day that the first of those tested molted into an adult. This proved that not only adult but also nymphal stages of *P. quadratum* can acquire and transmit the beef leafcurl virus (Proeseler, 1978b).

6.2.3.3 Virus Multiplication in the Vectors. Following a single acquisition-access feeding, *P. quadratum* remain infective throughout their lifespan. There is no evidence of transovarial transmission from females to offspring. Bugs that have acquired the virus in the field during the summer are poor vectors. After hibernation the same insects become efficient vectors. Transmission by individual insects can be erratic, with long intervals of time during which no transmission is effected (Völk and Krczal, 1957). These observations point to the possibility that the virus multiplies in its vectors. Experimental proof of multiplication was obtained in the same manner as has been done with leafhopper-borne viruses, that is, by means of insect injection (Proeseler, 1966b). In these tests the inoculated bugs were confined daily to virus-free plants for 8 hr and starved for the remaining 16 hr. The short inoculation feeding period prevented the uptake of virus from test plants. After 4 weeks the surviving insects were homogenized in a small blender, with phosphate buffer at pH 7, and used as inoculum for virus-free *P. quadratum*. After serial passages and a calculated dilution of the original inoculum of 10^{-14}, the injected bugs continued to transmit the virus. Since the dilution endpoint of the beet leafcurl virus is between 10^{-5} and 10^{-6}, the results provided evidence that the virus multiplied in the vectors.

In several instances in which the multiplication of the virus in *P. quadratum* was tested by serial passages, an increase in the mortality rate of inoculated bugs was observed (Table IV). This mortality occurred irrespective of whether the inoculum was derived from diseased plants or from viruliferous insects. These results were reminiscent of those obtained with *Macrosteles fascifrons* injected with the agent of aster yellows (Maramorosch, 1952). The reason for the increased mortality is as yet unknown. Perhaps the serial passages enhanced the virulence of the infectious agent and its pathogenicity towards *P. quadratum*. The injection technique was also applied to nymphal stages of *P. quadratum*. After 4 passages the calculated dilution of the original inoculum reached 10^{-11} and these findings provided direct evidence that the beet leafcurl virus can multiply in nymphs of the vector (Proesler, 1966b).

TABLE IV. Serial Passage of Beet Leafcurl Virus in Adult *P. quadratum* (Fieb.)[a]

Results	Number of Passage				
	1st	2nd	3rd	4th	5th
Theoretical dilution	10^{-2}	10^{-5}	10^{-8}	10^{-11}	10^{-13}
Number of infected plants per total tested	18/35	4/30	11/25	5/23	5/20
Number of survivors (from original 100 inoculated insects) 3 weeks after injection	60	53	38	12	2

[a] After Proeseler, 1966b.

6.2.3.4 Detection of the Virus in the Vector. The injection technique was also applied to detect the virus in the insect vector. It was found that virus is present in the midgut, the hemolymph, the salivary glands and feces. Interestingly, virus was recovered from the dissected midguts of insects that were collected in the beginning of March, at the end of hibernation in an area where the disease had been prevalent. Although the bugs did not feed for several months, the homogenate prepared from their intestinal tracts contained sufficient virus, in active form, to render inoculated vectors infective. This indicates that the virus survived in the cell wall of the gut, and not in the lumen, as the longevity in vitro, even at 5°C, does not exceed 4 days. After the successful localization of bacilliform virions in purified preparations and in thin sections of diseased plants, attempts were also made to detect virions in vectors. Numerous bacilliform particles were detected in salivary glands, especially in the vicinity of tracheoles (Fig. 5). They did not differ morphologically from the virions found in plant material. In longitudinal and cross sections the same measurements were established, and the electron dense bodies in the central core as well as the surrounding membrane were observed. It can be expected that further studies will reveal the presence of virions in other organs of the vector (Eisbein and Proeseler, 1976).

6.2.4 Vector and Virus Under Field Conditions

The natural occurrence of *P. quadratum* exceeds the geographic distribution of the virus. The insects occur in Europe from Great Britain to the Soviet Union and

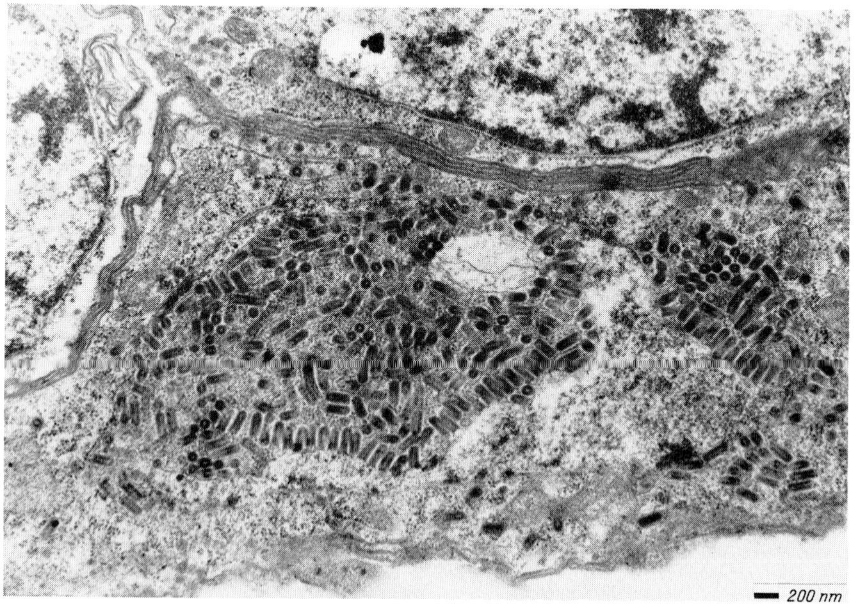

FIG. 5. Virus particles in the salivary gland of *P. quadratum*. (Original electron micrograph by Eisbein, Institute for Phytopathology, Aschersleben.)

from northern Scandinavia to the Alps. In addition, the bugs have been reported from large areas of Asia and North Africa. Outside the area of virus transmission, the insects are found in scattered halophilic biotopes. When bugs from such areas were collected and their vectorial capacity tested, it was found that they could acquire and transmit the virus (Lassack, 1955b). At present, economic losses due to the beet leafcurl virus have been reported from Poland, the German Democratic Republic, West Germany, and sporadic occurrences noted from the Soviet Union and Czechoslovakia. In the limited area where serious losses occur, climatic and soil conditions favor the development of the vectors. The apparent border of this area is the 18°C isothermic line for the month of July. Light sandy soils that warm up and dry rapidly provide favorable conditions. The insects are unable to develop in heavy clay, and they could hardly move when trapped in such areas after a downpour. The climate near the Atlantic ocean is also unfavorable. Although some of the ecological conditions are not yet completely understood, there is certainly a predominant ecological influence that governs the distribution of the virus and its vector. The disease has been known since 1903, and over the years the area of spread increased steadily in Central Europe. The outbreaks were especially disastrous in 1934-1936 and 1947-1954. An intensive program of surveillance and control had to be initiated, and through the use of insecticides not only *P. quadratum* but also other pests and vectors have been affected. Consequently, at present the beet leafcurl disease and the vector are of lesser importance in Central Europe (Lassack, 1966c; Klinkowski, 1964; Stachyra, 1968; Gersdorf, 1973).

6.2.5 Vector Control

Long before chemical insecticides became available for large-scale applications, control measures against *P. quadratum* were applied through the use of baiting strips. Throughout the disease area, beets were planted early in the season along the borders of fields, near the hibernating places of the bugs. Careful observation provided the necessary data so that it could be established when the insects invaded the beets and started ovipositing, at which time the beets were plowed under and the soil compressed. Only after completion of this operation could beets be planted on large areas. The delay in planting resulted in a decrease in yield, but this still was favorable compared to the losses that would have been incurred otherwise (Wille, 1929; Langenbuch, 1934). When broad spectrum insecticides became available, the baiting strips continued to be used, but instead of plowing them under, they were treated with the insecticides (Scheibe, 1951). At present, baiting strips are no longer used and there is no need for delaying the planting of beets. Instead, there is a thorough surveillance of the fields in early spring. As soon as the first bugs are observed to have entered, a 10-20 m strip along all sides of the field is treated with powdered parathion. Depending on the intensity of successive invasions, the procedure has to be repeated and sometimes extended to the entire field. Several other insecticides have been tested but the best results have constantly been obtained with parathion (Gersdorf, 1973).

6.3 *PIESMA QUADRATUM* (FIEB.) AND BEET LATENT ROSETTE DISEASE

In addition to the beet leafcurl virus, *P. quadratum* also acts as a vector of a different disease agent that causes the formation of numerous small leaves in affected beet plants. The spindly leaves are formed from growing points as well as from dormant buds, giving the plant a bushy appearance (Fig. 6A and B). The disease has been named "Latente Rosettenkrankheit" (beet latent rosette disease) (Schmutterer, 1976). In the German Democratic Republic, the disease has been observed as early as 1963 and 1964. Transmission through grafting and *P. quadratum* to beets (*Beta* sp.) as well as spinach (*Spinacia oleracea* L.) was accomplished, and it was assumed that the causative agent was a virus (Proeseler, 1965). When electron microscopy techniques were used and plant and vector material examined, rickettsia-like microorganisms were detected in the phloem of diseased beet and spinach plants. Companion cells, nearby parenchyma cells as well as xylem were free from the pathogen. The rickettsia-like organisms were also detected in the vector, especially in the vicinity of the intestinal tract (Nienhaus and Schmutterer, 1976). Adults and nymphs of *P. quadratum* can act as vectors. An incubation period, or circulation period of 10-30 days in nymphs, and 12-21 days in adults is required before the insects can transmit the rickettsia-like agents to plants. Following a single acquisition feeding, the bugs remain infective throughout their lives. The minimum acquisiton-access feeding period is less than 15 min. Increasing the infective feeding up to 72 hr gives optimal transmission results. *P. quadratum* transmits the rickettsia-like agent more

FIG. 6. Symptoms of beet latent rosette disease in beets. Note symptomatic spindly leaves formed from growing points (A) as well as from dormant buds (B). (Courtesy of Archives of the Institute of Phytopathology, Aschersleben.)

effectively than it does the beet leafcurl virus. The insects are able to transmit both agents simultaneously (Schmutterer, 1976).

6.4 *PIESMA CINEREUM* (SAY) AND SUGARBEET SAVOY VIRUS

The sugarbeet savoy disease has been known in the United States since 1890. Coons, Kotila and Stewart (1937) incriminated a virus as the causative agent and described the lace bugs or pigweed bugs, *P. cinereum* (Say) as the vector. Only the macropterous form of the lace bug is known. Color and markings are quite variable. The ground color is cinereous or stramineous, sometimes tinged with brownish or greenish, innotate or variegated with small spots, marks, or blotches of brown, fuscous, and black. The legs are stramineous or ochraceous, femora occasionally narrowly banded near middle with dark fuscous; the head is often largely black. Paranota are uniseriate, frequently with additional areoles anteriorly. The juga always surpass the tylus. The hemelytra are areolate. The bugs are 3.5 mm long (Drake and Davis, 1958).

The host range of *P. cinereum* is very broad. Primary hosts include *Amaranthus*, *Chenopodium, Salsola* and *Artriplex* spp. Among cultivated plants, sugarbeets, garden beets, lettuce, corn (*Zea mays*), clover, alfalfa, apple trees and many others can serve as hosts. The overwintering occurs in dry, protected places in patches of overwintering grass or fallen leaves. Contrary to other Piesma spp., *P. cinereum* occurs not only in North America but also in Central and South America (Drake and Davis, 1958). Virus-free insects are usually reared on *Amaranthus caudatum* L. Continuous rearing throughout the year can be carried out at 23.5°C and 18 hr/day illumination (Schneider, 1964).

The first symptoms of sugarbeet savoy virus disease are veinclearing and vein enlargement, resulting in a netlike leaf surface. Afterwards the edges of leaves, especially of younger ones, bend downwards. The overall growth of diseased plants is curtailed. Phloem necrosis and discoloration of roots occur, and the sugar content of beets is significantly lowered and the total yield seriously affected. Varieties VS 215 x 216, and hybrids VS 104 and VS 400 have a certain degree of resistance that is combined with resistance to *Cercospora beticola* Sacc., curly top virus and black root. The savoy virus overwinters in diseased plants or in the vectors (Coons, Kotila and Stewart, 1937; Coons *et al*., 1958).

Although the insect vector occurs throughout the Western Hemisphere, the savoy virus has been limited, so far, to North America, including Canada, and the states in the U. S. east of the Rocky Mountains where sugar beets are cultivated. The main areas of outbreaks have been recorded from Colorado, Wyoming, Montana, South and North Dakota, Texas, Nebraska, Kansas, Minnesota, Wisconsin, Indiana, Ohio and Michigan. In most years only a few plants in an affected field are infected and epidemics occur seldom.

Heavier outbreaks have been observed in fields that border on weedy lands, forests or pastures, where the vectors find favorable conditions for overwintering.

Outbreaks appear at the edges of fields near host plants that favor overwintering of the bugs. The rate of infection decreases with the distance from such spots (Hildebrand and Koch, 1942; Coons *et al.*, 1958). Only a small percentage of lace bugs caught under natural field condtions proved viruliferous. On the other hand, bugs caught as far as 5 miles from a beet planting were still infectious. Only one of 377 plants, exposed to lace bugs, became infected, indicating an unusually low efficiency of lace bug vectors. When several bugs were used in experimental transmission tests, the virus was transmitted serially to several plants (Schneider, 1964). It seems likely that the failure to observe virus particles in thin sections of diseased plants was due to the low concentrations and scanty transmission of the beet savoy virus (Arnott and Smith, 1967).

6.5 REFERENCES

Arnott, H. J., and Smith, K. M. (1967). Intracellular inclusions in the gut epithelial cells of *Piesma cinereum* Say. *J. Cell. Biol.* 34: 639-646.

Coons, G. H., Kotila, J. E., and Stewart, D. (1937). Savoy, a virus disease of beet transmitted by *Piesma cinerea. Phytopathology* 27: 125.

Coons, G. H., Stewart, D., Bockstahler, H. W., and Schneider, C. (1958). Incidence of savoy in relation to the variety of sugar beets and to the proximity of wintering habitat of the vector, *Piesma cinerea. Plant Dis. Reptr.* 42: 502-511.

Drake, C. J., and Davis, N. T. (1958). The morphology and systematics of the Piesmatidae (Hemiptera), with keys to world genera and american species. *Ann. Entomol., Soc. Amer.* 51: 567-581.

Ehrhardt, P., and Schmutterer, H. (1965). Untersuchungen zum Anstich- und Saugverhalten sowie zur Nahrungsaufnahme der Rübenblattwanze *Piesma quadrata* Fieb. (Heteropt., Piesmidae). *Z. angew. Ent.* 56: 41-55.

Eisbein, K. (1973) Weitere elektronenmikroskopische Untersuchungen des Rübenkräusel-Virus (*Beta* virus 3) in Ultradünnschnitten und nach Reinigung aus der Futterrübe. *Arch. Phytopathol. u. Pflanzenschutz* 9: 91-94.

Eisbein, K. (1976). Untersuchungen zum elektronenmikroskopischen Nachweis des Rübenkräusel- -Virus (*Beta* virus 3) in *Beta vulgaris* L. und *Piesma quadratum* Fieb., *Arch. Phytopathol. u. Pflanzenschutz* 12: 299-313.

Eisbein, K. und Proeseler, G. (1976). Nachweis der Partikel des Rübenkraüsel-Virus in seinem Vektor *Piesma quadratum* Fieb. (Kurze Mitteilung). *Arch. Phytopathol. u. Pflanzenschutz* 12: 227-228.

Gersdorf, E. (1973). Die Kräuselkrankheit der *Beta*-Rübe und ihr Überträger *I. I. R. B.* 6: 51-59.

Hildebrand, A. A., and Koch, L. H. (1942). Savoy disease of sugar beets in southwestern Ontario. *Phytopathology* 32: 328-331.

Klinkowski, M. (1964). Die Ausbreitung von Viruskrankheiten unserer Nutzpflanzen im europäischen Maßstab. *Sitzungsberichte Dt. Akad. Landwirtschaftswiss. Berlin* 13: 5-25.

Kotik-Królicka, Z. und Szulc, P. (1961). Wstepne badania nad porównaniem róznych metod zwalczania paszczyńca burakowego (*Piesma quadrata* Fieb.) i kedzierzawski wirusowej (*Beta* virus 3). *Biul. Inst. Hodowli Aklimatyzacji Roślin, Prace Zakladu Buraka i Innych Roślin Korzeniowych Nr*. 1/2: 81-93.

Krczal, H. und Völk, J. (1956). Über den Einfluß des Lichtes auf die Generationsfolge der Rübenblattwanze (*Piesma quadratum* Fieb.). *Nachrichtenbl Dt. Pflanzenschutzdienst, Braunschweig* 8: 145-147.

Langenbuch, R. (1934). Der Stand der Rübenwanzenfrage. *Mitt. Dt. Landwirtsch.-Ges.* 49-238-240.

Lassack, H. (1955a). Einige Beobachtungen über die Übertragung des Rübenkräusel-Virus durch *Piesma quadrata* Fieb. *Nachrichtenbl. Dt. Pflanzenschutzdienst, Braunschweig* 7: 60-62.

Lassack, H. (1955b). Ein Beitrag zur Rassenfrage der Rübenblattwanze (*Piesma quadrata* Fieb.). *Zucker* 8: 127-129.

Lassack, H. (1955c). Die Ausbreitung der Rübenkräuselkrankheit und ihres Vektors der Rübenblattwanze *Piesma quadrata* Fieb. (Heteropt. Piesmid.) in Niedersachsen. *Z. angew. Ent.* 38: 67-72.

Lee, P. E. (1968). Partial purification of wheat striate mosaic virus. *Virology* 34: 583-589.

Maramorosch, K. (1952). Direct evidence for the multiplication of aster-yellows virus in its insect vector. *Phytopathology* 42: 59-64.

Nienhaus, F., and Schmutterer, H. (1976). Rickettsialike organisms in latent rosette (witches' broom) diseased sugar beet (*Beta vulgaris*) and spinach (*Spinacia oleracea*) plants and in the vector *Piesma quadratum* Fieb. *Z. Pflanzenkrankh. Pflanzenschutz* 83: 641-646.

Proeseler, G. (1965). Die Blattwanze *Piesma quadrata* Fieb. und das Rübenkräusel-Virus (Savoia betae Holmes) unter besonderer Berücksichtigung der Beziehungen zwischen Virus und Vektor. Diss. Martin-Luther-Univ. Halle-Wittenberg, 106 S.

Proeseler, G. (1966a). Beziehungen zwischen der Rübenblattwanze *Piesma quadratum* Fieb. und dem Rübenkräusel-Virus. I. Virusübertragungsversuche und Zucht des Vektors. *Phytopath. Z.* 56: 191-211.

Proeseler, G. (1966b). Beziehungen zwischen der Rübenblattwanze *Piesma quadratum* Fieb. und dem Rübenkräuselvirus. II. Injektionsversuche. *Phytopath. Z.* 56: 213-237.

Proeseler, G. (1967). Ein Beitrag zur Nomenklatur der Piesmatidae (Heteroptera: Piesmatidae). *Beitr. Ent.* 17: 305-307.

Proeseler, G. (1978a). Entwicklungsdauer und Anzahl der Larvenstadien von *Piesma quadratum* Fieb., dem Vektor des Rübenkräusel-Virus. *Nachrichtenbl. Pflanzenschutz DDR* 32: 6-7.

Proeseler, G. (1978b). Larven von Piesma quadratum (Fieb.) als Vektoren des Rübenkräusel-Virus. *Nachrichtenbl. Pflanzenschutz DDR* 32: 254-256.

Proeseler, G. (1978c). Die Akquisitions- und Zirkulationszeit des Rübenkräusel-Virus. *Archiv Phytopathol. u. Pflanzenschutz* 14: 95-98.

Scheibe, K. (1951). Die Rübenblattwanze bedroht den Rübenanbau Westdeutschlands. *Gesunde Pflanzen* 3: 199-122.

Schmutterer, H. (1976). Die Rübenblattwanze *Piesma quadratum* Fieb. als Vektor eines neuen Pathogens bei Zucker und Futterrüben. *Z. Pflanzenkrankh, u. Pflanzenschutz* 83: 606-610.

Schmutterer, H. und Ehrhardt, P. (1963). Zur Übertragung des Rübenkräusel-Virus von infektiösen Blattwanzen (*Piesma quadrata* Fieb.) durch Injektion von Haemolymphe. *Z. angew. Ent.* 52: 425-428.

Schmutterer, H. und Ehrhardt, P. (1966a). Weitere Versuche zur künstlichen Umwandlung von Larven der Blattwanze *Piesma quadrata* Fieb. (Heteroptera) in Vektoren des Rübenkräusel-Virus. *Entomologia exp. appl.* 9: 278-286.

Schmutterer, H. und Ehrhardt, P. (1966b). Zur Kenntnis des Wirtspflanzenkreises beim Rübenkräusel-Virus (*Beta* virus 3). *Z. Pflanzenkrankh. (Pflanzenpath.) Pflanzenschutz* 73: 271-283.

Schneider, C. L. (1964). Studies on the transmission of sugar beet savoy virus by the vector, *Piesma cinereum* (Say). *Plant Dis. Reptr.* 48: 843-845.

Stachyra, T. (1968). Rytmika gradacji paszczyńca burakowego (*Piesma quadrata* Fieb.), oraz smiany nasilenia kedzierzawki paszczyńcowej w Europie środkowej wedug materiaow z lat 1903-1964. *Prace Nauko Instytutu Ochrony Roślin Poznań* 10: 57-89.

Stuben, M. (1958). Beobachtungen über den Einfluβ der Beleuchtung mit Leuchtstoffröhren auf die Verhinderung der Winterruhe bei *Piesma quadratum* (Fieb.). *Z. Pflanzenkrankh. (Pflanzenpath.) Pflanzenschutz* 65: 211-214.

Völk, J. und Krczal, H. (1975). Übertragungsversuche mit *Piesma quadratum* Fieb., dem Vektor der Kräuselkrankheit der Zucker- und Futterrübe. *Nachrichtenbl. Dt. Pflanzenschutzdienst, Braunschweig* 9: 17-22.

Wille, J. (1929). Die Rübenblattwanze *Piesma quadrata* Fieb. *Monographien zum Pflanzenschutz* Nr. 2, Springer-Verlag Berlin, 116 S.

Chapter 7

BEETLES

J. P. Fulton, H. A. Scott

Department of Plant Pathology
University of Arkansas
Fayetteville, Arkansas

Rodrigo Gamez

Centro de Investigación en Biología Celular y Molecular
Universidad de Costa Rica
San Jose, Costa Rica

7.1 INTRODUCTION . . . 116
7.2 THE VIRUSES . . . 116
7.2.1 The Comovirus Group . . . 116
7.2.2 The Tymovirus Group . . . 118
7.2.3 The Bromovirus Group . . . 118
7.2.4 The Southern Bean Mosaic Virus Group . . . 119
7.2.5 Other Viruses . . . 120
7.3 THE BEETLES . . . 121
7.3.1 Chrysomelidae . . . 121
7.3.2 Coccinellidae . . . 122
7.3.3 Curculionidae . . . 122
7.3.4 Other Beetles . . . 122
7.4 BEETLE VECTORING OF VIRUSES . . . 122
7.4.1 Virus Acquisition, Retention, and Transmission . . . 122
7.4.2 Vector Efficiency . . . 124
7.4.3 Epidemiology of Beetle-Transmitted Viruses . . . 125
7.5 ASSOCIATION OF VIRUS WITH BEETLES . . . 126
7.5.1 Hemolymph . . . 126
7.5.2 Regurgitant . . . 126
7.5.3 Feces . . . 127
7.6 SPECULATION ON MECHANISMS OF TRANSMISSION . . . 127
7.7 ACKNOWLEDGMENT . . . 128
7.8 REFERENCES . . . 128

ISBN 0-12-326450-2

7.1 INTRODUCTION

Leaf-feeding beetles have long been recognized as vectors of certain plant viruses (Smith, 1924). General aspects of beetle transmission have been reviewed (Selman, 1973; Walters, 1969) and some discussion presented regarding mechanisms of transmission of viruses by beetles. Because of the method of feeding and the stability of the viruses involved, the commonly held notion (Smith, 1965) suggests that transmission results from a simple mechanical process involving contamination of the beetle's mouthparts. More recent reviews (Fulton *et al.*, 1975; Fulton and Scott, 1977) have emphasized, however, that a complex biological phenomenon is involved. Although leaf-feeding beetles acquire virus by feeding and retain and transmit virus, in some cases for extended periods of time, most recent studies clearly suggest that the process is not adequately explained on the basis of contamination of mouthparts. This paper will review beetle vectoring of plant viruses, beetle-virus associations, and will discuss possible mechanisms of virus transmission by their vectors.

7.2 THE VIRUSES

Virus transmission by beetles is a very specific phenomenon. These viruses are not transmitted by other commonly recognized vectors, viz. aphids or leaf-hoppers. There are, however, a few reported exceptions to this generalization (Lana and Taylor, 1976; Whitney and Gilmer, 1974; Schuster, 1963) and there is a critical need for reconfirmation of these reports.

The beetle-transmitted viruses have many similar properties including being relatively stable, easily transmitted mechanically, and having high antigenicity. All are polyhedral RNA viruses ca 25 to 30 nm in diameter. The viruses are divided into several groups which are distinguished by their serological affinities and the number of sedimenting components. Because of high yields of purified virus particles, a great deal is known of their chemical and physical properties as well as their biological properties. Some of this information is summarized below.

7.2.1 The Comovirus Group

Viruses in this group are efficiently transmitted by several beetles (Table 1). Many strains have been reported from different parts of the world and found to infect a range of host plants, viz. members of the Leguminosae, Cruciferae, Solanaceae, and Cucurbitaceae. However, each virus strain described infects a relatively narrow host range. A few are transmitted through seeds. Although vectors are not determined for certain members of this group, it is logical to implicate beetles.

Purified virus preparations contain three centrifugal components designated top (T), middle (M), and bottom (B). Top components are empty protein shells, whereas, M and B are nucleoprotein components containing 24% RNA and 33% RNA, re-

TABLE I. Beetle Vectors of the Comovirus Group

Virus	Vector	References
Cowpea Mosaic, Arkansas	*Cerotoma arcuata* *Cerotoma atrofasciata* *Cerotoma ruficornis* *Cerotoma variegata* *Diabrotica adelpha* *Diabrotica undecimpunctata* *Diabrotica virgifera* *Diabrotica laeta* *Gynodiabrotica variabilis* *Acalymma vittatum* *Epilachna varivestis* *Epilachna* sp. *Diphaulaca meridae* *Systena sp.*	Smith, 1924; Walters & Barnett, 1964; Jansen & Staples, 1970a, b, 1971; Dale, 1949, 1953; Slack & Fulton, 1971; Sanderlin, 1973; Fulton & Scott, 1974; Debrot & de Rojas, 1967; Diaz, 1972; Valverde, Gonzales, Moreno, & Gamez (personal com.); Kvicala *et al.*, 1970, 1973
Cowpea Mosaic, Yellow	*Cerotoma trifurcata* *Diabrotica balteata* *Diabrotica undecimpunctata* *Diabrotica virgifera* *Acalymma vittatum* *Ootheca mutabilis* *Epilachna varivestis* *Paraluperodes quaternus* *Nematocerus acerbus*	Jansen & Staples, 1971; Chant, 1959; Whitney & Gilmer, 1974
Bean Pod Mottle	*Cerotoma trifurcata* *Diabrotica balteata* *Diabrotica undecimpunctata* *Colaspis lata* *Epicanta vittata*	Ross, 1963; Walters, 1964b; Horn *et al.* 1970; Patel & Pitre, 1971; Moore & Scott, 1971; Sanderlin, 1973
Bean Rugose Mosaic	*Cerotoma ruficornis* *Diabrotica balteata* *Diabrotica adelpha*	Gamez, 1972a
Quail Pea Mosaic and Bean Curly Dwarf Mosaic	*Cerotoma ruficornis* *Diabrotica undecimpunctata* *Epilachna varivestis*	Moore, 1973; Meiners *et al.*, 1977
Broad Bean Stain	*Apion vorax* *Apion aestivum* *Apion aethiops* *Sitona lineatus*	Cockbain *et al.*, 1975
Echtes Ackerbohnenmosaik	*Apion vorax* *Apion aethiops* *Sitona lineatus*	Cockbain *et al.*, 1975
Radish Mosaic	*Epitrix hirtipennis* *Diabrotica undecimpunctata* *Phyllotreta striolata*	Campbell & Colt, 1967; Tochihara, 1968; Stefanac & Mamula, 1972
Red Clover Mottle	*Apion apricans* *Apion varipes*	Gerhardson & Pettersson, 1974
Squash Mosaic	*Diabrotica undecimpunctata* *Diabrotica balteata* *Acalymma vittata* *Diabrotica longicornis* *Diabrotica virgifera* *Acalymma thiemei* *Acalymma trivittata* *Epilachna chrysomelina*	Freitag, 1941a, b, 1956; Lastra, 1968; Sitterly, 1960; Stoner, 1963; Cohen & Nitzany, 1963
Andean Potato Mottle	vector unknown	Fribourg *et al.*, 1977

spectively. The importance of the latter components has been demonstrated biologically. For example, cowpea mosaic virus requires a mixture of M and B components to initiate infections; neither component alone is infectious (Van Kammen, 1968).

Although antigenic relatedness among virus isolates may be quite distant all are related serologically. It is unclear, however, as to what constitutes the serological distinction that is required to specify separate members of the group. The situation is further aggravated by the fact that two members of the group are designated cowpea mosaic, the Arkansas and the yellow isolate. These two are very distantly related serologically (Swaans and Van Kammen, 1973). A suggestion was made (Fulton and Scott, 1977) that the various viruses be clustered into serogroups. Because of associated similarities within a serogroup, such as host susceptibility and resistance, vector relationships, and geographical distribution, serogrouping merits strong consideration. Based on Ouchterlony double diffusion tests, all recent isolates of cowpea mosaic virus from North America belong to the Arkansas isolate serogroup whereas isolates from Africa are in the yellow isolate serogroup (Fulton and Scott, 1977).

7.2.2 The Tymovirus Group

Beetle vectors have been recognized for most members of this group (Table II). Each virus exhibits a narrow host range but certain strains attack hosts from several families of seed plants and occur in widely separated geographical locations. Purified virus preparations typically contain two centrifugal components; a top or protein shell component and a bottom or nucleoprotein component containing 34% RNA. Virus isolates are easily identified by serological tests.

7.2.3 The Bromovirus Group

Although viruses in this group are transmitted by beetles the efficiency of transmission in all cases is low. Hobbs (1977) has suggested that with cowpea chlorotic mottle virus this may be due to a rapid reduction in specific infectivity in an in-

TABLE II. Beetle Vectors of the Tymovirus Group

Virus	Vector	References
Turnip Yellow Mosaic	*Phyllotreta spp.*	Markham & Smith, 1949
	Psylliodes spp.	
	Phaedon cochleariae	
Belladonna Mottle	*Epitrix atropae*	Weideman & Bode, 1973
Eggplant Mosaic	*Epitrix spp.*	Dale, 1954
Okra Mosaic	*Podagrica sjostedti*	Lana & Taylor, 1976;
	Podagrica uniforma	Lana *et al.*, 1974
	Syagrus calcaratus	
Scrophularia Mottle	*Cionus tuberculosis*	Bercks, 1973
	Cionus scrophulariae	
	Cionus hortulanus	
	Cionus alauda	
Wild Cucumber Mosaic	*Acalymma trivittata*	Freitag, 1941b
Desmodium Yellow Mottle	Vector unknown	Walters & Scott, 1972
Andean Potato Latent	*Epitrix* sp.	Fribourg *et al.*, 1975

TABLE III. Beetle Vectors of the Bromovirus Group

Virus	Vector	References
Broad Bean Mottle	*Acalymma trivittata* *Diabrotica undecimpunctata* *Colaspis flavida*	Walters & Surin, 1973
Cowpea Chlorotic Mottle and Bean Yellow Stipple	*Cerotoma trifurcata* *Diabrotica undecimpunctata* *Epilachna varivestis* *Certoma ruficornis* *Diabrotica balteata* *Diabrotica adelpha*	Walters & Dodd, 1969; Gamez, 1972b; Hobbs; 1977; Gamez, 1976
Brome Mosaic	*Oulema melanopa* *Diabrotica longicornis*	Nault (personal com.)

fected host. The bromovirus group and their beetle vectors are listed in Table III. Unlike viruses of the other groups, these exhibit a wide host range.

The viruses in the bromovirus group are single component nucleoproteins containing 22% RNA. However, four different-sized RNAs are distributed among the particles. The four RNA species have molecular weights of 1.1×10^6, 1.0×10^6, 0.7×10^6, and 0.3×10^6. The first three are required for infection.

7.2.4 The Southern Bean Mosaic Virus Group

These viruses are very efficiently transmitted by beetles (Table IV). In some cases virus is retained by beetles for extended periods of time (Walters and Henry, 1970; Slack and Scott, 1971). This group is represented by three biologically distinct strains, southern bean mosaic (Zaumeyer and Harter, 1943), the cowpea strain (Shepherd and Fulton, 1962) and the severe bean mosaic strain (Grogan and Kimble, 1964). The host range of each strain is narrow, infecting only limited numbers of species of Leguminosae. The virus is seed-borne in several susceptible hosts (Zaumeyer and Harter, 1943; Shepherd and Fulton, 1962, Uyemoto and Grogan, 1977) and this characteristic appears to be of significance in sporadic outbreaks of the virus.

The virus is a single-component nucleoprotein with 21% RNA and is very stable with a high thermal inactivation point and an extended longevity in vitro. All strains are serologically related but distinct.

TABLE IV. Beetle Vectors of Southern Bean Mosaic Virus Group

Virus	Vector	References
Southern Bean Mosaic	*Cerotoma trifurcata* *Epilachna varivestis* *Atrachya menetris* *Acalymma vittata*	Walters, 1964a; Fulton & Scott, 1974; Iizuka, 1974; McLaughlin Milbrath, 1978
Cowpea, Strain, Southern Bean Mosaic	*Cerotoma trifurcata* *Epilachna varivestis*	Walters, 1965; Fulton & Scott, 1974; Slack & Scott, 1971; Walters & Henry, 1970
Severe Bean Mosaic	*Cerotoma trifurcata*	Walters, 1969

7.2.5 Other Viruses

There is a growing list of beetle-transmitted, isometric viruses which are not related to viruses in the above groups. Included in this list are cocksfoot mottle (Catherall, 1970b), rice yellow mottle (Bakker, 1975), bean mild mosaic (Waterworth, *et al.*, 1977), blackgram mottle (Scott and Phatak, 1978), and maize chlorotic mottle (Nault, *et al.*, 1978) (Table V).

It should be noted that tobacco mosaic virus, the cowpea strain of tobacco mosaic virus, and tobacco ringspot virus are not listed even though there are reports of beetle transmissions (Chant, 1959; Orlob, 1963; Schuster, 1963). We (Scott and Fulton,

TABLE V. Beetle Vectors of Miscellaneous Viruses

Virus	Vector	References
Cocksfoot Mottle	*Oulema melanopa* *Oulema lichenis*	Serjeant, 1967; Catherall, 1968; A'Brook & Benigno, 1972; Benigno & A'Brook, 1972
Rice Yellow Mottle	*Oulema dunbrodiensis* *Cryptocephalus chalybeipennis* *Cryptocephalus nigrum* *Monolepta flaveola* *Monolepta hamatura* *Sesselia pusilla* *Chaetocnema abyssinica* *Chaetocnema kenyensis* *Chaetocnema pulla* *Dactylispa bayoni* *Dicladispa paucispina* *Dicladispa viridicyanae* *Trichispa sericea*	Bakker, 1971, 1974
Erysimum Latent	*Phyllotreta atra* *Phyllotreta nigripes* *Phyllotreta undulata*	Shukla *et al.*, 1975
Bean Mild Mosaic	*Diabrotica undecimpunctata* *Epilachna varivestis*	Waterworth *et al.*, 1977
Phleum Mottle	*Oulema melanopa* *Oulema lichenis*	A'Brook & Benigno, 1972; Benigno & A'Brook, 1972; Catherall, 1970a
Maize Chlorotic Mottle	*Oulema melanopa* *Chaetocnema pulicaria* *Systena frontalis* *Diabrotica undecimpunctata* *Diabrotica longicornis* *Diabrotica virgifera*	Nault *et al.*, 1978
Blackgram Mottle	*Cerotoma trifurcata* *Epilachna varivestis*	Scott & Phatak, 1978

1976, 1978) have not found a sufficiently strong association between virus and vector to classify these as being beetle vectored. Because these are stable viruses an occasional transmission may occur as a result of accidental contamination. The decision to exclude these viruses is arbitrary, particularly when we include the bromoviruses which have low levels of beetle transmission.

7.3 THE BEETLES

Most of the studies on beetle vectoring of plant viruses have utilized field-collected beetles. There are obvious disadvantages to using these for certain types of studies. Such field-collected beetles are only available during the growing seasons and may already be carrying a plant virus. Furthermore, one has little control over the age of the beetle or the proven efficiency of a known population. Some of the beetles used in vector studies can be reared in the laboratory, (Isely, 1930; Herzog *et al.*, 1974; Howe and George, 1966). These techniques have been adapted to rearing large populations for virus-vector studies (Holcomb and Fulton, 1978).

The plant-feeding beetles known to transmit viruses occur in several families as indicated below.

7.3.1 Chrysomelidae

Beetles most commonly recognized as virus vectors are in the family *Chrysomelidae*. These are the so-called leaf beetles and flea beetles. Beetles in this family usually have a very limited range of hosts on which they will feed and reproduce, and this is undoubtedly a factor determining the viruses each species vectors. Host preference is apparently determined by attractant substances in the plants. As was demonstrated by McLaughlin and Milbrath (1978), the striped cucumber beetle, *Acalymma vittata*, can be induced to feed on selected legumes by treating their leaves with an alcohol extract from a preferred host, the Butternut squash. Using this technique they were able to show that *A. vittata* could vector southern bean mosaic virus from bean to bean.

The larval stages of *Chrysomelidae* are often not as readily apparent, nor is their feeding damage as evident, as the adults. In many cases the larvae are root feeders. Transmission by larvae has been studied in only a few cases. Markham and Smith (1949), investigating the transmission of turnip yellow mosaic virus by *Phaedon cochleariae*, found that larvae were actually better transmitters than adults. Nault *et al.* (1978) recently showed that larvae of *Oulema melanopa* transmitted maize chlorotic mottle virus more efficiently than adults. However, with phleum mottle (Catherall, 1970a) and cocksfoot mottle viruses (Serjeant, 1967) the larval stage of *Oulema melanopa* transmitted these two viruses at much lower levels than the adults. There is no evidence with any beetle vectors that virus acquired by a larva can be carried through pupation to the adult.

Various genera containing species known to vector one or more viruses include the following: *Cerotoma, Diabrotica, Acalymma, Ootheca, Paraluperodes, Phaedon, Phyllotreta, Psylliodes, Epitrix, Systena, Podagrica, Syagrus, Colaspis, Sesselia, Chaetocnema, Oulema, Cryptocephalus, Monolepta, Dactylispa, Dicladispa, Trichispa,* and *Atrachya*. This list has been expanded substantially in recent years and as knowledge of beetle vectoring of plant viruses is increased the list will no doubt be expanded further.

7.3.2 Coccinellidae

Most members of this large family are predaceous on aphids and scale insects. Members of only one genus, *Epilachna*, are leaf-feeding beetles. The Mexican bean beetle, *Epilachna varivestis*, can transmit such legume viruses as cowpea mosaic (Jansen and Staples, 1970b), southern bean mosaic (Fulton and Scott, 1974), bean pod mottle (Fulton and Scott, 1974), bean mild mosaic (Waterworth *et al.* 1977), and bean curly dwarf mosaic (Meiners *et al.*, 1977). Larvae are leaf feeders, and they, as well as adults, are capable of transmitting many of these viruses. Transmission of the legume viruses mentioned above by *E. varivestis* has been accomplished experimentally. There is no evidence to suggest that this vector is a significant factor in the natural occurrence and spread of any of these viruses. Another species, *E. chrysomelina*, can transmit squash mosaic virus (Cohen and Nitzany, 1963).

7.3.3 Curculionidae

There are only a few examples of weevils in this family being recorded as virus vectors. Apparently they are as efficient in their ability to acquire, transmit and retain virus as are chrysomelid beetles (Cockbain *et al.*, 1975). Genera containing species capable of virus transmission include *Apion*, *Sitona*, and *Cionus*.

7.3.4 Other Beetles

Patel and Pitre (1971) demonstrated transmission of bean pod mottle virus by the striped blister beetle, *Epicanta vittata*, in the family Meliodae. The transmission levels were low. They were also able to demonstrate that low levels of beetles collected in the field were viruliferous.

7.4 BEETLE VECTORING OF VIRUSES

7.4.1 Virus Acquisition, Retention, and Transmission

In many of the experimental trials evaluating transmission, beetles have been given a feeding access to infected plants for periods of 24 or 48 hours. We have found, however, that beetles can acquire virus in much shorter periods of time and where attempts have been made to determine the minimum time required for virus acquisition it has been found that virus can be acquired on the first bite. The percentage of beetles that becomes viruliferous increases with increases in feeding time. In addition, beetles can become viruliferous when a purified preparation of virus is injected into the hemocoel (Slack and Scott, 1971; Sanderlin, 1973; Fulton and Scott, 1974). We have suggested that this is a common feature of all virus-beetle vector combinations (Fulton *et al.*, 1975). Beetles can also acquire virus by drinking from a purified preparation.

The studies of Dale (1953) and Walters and Henry (1970) show that the amount of tissue consumed during acquisition feeding will affect only slightly the percentage of beetles that will transmit virus during the first day following acquisition. It does, however, have a definite effect on the length of time the beetles will remain viruliferous, at least in the case of those viruses which can be retained by the beetles for extended periods of time. Jansen and Staples (1970a) demonstrated a similar phenomenon when different acquisition host plants were utilized. Using cowpeas or soybeans infected with cowpea mosaic virus as acquisition hosts, transmission did not seem to be affected during the first day but subsequent retention of virus was much greater for beetles which had acquired virus from cowpea. This may be due to a difference in virus titre in the two hosts so that beetles acquired a much larger charge of virus from infected cowpeas than from soybeans.

The length of time a beetle will retain and transmit virus varies greatly with virus, beetle, host, and environmental conditions. With some of the relatively stable viruses such as southern bean mosaic, squash mosaic, and cowpea mosaic when beetle vectors are tested daily the numbers of beetles transmitting virus the first day are high. On subsequent days the rates of transmission drop rapidly but occasional beetles will transmit virus for three weeks or longer. Transmission is erratic in that a beetle that transmits virus may have fed and failed to transmit virus on several previous days. Transmission of some viruses has been recorded for only one or two days following acquisition. Both Walters(1969) and Selman (1973) have divided the beetle-transmitted viruses into two groups: viruses which can be retained only one or two days and viruses retained for longer periods. At the present time, because of the variety of variables which can affect the time of retention, it seems best to avoid such groupings. When future studies reveal the manner in which virus is retained by a beetle and it can be shown that two different mechanisms are involved in long term and short term retention, a logical base for such groupings can be established.

The length of time a beetle remains viruliferous also depends on the species involved. The Mexican bean beetle can retain the various viruses it transmits for only short periods of time whereas the bean leaf beetle retains the same viruses for extended periods (Fulton and Scott, 1974). Gamez (1972a) and Cartin and Gamez (1973) evaluated the efficiency of *Cerotoma ruficornis* and *Diabrotica balteata* in transmitting bean rugose mosaic virus and found that *C. ruficornis* retained virus much longer (9 days) than did the species of *Diabrotica* (3 days).

In most studies on retention of virus, beetles were test-fed on a daily basis. It appears that beetles may retain virus for much longer periods when they are less active, since Walters *et al.* (1972) have shown that beetles emerging after overwintering under natural conditions may be viruliferous and able to transmit bean pod mottle virus. Long retention of virus by beetles might suggest the replication of virus in the bodies of beetles but there is no evidence supporting such an idea.

In most cases that have been studied in detail beetles can transmit virus immediately without evidence of a latent period. A'Brook and Benigno (1972) obtained evidence suggesting a latent period before transmission of cocksfoot mottle and phleum mottle viruses by *Oulema* spp. Their conclusion was based on a low level of transmission obtained during the first two hours as compared to a much higher rate of transmission in a subsequent 22-hour test feed. Dale (1953) obtained no evidence of a latent period in transmission of cowpea mosaic virus by *Cerotoma ruficornis.* A latent period is difficult to evaluate because on the basis of individual beetles most trials record cases where, following an acquisition feed, a beetle feeds for several days before transmitting virus. In the same trial the majority of beetles, however, will transmit virus immediately. Since the mechanism of transmission is not yet known it is difficult to understand why a viruliferous beetle fails to transmit virus every time it feeds.

At the present time evidence suggests that transmission is accomplished only during feeding. Another mechanism could involve entrance of the virus through injuries to epidermal cells made by claws of the beetle. The claws could become contaminated with virus from the acquisition host, feces, or by reflexive bleeding. We have attempted to evaluate this possible mechanism with several viruses and have failed to obtain any evidence that transmission occurs by the clawing of a beetle.

Effects of environmental variables such as temperature, humidity, and light on acquisition and transmission have not been evaluated in great detail. These conditions obviously affect the activities of beetles and in this way could affect transmission. Most efficient transmission of cowpea mosaic virus by *Cerotoma ruficornis* and *C. trifurcata* is recorded at temperatures of 25 to 28°C. Beetles are also most active at these temperatures. Preliminary results suggest that transmission is inhibited above 30°C even though beetles may feed. Lower temperatures adversely affect both feeding activity and transmission.

7.4.2 Vector Efficiency

Vector efficiency is usually thought of in terms of the relative number of a population of beetles which transmit virus following an acquisition feed. Efficiency may vary greatly between species of beetles. Cockbain *et al.* (1975) obtained 43% transmission of Echtes Ackerbohnenmosaik virus with *Apion vorax* as compared to 2% or less with *A. aethiops*, *A. aestivum*, or *Sitona lineatus*. Bakker (1974) studying transmission of rice yellow mottle virus with several species of chrysomelid beetles recorded efficiencies ranging from 25% for *Chaetocnema pulla* to 100% for *Apo-*

phylia sp. Gamez (1972a) and Cartin and Gamez (1973) found that *Cerotoma ruficornis* transmitted bean rugose mosaic virus at a level of near 80% whereas *Diabrotica balteata* and *D. adelpha* were much less efficient at levels near 20 and 10%, respectively.

The efficiency with which various viruses are transmitted by the same species of beetle may also vary greatly. Hobbs (1977) was able to obtain only low levels of transmission (20% or less) of cowpea chlorotic mottle virus by *Cerotoma trifurcata*, whereas this same beetle transmits cowpea mosaic virus or southern bean mosaic virus at levels near 100% (Sanderlin, 1973; Walters, 1964a).

7.4.3 Epidemiology of Beetle-transmitted Viruses

The epidemiology of beetle-transmitted viruses has received very little study. In several cases wild perennial hosts of the viruses have been recorded (Walters and Lee, 1969; Moore *et al.* 1969; Hobbs, 1977; Shukla *et al.*, 1975). In tropical areas *Phaseolus lathyroides* L. has been recognized as a reservoir of cowpea mosaic virus (Alconero and Santiago, 1973; Lima and Nelson, 1977). Walters *et al.* (1972) showed that bean pod mottle virus could survive in overwintering adult beetles. Many of these viruses are seed-borne in one or more of their hosts. The importance of these factors in subsequent virus spread, however, has not been evaluated in most cases. Uptone (1969), in his studies on cocksfoot mottle virus, concluded that although the beetle vectors *Oulema melanopa* and *O. lichenis* occur at only very low population levels in cocksfoot, they may be of significance in spreading the virus from grasses to cocksfoot and from cocksfoot back to interplanted grasses. Severe build-up of virus within cocksfoot, however, was attributed to cutting implements and grazing.

Moreno, Gonzalez and Gamez (personal communication) have made interesting observations on the relative amounts of cowpea mosaic virus and cowpea chlorotic mottle virus during wet and dry seasons and differing cropping situations in Costa Rica. Cowpea mosaic virus infected a much greater percentage (85%) of the plants during the rainy season than during the dry season (7%). The opposite was recorded with cowpea chlorotic mottle which infected 30% of the plants during the dry season as opposed to less than 1% during the rainy season. These differences could not be correlated with the beetles present or the viruses carried by field-collected beetles. Field-collected beetles carried cowpea mosaic virus, cowpea chlorotic mottle virus, or both viruses simultaneously. More individuals of *Cerotoma ruficornis* carried cowpea mosaic virus than cowpea chlorotic mottle virus; the reverse occurred with *Diabrotica balteata*. They also recorded more virus in cowpeas in monoculture than in multiple cropping of cowpeas and corn. This was true during both the rainy and dry seasons and for both cowpea mosaic virus and cowpea chlorotic mottle virus. Preliminary results suggest that there is a greater number of early infections in multiple cropping and more late season spread in monoculture. Also, considerable differences were recorded in the incidence of virus disease in different varieties of cowpea.

7.5 ASSOCIATION OF VIRUS WITH BEETLES

7.5.1 Hemolymph

Freitag (1956) first demonstrated that a beetle-transmitted virus, squash mosaic, could be found in the hemolymph of the beetle vector. Several other viruses have been recorded in hemolymph of beetle vectors (Slack and Scott, 1971; Slack and Fulton, 1971; Sanderlin, 1973). Beetles can be made viruliferous by injection of virus into the hemocoel (Slack and Scott, 1971; Sanderlin, 1973; Fulton and Scott, 1974; Scott and Fulton, 1976; 1978). Such information suggests that there is a correlation between virus in the hemolymph and the ability of beetles to vector virus. Recent studies, however, indicate that non-vectored viruses also appear in the hemolymph (Scott and Fulton, 1978).

Hemolymph is usually obtained from beetles by severing a metathoracic leg and collecting the material which exudes in a capillary. Our recent studies demonstrate that the heomlymph contains an inhibitor and this may have contributed to erratic results in earlier studies evaluating the presence of virus in hemolymph. Dilution of at least 200-fold of the hemolymph to overcome the inhibitor allows detection of virus much more readily. Ramirez and Gamez (personal communication) have been able to detect and quantitate virus in the hemolymph utilizing a passive hemagglutination technique.

Virus appears in the hemolymph very rapidly. In our studies with the cowpea strain of southern bean mosaic virus, bean leaf beetles, *Cerotoma trifurcata*, were fed on infected plants for 30 seconds. A metathoracic leg was then quickly severed and hemolymph collected; a procedure requiring no more than an additional 30 seconds. Virus was detected under these conditions. Virus was detected for seven days after a feeding period of 24 hours on infected plants. The concentration of virus is reduced with time, as is the ability of the beetle to transmit virus, and this suggests that multiplication of virus in the beetle does not occur (Slack and Scott, 1971; and Fulton, unpublished data).

Virus was found in the serum fraction of the hemolymph and also in the hemocytes. When a vectored virus, the cowpea strain of southern bean mosaic, and a non-vectored virus, the cowpea strain of tobacco mosaic, were injected into the hemocoel both were observed in phagocytic vacuoles in hemocytes when ultrathin sections were examined in the electron microscope. The cowpea strain of tobacco mosaic virus appeared to be in very high concentrations in such phagocytic vacuoles (Kim *et al.*, 1978).

7.5.2 Regurgitant

Early investigators of beetle transmission (Smith, 1924; Markham and Smith, 1949) suggested that the ability of these insects to transmit viruses is related to the regurgitation of infective virus from the foregut. There have been questions raised as to whether beetles regurgitate during feeding (Selman, 1973). Using several kinds of stimuli, beetles can be induced to regurgitate. Whether they regurgitate in feeding would seem to be a moot question, for our studies indicate that they deposit virus on the surface of the leaf during feeding. This is true for both a transmitted virus (the cowpea strain of southern bean mosaic) and a non-transmitted virus (the cowpea strain of tobacco mosaic) (Scott and Fulton, 1978). Both of these viruses appear in regurgitant following injection of purified preparations into the hemocoel.

Regurgitant from bean leaf beetles, *Cerotoma trifurcata*, fed on plants infected with southern bean mosaic virus contained virus activity greater than that in sap from infected tissues on which beetles have fed. The activity of southern bean mosaic virus or the cowpea strain of tobacco mosaic virus in regurgitant was maintained at the same level as that in plant sap when both were stored for seven days at room temperature indicating that there was no inactivation of virus by components of the regurgitant. There was no evidence of virus inhibitors in the regurgitant in amounts comparable to those in the hemolymph. Although the presence of virus in the regurgitant is undeniable, its relation to transmission is as yet not determined. A bean leaf beetle which has fed on a plant infected with the cowpea strain of tobacco mosaic virus will not be able to infect a healthy plant although that beetle will deposit virus on the surface of the leaf during feeding. If the same virus is injected into the hemocoele of a beetle, virus will appear in the regurgitant and be deposited on the leaf without producing infection (Scott and Fulton, 1978). There is also some evidence suggesting that southern bean mosaic virus is deposited on the surface of a leaf during feeding by a viruliferous beetle at times when transmission does not occur.

7.5.3 Feces

Virus can be recovered from fecal material from beetles which have fed on infected plants (Slack and Fulton, 1971). This is true for both a beetle-vectored and a non-beetle-vectored virus. A'Brook and Benigno (1972) obtained cocksfoot mottle and phleum mottle viruses from larvae feces. Nault (personal communication) also demonstrated the presence of maize chlorotic mottle virus in the feces of *Oulema melanopa* larvae. It has been suggested that contamination of susceptible plants with virus in this manner results in transmission attributed to beetle feeding. There is little direct evidence to support this idea.

7.6 SPECULATION ON MECHANISMS OF TRANSMISSION

As we indicated in the introduction, a complex biological phenomenon is involved in virus transmission by beetles. There is a high degree of specificity between beetle vectors and the viruses they transmit. The process is not simply one of contamination of the mouth parts, or, at the moment, this does not explain several anomalies. Such viruses as tobacco mosaic and tobacco ringspot should be readily transmitted by beetles if contamination of the mouth parts were the only factor involved. We know that these viruses are present in regurgitant of beetles which have fed on infected plants and that these viruses are deposited on the surface of leaves during feeding without infecting the plants. Some of these viruses are retained in the hemolymph. It is as difficult to explain why these viruses are not transmitted by beetles as it is to explain why other viruses are readily transmitted by beetles.

Our studies on the association of virus with the beetles has resulted in interesting observations: a rapid movement of virus into the hemolymph, an inhibitor in the hemolymph, and virus present in the regurgitant and feces. In spite of the association of some of these situations with transmission, such as beetles becoming viruliferous following the injection of virus into the hemocoele, none of these explain mechanisms of transmission or pathways of movement of virus within beetles to accomplish transmission. The presence of virus in the hemolymph following an acquisition feed suggests that the hemolymph is one of the places virus is stored during long-term retention by the beetle. We assume that virus moves into the hemocoele from the midgut. This movement must be somewhat selective and active since virus appears in the hemolymph very quickly. The pathway by which virus moves from the hemocoele to the regurgitant is not known. It can be assumed the movement could be from the hemocoele into the midgut and thus into the regurgitant. Although beetles which vector viruses do not have salivary glands, they do have gnathal glands opening into the buccal cavity (Srivastava, 1959). It is not known whether these glands are involved in virus transport. The indication that virus in the regurgitant is more concentrated than virus in the plant tissue being fed upon indicates that virus movement in the beetle is a selective process resulting in concentration of virus in the regurgitant.

As has been indicated, the association of virus with beetle hemolymph, regurgitant, or feces is not restricted to beetle-transmitted viruses but has also been observed with non-beetle-transmitted viruses. This suggests that the explanation of the specificity of beetle-transmitted viruses and an understanding of why certain viruses can be transmitted by beetles and why others cannot must be related to a virus-mediated difference. At the present time there is no information suggesting a basis for this difference.

7.7 ACKNOWLEDGMENT

This work was supported in part by grants from the U.S. National Science Foundation (No. 0IP75-05494) and Consejo Nacional de Investigaciones Cientificas y Technologicas de Costa Rica.

7.8 REFERENCES

A'Brook, J., and Benigno, D. A. (1972). The transmission of cocksfoot mottle and phleum mottle viruses by *Oulema melanopa* and *O. lichenis*. *Ann. Appl. Biol.* 72: 169-176.

Alconero, R., and Santiago, A. (1973). *Phaseolus lathyroides* as a reservoir of cowpea mosaic virus in Puerto Rico. *Phytopathology* 63: 120-123.

Bakker, W. (1971). Three new beetle vectors of rice yellow mottle virus in Kenya. *Neth. J. Pl. Path.* 77: 201-206.

Bakker, W. (1974). Characterization and ecological aspects of rice yellow mottle virus in Kenya. *Mededel. Landbouwhogeschool Wageningen.* 110: 152 pp.

Bakker, W. (1975). Rice Yellow Mottle virus. *CMI/AAB Descriptions of Plant Viruses* No. 149.

Benigno, D. A., and A'Brook, J. (1972). Infection of cereals by cocksfoot mottle and phleum mottle viruses. *Ann. Appl. Biol.* 72: 43-52.

Bercks, R. (1973). Scrophularia mottle virus. *CMI/AAB Descriptions of Plant Viruses* No. 113.

Campbell, R. N., and Colt, W. M. (1967). Transmission of radish mosaic virus. *Phytopathology* 57: 502-504.

Cartin, F., and Gamez, R. (1973). Transmission of two new strains of bean rugose mosaic virus by Chrysomelid beetles. *Abstracts XIII Annual Meeting, Caribbean Division, A. P. S.*, San Jose, C. R., Dec. 9-14, p. 12.

Catherall, P. L. (1968). Cocksfoot mottle virus. *Rep. Welsh Pl. Breed. Stn.* 1967: 120.

Catherall, P. L. (1970a). Phleum mottle virus. *Pl. Path.* 19:101-103.

Catherall, P. L. (1970b). Cocksfoot mottle virus. CMI/AAB *Descriptions of Plant Viruses* No. 23.

Chant, S. R. (1959). Viruses of cowpea, *Vigna unguiculata* L. (Walp.), in Nigeria. *Ann. Appl. Biol.* 47: 565-572.

Chant, S. R. (1962). Further studies on the host range and properties of Trinidad cowpea mosaic virus. *Ann. Appl. Biol.* 50: 159-162.

Cockbain, A. J., Cook, S. M., and Bowen, R. (1975). Transmission of broad bean stain virus and Echtes Ackerbohnenmosaik-virus to field beans (*Vicia faba*) by weevils. *Ann. Appl. Biol.* 81: 331-339.

Cohen, S., and Nitzany, F. E. (1963). Identity of viruses affecting cucurbits in Israel. *Phytopathology* 53: 193-196.

Dale, W. T. (1949). Observations on a virus disease of cowpea in Trinidad. *Ann. Appl. Biol.* 36: 327-333.

Dale, W. T. (1953). The transmission of plant viruses by biting insects, with particular reference to cowpea mosaic. *Ann. Appl. Biol.* 40: 384-392.

Dale, W. T. (1954). Sap-transmissible mosaic diseases of solanaceous crops in Trinidad. *Ann. Appl. Biol.* 41: 240-247.

Debrot, E., and de Rojas, C. E. B. (1967). El virus del mosaico del frijol, *Vigna sinensis* Endl. (Cowpea Mosaic Virus) en Venezuela. *Agronomia Tropical* 17: 3-15.

Diaz, A. J. (1972). Estudio y caracterizacion de un mosaico del frijol de costa (*Vigna sinensis*) en El Salvador. *Phytopathology* 62: 754 (abst.).

Freitag, J. H. (1941a). Insect transmission, host range, and properties of squash mosaic virus. *Phytopathology* 31: 8 (abst.).

Freitag, J. H. (1941b). A comparison of the transmission of four cucurbit viruses by cucumber beetles and by aphids. *Phytopathology* 31: 8 (abst.).

Freitag, J. H. (1956). Beetle transmission, host range, and properties of squash mosaic virus. *Phytopathology* 46: 73-81.

Fribourg, C. E., Hinostroza De Lekew, A. M., and Jones, R. A. C. (1975). Host range, some properties and beetle transmission of four isolates of Andean potato latent virus from Peru. Page 103 in *Proc. Sixth Triennial Conference of the European Association for Potato Research*. Wageningen. The Netherlands. 233 pp.

Fribourg, C. E., Jones, R. A. C., and Koenig, R. (1977). Andean potato mottle, a new member of the cowpea mosaic virus group. *Phytopathology* 67: 969-974.

Fulton, J. P., Scott, H. A., and Gamez, R. (1975). Beetle transmission of legume viruses. *In* "Diseases of Tropical Legumes" (Bird and Maramorosch, eds.), pp. 123-131. Academic Press, New York.

Fulton, J. P., and Scott, H. A. (1974). Virus vectoring efficiencies of two species of leaf-feeding beetles. *Proc. Amer. Phytopath. Soc.* 1: 159 (abst.).

Fulton, J. P., and Scott, H. A. (1977). Bean rugose mosaic and related viruses and their transmission by beetles. *Fitopatologia Brasileira* 2: 9-16.

Gamez, R. (1972a). Los virus del frijol en Centro America. II. Algunas propiedades y transmission por crisomelidos del virus del mosaico rugoso del frijol. *Turrialba* 22: 249-257.

Gamez, R. (1972b). Some properties and beetle transmission of bean yellow stipple virus. *Phytopathology* 62: 759 (abst.).

Gamez, R. (1976). Los virus del frijol en Centro America. IV. Algunas propiedades y transmission por insectos crisomelidos del virus del moteado amarillo del frijol. *Turrialba* 26: 160-166.

Gerhardson, B., and Pettersson, J. (1974). Transmission of red clover mottle virus by clover shoot weevils, *Apion* spp. *Swedish J. Agric. Res.* 4: 161-165.

Grogan, R. G., and Kimble, K. A. (1964). The relationship of severe bean mosaic virus from Mexico to southern bean mosaic virus and its related strain in cowpea. *Phytopathology* 54: 75-78.

Herzog, D. C., Eastman, C. E., and Newsom, L. D. (1974). Laboratory rearing of the bean leaf beetle. *J. Econ. Entomol.* 67: 794-795.

Hobbs, H. A. (1977). Beetle transmission of cowpea chlorotic mottle virus. M. S. Thesis, University of Arkansas, 18 pp.

Holcomb, C. E., and Fulton, J. P. (1978). Rearing the bean leaf beetle for plant virus vector studies. *Plant Dis. Rep.* 62: 12-14.

Hollings, M. (1973). Turnip rosette virus. CMI/AAB *Descriptions of Plant Viruses* No. 125.

Horn, N. L., Newsom, L. D., Carver, R. G., and Jenson, R. L. (1970). Effects of virus diseases on soybeans in Louisiana. *Louisiana Agr.* 13: 12.

Howe, W. L., and George, B. W. (1966). Corn rootworms. *In* "Insect Colonization and Mass Production" (Carroll N. Smith, ed.), pp. 367-383. Academic Press, New York.

Iizuka, N. (1974). Southern bean mosaic virus in soybean. *Shokubutsu-Boeki* (Plant Protection) 28: 471-474.

Isely, D. (1930). The biology of the bean leaf beetle. *Arkansas Agric. Exp. Sta. Bull.* 248: 1-20.

Jansen, W. P., and Staples, R. (1970a). Effect of cowpeas and soybeans as source or test plants of cowpea mosaic virus on vector efficiency and retention of infectivity of the bean leaf beetle and spotted cucumber beetle. *Plant Dis. Rep.* 54: 1053-1054.

Jansen, W. P., and Staples, R. (1970b). Transmission of cowpea mosaic virus by the Mexican bean beetle. *Jour. Econ. Ent.* 65: 1719-1720.

Jansen, W. P., and Staples, R. (1971). Specificity of transmission of cowpea mosaic virus by species within the subfamily Galerucinae, family Chrysomelidae. *Jour. Econ. Ent.* 64: 365-367.

Kim, K. S., Scott, H. A., and Robison, M. D. (1978). Ultrastructural responses of bean leaf beetle hemocytes to beetle-transmitted and non-transmitted plant viruses. *Proc. Am. Phytopath Soc.* 4: 130 (abst.)

Kvicala, B. A., Smrz, J., and Blanco, N. (1970). Some properties of cowpea mosaic virus isolated in Cuba. *Phytopath. Z.* 69: 223-235.

Kvicala, B. A., Smrz, J., and Blanco, N. (1973). A beetle-transmitted virus disease of cowpea in Cuba. *FAO Plant Protection Bull.* 21: 27-29.

Lana, A. O., Gilmer, R. M., Chheda, H. D., and Fatokum, D. O. (1974). A virus-induced mosaic of okra (*Hibiscus esculentus*) in Nigeria. *Plant Dis. Rep.* 58: 616-619.

Lana, A. O. and Taylor, T. A. (1976). The insect transmission of an isolate of okra mosaic virus occurring in Nigeria. *Ann. Appl. Biol.* 84: 361-364.

Lastra, R. (1968). Occurrence of cucurbit viruses in Venezuela. *Plant Dis. Rep.* 52: 171-174.

Lima, J. A. A., and Nelson, M. R. (1977). Etiology and epidemiology of mosaic of cowpea in Ceara, Brazil. *Plant Dis. Rep.* 61: 864-867.

McLaughlin, M. R., and Milbrath, G. M. (1978). Transmission of southern bean mosaic virus by the striped cucumber beetle, *Acalymma vittata. Proc. Am. Phytopath. Soc.* 4: 129-130 (abst.).

Markham, R., and Smith, K. M. (1949). Studies on the virus of turnip yellow mosaic. *Parasitology* 39: 330-342.

Meiners, J. P., Waterworth, H. E., Lawson, R. H., and Smith, F. F. (1977). Curly dwarf mosaic disease of beans from El Salvador. *Phytopathology* 67: 163-168.

Moore, B. J. (1973). Quail pea mosaic virus: a new member of the comovirus group. *Plant Dis. Rep.* 57: 311-315.

Moore, B. J., and Scott, H. A. (1971). Properties of a strain of bean pod mottle virus. *Phytopathology* 61: 831-833.

Moore, B. J., Scott, H. A. and Walters, H. J. (1969). *Desmodium panicum*, a perennial host of bean pod mottle virus in nature. *Plant Dis. Rep.* 53: 154-155.

Nault, L. R., Styer, W. E., Coffey, M. E., Gordon, D. T., Negi, L. S., and Niblett, C. L. (1978). Transmission of maize chlorotic mottle virus by chrysomelid beetles. *Phytopathology* 68: 1071-1074.

Nemoto, M., Yoshida, K. and Goto, T. (1974). Studies on virus diseases of melon; squash mosaic virus. *Ann. Phytopath. Soc. Japan* 40: 117-118.

Orlob, G. B. (1963). Reappraisal of transmission of tobacco mosaic virus by insects. *Phytopathology* 53: 822-830.

Patel, V. C., and Pitre, N. H. (1971). Transmission of bean pod mottle virus to soybean by the striped blister beetle, *Epicanta vittata. Plant Dis. Rep.* 55: 628-629.

Ross, J. P. (1963). Transmission of bean pod mottle virus in soybeans by beetles. *Plant Dis. Rep.* 47: 1049-1050.

Sanderlin, R. S. (1973). Survival of bean pod mottle and cowpea mosaic virus in beetles following intrahemocoelic injections. *Phytopathology* 63: 259-261.

Schuster, M. F. (1963). Flea beetle transmission of tobacco ringspot virus in the lower Rio Grande Valley. *Plant Dis. Rep.* 47: 510-511.

Scott, H. A., and Fulton, J. P. (1976). Behavior of vectored and nonvectored viruses during feeding activity of the bean leaf beetle. *Proc. Amer. Phytopath. Soc.* 3: 249 (abst.).

Scott, H. A., and Fulton, J.P. (1978). Comparison of the relationship of southern bean mosaic virus and the cowpea strain of tobacco mosaic virus with the bean leaf beetle. *Virology* 84: 197-199.

Scott, H. A., and Phatak, H. C. (1978). A beetle-transmitted virus from *Phaseolus mungo. Proc. Am. Phytopath. Soc.* 4: 131 (abst.).

Selman, B. J. (1973). Beetles-phytophagus Coleoptera. *In* "Viruses and Invertebrates" (A. J. Gibbs, ed.), pp. 157-177. North-Holland Publishing Co., London.

Serjeant, E. P. (1964). Cocksfoot mottle virus. *Pl. Path.* 13: 23-24.

Serjeant, E. P. (1967). Some properties of cocksfoot mottle virus. *Ann. Appl. Biol.* 59: 31-38.

Shepherd, R. J., and Fulton, R. W. (1962). Identity of a seed-borne virus of cowpea. *Phytopathology* 52: 489-493.

Shukla, D. W., Proeseler, G., and Schmelzer, K. (1975). Studies on viruses and virus diseases of cruciferous plants XVIII. Beetle transmission and some new natural hosts of Erysimum latent virus. *Acta Phytopathologica Academiae Scientiarum Hungaricae* 10 (3/4): 211-215.

Sitterly, W. R. (1960). A new insect vector of squash mosaic virus. *Plant Dis. Rep.* 44: 134-135.

Slack, S. A., and Fulton, J. P. (1971). Some plant virus-beetle vector relations. *Virology* 43: 728-729.

Slack, S. A., and Scott, H. A. (1971). Hemolymph as a reservoir for the cowpea strain of southern bean mosaic virus in the bean leaf beetle. *Phytopathology* 61: 538-540.

Smith, C. E. (1924). Transmission of cowpea mosaic by bean-leaf beetle. *Science* 60: 268.

Smith, K. M. (1965). Plant virus-vector relationships. *Adv. Virus Res.* 11: 61-95.

Srivastava, U. S. (1959). The maxillary glands of some Coleoptera. *Proc. R. Ent. Soc. Lond.* (A) 34: 57-62.

Stefanac, Z., and Mamula, D. (1971). A strain of radish mosaic virus occurring in turnip in Yugoslavia. *Ann. Appl. Biol.* 69: 229-234.

Stoner, W. N. (1963). A mosaic virus transmitted by beetles and a grasshopper. *Phytopathology* 53: 890 (abst.).

Swaans, H., and Van Kammen, A. (1973). Reconsideration of the distinction between the severe and yellow strains of cowpea mosaic virus. *Netherlands Jour. of Plant Pathology.* 79: 257-265.

Tochihara, H. (1968). Radish enation mosaic virus. *Ann. Phytopath. Soc. Japan* 34: 129-136.

Upstone, M. C. (1969). Epidemiology of cocksfoot mottle virus. *Ann. Appl. Biol.* 64: 49-55.

Uyemoto, J. K., and Grogan, R. G. (1977). Southern bean mosaic virus: evidence for seed transmission in bean embryos. *Phytopathology* 67: 1190-1196.

Van Kammen, A. (1968). The relationship between the components of cowpea mosaic virus I. Two ribonucleoprotein particles necessary for the infectivity of CPMV. *Virology* 34: 312-318.

Walters, H. J. (1964a). Transmission of southern bean mosaic virus by the bean leaf beetle. *Plant Dis. Rep.* 48: 935.

Walters, H. J. (1964b). Transmission of bean pod mottle virus by bean leaf beetles. *Phytopathology* 54: 240.

Walters, H. J. (1965). Transmission of the cowpea strain of southern bean mosaic virus by bean leaf beetles. *Phytopathology* 55: 1080.

Walters, H. J. (1969). Beetle transmission of plant viruses. *Adv. Virus Res.* 15: 339-363.

Walters, H. J., and Barnett, O. W. Jr. (1964). Bean leaf beetle transmission of Arkansas cowpea mosaic virus. *Phytopathology* 54: 911.

Walters, H. J., and Dodds, N. L. (1969). Identification and beetle transmission of an isolate of cowpea chlorotic mottle virus from Desmodium. *Phytopathology* 59: 1055 (abst.).

Walters, H. J., and Henry, D. G. (1970). Bean leaf beetle as a vector of the cowpea strain of southern bean mosaic virus. *Phytopathology* 60: 177-178.

Walters, H. J., and Lee, F. N. (1969). Transmission of bean pod mottle virus from *Desmodium paniculatum* to soybean by the bean leaf beetle. *Plant Dis. Rep.* 53: 411.

Walters, H J., and Scott, H. A. (1972). Host range and some properties of Desmodium yellow mottle virus. *Phytopathology* 62: 125-128.

Walters, H. J., and Surin, P. (1973). Transmission and host range studies of broad bean mottle virus. *Plant Dis. Rep.* 57: 833-836.

Walters, H. J., Lee, F. N., and Jackson, K. E. (1972). Overwintering of bean pod mottle virus in bean leaf beetles. *Phytopathology* 62: 808 (abst.).

Waterworth, H. E., Meiners, J. P., Lawson, R. H., and Smith, F. F. (1977). Purification and properties of a virus from El Salvador that causes mild mosaic in bean cultivars. *Phytopathology* 67: 169-173.

Weidemann, H. L., and Bode, O. (1973). Untersuchungen über ein neues "isometric virus" aus *Atropa belladonna* L. IV. Versuch zur Übertragbarkeit des belladonna mottle virus. *Phytopath. Z.* 76: 6-10.

Whitney, W. K., and Gilmer, R. M. (1974). Insect vectors of cowpea mosaic virus in Nigeria. *Ann. Appl. Biol.* 77: 17-21.

Zaumeyer, W. J., and Harter, L. L. (1943). Two new virus diseases of beans. *Jour. Agr. Res.* 67: 305-327.

Chapter 8

BARK BEETLES, *CERATOCYSTIS ULMI* AND DUTCH ELM DISEASE*

Francis W. Holmes

Shade Tree Laboratories
University of Massachusetts
Amherst, Massachusetts

8.1 INTRODUCTION 134
8.1.1 Disease Discovery 134
8.1.2 Pathogen Discovery 134
8.1.3 Vector Discovery 135
8.1.4 Other Potential Vectors 135
8.1.5 Disease Spread and Distribution 135
8.2 TRANSMISSION 136
8.2.1 Vectors, Spread and Distribution 136
8.2.2 Means of Disease Local Transmission 136
8.2.3 Efficiency of Disease Transmission 137
8.2.4 Means of Disease Distant Transmission 137
8.3 ELM BARK BEETLE CHARACTERISTICS: CONTRASTING *SCOLYTUS MULTISTRIATUS* AND *HYLORGOPINUS RUFIPES* 138
8.3.1 Appearances 138
8.3.2 Brood Galleries 139
8.3.3 Life Cycles in Relation to Transmission 139
8.3.4 Feeding Pheromone Released when Brood Galleries are Started 140
8.3.5 Proposed Consequences of Partial Versus Thorough Sanitation 140
8.4 DISEASE CONTROL BY INSECT CONTROL 141
8.4.1 Insecticides 141
8.4.2 Sanitation 141
8.4.3 Repellants 142
8.4.4 Biological Control 142
8.4.5 Trapping 142
8.5 CLOSING REMARKS 143
8.6 ACKNOWLEDGMENTS 143
8.7 REFERENCES 143

*Paper No. 2210, Massachusetts Agricultural Experiment Station, University of Massachusetts at Amherst, including research supported (in part) from Experiment Station Project 110.

ISBN 0-12-326450-2

8.1 INTRODUCTION

8.1.1 Disease Discovery

One must know something of the history and background of Dutch elm disease to see why the elm bark beetle vectors are so very important in this disease. One way that this disease is unusual is that it is so new. It was never seen in the world until it was pointed out by Dr. Dina Spierenburg of the Dutch Plantenziektekundige Dienst as of 1919 (Spierenburg, 1921), at which point in time it appeared to have existed in the Benelux area of Europe for only the previous two or three years. Most of the research on this disease in the early years was done in the Netherlands. Because nearly all the early research publications therefore came from Dutch sources, the disease came to be known as the "Dutch elm disease."

8.1.2 Pathogen Discovery

The Dutch elm disease is caused by a fungus, an unusual member of the group Ascomycetes, being primitive in some ways and very advanced in other ways (Rosinski, 1961). This fungus today is called *Ceratocystis ulmi* by most people and *Ophiostoma ulmi* by the remainder. Most of the time, *C. ulmi* is buried in the sapwood vessels where it is hard to get at.

There was some early suspicion by Dr. John Boyce that *C. ulmi* might have come from China to Europe, with Asiatic troops engaged in World War I, especially since Chinese Elm and Siberian Elm are resistant, but Dr. Aoshima could not find it in China in the late 1930's. Everywhere that the fungus went, the disease was recognized as something new. So, in spite of these early suspicions, we now believe that the disease fungus actually originated about 1917 or 1918 in the Benelux area.

Possibly it arose from some mutation in a fungus that was rather similar but not able to cause the devastation that the present fungus can do. One example of such a mutation would be the ability to form a sticky substance on the spores, whereby the spores would adhere to bark beetles. After all, mustard gas—a well-known and powerful mutagen—was in widespread use by the armies fighting in that general area at that time.

The 1920's were spent in a prolonged struggle to determine what caused the Dutch elm disease. Dr. Marie Schwarz had found the real cause at the very outset of the study as part of her doctoral thesis. She called the non-sexual stage of this fungus *Graphium ulmi* (Schwarz, 1922). At first few believed her, and the trouble was ascribed variously to drought, to poison gas from World War I, to bacterial infections, etc. However, in 1928-30 Dr. Johanna Westerdijk, Directrice of the Phytopathlogisch Laboratorium "Willie Commelin Scholten," assigned Dr. Christine Buisman to test Schwarz's conclusions. Buisman's experiments convincingly vindicated Schwarz (Buisman, 1928).

8.1.3 Vector Discovery

Naturally it would be very difficult to show how any disease was transmitted before one even knew what caused it. For example, drought would not require any agent of transmission. But beginning in the late 1920's, and continuing into the early 1930's, at the Agricultural College in Wageningen, Netherlands, the doctoral thesis work of Fransen, assisted by Buisman, showed that the disease was indeed carried from tree to tree by elm bark beetles (Fransen, 1931; Fransen and Buisman, 1935). This was promptly confirmed elsewhere (Middleton *et al.*, 1935; Walter, 1935). In Europe the Dutch elm disease is carried by *Scolytus scolytus* Fabricius and by *Scolytus multistriatus* (Marsh.). (In Scandinavia *Scolytus laevis* is the additional vector but does not seem very effective.)

8.1.4 Other Potential Vectors

Several other insects, which feed on elms and might be supposed to be likely to carry the fungus from tree to tree, are discussed in the forthcoming Dutch elm disease bulletin of the Cooperative Regional Research Project on Wilt Diseases of Trees (Project NE-99, formerly NE-25), a bulletin which cites 460 references (Sinclair and Campana, 1978). These insects are (1) weevils in the genus *Magdalis*, (2) the long-horned beetle *Saperda tridentata* and (3) the carpenter worm, *Zeuzera pyrina* (Collins *et al.*, 1936; Readio, 1925; Tyler *et al.*, 1939). However much these insects may occasionally make contact with spores of the fungus, their life-cycles and their procedures for feeding and egg-laying probably give them little chance to transmit. The only other significant transmission from tree to tree is by root grafts between very closely adjoining trees which is important only in hedgerows or along densely-elm-planted streets.

8.1.5 Disease Spread and Distribution

Elm bark beetles naturally were already well distributed in Europe and the record of disease spread therefore is a record of spread by the fungus. The Dutch elm disease reached England in 1926, it spread southwestward to Italy and the Balkans, westward to Poland and Russia to the Caucasus and southeastward into Spain. It has gone as far as Tashkent, Iran, Turkey and even India. In 1930 the disease reached the United States. The fungus and the beetles were brought to this country on logs of European species of elm which, having a finer grain, are much more suitable for furniture veneer than the rather coarse-grained American elm species.

Dutch elm disease seemed to have been wiped out in the USA in 1931 but in 1932 it was found here again in several places. Indeed it was found entering nearly every port on the Atlantic and Gulf seaboards, from the St. Lawrence River to New Orleans. Since that time it has been spreading in this country (Barger and Hock, 1971) until now it has reached the Pacific Ocean in both Oregon and California and is as far south as the northern parts of Gulf states like Texas and Alabama. The disease does not yet occur in the whole range of the American elm in this country; for example I have not yet heard of it in Florida or Georgia. Also it has not been reported in Japan, and quite likely it is not in China either.

8.2 TRANSMISSION

8.2.1 Vectors, Spread and Distribution

In transatlantic travel in the 1930's, *S. multistriatus* came with *C. ulmi* (Collins, 1936, 1941), but it had also come ahead and indeed was already established in this country (Chapman, 1910) before the disease even existed! *S. scolytus*, on the other hand, remained in Europe. When the fungus reached North America, it found another elm bark beetle species also already here, namely *Hylurgopinus rufipes* (Eichh.), which had been found here as early as the turn of the century and probably had always been a North American beetle since it is not known to occur in Europe (Parker *et al.*, 1948). *H. rufipes* probably occurs wherever elms naturally grow in North America (Baker, 1972). It was not long before both these beetles were carrying the fungus here (Parker *et al.*, 1947, 1948). *H. rufipes* could survive farther north than *S. multistriatus* (Pomerleau, 1965a, 1965b; Thomas, 1971; Pantyukhov, 1958). However, where both beetles occurred together, the lesser European elm bark beetle seemed to take over and it sometimes became difficult to find the North American one (Thompson and Matthysse, 1972).

In Europe the greater European elm bark beetle, *S. scolytus*, is considered the most important vector of the disease and we are very fortunate that it has not yet come to our country. The Europeans on the other hand, are also fortunate that *H. rufipes* has not become established in Europe, because it might cause considerable destruction of elms by transmitting the disease fungus more readily in northern areas like Denmark and Scandinavia.

A U.S. quarantine against elm was proposed in 1928, instituted in 1935, and removed again about 1965. In the dozen years since its removal *S. scolytus* does not seem to have come to North America. Recently the North American elm bark beetle and *C. ulmi* were found entering England (Walker, 1973) from Canada on logs of the rock elm, *Ulmus thomasii*, but the beetle does not seem to have become established there.

8.2.2 Means of Disease Local Transmission

Compared to insects that transmit viruses and insects that transmit mycoplasma-like organisms—insects which pick up the plant pathogen by sucking juices out of the plant—the situation with the Dutch elm disease is somewhat crude. The spores of the fungus are sticky. In the case of the ascospores, the slime seems to be added in the neck of the perithecium; it is not present on them when the ascus breaks open.

In the case of conidia it appears to be made at the tip of the conidiophore soon after the first two spores are fully formed. The fungus grows in the bark of the dead tree. As the young adult insect moves about the pupal gallery and bores an emergence hole in the bark of dead elm trees, it picks up these sticky spores on the outside surface of its body. The spores remain attached when it flies to feed. Generally the elm bark beetles feed on healthy elm trees and they lay their eggs under the bark of dying or very recently killed elm trees (Collins *et al.*, 1940), but only on elms. The lesser European elm bark beetle chews through the thin bark in the crotches of very young healthy elm twigs and the North American elm bark beetle burrows into the bark of moderately small branches of healthy elms, where it feeds on the inner bark next to the cambium (Thompson and Matthysse, 1972). In both cases the beetles chew through the cambium and into the xylem where they cut across xylem vessels. The beetles move very actively in the wound and rub off the sticky spores. If the wound remains moist enough, the spores germinate there.

8.2.3 Efficiency of Disease Transmission

The result is a very inefficient transmission. Often the beetle is bearing no spores. Often the spores are not rubbed off at all, or are rubbed off in the wrong place, or do not germinate, or germinate in such a way as not to lead to growth of the fungus into functional xylem vessels.

8.2.3.1 Aggregating Pheromones. However, another aspect of elm bark beetle flight tends to compensate for a low degree of vector efficiency by an individual beetle. There is an entire symposium in this very Conference on that matter. Elm bark beetles fly in little clouds, attracted to each other by aggregating pheromones. Therefore some susceptible trees chance to escape inoculation for many years, while others die. (Naturally this also gives rise to the concept of controlling the disease by trapping the beetles. We shall discuss that later when we come to control measures.)

The advantage of the aggregating pheromones to the Dutch elm disease is that one does not have just a single beetle arriving at a tree. One could expect that the single beetle usually would fail to transmit the disease, although occasionally, of course, it succeeds. We do not know the exact percentages that transmit under various circumstances, but there certainly are cases where an individual tree is fed upon by hundreds of beetles.

8.2.4 Means of Disease Distant Transmission

These beetles, of course, are not transported only by their own wing power. Sometimes they are blown long distances by wind. They can be carried by automobiles; the beetles are so small that they attract little attention inside a car window. As the disease was first moving into Wisconsin, Hafstad (1958) showed that all the early cases occurred either at gasoline service stations or at roadside rest stops along the major highways leading into the state, presumably brought there by automobiles.

8.3 ELM BARK BEETLE CHARACTERISTICS: CONTRASTING *SCOLYTUS MULTISTRIATUS* AND *HYLURGOPINUS RUFIPES*

8.3.1 Appearances

Both North American and lesser European elm bark beetle larvae are legless white grubs with amber head capsules, the former ½ the body width, the latter ¾ (Wallace and Beard, 1942). Both North American and lesser European adult elm bark beetles are 2 to 3 mm long. The lesser European beetle is shiny with red-brown elytra covering its wings and a black pronutum or thorax partly covering its head; the North American beetle is uniformly grey-brown with a rough texture (Fig. 1). The lesser European male beetle is distinguished by a brush of golden setae on the front of its head; the North American beetle shows no such convenient difference between the sexes. The lesser European elm bark beetle's lower abdomen is concave; that of the North American elm bark beetle is more dome-like.

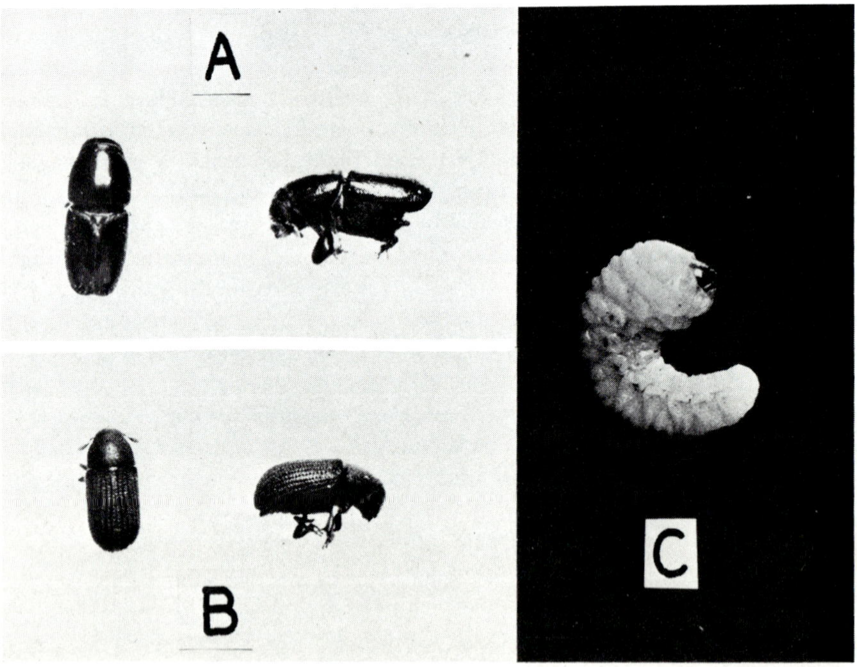

Fig. 1. Top and side views of adult elm bark beetles. Above (A), Lesser European Elm Bark Beetle, *Scolytus multistriatus* (Marsh.); below (B), our native North American Elm Bark Beetle, *Hylurgopinus rufipes*. (Eichh.). At right (C), larva of *S. multistriatus*. Actual length of beetles is 2-3½ mm.

8.3.2 Brood Galleries

The lesser European mother beetle bores an egg gallery parallel to the grain of the wood and the larval tunnels radiate in a fan crossing the wood grain. The North American mother beetle's egg gallery crosses the wood grain and her larvae bore tunnels more or less parallel to the grain. Both species, of course, make their galleries half in the outer layer of wood and half in the inner layer of bark.

The lesser European beetle seems to prefer elms in drier locations such as along streets, and the North American beetle does better in woodlots and forests which are somewhat moister. By the same token the lesser European beetle colonizes the sunny portions of the crown of a tree and, when both species occur in the same tree, the North American beetles attack shaded limbs and lower trunk. The lesser European beetle breeds only in elms, except that it can breed also in the zelkova (*Zelkova serrata*), which is in the elm family (Baker, 1972). It prefers these two for feeding (Becker and Mankowsky, 1965) although it can feed on other hardwoods (Dixon, 1964). The North American beetle is reported to breed also in basswood, ash and wild cherry (Baker, 1972). Since the zelkova, basswood, ash and wild cherry seem not to contract natural infections of the Dutch elm disease, it may be inferred that they cannot be infected by *Ceratocystis ulmi* borne by elm bark beetles. Indeed at the Shade Tree Laboratories we tried inoculating trees of 70 species with the fungus and got no infection, although we found that it will grow, and fruit both sexually and asexually, on many dead substrata.

8.3.3 Life Cycles in Relation to Transmission of Disease

In the northern part of the United States, like Massachusetts, the lesser European elm bark beetle completes two generations per year and in somewhat warmer areas, like southern Caifornia and New Jersey, it is reported to complete part of a third (Brown, 1965). Thus, in general, the larvae of alternate generations overwinter in the bark. They are the more significant insects for transmission to the healthy elm trees since as adults they are feeding there in June and July, when the xylem vessels seem to be most suitable for the growth of the fungus. By August even our artificial inoculations usually yield very poor results. The generation of beetles that develops in late summer and emerges in warm days in September and October cannot cause cases of Dutch elm disease at that time, since elms are almost insusceptible by then. However, these beetles can lead to a great increase in the number of young, overwintering larvae, ready to emerge as young adults the following spring.

By contrast, the North American elm bark beetle generally completes only 1½ generations per year and therefore overwinters each year *both* as adults *and* as immature larvae (Thompson and Matthysse, 1972). The result is that in Massachusetts these adults can emerge in April or May, about 2 weeks earlier than the smaller European elm bark beetle. Thompson's test inoculations of elm trees with the Dutch elm disease fungus during the various months of the spring, however, showed that the fungus is able to persist in the wound, and then to invade the elm tissues and cause outward symptoms of disease at about the same time as inoculations in June (Thompson and Matthysse, 1972).

The overwintering larvae of our native elm bark beetle finish their development in the late spring. In Massachusetts they emerge in June and July, a time when elm trees are still highly susceptible to inoculation. Since elm susceptibility occurs later in the season in northern areas like Maine and Quebec (Pomerleau, 1965a), the native beetle that overwintered as a larva can be very important in those areas (Pomerleau, 1965b). In more southern areas, the European beetles and the overwintering adult native beetles are the more important.

In all cases the transmission occurs during feeding by the beetles. Egg-laying is carried out only on dying or on recently killed elm trees or parts of elm trees, but of course it is not necessary that these trees or these parts of trees should have been killed by Dutch elm disease. Eggs can be laid in elm trees that were killed by bulldozers or by ice-storm breakage or by anything else. The fungus brought by the parent beetles ordinarily will grow in the beetle galleries in such trees, so that the emerging young adults will be carrying the fungus even though the tree never had the Dutch elm disease.

8.3.4 Feeding Pheromone Released When Brood Galleries Are Started

Aggregating pheromones were mentioned earlier. But another pheromone is released by the elm bark beetles when the virgin female beetle is feeding upon the inner bark, making a brood gallery (Peacock *et al*., 1971). This will bring large numbers of other beetles to the same weakened or diseased elm so that the whole tree is quickly colonized by thousands of females about to chew "maternal galleries" in which to lay their eggs. Males are also attracted to this area. They search over the bark for the females' tunnels where they may mate at the entrance or even within the tunnel. A male beetle may inseminate only one female or many; the average is reported to be 17 (Bartels and Lanier, 1974). It would be useful to learn whether the female mates with only one or with several males. After mating, the female no longer releases that pheromone that attracted other beetles to the tree (Elliott *et al*., 1975). Most of the pheromone research has been done with the lesser European elm bark beetle and the whole story of pheromones remains to be worked out for the native elm bark beetle.

8.3.5 Proposed Consequences of Partial Versus Thorough Sanitation

Since the emerging young adult elm bark beetles may go *directly* to breeding sites, the feeding on the twigs or branches of healthy trees may turn out to be largely a phenomenon of scarcity of suitable places to breed. Therefore a *partially* effective program of cleaning up the dead trees,where elm bark beetles could lay their eggs, might be just as bad as no sanitation program at all, since it might merely divert the beetles from "brood wood" to healthy trees for their feeding. The answer may lie in a fanatical thoroughness of sanitation, if sanitation be attempted.

8.4 DISEASE CONTROL BY INSECT CONTROL

8.4.1 Insecticides

The Dutch elm disease fungus is very difficult to get at, since most of the time it is in the xylem vessels, under the bark, fairly deep within the tree. Moreover, the tree does not have a circulating system and consequently will not bring injected chemicals repeatedly to the same place in its vascular system. The result has been that Dutch elm disease control has always depended upon insect control.

Except for early efforts at eradication, the first effective control occurred after World War II, when DDT became available for public use and was applied to elm trees (Whitten and Parker, 1945; Plumb, 1950; Matthysse *et al.*, 1954). Losses in towns where a large amount of DDT was carefully and methodically sprayed onto each tree sometimes dropped to as little as a half of one percent per year, a level that was considered acceptable. Nevertheless Becker's studies of Massachusetts towns showed that control seldom was obtained, usually because of inadequate coverage of twigs high in the trees.

After DDT was banned, losses became much higher in spite of the fact that methoxychlor (Matthysse *et al.*, 1954) had been claimed to have as high an effectiveness as DDT. (For those who found DDT ineffective, of course, it *did* have as high an effectiveness.) Some think that methoxychlor is essentially worthless for Dutch elm disease control; others, that methoxychlor could be quite satisfactory although it is not properly used in most current practice (Barger, 1976).

Many other insecticidal chemical treatments have been tried on or in elms (Al-Azawi and Casida, 1958; Al-Azawi and Norris, 1959; Williams and Brown, 1969). Perhaps the most famous one was the injection of elm trees about the mid-1960's with capsules of Bidrin® (Doane, 1962; Norris, 1967). At the Shade Tree Laboratories, we tried out this injection procedure. Our results were expressed in terms of whether or not there was greater loss among the untreated control trees than among those treated. In Massachusetts, over a 4-experiment program in a period of 2 years, we found in each of the 4 experiments essentially the same loss whether or not Bidrin was applied to the tree (Holmes, 1966). We consider that this was the result of our very great care to intersperse the treated trees alternately with control trees throughout the area of the experiment, since no one knew what sources, in what locations, would yield what numbers of elm bark beetles bearing spores of *C. ulmi*. (Between every treated tree and every control tree there was also an elm that was not part of the experiment, in order that there should be no question of chemical transmission through any root graft.)

8.4.2 Sanitation

Very recent work in Detroit by Barger, Worley and Cannon (1978) has shown that Matthysse was quite right in his finding in the Lake George area of New York state in the early 1950's, namely that thorough elm sanitation over the area of a whole town *can* lead to satisfactory control of Dutch elm disease in that town (Matthysse, 1958). However, the moment that sanitation is abandoned, there can be a great outbreak of the disease (Marsden, 1953). The Detroit study shows that it is actually cheaper to spend the extra money required for 3 surveys per summer looking for Dutch elm disease, provided that all diseased elms be *very* promptly removed–within two weeks (Barger, 1977; Cannon *et al.*, 1977). It was cheaper because money was saved through fewer costs of tree removals (Cannon and Worley, 1976).

8.4.3 Repellants

Efforts have been made to repel elm bark beetles from selected trees and to prevent them from reproducing in or emerging from logs of killed elms, by the injection of various chemicals (Doane, 1958; Whitten, 1941; Touhey and Bray, 1961; Himelick and Neeley, 1961, 1963). However, since xylem transport is so much less effective in trees that have become diseased with Dutch elm disease than it is in healthy trees, these efforts did not prove practical enough for general adoption.

8.4.4 Biological Controls

Fungi (Doane, 1959), bacteria (Doane, 1960), nematodes (Hunt, 1972; Hunt and Hague, 1974) and other insects (Bosman and Meijeraan, 1969; Galford, 1967; Kennedy and Roberts, 1974; Truchan and Butcher, 1970) attack the elm bark beetles, but not severely enough to reduce the population to the point of affecting the amount of Dutch elm disease.

8.4.5 Trapping

Exciting, recent research has been the trapping of millions of beetles by attraction to pheromones (Peacock *et al.*, 1971; Pearce *et al.*, 1975; Jones and Lanier, 1977). You can imagine what good news it was when the pheromones proved to be pure chemicals which could be made synthetically (Lanier *et al.*, 1977)! Although the researchers have stated that the trapping does not *of itself* seem to reduce the amount of Dutch elm disease (Peacock, 1975), it does not interfere with simultaneous use of either sanitation or insecticidal sprays. Trapping ought to be a very effective *supplement* to a sanitation effort, where most of the dead or dying elm wood was successfully removed from an area, but where beetles had to be mopped up as they blew in from other areas or as they emerged from the few remaining broken branches that had been overlooked in the sanitation program. Yet one of the study areas had indeed been subject to thorough sanitation and *still* the trapping remains to be reported to have been of much help. Further research is needed to explain this.

8.5 CLOSING REMARKS

Although we have been unfortunate that both DDT spraying of elms and burning of elm logs (under certain circumstances) were banned, we have at least the compensation that the present methods of control do not interfere with one another. We should use a combination of all available methods (Schreiber and Peacock, 1974). In my view of the near future, we ought to be able (1) to hold down the number of elm bark beetles by promptly and thoroughly destroying their breeding places, (2) *further* to reduce their population by attracting them to traps containing pheromones, (3) *yet further* to reduce their population by having an insecticide present on the healthy tree when they arrive to feed there, (4) *still* to watch healthy elm trees sharply and daily in order (5) to prune out infected branches instantly when the wilt first appears, thereby removing the disease entirely from the tree by a radical surgery despite a risk of attracting bark beetles (Hart *et al.*, 1967), (6) to cut root grafts between an infected elm and its neighboring elm, and also beyond its neighbors, and (7) to try to inject the most valued elms, via small roots, with one of the very recently registered systemic fungicides. Incidentally, the program of daily watching seems susceptible to the use of citizen volunteers. Research ought to be applied to the possibility of including a compatible systemic insecticide with a compatible systemic fungicide in the same injection mixture.

The other major approach to disease control, and one in which the Shade Tree Laboratories have been deeply involved, has to do with the production of resistant elm trees, but this has been strictly limited to resistance to the fungus once it has entered the tree. I know of no studies on the resistance of each of the many elm clones to each species of elm bark beetle. This, too, deserves research attention. However, resistance does not offer help to save the present stand of large elms.

Those who wish to study Dutch elm disease further may consult the list of 2,023 references published in 1976 (supplement in 1977) by Laut and Schomaker.

8.6 ACKNOWLEDGMENTS

I thank W. B. Becker, R. A. Mankowsky, G. N. Lanier, J. W. Peacock, and J. H. Barger for advice, ideas and unpublished results that helped me to prepare this text. As an erstwhile pathologist now specializing in urban shade trees from every point of view, I readily accept responsibility for errors in expression or interpretation of this information.

8.7 REFERENCES CITED

Al-Azawi, A. F., and Casida, J. E., (1958). The efficiency of systemic insecticides in the control of the smaller European elm bark beetle. *J. Econ. Entomol.* 51: 789-790.

Al-Azawi, A. F., and Norris, Jr., D. M. (1959). Experimental prevention of bark beetle transmission of *Ceratocystis ulmi* (Buis.) Moreau with the systemic insecticide Chipman R-6199. *J. Econ. Entomol.* 52: 902-904.

Baker, W. L. (1972). Eastern forest insects. *U. S. Dep. Agric. Misc. Publ.* 1175. 642 pp.

Barger, J. H. (1976). Dutch elm disease and methoxychlor. *U. S. Dep. Agric., For. Serv., Res. Pap.* NE-353: 1-5.

Barger, J. H. (1977). Improved sanitation practice for control of Dutch elm disease. *U. S. Dep. Agric., For. Serv., Res. Pap.* NE-386: 1-12.

Barger, J. H., and Hock, W. K. (1971). Distribution of Dutch elm disease and the smaller European elm bark beetle in the United States as of 1970. *Plant Dis. Rep.* 55: 271-272.

Barger, J. H., Worley, D. P., and Cannon, Jr., W. N. (1978). Effective Dutch elm disease control by early discovery and quick removal. *Proc. 32nd Mass. Annu. Dutch Elm Disease Conf.* (Oct. 5, 1977): 6-12, fig. 1-6.

Bartels, J. M., and Lanier, G. N. (1974). Emergence and mating in *Scolytus multistriatus* (Coleoptera: Scolytidae). *Ann. Entomol. Soc. Amer.* 67: 365-370.

Becker, W. B. and Mankowsky, R. A. (1965). Twig feeding by the smaller European elm bark beetle on different kinds of trees. *J. Econ. Entomol.* 58: 132-134.

Bosman, B. T., and Meijeraan, J. W. (1969). [New data on some Hymenoptera parasitic on elm bark beetles in the Netherlands.] *Nederl. Bosb. Tijdschr.* 41: 325-330. (For. Abstr. 31: 357. 1970)

Brown, L. R. (1965). Seasonal development of the smaller European elm bark beetle in southern California. *J. Econ. Entomol.* 58: 1176-1177.

Buisman, C. J. (1928). De oorzaak van de iepenziekte. *Tijdschr. Ned. Heidemaatsch.* 40: 338-345.

Cannon, W. N., Jr., and Worley, D. P. (1976). Dutch elm disease control: performance and costs. *U. S. Dep. Agric., For. Serv., Res. Pap.* NE-345: 1-7.

Cannon, W. N., Jr., Barger, J. H., and Worley, D. P. (1977). Dutch elm disease control: the economics of intensive santiation. *U. S. Dep. Agric., For. Serv., Res. Pap.* NE-387: 1-28.

Chapman, J. W. (1910). The introduction of a European Scolytid, the smaller elm bark beetle, *Scolytus multistriatus* (Marsh.) into Massachusetts. *Psyche* 17: 63-68.

Collins, C. W. (1936). Insect vectors of the Dutch elm disease caused by the fungus *Ceratostomella ulmi* (Schwarz) Buisman. *Nat. Shade Tree Conf. Proc.* 11: 127-132.

Collins. C. W. (1941). Studies of elm insects associated with Dutch elm disease fungus. *J. Econ. Entomol.* 34: 369-372.

Collins, C. W., Buchanan, W. D., Whitten, R. R., and Hoffman, C. H. (1936). Bark beetles and other possible insect vectors of the Dutch elm disease *Ceratostomella ulmi* (Schwarz) Buisman. *J. Econ. Entomol.* 29: 169-176.

Collins, D. L., Parker, K. G., and Dietrich, H. (1940). Uninfected elm wood as a source of the bark beetle (*Scolytus multistriatus* Marsham) carrying the Dutch elm disease pathogen. *Cornell Agric. Exp. Stn. Bull.* 740. 14 pp.

Dixon, E. B. (1964). Attack response of the smaller European elm bark beetle, *Scolytus multistriatus* in confinement. *J. Econ. Entomol.* 57: 170-172.

Doane, C. C. (1958). Insecticides to prevent the emergence of *Scolytus multistriatus*. *J. Econ. Entomol.* 51: 469-471.

Doane, C. C. (1959). *Beauvaria bassiana* as a pathogen of *Scolytus multistriatus*. *Ann. Entomol. Soc. Amer.* 52: 109-111.

Doane, C. C. (1960). Bacterial pathogens of *Scolytus multistriatus* Marsham as related to crowding. *J. Insect Pathol.* 2: 24-29.

Doane, C. C. (1962). Toxicity of translocated insecticides to *Scolytus multistriatus*. *Conn. Agric. Exp. Stn. Rep. Prog.* #6.

Elliott, E. W., Lanier, G. N., and Simeone, J. B. (1975). Termination of aggregation by the European elm bark beetle. *Scolytus multistriatus*. *J. Chem. Ecol.* 1: 283-289.

Fransen, J. J. (1931). Enkele gegevens omtrent de verspreiding van de door *Graphium ulmi*

Schwarz veroorzaakte iepenziekte door de iepenspintkevers, *Eccoptogaster (Scolytus) scolytus* F. and *E. (Scolytus) multistriatus* Marsh. in verband met de bestrijding dezer ziekte. *Tijdschr. Plantenziekten* 37: 49-62.

Fransen, J. J., and Buisman, C. J. (1935). Infectieproeven op verschillende iepensoorten met behulp van iepenspintkevers. *Tijdschr. Plantenziekten* 41: 221-239.

Galford, J. R. (1967). Emergence of *Entedon leucegramma* from smaller European elm bark beetle larvae reared on artificial media. *J. Econ. Entomol.* 60: 1482-1483.

Hafstad, G. E. (1958). Possible long distance spread of Dutch elm disease by transported beetles. *Plant Dis. Rep.* 42: 893-894.

Hart, J. H., Wallner, W. E., Caris, M. R., and Dennis, G. K. (1967). Increase in Dutch elm disease associated with summer trimming. *Plant Dis. Rep.* 51: 476-479.

Himelick, E. B., and Neely, D. (1961). Prevention of bark beetle development in undesirable elms for control of Dutch elm disease. *Plant Dis. Rep.* 45: 180-184.

Himelick, E. B., and Neely, D. (1963). Systemic control of bark beetle colonization of diseased American elms. *Plant Dis. Rep.* 47: 87-88.

Holmes, F. W. (1966). Research with Bidrin. *Mass. Tree Wardens, Arborists & Util. Conf. Proc.* (Mar. 14-15): 72-76.

Hunt, D. J. (1972). Nematodes associated with the bark beetle vectors of Dutch elm disease. *Intern. Sympos. Nematol., Europ. Soc. Nematol.*, Reading, UK, (Sept.) Proc. 11: 30-31 (Abstr.).

Hunt, D. J., and Hague, N. G. M. (1974). The distribution and abundance of *Parasitaphelenchus olhami*, a nematode parasite of *Scolytus scolytus* and *S. multistriatus*, the bark beetle vectors of Dutch elm disease. *Plant Pathol.* 23: 133-135.

Jones, W. W., and Lanier, G. N. (1977). Termination of larval diapause in the European elm bark beetle. *Ann. Entomol. Soc. Amer.* 70(3): 387-390.

Kennedy, B. H., and Roberts, N. H. (1974). *Dendrosoter protuberans* (Nees) an introduced parasite of *Scolytus multistriatus* (Marsham) is established in Detroit, Michigan. *Entomol. Soc. Amer., North Central Branch, Annu. Mtg.* 29: 177-178.

Lanier, G. N., Gore, W. E., Pearce, G. T., Peacock, J. W., and Silverstein, R. M. (1977). Response of the European elm bark beetle, *Scolytus multistriatus* (Coleoptera: Scolytidae), to isomers and components of its pheromone. *J. Chem. Ecol.* 3(1): 1-8.

Laut, J. G., and Schomaker, M. E. (1976). Dutch elm disease, a bibliography. Colorado State For. Serv., Colorado State Univ. (with 1977 supplement). 153 pp.
references.

Marsden, D. H. (1953). Dutch elm disease: an evaluation of practical control efforts. *Plant Dis. Rep.* 37: 3-6.

Matthysse, J. G. (1958). An evaluation of mist blowing and sanitation in Dutch elm disease control programs. *Cornell Univ. Misc. Bull.* 30. 16 pp.

Matthysse, J. G., Miller, H. C., and Thompson, H. E. (1954). Insecticide deposits for control of elm bark beetles. *J. Econ. Entomol.* 47: 739-746.

Middleton,Buchanan, W. D., May, C., and Walter, J. M.(1935). *Ceratostomella (Graphium) ulmi,* the cause of Dutch elm disease, transmitted by *Scolytus multistriatus. J. Econ. Entomol.* 28: 138.

Neely, D. (1975). Sanitation and Dutch elm disease control. pp. 76-87 *in*: D. A. Burdekin and H. M. Heybroek, compilers. Dutch elm disease. *Proc. IUFRO Conf.*, Minneapolis, St. Paul, USA, Sept. 1973. (U. S. Dep. Agric., For. Serv., Northeast. For. Exp. Stn.) 94 pp.

Norris, D. M., Jr. (1967). Systemic insecticides in trees. *Annu. Rev. Entomol.* 12: 127-148.

Pantyukhov, G. A. (1958).[Cold-hardinessof the larvae of striated bark beetle (*Scolytus multistriatus* Marsh.)]. *Zool. Z. Moskva* 37: 1339-1344. Transl. Canada Dep. For. No. 263. 1968. 14 pp. (For. Abstr. 30: 508. 1969).

Parker, K. G., Collins, D. L., Tyler, L. J., Connola, D. P., Ozard, W. E., and Dietrich, H. (1947). The Dutch elm disease–association of *Ceratostomella ulmi* with *Scolytus multistriatus*, its advance into new areas, and control of the disease. *Cornell Agric. Exp. Stn. Mem.* 275. 44 pp.

Parker, K. G., Hagemann, L. E., Collins, D. L., Tyler, L. J., Dietrich, H., and Ozard, W. E. (1948). The association of *Hylurgopinus rufipes* with the Dutch elm disease. *J. Agric. Res.* 76: 175-183.

Peacock, J. W. (1975). Research on chemical and biological control of elm bark beetles. pp. 18-49 *in* D. A. Burdekin and H. M. Heybroek, compilers. Dutch elm disease. *Proc. IUFRO Conf.*, Minneapolis-St. Paul, USA, Sept. 1973. (U. S. Dep. Agric., For. Serv. Northeast. For. Exp. Stn.). 94 pp.

Peacock, J. W., Lincoln, A. C., Simeone, J. B., and Silverstein, R. M. (1971). Attraction of *Scolytus multistriatus* (Coleoptera: Scolytidae) to a virgin-female-produced pheromone in the field. *Ann. Entomol. Soc. Amer.* 64: 1143-1149.

Pearce, G. T., Gore, W. E., Silverstein, R. M., Peacock, J. W., Cuthbert, R., Lanier, G. N., and Simeone, J. B. (1975). Chemical attractants for the smaller European elm bark beetle *Scolytus multistriatus* (Coleoptera: Scolytidae). *J. Chem. Ecol.* 1: 115-124.

Plumb, G. H. (1950). Control of Scolytus and Dutch elm disease by concentrated DDT sprays. *J. Econ. Entomol.* 43: 110-111.

Pomerleau, R. (1965a). The period of susceptibility of *Ulmus americana* to *Ceratocystis ulmi* under conditions prevailing in Quebec. *Canad. J. Bot.* 43: 787-792.

Pomerleau, R. (1965b). Seasonal transmission of the elm disease by *Hylurgopinus rufipes* in Quebec. *Canad. J. Bot.* 43: 1592-1595.

Readio, P. A. (1935). The entomological phases of the Dutch elm disease. *J. Econ. Entomol.* 28: 341-352.

Rosinski, M. A. (1961). Development of the ascocarp of *Ceratocystis ulmi. Amer. J. Bot.* 48: 285-293.

Schreiber, L. R., and Peacock, J. W. (1974). Dutch elm disease and its control. *U. S. Dep. Agric., Agric. Inf. Bull.* 193. 15 pp.

Schwarz, M. B. (1922). Das Zweigsterben der Ulmen, Trauerweiden und Pfirsichbäume. *Meded. Phytopath. Lab. "Willie Commelin Scholten," Baarn*, 5: 1-73.

Sinclair, W. A., and Campana, R. J., eds. (1978). Dutch elm disease: perspectives after 60 years. *New York Agr. Exp. Sta. Cornell Univ., Ithaca, N. Y., Search-Agr.* 9: 52 pg. 460 references.

Spierenburg, D. (1921). Een onbekende ziekte in de iepen. *Tijdschr. Plantenziekten* 27: 53-60.

Thomas, J. B. (1971). Smaller European elm bark beetle found in Sault Ste. Marie, Ontario. *Bi-mon. Res. Notes, Canad. For. Serv.* 27:3.

Thompson, H. E., and Matthysse, J. G. (1972). Role of the native elm bark beetle, *Hylurgopinus rufipes* (Eichh.), in transmission of the Dutch elm disease pathogen, *Ceratocystis ulmi* (Buisman) C. Moreau. *Search Agric. (Cornell Univ. Agric. Exp. Stn.)* 2(1): 1-16.

Touhey, J. G., and Bray, D. E. (1961). An evaluation of certain compounds as feeding deterrents against the smaller European bark beetle *Scolytus multistriatus. J. Econ. Entomol.* 54: 195-196.

Truchan, J., and Butcher, J. W. (1970). Cold hardiness of *Dendrosoter protuberans. J. Econ. Entomol.* 63: 328-330.

Tyler, L. J., Parker, K. G., and Pechuman, L. L. (1939). The relation of *Saperda tridentata* to infection of American elm by *Ceratostomella ulmi. Phytopathology* 29: 547-549.

Walker, C. (1973). Interception of the North American native elm bark beetle (*Hylurgopinus rufipes* (Eichh.)) imported from Canada into Britain in logs of rock elm (*Ulmus thomasii* Sar.). *Plant Pathol.* 22: 147.

Wallace, P. P., and Beard, R. L. (1942). Larval characteristics of certain elm bark infesting Coleoptera. *Canad. Entomol.* 74: 86-88.

Walter, J. M. (1935). Technique advantageous for the isolation of *Ceratostomella ulmi* from bark beetles. *Phytopathology* 25: 37-38 (Abstr.).

Whitten, R. R. (1941). The internal application of chemicals to kill elm trees and prevent bark beetle attack. *U. S. Dep. Agric. Circ.* 605. 11 pp.

Whitten, R. R. (1942). Toxic and repellent sprays for the control of elm bark beetles 1942. *U. S. Dep. Agric. Circ*. 647. 12 pp.

Whitten, R. R., and Parker, D. E. (1945). Experimental control of shade tree insects with DDT. *Nat. Shade Tree Conf. Proc*. 21: 13-17.

Williams, L. H., and Brown, H. E. (1969). Some biological investigations of the smaller European elm bark beetle in Missouri with reference to systemic insecticidal control. *J. Econ. Entomol.* 62: 1381-1386.

Chapter 9

THRIPS

T. N. Ananthakrishnan

Zoological Survey of India
Calcutta, India

9.1	INTRODUCTION	149
9.2	THRIPS AND GALL FORMATION	150
9.3	THRIPS AS VECTORS OF BACTERIAL, FUNGAL AND VIRAL PATHOGENS	156
	9.3.1 Bacteria	156
	9.3.2 Fungi	157
	9.3.3 Viruses	158
9.4	ACKNOWLEDGMENT	161
9.5	REFERENCES	161

9.1 INTRODUCTION

Thrips or fringe wings belonging to the order Thysanoptera include mostly phytophagous and mycophagous species which are often known to influence the development of disease in plants by acting as causal agents of dissemination of pathogens like bacteria, fungi and viruses. Any disease is a biological process resulting from the interaction of two or more organisms influenced by environmental factors. This naturally involves the number, population periodicity or occurrence of vectors, as well as the favorable or unfavorable conditions for multiplication, which are vital for an assessment of the seriousness of the disease from the economic view point, not overlooking the fact that insects can cause sudden outbreaks of diseases. The causation of these diseases may be directly through the influence of salivary secretions which may be toxic to the plant tissues resulting in physiological disturbances in plants or indirectly through pathogen dissemination, inoculation and ingression (Leach, 1940). Phytotoxaemias, as the former are often designated, frequently result in local lesions at the insect's feeding point, with or without local dif-

ISBN 0-12-326450-2

ferentiation. All the cells in the vicinity of the feeding punctures will, to a limited extent, be subject to the influence of the components of salivary secretions. And some of these components may be translocated, thus leading to considerable physiological disturbances in the plant and consequential disease symptoms. The incidence and severity of these disease symptoms appear invariably directly correlated with the number of insects feeding (Carter, 1973). Several thrips are known to cause direct injury to several crop plants and other vegetation without dissemination of pathogens. Serious damage may result through both feeding and ovipositing by the terebrantian thrips in particular, with consequent mechanical injury to growing tissues as well as chemical injuries to adjacent cells, possibly due to the toxic effects of salivary substances. One of finest examples of direct injuries by thrips is provided by the cardamom thrips (*Sciothrips cardamomi* Ramk.). The tender panicle stalks subject to attack by heavy populations of thrips become stunted and do not bear the full complement of flowers; the feeding injuries result in flower and immature pod sheddings, reduction in pod number and scabbing of pods. Instances of damage to crop plants by such direct attacks of thrips are many (Ananthakrishnan, 1969, 1971, 1973; Bournier, 1970; Lewis, 1973). In view of the emphasis herein laid on typical phytotoxaemic effects as illustrated by such malformations as galling, the relevant information on the direct damage potential by thrips feeding and oviposition is presented as a tabular statement in Table I.

Besides the role of thrips in gall formation, instances of transmission of bacterial and fungal pathogens by thrips are reported, the pathogens being mechanically transmitted, in many cases without any specificity. However, distinct vector-pathogen relationships do exist in the case of transmission of viruses. The two principal aspects discussed here pertain to (a) thrips in relation to gall formation in plants and (b) thrips as vectors of phytopathogenic bacteria, fungi, and viruses.

9.2 THRIPS AND GALL FORMATION

Growth abnormalities characterized by eccentric cell growth (hypertrophy) and multiplication (hyperplasia) are typical of galled tissues. Gall formation is a phase in the induction of plant disease: the plants reacting to the stimulus induced by the feeding of thrips; the tissues responding to the stimulus only at a time when tissue differentiation has not yet commenced. The degree of hyperplasia and hypertrophy vary with different plants. The form of the gall and its position on the host plant are characteristic of different thrips species. Invariably the species may be identified by the shape of the gall. A large proportion of thrips galls are produced in the bud condition or in tissues developing from buds. Though thrips galls are always associated with the development of the immature stages of the concerned species, heavy population of both adult and immature stages are invariably present during gall development. Tissue changes associated with gall development may be *kataplasmatic* with little or no differentiation, or *prosoplasmatic* with well defined tissues, with distinct protective, mechanical and nutritive tissues. The protective tissues often de-

TABLE I. Some Major Thrips Pests and Effects of Infestation

Species	Common Name	Effects	Distribution
Limothrips cerealium Haliday	cereal thrips	failure of florets to develop	U.K., Europe and U.S.A.
Limothrips denticornis Haliday	cereal thrips	gooseneck-like stalks in barley	U.K. and Europe
Anaphothrips obscurus (Muller)	oat thrips	sterility in oats	Europe
Haplothrips tritici Kurdj.	wheat thrips	damage to developing ears	Europe
Kakothrips robustus Uzel	pea thrips	prevents pod formation or with small sickly undersized pods	Europe
Taeniothrips meridionalis Priesner	apple, peach thrips	fruit malformation	Europe
Taeniothrips inconsequens (Uzel) *Thrips imaginis* Bagnell *Thrips madroni* Moulton	pear/apple blossom thrips	prevents fruit formation	U.S.A.
Caliothrips fasciatus (Perg)	pear thrips	dropping of leaves and scars on fruit	U.S.A.
Scirtothrips aurantii Faure	citrus thrips	scarring of fruits, rat tailing of twigs, and failure of buds to develop	Africa
Thrips angusticeps Uzel	flax thrips	growth retardation and poor fiber source	Europe
Selenothrips rubrocinctus (Giard)	cocoa thrips	serious damage to leaves and pods	Trinidad, Gold Coast and India
* *Thrips tabaci* Lind.	onion/cotton thrips	retardation of growth	cosmopolitan
* *Scirtothrips dorsalis* Hood	chilli thrips	curling and shedding of leaves, fresh buds become brittle and drop down	India
Scirtothrips bispinosus (Bagnall)	coffee thrips	leaves turn leathery, brittle and crinkled; buds badly damaged; rosette growth	India
Diarthrothrips coffeeae Bagnall	coffee thrips	serious defoliation	Africa
Sciothrips cardamomi (Ramk.)	cardamom thrips	shedding of flowers and immature pods and scabbing of pods	India
* *Rhipiphorothrips cruentatus* Hood	grapevine thrips	withering, curling and dropping of leaves	India
Baliothrips biformis (Bagnall)	paddy thrips	sickly, blighted, faded seedlings with inward curling of leaves	India and S. E. Asia
* *Caliothrips indicus* (Bagnall)	groundnut thrips cotton thrips	tender foliage becoming pale, spotted and sickly	India, Egypt
Caliothrips sudanensis Bagnall	cotton thrips	damage to seedlings	Egypt, India

* indicates highly polyphagous species.

velop trichomes, glands, hairs, etc., in excessive quantities. The mechanical tissues with sclerenchyma cells, heavily lignified fibers and nutritive zones with thin-walled cells and dense protoplasmic contents are characteristic of many galls (Meyer, 1969). Most thrips form *organoid* galls retaining structurally and anatomically very close similarity with the affected organs. Complex galls lose identity with the affected organ and are known as *histoid* galls.

Malformations on *Galium verum, Stellaria media, Vernonia chamaedrys* caused by *Physothrips basicornis* (Reuter), *Frankliniella intonsa* (Trybom) and *Thrips tabaci* Lind., respectively, were reported at the turn of this century. Grevillius (1910) described the anatomical peculiarities of the galls on *Stellaria media* in Europe caused by *Tmetothrips subapterus* Haliday, *Taeniothrips atratus* (Haliday) and *Thrips tabaci*, and considered them as pseudocecidia. Karny *et al.* (1913, 1914-16) and van Leeuwen-Reijnvann *et al.* (1911, 1913) reported galls from 31 natural orders and 56 species of thrips from about 68 host plants from Java, primarily among Moraceae, Urticaceae, Piperaceae, Euphorbiaceae, Vitaceae and Myrtaceae. Though most of these are dicots, the Liliaceae, Araceae and Gramineae have also been implicated. Galls induced by *Physothrips* species on Pteridophyta in the Netherlands Indies, cause rolling of leaf margin on *Asplenium nidus* and *Pleopeltis pteropus*. Karny (1921) cites gall-like anomalies on *Thunbergia fragrens* by *Anaphothrips involvens* Karny and on tea leaves by *Danothrips theivorus* (Karny). Bagnall (1928, 1929) reported gall thrips in *Ficus* from Samoa, Tonga and Australia. Baudys and Kratochvil (1941) report numerous terebrantian galls from Czechoslovakia implicating *Taeniothrips pini* (Uzel) on different species of pines, *Taeniothrips salacis* on *Salix alaba, Taeniothrips atratus* on *Stellaria* spp., *Haplothrips aculeatus* (Fabricius) on *Vernonica chamaedrys*, and *Taeniothrips firmus* (Uzel) on *Vicia cracca*. Wahlgren (1945) reports similar galls from Europ. Mani (1948, 1964, 1973) reviewed available information on thrips galls and gall thrips and provided further information on some host plants and gall anatomy. Doctors W. and J. van Leeuwen-Reijnvann (1956) discuss the aetiology of ceratonean thrips galls found on leaves of the Malaysian *Schefflera* species *S. scandens* (Bl.), *S. divaricata* (Bl.), *S. elliptica* (Bl.), *S. polybotrya* Miq., and *S. lucescens* (Bl.) and implies that these galls result from chemomorphosis. Knechtel (1951) considers the anomalies caused by *Taeniothrips atratus* (Haliday) on *Silene vulgaris*, *Taeniothrips nicipes* (Zetter-st.) on *Digitalis ambigua* and *Thrips tabaci* on *Nicotiana tabacum* as galls. Franssen and Mantel (1967) record well developed leaf roll galls caused by *Anaphothrips euphorbiae* Uzel on *Euphorbia cyparissias* and by *Anaphothrips silvarum* Priesner on *Galium verum*. Over the last quarter of a century our knowledge of Indian gall thrips has been considerably enhanced due to investigations by Ananthakrishnan (1949-1978) and associates who described more than a hundred species, the dominant gall families being Moraceae (*Ficus* sp), Melastomaceae (*Memecylon* sp), Araliaceae (*Schefflera* sp) and to some extent Combretaceae, Loranthaceae, Oleaceae, Myrtaceae and Rubiaceae. Mound (1971) provides detailed studies on the galls and gall thrips of *Acacia* in Australia.

Thrips galls are almost exclusively leaf galls, very rarely bud, flower, and stem galls. Though most galls conform to the leaf roll and leaf fold types, simple sac-like structures to highly convolute pouch galls also occur. The complexity invariably depends on the thrips population within the gall, the populations often varying with seasons, types of galls and stage of development of the gall. Simple tubular or siphonous galls of *Schefflera racemosa* may have one to two individuals in each gall, but a complex gall of *Calycopteris floribundus* harbors anywhere between 200-5000 individuals in galls measuring 2-8 cm wide. The largest thrips gall so far on record is 15 cm in diameter, from Solomon Islands, produced by *Tolmetothrips granti* with a population of over 10,000 individuals (Mound, 1970). Besides these, some of the more striking thrips galls include those of *Terminalia chebula*, *Acacia leucophloea*

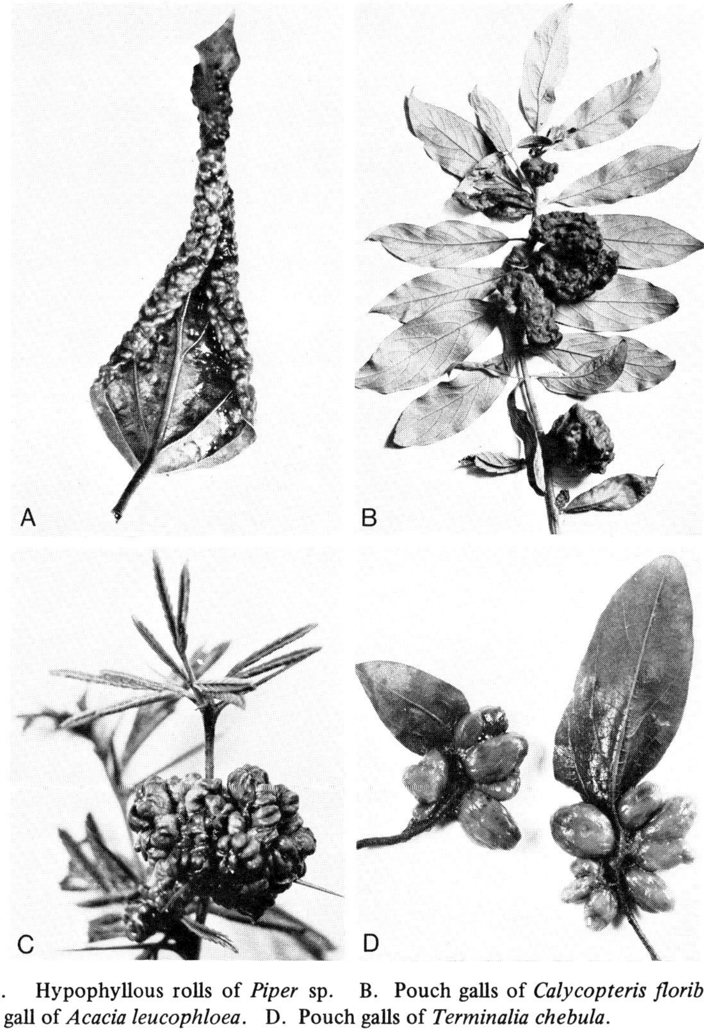

FIG. 1. A. Hypophyllous rolls of *Piper* sp. B. Pouch galls of *Calycopteris floribundus*. C. Rosette gall of *Acacia leucophloea*. D. Pouch galls of *Terminalia chebula*.

and other species of *Acacia*, *Piper* spp., *Memecylon* spp., horn galls on various species of *Schefflera*, and stem galls on *Casuarina stricta* (Figs. 1 and 2).

Detailed classification aspects of thrips galls are discussed by Karny *et al.* (1913), Mani (1964) and Ananthakrishnan (1969; 1978). The leaf roll-leaf fold galls with their diverse modifications and the nonleaf roll-fold galls involving several categories from the small epidermal emergence galls to large bladder or pouch galls are indicated in Table II. About 50% of the leaf galls are leaf rolls of one kind or another, while 25% are leaf folds. Leaf rolls, folds and curls are associated with more or less pronounced swellings of the lamina, and increased thrips populations often result in surface tuberculations of the galls.

Maresquelle (1937) divided the process of gall development into three fundamental phases, viz., the inhibition of normal development, cellular hypertrophy and adaptive phenomena. Thrips galls do not reveal striking abnormalities as do galls induced by other insects, yet they do seem to involve inhibition of normal development and cellular hypertrophy; and the adaptive phenomena appear to be well expressed as evidenced by the morphological manifestation of thrips galls.

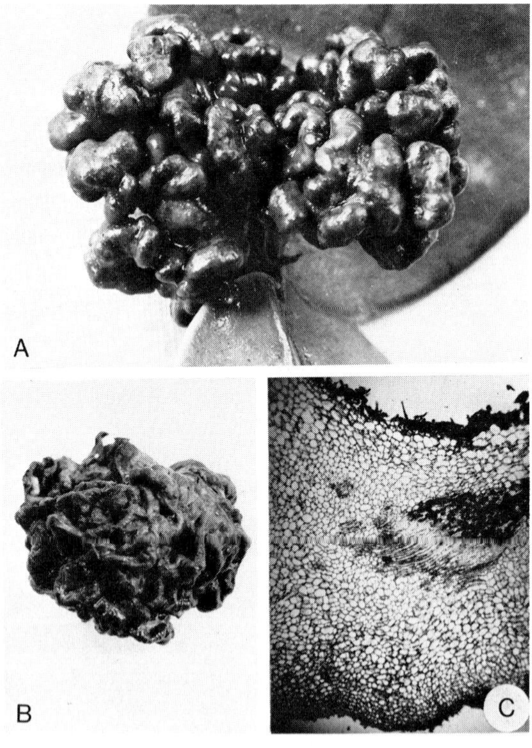

FIG. 2. A. Gall of *Memecylon lushingtoni* (highly enlarged). B. A single enlarged pouch gall of *Calycopteris floribundus*. C. Section of *Calycopteris* gall showing increase in parenchymatous layers and scattered vessels.

TABLE II. Classification of Major Thrips Gall Types

Gall Types	Host Plants	Thrips Species	Distribution
A. Leaf Roll-Fold Galls			
marginal roll/fold (partial folds)	*Maytenus senegalensis* (Lam.)	*Alocothrips hadrocerus* (Karny)	India
complete marginal folds	*Piper nigrum* Linn.	*Liothrips karnyi* (Bagnall)	India
	Anogeissus sp.	*Lygothrips jambuvasi* (Ramk.)	India
typical leaf fold (two margins meeting at middle)	*Mimusops elengi* Linn.;	*Arrhenothrips ramakrishnae* Hood	India
	Ficus retusa Linn.	*Arrhenothrips dhumrapaksha* Ramk.	India
	Ficus benjamina Linn.	*Gynaikothrips uzeli* Zimm.	India, S. E. Asia
		Mesothrips jordani Uzel	India, Java, Malaya, Hong Kong
leaf roll (one edge rolled over other)	*Casearia tomentosa* Roxb.	*Gynaikothrips flaviantennatus* Moulton	India
epiphyllous roll (leaf blade on either side of midrib rolling inwards)	*Ventilago maderaspatana* Gaerten	*Schedothrips orientalis* Ananthakrishnan	India
	Memecylon lushingtoni Gamble	*Byctothrips ayyari* Ananthakrishnan	India
	Eugenia axillaris Vell.	*Mixothrips craigheadi* Stannard	Florida
hypophyllous roll	*Pavetta hispidula* Linn.	*Teuchotrips longus* Schmz.	India, Ceylon
spiral leaf roll	*Jasminum multiflorum* (Burm-f.)T. Anders	*Eothrips coimbatorensis* Ramakrishna	India
B. Nonleaf Roll-Fold Gall			
bladder /pouch galls	*Calycopteris floribundus* Lamk.	*Austrothrips cochinchinensis* Karny	India
	Terminalia chebula	*Dixothrips onerosus* Ananthakrishnan	India
	Acacia pendula A. Cunn. Ex G. Don	*Kladothrips rugosus* Froggatt	Australia
phylloclade galls	*Acacia harpophylla* Acacia melanoxylon Br.	*Kladothrips acaciae* (Moulton)	Australia
		Koptothrips zelus Mound	Australia
rosette gall	*Acacia leucophloea* (Roxb.) Willd	*Thilakothrips babuli* Ramakrishna	India
blister galls	*Memecylon* spp.	*Crotonothrips gallarum* Ananthakrishnan	India
		Crotonothrips coorgensis Ananthakrishnan	India
large bud galls	*Myrcia* sp.	*Phrasterothrips conducans* Priesner	Brazil
horn galls-ceratonean galls	*Schefflera racemosa* Harms.	*Liothrips associatus* Anan. and Jag.	India
		Liothrips ramakrishnae Anan and Jag.	India
stem galls	*Casuarina stricta* Miq	*Thaumatothrips froggatti* Karny	Australia
	Microgyne macrophylla	*Eurynchothrips flavicornis* Bagnall	Sierra Leone
small subspherical stem galls	*Acacia aneura* F.	*Onychothrips tepperi* (Uzel)	Australia

Inhibition of normal development or hypoplasia has been known in the thrips galls of *Ficus benjamina* and *Fagraea litoralis* (Karny *et al.*, 1913). Thrips galls are induced during the process of differentiation of the concerned organ (in many cases, the leaf0. The feeding injury induces meristematic activity in the differentiating tissue, resulting in a gall usually comprised of isodiametric, closely arranged, parenchymatous cells. Mature galls induced by thrips, particularly as in the epiphyllous roll galls on the leaves of *Casearia tomentosa* made by *Gynaikothrips flaviantennatus*, significantly differ from the normal leaf structure, lacking the routine developmental pattern indicating the inhibition of normal development. Further, with continued and differential feeding action of the cecidogenous thrips, it has been observed that in the gall laminae of *Casearia tomentosa*, cell divisions are predominantly periclinal in the mesophyll adjoining the inner lining of the gall chamber contributing to the bulk of the gall; whereas the divisions inside appear predominantly anticlinal, thus facilitating the inrolling of the lamina. This aspect is inferred to be a unique condition in thrips galls involving considerable reorientation of the morphogenetic activity. In the galls induced by *Alocothrips hadrocerus* on *Maytenus senegalensis*, *Arrhenothrips ramakrishnae* on *Mimusops elengi*, *Aneurothrips* sp. on *Cordia obliqua*, and *Cordia suaveolens* and *Teuchothrips longus* on *Pavetta hispidula* moderate epidermal hypertrophy (a feature restricted to thrips galls) has been known. An enormous increase in the mesophyll cells from 45-75 layers in the galls of *Calycopteris floribundus*, 25-40 layers in the pouch gall of *Terminalia chebula*, 17-60 layers in *Pavetta hispidula* is indicated in gall tissues. Curiously enough, in the ceratonean galls of *Schefflera racemosa*, the simple galls show 7-12 layers, whereas the more fleshy complex galls have 23-35 layers of closely arranged parenchyma cells. Although, as mentioned before, adaptive phenomena such as the differentiation of a highly specialized nutritive tissue and irrigating vascular strands are not frequent, occasional instances of sclereid differentiation and tannin accumulation in the mature galls of *Cordia* and *Maytenus* have been recorded. Development of small papillose hairs on the galled leaves which are otherwise absent on the normal leaves of *Cordia* appear significant. Variations in stomatal shape, size, distribution and frequency have been observed in the ceratoneon pouch galls of *Schefflera racemosa* induced by *Liothrips* spp.

9.3 THRIPS AS VECTORS OF BACTERIAL, FUNGAL AND VIRAL PATHOGENS

9.3.1 Bacteria

The role of thrips as vectors of bacterial pathogens is very limited (Bailey, 1935; Sakimura, 1947). Mention was made by Pussard-Radulesco (1930) that thrips act as disseminators of pathogens, particularly *Bacterium dianthi* Arthur and Bailey on carnations and *Alternaria* sp. on tobacco. Buchanan (1932) reported *Hercinothrips femoralis* (Reuter) associated with a bacterial disease of plants on which the insects had caused feeding injuries.

Larvae of *Frankliniella moultoni* Hood and *Taeniothrips inconsequens* (Uzel) have been known to carry *Erwinia amylovora* (Burrill) (=*Bacillus amylovorus*) responsible for fire blight of pears (Thomas and Ark, 1934; Bailey, 1935). Buchanan (1932) conclusively demonstrated that the bacterial haloblight of beans (*Pseudomonas medicaginis* var., *Phaseolicola* [Burkh.] Stappe and Kotte) was transmitted by *Hercinothrips femoralis* (Reuter), since the lesions were always associated with feeding injuries. However, Leach (1940) thought that transmission was mechanical. Bailey (1935) noted spores of the black mold *Aspergillus niger* Van Tieg on the body of adults and larvae of onion thrips, *Thrips tabaci*. Johnson (1956) observed that pod twist of beans caused by *Pseudomonas flectans* John, was transmitted by *Taeniothrips nigricornis* Schmutz. Large populations of these thrips were observed on wild and cultivated hosts, the important reservoir of the pathogen being the wild *Phaseolus lathyroides* L., so that natural infestation was widespread. Transmission tests resulted in a distortion of bean pods.

9.3.2 Fungi

Very few thrips-borne fungal diseases have been experimentally demonstrated, even though casual, purely observational records exist of association of thrips with fungal pathogens. As in the case of bacteria, thrips transmit fungal pathogens mechanically. Leach (1940), Bailey (1935), Sakimura (1941), Carter (1973) and Ananthakrishnan (1973) provided some relevant information about the vector role of thrips.

Johnson (1911) observed that the rust spores of *Puccinia graminis tritici* Eriks and Henn., *P. rubigo-vera tritici* [Eriks] (Carlton), were commonly carried by *Frankliniella tritici* (Fitch). Howard (1923) first reported *Cryptothrips floridensis* Watson disseminating spores of *Pestalozzia* sp., causing disease on raw camphor. Ramakrishna Ayyar (1928) recorded the association of *Euphysothrips minozii* Bagnall (= *E. fungivora* Ramk.) with wheat rust. This was subsequently confirmed by Ananthakrishnan (1973) who experimentally transferred *E. minozii* from rust infested to healthy plants. Hansen (1929) reported that *Aspergillus niger* was carried by thrips into green figs. Subsequently Hansen and Davey (1932) showed that thrips entered the figs and carried the organisms of smut.

Culturing of thrips on infested figs revealed the presence of various species of fungi, bacteria and yeasts (Hansen, 1929). Experiments focusing on the individual disease potential by thrips alone indicated that nearly 16.6% of the 1557 experimentally sealed cyconia showed the accumulation/occurrence of smuts and molds, ruling out the possibility of larger insects as possible vectors. Thrips appeared as powerful vectors along with mites, as nearly 60% of the population carried spores (Hansen and Davey, 1932). The order of the frequency of occurrence as observed from cultures was as follows: miscellaneous fungi, bacteria, *Hormodendrium*, *Aspergillus*, *Pencillium*, *Alternaria*, *Rhizopus*, *Acrostalgamus*, and a few yeasts. The causal organisms were transmitted by thrips according to Caldis (1927), Hansen (1929), Smith and Hansen (1931), Hansen and Davey (1932), and Davey and Smith (1933).

The following species of thrips were involved: *Thrips tabaci* Lind., *Frankliniella tritici* (Fitch), *F. moultoni* Hood., *F. occidentalis* (Perg.) and *Leptothrips mali* (Fitch). *Puccinia graminis* (Pers.) can be experimentally transmitted by *Hercinothrips femoralis* after the insects contact uredospores on infected plants (Granovsky and Levine, 1932).

Experimental transmission of azalea flower spot, *Ovulina azaliae* Weiss, by *Heterothrips azaliae* Hood, was reported by Weiss and Smith (1940) and Smith and Weiss (1942). Yarwood (1943) reported *Thrips tabaci* in heavily mildewed leaves of grapes, but the absence of thrips on leaves with dead mycelia. This association appears to be widespread, and thrips are known to feed on and destroy established mildew colonies and to aid in the mechanical dispersal of the fungus. Thrips are also implicated as vectors of ergot, visiting infected flowers and carrying the fungus to healthy flowers (Langdon and Champ, 1954). *Euphysothrips (Megaphysothrips) subramanii* (Ramk.) and *Taeniothrips xanthoceros* are known to aid in the dispersal of the uredospores of *Hamileia vastatrix*, the coffee leaf rust, in India and South Africa, respectively (Ananth and Chokkanna, 1961; Crowe, 1963; Ananthakrishnan, 1971, 1973). *Frankliniella intonsa* is reported to carry the spores of *Ustilago maydis* Do. Tul., in Italy (Grancini, 1963).

9.3.3 Viruses

Our knowledge of the vector-virus relationships of thrips appears to be more complete than that of the bacterial and fungal diseases; and, as earlier mentioned, the relationship is biologically more intimate and specific. Bailey (1935), Sakimura (1947, 1962a and b, 1963, 1970), Best (1968) and, to a limited extent, Mound (1973) provide excellent reviews on varied aspects of the vector role of thrips. Positive confirmation pertains only to one virus, the tomato spotted wilt virus (TSWV). Evidence appears incomplete regarding the vector role of *Scirtothrips manihoti* (Bondar), *Liothrips pistachiae* Krentazb and *Thrips tabaci* in relation to the manihot mosaic virus of Brazil, the pistachio virus of Southern Russia, and the sunflower mosaic virus of Argentina (Travesi, 1949; Sakimura, 1962). Negative cases of transmission have been equally well analyzed by Sakimura (1947); more than forty different plant viruses have been tested for transmission by thrips with negative results. *Thrips nigripilosus* Uzel, *Chaetanaphothrips orchidii* (Moulton), *Hercinothrips femoralis* (Reuter), *Kurtomathrips morrilli* Hood., *Parthenothrips dracaenae* (Heeger) and *Frankliniella intonsa* (Trybom) proved to be non-vectors (Sakimura 1947). In contrast to aphid or leafhopper-borne viruses, the vector specificity of TSWV is very narrow.

Phytophagous insects, in general, show a distinct tendency towards systematic oligophagy feeding on a number of related species of host plants, and this is equally true of many species of thrips. Strict monophagy is not uncommon, as well as polyphagy, and there can be no better examples than *Thrips tabaci* Lind., occurring all over the tropical and temperate zones, and the equally cosmopolitan *Frankliniella schultzei* which are recognized vectors of the TSWV and known to feed on a wide variety of host plants all over the world. Different thrips species show variability regarding host preferences and feeding habits. In transmission tests, good source plants are essential, and often the natural host plants are best suited. The essential criterion for a good virus source is provided by plants supplying good nutrition for larvae, non-necrotic reaction from infection and systematic, prolonged infection with high virus titer (Sakimura 1961). *Emilia sonchifolia* (Compositae), widely distributed, appears to be the most favorable host plant for several thrips species. Sakimura (1961) has discussed in detail the techniques for handling thrips in transmission experiments with TSWV.

The nature of the mouth parts of thrips appears eminently suited for sucking. Mound (1971) demonstrated that the maxillary stylets, which are "C-like" in section, fit together to form a tube with a subapical aperture. While feeding, the mesophyll tissues alone, not the vascular tissues, are affected. Two basic types of feeding by thrips are recognized–the shallow type restricted to epidermal tissues or a few layers of mesophyll, and the penetrating type into deeper mesophyll tissues. Thrips feed on a wide range of host plants. Therefore, the seasonal incidence of the vector species is of importance in view of the proven ability of certain host plants to thrive in different seasons. This has its impact on the vector species, in particular on the length of the life cycle as well as the production of color variation. For example, the summer forms of *Thrips tabaci* are more yellow and winter forms darker, with intergrades. The length of the life cycle on an average is 17-18 days in summer, 38 days in winter, and adults may vary from 22-123 days (Sakimura 1937). Further, geographic strains also exist, resulting in slightly different biotic potentials. A knowledge of the food habits, life cycle, population density and host range are important aspects in the study of vector potential of thrips species in transmitting viruses. The virus, vector and the host plant constitute an inseparable ecological trinity (Carter, 1939), and the interactions between these and the physical and biotic environment are naturally complex. The size of a population may depend on climatic factors and nutritional status. The variability of vector species in relation to biotic potential, color, nature of wing, i.e., macropterous/brachypterous, or larger or smaller forms, as well as the comparative ability of both sexes to transmit the virus, are important aspects in vector-virus studies. The role of thrips as vectors of TSWV has been very well established. The known four species belong to the genera *Thrips* and *Frankliniella* i.e., *Thrips tabaci* Lind, *Frankliniella schultzei* (Trybom), *Frankliniella fusca* (Hind), and *F. occidentalis* (Perg.). All these species are known to have color variants (Sakimura, 1962, 1963). Whereas the dark and light forms of *Thrips tabaci* and *F. occidentalis* do not differ in their ability to transmit viruses, in *Frankliniella schultzei* the pale forms are unable to transmit the virus, and in the other species the comparative efficiency of color variants are unknown (Sakimura, 1969; Mound, 1973). In *Frankliniella occidentalis*, the males of both color forms are as effective as females in transmitting the virus (Sakimura, 1962b). The vector-virus relationships of these four vector species have common characteristics: (a) the inability of adults to acquire the virus, (b) a well defined latent period, and (c) long retention period. Host preference, reproductive potential and resistance to ecological factors vary with the different vector species.

Thrips tabaci has been reported as (a) vector of tobacco ring spot virus under laboratory conditions (Messieha, 1969). The TSWV was recognized in 1906. Since then, various strains of the virus have been reported from all over the world as being responsible for diseases of crops and ornamental plants, and, when occurring in weeds, as serving as sources of infection to cultivated crops. *Frankliniella fusca*, a well-known pest on tobacco, peanut and cotton in parts of Canada, United States and Mexico, has a very wide host range among cultivated plants and weed species. This thrips has been reported as a potential vector of spotted wilt disease of tobacco (Sakimura, 1963). Successful virus acquisition is accomplished following shallow feeding by *Thrips tabaci*, involving direct application of the mouth cone on the leaf surface after maceration of leaf surface. TSWV has been known under different names such as the tomato bronzing virus, *Lycopersicum* virus 7, and Makhorka tip chlorosis virus (Sakimura, 1962). Of the four vector species, the proven major vector in field transmission is *Thrips tabaci*, followed by *F. schultzei*, *F. occidentalis* and *F. fusca* in the order cited. Both *tabaci* and *schultzei* are known to transmit every strain of TSWV all over the world. The TSWV has a wide host range comprising 166 species (mostly dicotyledons) from 34 families. Best (1968) described the symptoms of TSWV necrotic lesions (primary and systemic), necrotic ring spotting, non necrotic ring spotting, vein or net necrosis, non necrotic vein cleaning, yellowing and chlortic mottling. Carter (1939) provided interesting information on the migration of *Thrips tabaci* in relation to the incidence of infection. He demonstrated that populations of *Thrips tabaci* on TSWV diseased weeds, such as *Emilia sonchifolia*, are consistently higher than on healthy plants, so that the infected plants are more suitable hosts than healthy ones. An interesting pattern has been the movement of *Thrips tabaci*, the major vector: brought on *Emilia sonchifolia* to Hawaii, *Portulaca* to Brazil, chicory and *Dahlia* tubers to India. The thrips vector-virus relationship appears to be the result of a long period of evolution and adaptation.

Some of the economically important plants susceptible to TSWV are spinach, broad bean, potato, eggplant, bell pepper, tomato and chicory. TSWV is readily juice transmissible, and purified viral particles are isometric, 85 μm in diameter (Black, 1955).[1] Moreover, artificial transmission is obtained by a slight rubbing movement. Virus transmission occurs only when thrips feed on infested plants in the larval stage, and only such larvae can transmit the virus as adults. It is a general rule in insect transmission that the younger the plant, the more readily it is inoculated and the shorter the incubation period (Black, 1954). Generally there is an incubation period of several days between the time of virus acquisition and its transmission to healthy plants (Bald and Samuel, 1931). In other words, the adults cannot acquire the virus unless they had fed on infected plants as larvae. The virus is often retained by the vector throughout life, and transmission may be erratic. The multiplication of the virus in the vector has not been established. Retention is erratic or inconsistent, and this appears to be related to the amount of virus acquired by the insect. It is this inconsistency in the pattern of retention that has given thrips-borne viruses the status of persistent viruses. The maximum period of retention is 22-24 days for *F. schultzei*, 30 days for *F. occidentalis* and *T. tabaci*, and 43 days

[1] Actual dimensions, according to more recent measurements, are 70-80 μm.

for *F. fusca*. The only difference between the persistent thrips-borne virus and most persistent aphid-borne viruses is that TSWV is readily sap transmissible and there is no involvement of vascular tissues (Sakimura, 1962). Retention periods appear variable in relation to the length of the post infection period as well as of transmission. The inconsistant pattern seems analogous to those of circulative non-propagative viruses (Sakimura, 1963). Smith (1932) and Linford (1932) record 5 and 10 days as minimal for *Thrips tabaci*, whereas Bailey (1935) observed a period of about 10 days for *F. occidentalis*. The retention period of the virus differs among species. Sometimes, but not always, the four species can retain infectivity throughout life. Infectivity persists throughout the population, but is not transovarial (Best, 1968).

One of the earlier reasons put forward for nonacquisition of the virus by adult thrips has been the inability of the virus to penetrate the gut wall (Sakimura, 1947). Recent experiments on gut penetration have shown that the adults and larvae behave alike and that the permeability of the midgut wall does not differ (Best, 1968). Day and Irzykiewicz (1954) reported that the pH value of the midgut contents of both larvae and adults is between 5.0 and 5.6, a range that inactivates the virus. These findings require further experimental confirmation. With the increasing tendency towards polyphagism in many species such as *Scirtothrips dorsalis*, *Frankliniella schultzei*, *Thrips tabaci*, and *Megalurothrips distalis*., increased chances of their vector role cannot be ignored. Detailed systematic studies based both on field and experimental investigations of potential vectors might yield interesting results.

9.4 ACKNOWLEDGMENT

I am grateful to Dr. K. F. Harris for preparing composites of the photographs.

9.5 REFERENCES

Ananth, K. C., and Chokanna, N. G. (1961). A contribution to the knowledge of spore dispersal of *Hamileia vastatrix* by certain species of thrips occurring in *Coffea arabica* in South India. *Indian coff.* 25(1): 337-338.

Ananthakrishnan, T. N. (1969). Indian Thysanoptera. *CSIR Zool. Monograph*, 1, 171 pp. New Delhi.

Ananthakrishnan, T. N. (1971). Thrips (Thysanoptera) in Agriculture, Horticulture and Forestry–Diagnosis, Bionomics and Control. *J. Scient. and Ind. Res.* 30(3): 113-146.

Ananthakrishnan, T. N. (1971). Further studies on Indian gall thrips. *Marcellia* 37(3): 111-127.

Ananthakrishnan, T. N. (1973a). "Thrips: Biology and Control." Macmillan, India. 120 pp.

Ananthakrishnan, T. N. (1973b). Mycophagous Tubulifera of India (Thysanoptera: Insecta). *Occasional Publication No. 2, Entomology Research Unit, Loyola College, Madras.* 144 pp.

Ananthakrishnan, T. N., and Jagadish, A. (1969). Studies on Indian gall thrips (Thysanoptera) *Marcellia* 36: 7-104.

Ananthakrishnan, T. N. (1978). Thrips galls and gall thrips. Tech. *Mangr.* No. 1: 1-69. Zool. Surv. India.

Bailey, S. F. (1935). Thrips as vectors of plant diseases. *J. Econ. Entomol.* 28: 856-863.

Bald, J. G., and Samuel, G. (1931). Investigation on 'Spotted Wilt' of tomatoes. II. *Aust. Coun. Sci. Ind. Research Bull.* 54: 2411.

Bagnall, R. S. (1928). On some Samoan and Tongan Thysanoptera, with special reference to ficus gall causers and their inquilines. Insects of Samoa. 7: 55-76, *British Museum (N. H.)*, London.

Bagnall, R. S. (1929). On some new genera and species of Australian Thysanoptera (Tubulifera) with special reference to gallspecies. *Marcellia* 25: 184-204.

Baudys, E., and Kratochvil, J. (1941). Prispevek k poznani nasich Thysanopterocecidii. *Sborn ent. Odd. zemsk. Mus. Praze.* 19: 136-147.

Best, R. J. (1968). Tomato spotted wilt virus. *In* "Advances in Virus Research," Vol. 13 (K. M. Smith and M. A. Lauffer, eds.) pp. 65-146. Academic Press, New York.

Black, L. M. (1954). Parasitological reviews: arthropod transmission of plant viruses. *Exptl. Parasitol.* 3: 72-104.

Bournier, A. (1970). Principaux types de dégâts de Thysanoptères sur les plantes cultivées. *Ann. Zool. Ecol. Anim.* 2: 237-259.

Buchanan, D. (1932). A bacterial disease of beans transmitted by *Heliothrips femoralis* Reut. *J. Econ. Entomol.* 25: 49-53.

Caldis, P. D. (1927). Etiology and transmission of endosepsis (internal rot) of the fruit of the fig. *Hilgardia* 2: 319.

Crowe, T. J. (1963). Possible insect vector of the uredospores of *Hamileia vastatrix* Kenya. *Trans. Br. mycol. Soc.* 46(1): 24-26.

Davey, A. E., and Smith, R. E. (1933). The epidemiology of the fig spoilage. *Hilgardia* 7(13); 537.

David, B. V. *et al.* (1973). On the host range of the mycophagous thrips *Euphysothrips minozii* Bagnall. *Madras Agric. J.* 60(5); 337.

Day, M. F., and Irzykiewicz, H. (1954). Physiological studies on thrips in relation to transmission of tomato spotted wilt virus. *Australian J. Biol. Sci.* 7: 274-281.

Docters Van Leeuwen, W. J., and Karny, H. (1911). Eiringe gallen aus Java Beitrage. *Marcellia* 10: 6593.

Docters Van Leeuwen, W. J., and Karny, H. (1913). Beitrage zur Kenntnis der Gallen von Java, 5, Uber die Javanischen Thysanopterocecidien. *Bull. Jardin Bot. Buitenzorg*, (2), 10: 1-26.

Docters Van Leeuwen, W. W. (1956). The aetiology of some thrips galls found on leaves of Malaysian *Schefflera* sp. *Acta Botanica Neerlandica* 5(1): 80-89.

Frannsen, C. J. H., and Mantel, W. P. (1967). Deux galles, nouvelles pour la Hollande, produite par les piquires d'*Anaphothrips. Entom. Bericht.* 27: 195-197.

Gardner, M. W., and Whipple, O. C. (1934). Spotted wilt of tomatoes and its transmission. *Phytopath.* 24: 11-35.

Grancini, P. (1963). Danni da tripidi al mais. *Maydica* 8: 90-94.

Granovsky, A. A., and Levine, M. N. (1932). The dissemination of cereal rust spores in the greenhouse by terrestial invertebrates. *Phytopath.* 22: 9-10.

Grevillius, A. Y. (1910). Notizen über Thysanopterocecidien auf *Stellaria media* cyr., und *Polygonum Convolvulus* L. *Marcellia* 9: 161.

Hansen, H. N. (1929). Thrips as carriers of fig decaying organisms. *Science* LXIX: 356-357.

Hansen, H. N., and Davey, A. E. (1932). Transmission of smut and molds in fig. *Phytopath.* 22: 247-253.

Howard, N. O. (1923). The relation of an undescribed sp. of *Pestalozzia* to a disease of *Cinnamom camphora* Nee and Eberm. *Phytopath.* 13: 47-48.

Ie, T. S. (1970). Tomato spotted wilt virus. *CMI/AAB Descriptions of Plant Viruses* No. 39.

Johnson, J. C. (1956). Pod twist: a previously unrecorded bacterial disease of French bean *Phaseolus vulgaris* (L). *Queensland J. Agric. Sci.* 13: 127-158.

Karny, H. H. (1911). Über Thrips-Gallen and Gallen-Thrips. *Zentralbl. Bakt. Jena.* (2) 30: 556-572.

Karny, H. H. (1913). Über Gallen bewohende Thysanopteren. *Verh. Zool. Bot. Ges. Wien.* 63: 6-12.

Karny, H. H. (1921). Malaysian Thysanoptera V. Thysanoptera on tea. *Treubia* 2(1): 66-83.

Karny, H., van Leeuwen-Reijnvann, W., and van Leeuwen-Reijnvann, J. (1913). Beiträge zur Kenntnis der Gallen von Java. 5. Über die javanischen Thysanoptero-cecidien und deren Bewohner. *Bull. Jardin bot. Buitenzorg.* (2) 10: 1-26, figs. 1-86.

Karny, H., van Leeuwen-Reijnvann, W., and van Leeuwen-Reijnvann, J. (1914, 1915, 1916). Beiträge zur Kenntnis der Gallen von Java. 2te Mitteilung über die javanischen Thysanopterocecidien und deren Bewohner. *Z. wiss. Insektenbiol.* 19: 210-208, 288-297, 355-369 (1914); 20: 32-39, 85-90, 138-147, 203-210, 249-256, 324-331 (1915); 21: etc. (1916).

Knechtel, N. (1951). Fauna Republicii popular Romane. (Insecta-Thysanoptera). 8(1): 39-42.

Krishnamurthy, K. V., A. Raman., and T. N. Ananthakrishnan (1973). On the morphology of the ceratoneon thrips galls of *Schefflera recemosa* Haras. *Marcellia* 38: 179-184.

Leach, J. G. (1940). "Insect transmission of plant diseases." McGraw-Hill Book Company Inc., New York and London. 615 pp.

Lewis, T. (1973). "Thrips: their biology, ecology and economic importance." Academic Press, Lond. and N. Y. 349 pp.

Linford, M. B. (1932). Transmission of the pineapple yellow spot virus by *Thrips tabaci, Phytopath.* 22: 301-324.

Langdon, R. F. N., and Champ. B. R. (1954). Insect vectors of *Claviceps paspali* in Queensland. *J. Australia Inst. Agri. Sci.* 20: 115-118.

Mani, M. S. (1948). Cecidozoa and zoocecidia from India. *J. Royal Asiatic Soc. Bengal. (sci.)* 14(2): 29-195.

Mani, M. S. (1964). "The Ecology of Plant Galls." Dr. W. Junk Publishers. The Hague. 434 pp.

Mani, M. S. (1973). "Plant Galls of India." Macmillan India, 353 pp.

Miles, P. W. (1968'. Insect secretions in plants. *Ann. Rev. Phytopathol.* 6: 137-164.

Miles, P. W. (1972). Saliva of Hemiptera. *Adv. Insect Physiol.* 9: 183-256.

Maresquelle, H. J. (1937). Les processus fondamentaux de la Cécidogenèse *Ann. Sci. Nat. Bot.* 10e serie. 19: 379-392.

Meyer, J. 1969). Problèmes actuels de Cécidologie. *Bull. Soc. bot. Fr.* 116-445-481.

Messieha, M. (1969). Transmission of tobacco ringspot virus by Thrips. *Phytopathol.* 59: 943-945.

Mound, L. A. (1970). Thysanoptera from the Solomon Islands. *Bull. Br. Mus. (Nat. Hist.) Ent.* 24: 83-125.

Mound, L. A. (1971). Gall forming thrips and allied species (Thysanoptera: Phlaeothripinae) for *Acacia* trees in Australia. *Bull. Brit. Mus. (Nat. Hist) Ent.* 35(9): 387-466.

Mound, L. A. (1973). Thrips and whiteflies. *In* "Viruses and Invertebrates" (A. J. Gibbs, ed.). Vol. 31, pp. 232-242. North-Holland Publ. Co., Amsterdam and London.

Pitman, H. A. (1927). Preliminary note concerning the transmission of 'spotted wilt' of tomatoes by an insect vector (*Thrips tabaci* L.) *J. Council. Sci. & Ind. Res.* (Australia) 1: 73-77.

Pussard-Radulesco, E. (1930). Recherches biologiques et cytologiques sur quelques thysanoptères. *Annales des Epiphytics* 3-4: 179-180.

Ramakrishna Ayyar, T. V. (1928). A contribution to our knowledge of the Thysanoptera of India. *Mem. Dept. Agr. India* 10: 215-316.

Sakimura, K. (1937). The life and seasonal histories of *Thrips taaci* in the vicinity of Tokyo, Japan. *Oyo-Dobutsugaku Zasshi* 9(1): 1-24.

Sakimura, K. (1940). Evidence for the identity of the yellow-spot virus with the spotted wilt virus. Experiments with the vector, *Thrips tabaci, Phytopath.* 30: 281-299.

Sakimura, K. (1947). Thrips in relation to gall forming and plant diseases transmission: a review. *Proc. Hawaii Entomol. Soc.* 13: 59-96.

Sakimura, K. (1961). Techniques for handling thrips in transmission experiements with the tomato spotted wilt virus. *Plant disease Reporter* 45, 10: 766 pp.

Sakimura, K. (1962a). The present status of thrips-borne viruses. *In* "Biological Transmission of Disease Agents" (K. Maramorosch, ed.), pp. 33-40. Academic Press, N. Y.

Sakimura, K. (1962b). *Frankliniella occidentalis* (Thysanoptera: Thripidae) vector of the tomato spotted wilt virus with special reference to colour forms. *Ann. Ent. Soc. Am.* 55: 387-389.

Sakimura, K. (1963). *Frankliniella fusca* an additional vector for the tomato spotted wilt virus with notes on *Thrips tabaci*, another vector, *Phytopath.* 53: 412-415.

Sakimura, K. (1969). A comment on the color forms of *Frankliniella schultzei* (Thysanoptera: Thripidae) in relation to transmission of the tomato spotted wilt virus. *Pacif. Insects* 11: 761-762.

Smith, K. M. (1932). Studies on plant virus disease XI. Further experiments with a ring spot virus and its identification with spotted wilt of tomato. *Ann. appl. Biol.* 19: 305-330.

Smith, R. E., and Hansen, H. N. (1931). Fruit spoilage disease in figs. *Univ. Calif. Agr. Exp. Sta. Bull.* 506: 1-84.

Thomas, H. E., and Ark, P. A. (1934). Fire blight of peas in related plants. *Univ. Calif. Agr. Exp. Sta. Bull.* 586: 18.

Traversi, B. A. (1949). Estudio initial sobre una enfermedad del, Girsol (Helianthus annus). *Argentina Rev. invest. agr. Buenos Aires* 3: 345-351.

van Leeuwen-Reijnvann, W., van Leeuwen-Reijnvann, J., and Karny, H. (1911). Eiringe gallen aus Java Beiträge. *Marcellia* 10: 65-93.

van Leeuwen-Reijnvann, W., van Leeuwen-Reijnvann, J., and Karny, H. (1913). Beiträge zur Kenntnis der Gallen von Java, 5. Über die Javanischen Thysanopterocecidien. *Bull. Jardin Bot. Buitenzorg* (2) 10: 1-26.

van Leeuwen-Reijnvann, W., and van Leeuwen-Reijnvann, J. (1956). The aetiology of some thrips galls found on leaves of Malaysian *Schefflera* sp. *Acta Botanica Neerlandica* 5(1): 80-89.

Wahlgren, E. (1945). Gallibildande Thysanoptera. *Opuscula Ent.* 10: 119-126.

Yarwood, E. E. (1943). Association of thrips with powdery mildrews. *Mycologia* 35: 189-193.

Chapter 10

FLIES

T. A. Zitter*

Agricultural Research and Education Center
University of Florida
Belle Glade, Florida

J. H. Tsai

Agricultural Research Center
University of Florida
Fort Lauderdale, Florida

10.1 INTRODUCTION 165
10.2 DIPTERA AS VECTORS OF DISEASE AGENTS 167
10.3 BIOLOGY OF *LIRIOMYZA* FLIES 170
10.4 MODE OF TRANSMISSION AND NATURAL SPREAD 172
10.5 REFERENCES 174

10.1 INTRODUCTION

The Diptera constitute one of the largest orders of class Insecta. The name "Diptera" derives from their having only a front pair of wings, while the hind wings are reduced and modified to form specialized organs known as halters, serving to maintain equilibrium during flight. The strong powers of controlled flight governed by these organs enable the Diptera to play a unique role in the epidemiology of plant and animal diseases. The vast majority of the Diptera are relatively small, but many are of great economic and medical importance.

The Diptera occupy a wide range of habitats which include crude petroleum pools, numerous aquatic habitats, and decaying organic matters. They also live within the tissues of both insects and plants. This order is divided into three suborders: the Nematocera, Brachycera, and Cyclorrhapha. The Nematocera are primitive and have long and many segmented antennae. They include the midges, gnats, and mosquitoes. The Brachycera and Cyclorrhapha are highly evolved and have

*Present address: Department of Plant Pathology, Cornell University, Ithaca, New York.

ISBN 0-12-326450-2

three to five antennal segments. The adults of Brachycera (straight-seamed flies) emerge through a T-shaped opening of the puparium, whereas the Cyclorrhapha emerge through a circular opening. Other characteristics such as legs, wing venation, chaetotaxy, and head and thoracic sutures are also used in the classification and identification of Diptera. The Cyclorrhapha are further divided into two divisions: Aschiza and Schizophora. The former can be distinguished from the latter by lack of a frontal suture in the shape of an inverted U on the face. The vegetable leafminer *Liriomyza sativae* (Blanchard) (Agromyzidae) belongs to Schizophora, which comprise nearly half the species in the order Diptera.

Many of the blood sucking flies and some of the scavenging flies are important vectors of disease agents of man and animals such as malaria, yellow fever, typhoid fever, filariasis, dysentery, sleeping sickness, and dengue. Some flies attack both cultivated and wild plants. Several species such as hessian fly, apple maggot, fruit flies, root gnats, gall midges, and leafminer flies are important pests of cultivated crops. *L. sativae*, as an example, damages its hosts by stippling and mining cotyledons and leaves. The fly larvae tunnel through the mesophyll layers of the leaf which greatly reduced the surface area available for photosyntheses. Plants such as cucurbits that are severely attacked in the seedling stage may be killed outright. Infested vegetable crops, such as celery, tomato, and lettuce may not only have their yields reduced, but also require costly culling and trimming. Infested ornamentals usually are not saleable due to loss of quality and vigor (Poe and Short, 1975).

The current leafminer crises facing the Florida celery industry clearly demonstrate how quickly this insect rose to prominence. In 1951, Wilson and Hayslip (1951) referred to the serpentine leafminer as a new pest of celery, but by 1973, Guzman *et al.* (1973) considered this leafminer to be the most important celery pest. It remains so to the present. Leafminers were considered minor on all crops prior to the advent of chlorinated hydrocarbon insecticides exemplified by DDT (dichloro-diphenyl-trichloroethane). DDT was particularly severe on the braconid and Chalcidoidea parasites that normally kept the agromyzids under control. By 1956, leafminers had become resistant to toxaphene (Genung, 1957) and by 1957, the flies were showing at least partial resistance to parathion (Wolfenbarger, 1958). By 1961, workers showed resistance to both parathion and diazinon (Genung and Harris, 1961). Tests by Janes (1974) and Janes and Genung (1975) showed resistance to Naled (Dibrom®) and azinphos-methyl (Guthion®). So it has continued to the present day, with researchers wondering how long the pyrethroids will be effective.

Although the leafminers developed resistance to various insecticides, their parasites failed to do so and as a consequence the leafminer problem became more serious. Leafminer immatures are attacked by at least nine species of parasitic Hymenoptera (Stegmaier, 1966), and the most numerous parasitic species is the braconid, *Opius dimidiatus* (Ashmead); most of the other Hymenoptera are representatives of the Chalcidoidea (C. Musgrave, Department of Entomology and Nematology, University of Florida, personal communicaton).

In addition to insecticide applications, several other control measures for serpentine leafminers have been investigated. The possibilities of varietal resistance to adult attack and larval development have been studied in lima beans, tomatoes, celery, and chrysanthemums (Kelsheimer, 1963; Webb and Smith, 1969; Musgrave *et al.*, 1975). Host plant nutrition and cultural methods also seem to affect susceptibility to leafminer attack (Woltz and Kelsheimer, 1958; Price and Poe, 1976).

Since 1974, celery integrated pest management (IPM) has been investigated as an alternative method for leafminer control. This has been a large scale interdisciplinary research effort working with crops of celery (Guzman *et al.*, 1977). By relying on leafminer parasites and predators, and the judicious use of baits or *Bacillus thuringiensis* preparations for lepidopterous larvae control, leafminer control in IPM plots was the same or superior to that achieved in commercial crops.

Apart from the damage inflicted by leafminer larvae and adults, leaf wounding by these insects has proven to transmit several viral agents (Costa *et al.*, 1958; Zitter and Tsai, 1977), and their role in the possible transmission of other plant pathogens such as bacteria is strongly suspect (Zitter, unpublished). Association of damaged tissue with foliar pathogens and organisms that are normally weakly pathogenic is fairly common.

Nearly 97% of the known arthropod-borne plant pathogens including viruses and prokaryotes are transmitted by insects (Harris, 1979). The vast majority of these pathogens (67%) are transmitted by Stenorrhyncha homopteran (aphids, whiteflies, and psyllids), especially aphids, which are the largest group of insect vectors both in terms of species described (Eastop, 1977) and viruses transmitted (Harris, 1977, 1978, 1979; Harris and Maramorosch, 1977). About 30% of these pathogens are transmitted by Auchenorrhyncha homopterans (leafhoppers and planthoppers) (Harris, 1979; Maramorosch and Harris, 1979; Tsai, 1979). The remaining 3% of these are transmitted by insects such as beetles, thrips, and leafminer flies (Harris, 1979; Harris and Maramorosch, 1980). No doubt a greater percentage of plant disease agents transmitted by insects other than the Homoptera will be realized as more study is initiated. This chapter discusses the nature and potential of plant viruses transmitted by leafminer flies.

10.2 DIPTERA AS VECTORS OF DISEASE AGENTS

The intimate associations of haematophagous Diptera with viruses and vertebrates are similar to vector-virus-plant relationships. Many of the animal viruses have evolved to such a degree that they almost completely rely on the flies for transmission between vertebrate hosts. The mechanisms of transmission can be either nonpersistent or persistent. Relatively few animal viruses are dependent on mechanical transmission by insect vectors for survival, and most of these nonpersistent viruses reach high titer in skin tumors and nodules and are readily accessible to the probing mouth parts of the feeding insects. The viruses are usually highly resistant to inactivation.

Nematocera are mostly minute and fragile insects. Among the various blood feeding families, the Simuliidae are important vectors of protozoan and nematode blood parasites. The Psychodidae, Ceratopogonidae, and Culicidae include known virus vectors. The transmission of reovirus type 3 by *Culex quinquefasciatus* (Say) has been reported (Miles and Stenhouse, 1969). The Shope's rabbit Papillomavirus is transmitted mechanically by mosquitoes (Marshall, 1973). The two strains of myxoma virus of rabbit are transmitted mechanically by *Culex annulirostris* (Skuse) and *Anopheles annulipes* (Walker) in Australia (Marshall and Douglas, 1961). In Brachycera, bloodfeeding is mostly found in the family Tabanidae and a few members of the family Rhagionidae. Tabanids vector trypanosomes and the tularemia bacterium. Cyclorrhapha are large flies. Many serve as mechanical vectors of viruses. In a few of them the mouthparts are equipped with small teeth capable of drawing blood. Others with piercing, sucking mouthparts include the tsetse flies, hornflies, and stable flies. The latter serve as mechanical vectors of vesicular stomatitis virus.

Diptera have been incriminated as vectors in the persistent transmission of animal viruses that replicate in alternate cycles in the cells of invertebrates and vertebrates. The viruses known as arboviruses include togaviruses, all the bluetongue-like viruses, some of the rhabdoviruses, and one picornavirus. The persistent viruses have a higher degree of vector specificity than nonpersistent viruses. An extrinsic incubation period of arbovirus in its vector is evident in the persistent transmission, but the length of incubation period varies with the virus isolate, vector species, dosage, and environmental factors. The vectors generally retain high infectivity, which slowly declines over the life of the insect. Yellow fever virus is persistently transmitted by *Aedes aegypti* (L.). Several encephalitis viruses are transmitted by a number of *Culex* spp. Dengue virus is transmitted by at least five species of *Aedes* in the persistent manner.

Only limited information is available on the involvement of dipterous insects in the transmission of plant pathogens. Two species of leafmining diptera (*Scaptomyza graminum* Fall and *Elachiptera costata* Loew) have been implicated in the spread of celery heart rot caused by *Erwinia carotovora* (Jones) Bergey, *et al.* (Leach, 1927). The olive fruit fly [*Dacus oleae* (Gmelin)] transmits *Pseudomonas savastanoi* (Erw. Smith) Steven (Breed *et al.*, 1957), the causal agent of olive knot, through the oviposition process (Petri, 1910).

The only Diptera demonstrated to serve as vectors of plant viruses are *Liriomyza* leafminers. Costa *et al.* (1958) reported that adult leafminer flies of the species *Liriomyza langei* (Frick) were capable of transmitting tobacco mosaic virus (TMV) and sowbane mosaic virus (SMV). Eight of 25 petunia plants (*Petunia hybrida* Vilm) were infected with TMV, while only 3 of 61 sowbane plants (*Chenopodium murale* L.) were infected. Sixty sowbane plants were infected with SMV out of 166 attempts, when sowbane also served as the virus source plant. Attempts to infect *C. amaranticolor* (Coste and Reyn.) with this same virus were negative. Since both TMV and SMV are highly infectious viruses, the mechanical movement of these viruses by insects can be expected. Such is the case with SMV, which is transmitted by leafminer flies in addition to the beet leafhopper [*Circulifer tenellus* (Baker)], fleahopper [*Holticus citri* (Ashmead)], and the green peach aphid [*Myzus persicae* (Sulz)] (Bennett and Costa, 1961), although the latter may be in doubt (Dias and Waterworth, 1967). Although not normally transmitted by arthropods, TMV has experimentally been transmitted by grasshoppers (Hollings and Huttinga,

1976), and aphids (Lojek and Orlob, 1969). More recent data indicate that TMV is transmitted by means of aphids' claws (Bradley and Harris, 1972; Harris and Bradley, 1973a, 1973b) or via the mouthparts only, with the aid of a helper agent (Harris, 1977; Pirone, 1977; Pirone and Harris, 1977).

In the same experiments (Costa *et al.*, 1958), leafminer flies failed to transmit nonpersistent, aphid-transmitted viruses which included beet mosaic virus, lettuce mosaic virus and tobacco etch virus. The principal test plants for each of these viruses were sugar beet (*Beta vulgaris* L.), lettuce (*Lactuca sativa* L.), and tomato (*Lycopersicon esculentum* Mill.), respectively.

In more recent experiments (Zitter and Tsai, 1977), leafminer flies were released in large numbers in cages where source and test plants were maintained together for more than a week. Successful transmission of celery mosaic (CeMoV) from celery to celery, and the two strains of watermelon mosaic virus (WMV-1 and WMV-2) from squash to squash was achieved. Both celery (Fig. 1) and squash (Fig. 2) can be extensively stippled by the female fly and mined by the developing larvae. The source

FIG. 1. Typical leafminer injury to celery leaves.

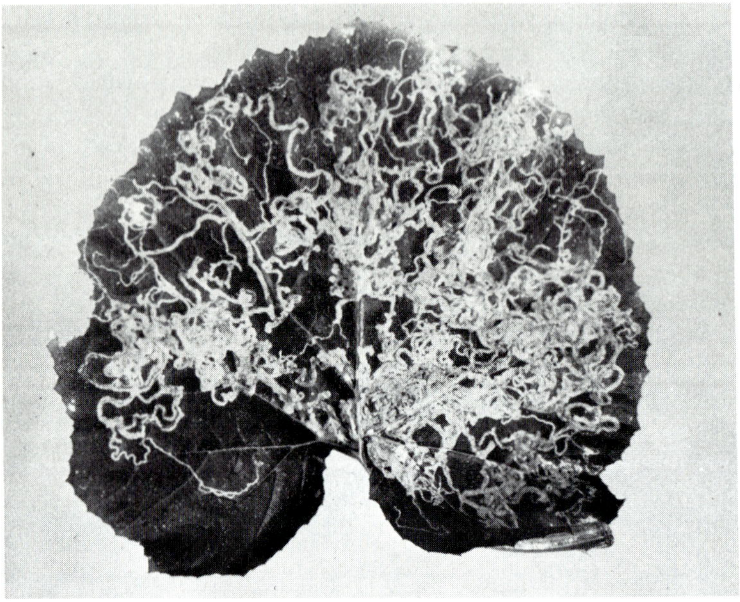

FIG. 2. Typical leafminer injury to squash leaves.

of the leafminer flies in all experiments was celery leaves collected from local farms. The flies were reared by two methods: 1) Initially the larvae-infested leaves were loosely packed into sealed plywood boxes to which a glass jar was attached to collect the emerging flies. The flies continued to emerge over a week's time and were released periodically into the inoculation cages. 2) Flies were also reared by allowing the celery leaves to desiccate over a period of 6 days after which time large numbers of pupae were collected and held in polyethylene bottles until the first flies appeared. The lids were then removed and the bottles placed inside the inoculation cages. Light intensities where the caged plants were held during the inoculation period had a decided effect on fly activity. Highest transmission rates were achieved when light intensities inside the cages measured 6,000 lux in an insectary and 9,469 lux in the greenhouse.

10.3 BIOLOGY OF *LIRIOMYZA* FLIES

L. sativae is commonly found in the American tropics and southern temperate regions (Spencer and Stegmaier, 1973; Musgrave *et al*., 1975). This species is a small insect ranging from 1.5 to 1.75 mm in length. The female is always larger than the male. It is bright yellow in the frons, the top of the head, and the femora. The head has a rather long arista arising from the last segment of antenna and usually has a pair of postvertical bristles standing divergently on the top. The wing has a distinct

FIG. 3. Female *Liriomyza sativae* adult. Note chitinized ovipositor.

anal cell. Most species of leafminer flies are more readily recognized by their characteristic mines than by the insects. The leafminer *Phytomyza minuscula* (Goureau) is known to make a serpentine mine. Most agromyzid leafminers make serpentine mines, and *Agromyza parvicornis* (Loew) has a blotch mine (Borror *et al.*, 1977). However, the only reliable taxonomic characteristics are based on the specific morphology of the male genitalia. The adults are often found on leaves of host plants. The adult female has a sharp and chitinized ovipositor (Fig. 3) which can be extended to the anal end with plunging actions to lacerate the leaf surface forming numerous stipplings (Fig. 1) from which plant sap exudes. The female may immediately either back up or turn around to feed on the sap. The punctures serve as ovipositional sites.

However, only less than 1% of the punctures are reported to contain eggs (Wolfenbarger, 1947). The male lacks the puncturing apparatus (Fig. 4) and is observed to feed on the same wound made by the female, but in the laboratory, male feeding frequency is much less than the female. In one 90 min. period, a female was noted to make 18 stipplings, and 14 feedings were made on those wounds. Both sexes are also found to feed on water drops on leaf surfaces. The adult mouthparts are a sponging or lapping type, which include a pair of lobe-like labella attaching to the end of the haustellum. The larvae are legless with a reduced head. A pair of sclerotized mandibles move in a vertical plane to rasp the mesophyll of the leaf. The

FIG. 4. Male *Liriomyza sativae* adult.

mine increases in width as the size of the larva increases. The mature larva can pupate either inside or outside the leaf mine. The pupae are of coarctuate type while the pupal stage is passed inside the puparium (Fig. 5). The average length of pupation is about 10 days, and the average life cycle is 24 days in the laboratory. However, it may vary greatly with the environmental conditions. In one test, about 5% of pupae remain viable after 8 weeks storage at 4.5°C.

10.4 MODE OF TRANSMISSION AND NATURAL SPREAD

The subject of transmission of disease agents has received continuous attention from both researchers and students in such fields as plant pathology, entomology, medical and veterinary medicine, microbiology, botany, and other biological sciences. The relationship between plant disease agents and their vectors can be simple or complicated in nature. Plant viruses are more dependent on their vectors for transmission than other pathogens such as bacteria and fungi. The recent rewakening of interest in the mechanisms of transmission of plant viruses has led to a re-evaluation and revision of traditional classification systems, especially with respect to transmission by aphid, leafhopper and planthopper vectors (Harris, 1977, 1978,

FIG. 5. Pupae of vegetable leafminer *Liriomyza sativae*.

1979). Such transmissions depend on vector probing and/or feeding behavior and seem more than mechanical processes. Transmission of potyviruses, a tobamovirus, and sowbane mosaic virus by leafminer flies may involve both ovipositor and mouthparts. Whether or not the transmission is solely a mechanical process has yet to be determined.

During the fall growing season and before aphid flights occurred, an attempt was made to create conditions favorable to test the natural spread of virus by leafminers (Zitter, unpublished). Zucchini squash (*Cucurbita pepo* L. 'Senator') were directly seeded into 30 ft. rows and were surrounded on either side by transplanted celery in two rows of comparable length. Four rows of celery in two treatments were intentionally sprayed weekly with methomyl (Lannate®), an insecticide known to increase leafminers by eliminating their parasites. In a third treatment the celery was sprayed with permethrin (Ambush®) which repels leafminer flies (Fig. 6). Squash source plants infected with WMV-2 (Zitter and Tsai, 1977) were spaced every 7 ft. in each row of squash. The squash were left unsprayed or sprayed with permethrin as a control. Although some mining of squash occurred in most of the plots, leafminers showed a strong preference for celery which negated efforts to achieve spread.

In a different experiment conducted on Florida's east coast about the same time, squash plants were severely stippled and mined by leafminers, but this occurred in the absence of nearby virus source plants and no infection was noted. The attraction

FIG. 6. The effect of permethrin on *Liromyza sativae* when applied to celery. Left, sprayed. Right, unsprayed control.

of leafminers to celery over their normally preferred weed hosts has previously been recorded (Genung and Janes, 1975).

The likelihood of achieving natural spread of potyviruses by leafminers is at best remote. In the laboratory this occurred when virus source plants and test hosts were held in cages and large numbers of flies were released. These exact conditions do not exist in nature. However, the celery-leafminer-celery mosaic relationship does exist and would appear to offer the potential for spread to occur. Leafminer populations can reach epidemic proportions in this closely spaced crop, and the titer of CeMoV is known to remain high in celery (Zitter, 1970).

10.5 REFERENCES

Bennett, C. W., and Costa, A. S. (1961). Sowbane mosaic caused by a seed transmitted virus. *Phytopathology* 51: 546-550.

Borror, D. L., Delong, D. M., and Triplehorn, D. A. (1977). "An Introduction to the Study of Insects" (3rd ed.) Holt, Rinehart and Winston, New York. 852 pp.

Bradley, R. H. E., and Harris, K. F. (1972) Aphids can inoculate plants with tobacco mosaic virus by clawing. *Virology* 50: 615-618.

Breed, R. S., Murray, E. G. D., and Smith, N. R. (1957). "Bergey's Manual of Determinative Bacteriology." (7th ed.) Williams and Wilkins Co., Baltimore. 1094 pp.

Costa, A. S., De Silva, D. M., and Duffus, J. E. (1958). Plant virus transmission by a leafminer fly. *Virology* 5: 145-149.

Dias, H. F., and Waterworth, H. E. (1967). The identity of a seed-borne mosaic virus of *Chenopodium amaranticolor* and *C. quinoa*. *Can. J. Bot.* 45: 1285-1295.

Eastop, V. F. (1977). Worldwide importance of aphids as virus vectors. *In* "Aphids as Virus Vectors" (K. Harris and K. Maramorosch, eds.), pp. 3-62. Academic Press, New York.

Genung, W. G. (1957). Some possible causes of insect resistance to insecticides in Florida. *Proc. Florida State Hort. Soc.* 70: 148-152.

Genung, W. G., and Harris, E. D. (1961). Notes on the biology and control of serpentine leaf-miner(s) in the Everglades. *Proc. Florida State Hort. Soc.* 74: 137-143.

Genung, W. G., and Janes, M. J. (1975). Host range, wild host significance, and infield spread of *Liriomyza trifolii* and population build-up and effects of its parasites in relation to fall and winter-celery (Diptera: Agromyzidae) *Univ. of Fla., Belle Glade AREC Res. Rept.* EV-1975-5, 18 pp.

Guzman, V. L., Burdine, H. W., Harris, E. D. Jr., Orsenigo, J. R., Showalter, R. K., Thayer, P. L., Winchester, J. A., Wolf, E. A., Berger, R. D., Genung, W. G., and Zitter, T. A. (1973). Celery production on organic soils of south Florida. *Univ. of Fla. Fla. Agr. Expt. Sta.* Bul. No. 757, 79 pp.

Guzman, V. L., Genung, W. G., Janes, M. J., and Zitter, T. A. (1977). Integrated production systems in celery. Preliminary reports I-IV. *Univ. of Fla., Belle Glade AREC Mimeo Series.*

Harris, K. F. (1977). An ingestion-egestion hypothesis of noncirculative virus transmission. *In* "Aphids as Virus Vectors" (K. F. Harris and K. Maramorosch, eds.), pp. 164-220. Academic Press, New York.

Harris, K. F. (1978). Aphid-borne viruses: ecological and environmental aspects. *In* "Viruses and Environment" (E. Kurstak and K. Maramorosch, eds.), pp. 311-337. Academic Press, New York.

Harris, K. F. (1979). Leafhoppers and aphids as biological vectors: vector-virus relationships. *In* "Leafhopper Vectors and Plant Disease Agents" (K. Maramorosch and K. F. Harris, eds.), pp. 217-308. Academic Press, New York.

Harris, K. F., and Bradley, R. H. E. (1973a). Importance of leaf hairs in the transmission of tobacco mosaic virus by aphids. *Virology* 52: 295-300.

Harris, K. F., and Bradley, R. H. E. (1973b). Tobacco mosaic virus: can aphids inoculate it into plants with their mouthparts? *Phytopathology* 63: 1343-1345.

Harris, K. F., and Maramorosch, K. (1977). Preface. *In* "Aphids as Virus Vectors" (K. Harris and K. Maramorosch, eds.). Academic Press, New York. 559 pp.

Harris, K. F., and Maramorosch, K., eds. (1980). "Vectors of plant pathogens." Academic Press, New York. In press.

Hollings, M., and Huttinga, H. (1976). Tomato mosaic virus, C. M. I./A. A. B. Descriptions of Plant Viruses. No. 156.

Janes, M. J. (1974). Insecticide evaluation for control of leafminers on celery. *Univ. of Fla., Belle Glade AREC Res. Rept.* EV-1974-14, 2 pp.

Janes, M. J., and Genung, W. G. (1975). Insecticide evaluation for control of leafminers and beet armyworms on celery, leafminers and cutowrms on lettuce. *Univ. of Fla., Belle Glade AREC Res. Rept.* EV-1975-1, 6 pp.

Kelsheimer, E. G. (1963). Tomato varietal resistance to leafminer attack. *Proc. Florida State Hort. Soc.* 76: 134-135.

Leach, J. G. (1927). The relation of insects and weather to the development of heart rot of celery. *Phytopathology* 17: 663-667.

Lojek, J. S., and Orlob, G. B. (1969). Aphid transmission of tobacco moxaic virus. *Science* 164: 1407-1408.

Maramorosch, K., and Harris, K. F. (1979). "Leafhopper Vectors and Plant Disease Agents." Academic Press, New York. 654 pp.

Marshall, I. D. (1973). Viruses and Diptera. *In* "Viruses and Invertebrates" (A. J. Gibbs, ed.) pp. 406-427. North Holland Publishing Co., Amsterdam.

Marshall, I. D., and Douglas, G. W. (1961). Studies in the epidemiology of infectious myxomatosis of rabbits. VIII. Further observations of changes in the innate resistance of Australian wild rabbits exposed to myxomatosis. *J. Hyg. Camb.* 59: 117-122.

Miles, J. A. R., and Stenhouse, A. C. (1969). Observations on reovirus type 3 in certain culicine mosquitoes. *Am. J. Trop. Med. Hyg.* 18: 427-432.

Musgrave, C. A., Poe, S. L., and Weems, H. V., Jr. (1975). The vegetable leafminer, *Liriomyza sativae* Blanchard (Diptera: Agomyzidae) in Florida. *Fla. Dept. Agric. and Consumer Services, Div. Plant Industry, Ent. Circ.* 162, 4 pp.

Petri, L. (1910). Untersuchungen uber die Darmbakterien der Olivenfliege. *Zentralbl. fur Bakt.* II 26: 357-367.

Pirone, T. P. (1977). Accessory factors in nonpersistent virus transmission. *In* "Aphids as Virus Vectors" (K. F. Harris and K. Maramorosch, eds.), pp. 221-235. Academic Press, New York.

Pirone, T. P., and Harris, K. F. (1977). Nonpersistent transmission of plant viruses by aphids. *Ann. Rev. Phytopathol.* 15: 55-73.

Poe, S. L., and Short, D. E. (1975). Pest populations on floral crops: Part IV. Leafminers. *Florida Flower Grower* 12: 2-4.

Price, J. F., and Poe, S. L. (1976). Response of *Liriomyza* (Diptera: Agromyzidae) and its parasites to stake and mulch culture of tomatoes. *Florida Entomologist* 59: 85-88.

Spencer, K. A., and Stegmaier, C. E., Jr. (1973). Agromyzidae of Florida with a supplement on species from the Caribbean. Arthropods of Florida and neighboring land areas. *Fla. Dept. Agr. and Consumer Services, Div. of Plant Industry.* Vol. 7, 205 pp.

Stegmaier, C. E., Jr. (1966). Host plants and parasites of *Liriomyza trifolii* in Florida (Diptera: Agromyzidae). *Florida Entomologist* 49 (2): 75-80.

Tsai, J. H. (1979). Vector transmission of mycoplasmal agents of plant diseases. *In* "The Mycoplasmas" (M. F. Barile, S. Razin, J. G. Tully, and R. F. Whitcomb, eds.), Vol. 3, pp. 265-307. Academic Press, New York.

Webb, R. E., and Smith, F. F. (1969). Effect of temperature on resistance in lima bean, tomato and chrysanthemum to *Liriomyza munda. J. Econ. Entomol.* 62: 458-462.

Wilson, J. W., and Hayslip, N. C. (1951). Insects attacking celery in Florida. *Univ. of Fla., Fla. Agr. Expt. Sta.*, Bul. No. 486.

Wolfenbarger, D. O. (1947). The serpentine leafminer and its control. *Univ. of Fla., Agr. Exp. Sta.* Press Bull. No. 639.

Wolfenbarger, D. O. (1958). Serpentine leafminer: brief history and summary of a decade of control measures in south Florida. *J. Econ. Entomol.* 51: 357-359.

Woltz, S. S., and Kelsheimer, E. G. (1958). Effect of variation in nitrogen nutrition of chrysanthemum on attack by serpentine leafminer. *Proc. Florida State Hort. Soc.* 71: 404-405.

Zitter, T. A. (1970). Titers of two virus diseases of celery affecting field spread. *Phytopathology* 60: 1321 (Abstr.).

Zitter, T. A., and Tsai, J. H. (1977). Transmission of three potyviruses by the leafminer, *Liriomyza sativae* (Diptera: Agromyzidae). *Plant Dis. Reptr.* 61: 1025-1029.

Chapter 11

LETHAL YELLOWING OF COCONUT PALM: SEARCH FOR A VECTOR

James H. Tsai

Agricultural Research Center
University of Florida
Fort Lauderdale, Florida

11.1	INTRODUCTION	177
11.2	PATTERNS OF SPREAD	179
11.3	SYMPTOMATOLOGY	179
11.4	CAUSAL AGENT	179
11.5	HISTORICAL PERSPECTIVES IN THE SEARCH FOR A VECTOR(S) OF LETHAL YELLOWING AGENT	180
11.6	VECTOR RESEARCH IN THE UNITED STATES	181
11.7	ATTEMPTS TO MECHANICALLY TRANSMIT LY AGENT	190
11.8	DETERMINATION OF INCUBATION PERIOD IN PALM	190
11.9	ATTEMPTS TO CULTURE COCONUT EMBRYO AND TISSUE IN VITRO	190
11.10	DISCUSSION	191
11.11	ACKNOWLEDGMENTS	197
11.12	REFERENCES	197

11.1 INTRODUCTION

Revelations 22:2 refers to the coconut as "the tree of life which bears twelve manner of fruits and yieldeth her fruit every month." Scientists generally believe that coconut came from the Indian Archipelago or Polynesia. It was dispersed in the sea by ocean currents. The triangular shape of the coconut fruit enables it to float up to 3,000 miles in 120 days, and still maintain its viability. Also, man has moved it from island to island and from one continent to another.

Of the 2,000 species of palm, the coconut, date and oil palms are important to world economy. Coconut palm (*Cocos nucifera* L.) is one of the most valuable trees in the world. Over 25 million tons of coconuts are produced each year on 6.5

ISBN 0-12-326450-2

million hectares of land throughout the tropics. The U.S., Western Europe and Japan import 75% of the world's production of coconut oil and copra. Millions of people in the tropics depend on coconut palms for food, shelter, fiber, and clothing (Woodroof, 1970).

Most parts of the coconut tree have economic value. The waxlike inflorescence, the spathes of inflorescence and the leaves of coconut palm are used for decorations in many countries. Coconut wood is used for cabinets, fish spears, and building materials. Leaves are woven into screens, hats, baskets, and fans; they are also used to thatch houses. The midribs of leaves are used for brooms whereas the sheath of the leaf base is woven into sandals and strainers. The shells of the nuts are used for lamp stands, buttons, containers, and other ornamentals. More importantly, they are processed for recovery of valuable activated charcoal and shell chemicals. Coconut husks are made into mats, brushes, upholstery stuffings, cordage, carpets, rope and various types of rubberized coir products.

Different parts of the trees also provide important foodstuffs. The palm hearts are eaten raw. Sap from inflorescences is used as a soft drink or can be made into wine. Coconut meat contains 20% protein and nutritious food for man and livestock, and it is also used for oil, ointment, and toiletries. The coconut water is used as a drink. Coconut milk (endosperm) is collected and dried under high vacuum to produce coconut honey, which is a well known constituent of media for growing plant tissues. Coconut oil has special qualities that differ in composition from animal fats originating in colder climates. The oil retains a distinctive advantage for certain uses and consumers prefer to pay a premium for the qualities. The U.S. imports 25% of the world's coconut oil for manufacturing soaps, margarine, detergents, candies and other confectionaries.

In Florida, coconut palms are grown solely for their aesthetic qualities. In a state which depends heavily on the tourist industry for revenue, the maintenance of a tropical landmark and beauty is of vital importance. The coconut palm has served this need well but is being threatened by a disease called lethal yellowing.

Lethal yellowing (LY) disease of coconut palm was first reported in Jamaica (Fawcett, 1891). It remained on the western part of the island for many years, then moved eastward across the mountain barrier against prevailing winds to the eastern coconut growing areas of the island in the 1960's (Grylls and Hunt, 1971a). In a short period, LY has demonstrated a tremendous capacity for destroying coconut palms. More than 50% of 4.3 million coconut palms in Jamaica alone have been killed by this disease. The annual death rate is estimated at 1,000,000 productive palms. This devastating disease has also been reported in the Bahamas (Leach, 1946), Cayman Islands (Seal, 1928), Cuba (DeLa Torre, 1906; Mijailova, 1967), Dominican Republic (Schieber and Hichez-Frias, 1970), and Haiti (Leach, 1946). In West Africa, this same devastating disease is known as Kaincope in Togo (Dring, 1963; Ollagnier and Weststeijn, 1961; and Steiner, 1976) or as Cape St. Paul Wilt in Ghana (Leather, 1959; Westwood, 1953).

LY was reported in Key West, Florida in 1965 (Martinez, 1965), killing 75% of 20,000 coconut palms before disappearing from the island in the late 1960's. In 1971, it was reported in mainland Florida. Since then LY has killed approximately

50% of the estimated 1.0 million coconut palms throughout the coconut-growing areas of Florida. The cost of cutting and removing each dead tree is about $50-70 and the cost of replacing a mature palm ranges from $150-300. However, the aesthetic value and environmental impact of the coconut palms in south Florida are far greater than the simple monetary value.

11.2 PATTERNS OF SPREAD

Two types of spread of LY are found in both Jamaica and Florida. The common type involves a local center of infection that appears on one or two palms. Subsequently, new cases appear at random around the initial focus to provide further local spread. The second type is a jump spread followed by the pattern of local spread. The distance of such jumps varies with locality, ranging from a few kilometers to 70 km (Carter, 1966).

11.3 SYMPTOMATOLOGY

The LY symptoms in immature coconut palms start with yellowish discoloration from the distal ends of leaf fronds, and are terminated by a final stage of decay of the spear leaves and palm hearts. The symptoms in fruit bearing palms are characterized by premature nut fall in all stages of development, blackening and necrosis of inflorescences in opened and unopened spathes, and root rot. A yellow flagleaf may also appear in the middle of the crown at this time. Later, as the disease progresses, the lower fronds turn yellow; this discoloration moves upwards into the crown over a period of 1-3 months. Finally, the decay of the spear leaves and hearts is always accompanied by a typical rank odor, and the crown topples from the top of the tree leaving a telephone pole-like naked trunk. The syndrome is complete in 3-5 months.

11.4 CAUSAL AGENT

Similarities of LY symptoms to those of bud rot and red ring disease of coconut palms include early nutfall, and the terminal death of the heart and crown. This led to the early studies of fungi and bacteria as possible LY causal agents (Grylls *et al.*, 1968; Leach, 1946; Martinez, 1965; Martyn 1945; Nutman and Roberts, 1955). More than 30 years were required for the recognition that bud rot is caused by *Phytophthora* fungus and red ring is caused by nematodes (Martyn, 1945, 1949).

Nutrient deficiencies and toxicities in the soil were also investigated, but none were implicated (Innes, 1949; Chen, 1966; Fiskell *et al.*, 1959). Martinez (1965) and Latta (1966) studied the possibility that nematodes were causal agents or vectors, but their studies proved negative. Subsequently, a virus was thought to be the causal agent (Carter, 1966; Grant, 1967). These authors conducted a series of unsuccessful experiments using dodder and mechanical means of transmission.

In 1972, Plavsic-Banjac *et al*. observed the presence of mycoplasma-like organisms (MLO) in tissues of a coconut palm infected with LY but not in healthy coconut palms. Since no other disease agents were observed in the infected materials, they suggested that MLO might be the causal agent. This hypothesis was further supported by the evidence of symptom remission following antibiotic treatments (Hunt *et al*., 1974; McCoy, 1972). Subsequently, several international laboratories have reported the association of MLO with diseased coconut and other palms (Beakbane *et al*., 1972; Heinze, 1972; Parthasarathy, 1973, 1974; Thomas, 1974, 1976, 1978; Waters and Osborne, 1978).

11.5 HISTORICAL PERSPECTIVES IN THE SEARCH FOR LY VECTOR(S)

Nutman and Roberts (1955) assumed that only large palms were susceptible to LY. They considered that study of an insect vector was impracticable. Carter (1966) conducted 300 tests using gross by mixed colonies from sweep net samples from the ground cover of diseased coconut plantations. The insects, consisting of Cicadellidae and Fulgoridae, were caged over healthy coconut seedlings; no transmission resulted. He then proposed that insects such as thrips [*Heliothrips haemorrhoidalis* (Bouche) and *Chaetanaphothrips orchidii* (Moulton)], mealybugs *Pseudococcus nipae* (Mask), *P. adonidum* (L.), and *Ferrisia virgata* (Ck 11), or aphids, *Cerataphis* sp., might be vectors. However, transmission trials yielded negative results. Grylls and Bor (1968) collected insects on or near diseased palms and separated them into family groups, then caged them on susceptible coconut palms; however, their efforts were without any success. Martinez (1965) conducted controlled experiments at Key West, Florida with a whitefly (*Aleurodicus* sp.), nematodes (*Rotylenchulus* sp. and *Xiphinema* sp.) and a fungus (*Fusarium* sp.). None of those agents could be shown to transmit the LY pathogen.

Heinze (1972) reported that a flatid, *Psenoflata brevis* (Van Duzee) (Homoptera: Flatidae) and a tingid bug, *Hybopharsa* sp. (Hemiptera: Tingidae) were able to complete the life cycle on coconut palm. These two species and other insects including *Cedusa* sp. (Homoptera: Derbidae), a *Colpoptera* sp. (Homoptera: Issidae), *Omalicna* spp. (Homoptera: Derbidae) and *Delphacodes* spp. (Homoptera: Delphacidae) were aspirated from coconut palms or by sweeping and suctioning the insects from the undergrowths. Many thousands of insects were caged on coconut palms over an 18-month period; two out of 60 caged palms receiving mixed insect species developed the LY syndrome. Heinze (1972) later used suction traps to collect insects in two areas in Jamaica that had high disease incidence. A large number of cicadellid leafhoppers, *Chlorotettix* spp., *Deltocephalus* spp., *Graminella* spp., *Tylozygus fasciatus*, *Xestocephalus* spp., *Empoasca* spp., and *Typhlocyba* spp. was separated and introduced into different cages containing healthy palms. Three suspected transmissions were recorded as a result of the introduction of the insects, but no follow-up study was conducted to pursue the lead. Other insects used in the earlier tests (Carter, 1966) such as mealybugs, *Nippaecoccus nipae* (Mask.) and two *Pseudococcus* spp., and the palm aphid, *Cerataphis variabilis* (H.R.L.) were also tested in various transmission experiments; no LY syndrome developed in any of the tests. Negative results in transmission experiments with the whiteflies, *Aleurodicus jamaicensis* (Ck11.), *A. ornatus* (CRH), three *Paraleyrodes* spp., a *Aleurotrachelus* sp., and a *Aleuroplatus* sp., led to the conclusion that whiteflies were not involved in transmission of the LY agent (Heinze, 1972). Eriophid mites such as *Notostrix jamaicae* (Keifer), *Epitrimerus* sp., and *Scolocenus* sp. were also tested as vectors of LY by Heinze and Schuiling (1970). A total of 91 young palms were colonized with those gallmites after 2-3 transfers from the diseased palms, but all test plants remained healthy.

Grylls and Hunt (1971b) employed large screen cages with aluminum angle frames to introduce the test insects into the cages containing palm seedlings. Groups of 154 to 4,523 unidentified insects belonging to Aleyrodidae, Aphididae, Fulgoridae, Cixiidae, and Cicadellidae were tested with no positive transmissions.

Johnson (1976) collected between 250,000 to 500,000 insects of mixed species in the field and tested them on palms. There were definitely four and possibly seven transmissions in the 358 palms used for testing leafhoppers and planthoppers. These results have been considered as accidental. Schuiling *et al*. (1976) also tested the adults of *Haplaxius crudus*, (Van Duzee), *Antillixius* sp., two *Omolicna* spp., *Colpoptera elevans* (Walker) and *Psenoflata brevis* (Van Duzee). Nymphs of *C. elevans* and *P. brevis* were also tried for transmission of the LY agent. Several thousand insects were tried but none of the 132 test palms became infected.

Eden-Green and Schuiling (1976) attempted to transmit LY agent with nymphs of *H. crudus* and *Proarna hilaris* (Germ) but with no positive results. Eden-Green (1978) reported that attempts were made to transmit the LY agent by *Haplaxius* nymphs feeding on diseased roots and by direct introduction of *P. hilaris*. About 900 adult *Haplaxius* were tested on 7 palms and 2,460 adult *Proarna* were introduced to 38 test palms. No transmissions were noted. Other acquisiton feeding tests were conducted in the field with *Chlorotettix minimus* (Baker), *C. viridius* (Van Duzee), *Graminella* spp., *Hortensia similis* (Walker), and *Metascarta histrio* as well as adult *Haplaxius*. After acquisition access feeds of 3-4 days on the attached fronds of diseased palms, the insects were kept on grasses for 10-14 days before transfer to test palms. The nymphs of *Haplaxius* sp. and the adults of *Chlorotettix* spp. and *Spangbergiella vulnerata* (Wolcott) were also injected with the extracts from diseased palms. None of the test palms showed LY symptoms.

Other insects such as *Hortensia similis* (Walker), *Tylozygus fasciatus* (Walker), *Carneocephala sagittifera* (Uhler), *Chlorotettix* spp., *Acinopterus* sp., *Xerophloea* sp. are currently being tested as possible vectors of LY agent in Jamaica (Dabek, 1978).

11.6 VECTOR RESEARCH IN THE U. S.

The research on the LY vector transmission conducted at the Agricultural Research Center, University of Florida, includes three major divisions. These divisions include toxins which may be induced by invertebrates or pathogen(s), dodder transmission and pathogen-vector-plant relationships. The pathogen-vector-plant relationships are based on the premise that LY disease is caused by a MLO and, like most plant diseases of mycoplasmal etiology, is therefore probably transmitted by a leafhopper or planthopper (Maramorosch, 1974; Maramorosch *et al.*, 1970; Tsai, 1979; Whitcomb and Davis, 1970). The emphasis of the vector transmission studies (Fig. 1) has been placed on homopteran insects which have been experimentally shown to feed on coconut and other palms.

Extensive sampling of insects by means of rotary flight traps and visual counts indicates that *Haplaxius crudus* (Van Duzee) is the only predominant homopterous species consistently associated with coconut and other palms (Howard, 1978; Woodiel and Tsai, 1978; Tsai and Kirsch, 1978; Tsai *et al.*, 1979; Tsai and Mead, 1979). Initial emphasis was therefore placed on testing *H. crudus* (Tsai, 1975, 1977a) and studying the biology and ecology of this insect (Tsai *et al.*, 1976; Tsai and Kirsch, 1978). Cicadellid leafhoppers are common on the undergrowth especially on the grasses. Many of these species are occasionally found to feed on palm (Fisher and Tsai, 1978a; Tsai, 1975, 1977b). One experiment was conducted from October of 1975 to September of 1976 by means of rotary flight traps and visual counts to compare the species compositions of the assemblages of auchenorrhynchous homopterans in coconut and sabal (*Sabal palmetto* W.) groves. The following insects were found in this study (Tsai and Mead, 1979):

Cicadellidae–*Aceratagallia gillettei* (Osborn and Ball), *A. sanguinoleta* (Provancher), *Agallia deleta* (Van Duzee), *A. constricta* (Van Duzee), *A. albidula* (Uhler), *Balclutha hebe* (Kirkaldy), *B. incisa* (Matsumura), *B. flavescens* (Baker), *B. guajanae* (DeLong), *Carneocephala sagittifera* (Uhler), *C. flaviceps* (Riley), *Chlorotettix viridius* (Van Duzee), *C. minimus* (Baker), *C. rugicollis* (Ball), *Deltocephalus sonorus* (Ball), *D. flavicosta* (Stal), *Draeculacephala producta* (Walker), *D. portola portola* (Ball), *D. inscripta* (Van Duzee), *D. bradleyi* (Van Duzee), *Empoasca plebeia* (DeLong and Davidson), *E. kraemeri* (Ross and Moore), *Erythroneura* sp., *Exitianus exitiosus* (Uhler), *Graminella villica* (Crumb), *G. cognita* (Caldwell), *G. nigrifrons* (Forbes), *G. albovenosa* (Sanders and DeLong), *Homalodisca insolita* (Walker), *Hortensia similis* (Walker), *Macrosteles fascifrons* (Stal), *Oncometopia nigricans* (Walker), *Paraphlepsius continuus* (DeLong), *Parabolocratus flavidus* (Signoret), *Planicephalus flavicosta* (Stal), *Polyamia obtecta* (Osborn and Ball), *Protalebra brasiliensis* (Baker), *Sanctanus fasciatus* (Osborn), *Spangbergiella vulnerata* (Wolcott), *Stragania robusta* (Uhler), *Texananus* sp., *Tylozygus bifidus* (Say), *T. geometricus* (Signoret), *Typhlocyba* sp., *Xerophloea viridis* (Fab.), *Xestocephalus tessellatus* (Van Duzee), *Xestocephalus* sp.;

Cixiidae–*Haplaxius crudus* (Van Duzee), *Bothriocera* sp., *Pintalia* sp.;

Delphacidae–*Chloriona slossonae* (Ball), *Delphacodes propinqua* (Fieber), *Megamelus gracilis* (Beamer), *Pissonotus piceus* (Van Duzee), *Euides weedi* (Van Duzee), *Neomegamelanus elongatus* (Ball), *Sogatella kolophon meridiana* (Beamer), *S. cubuanus* (Crawford), *S. molinus* (Fennah), *Bostaera balli* (Penner), *Nilaparvata wolcotti* (Muir and Giffard), *Pygospina spinata* (Caldwell), *Strobaera* sp.;

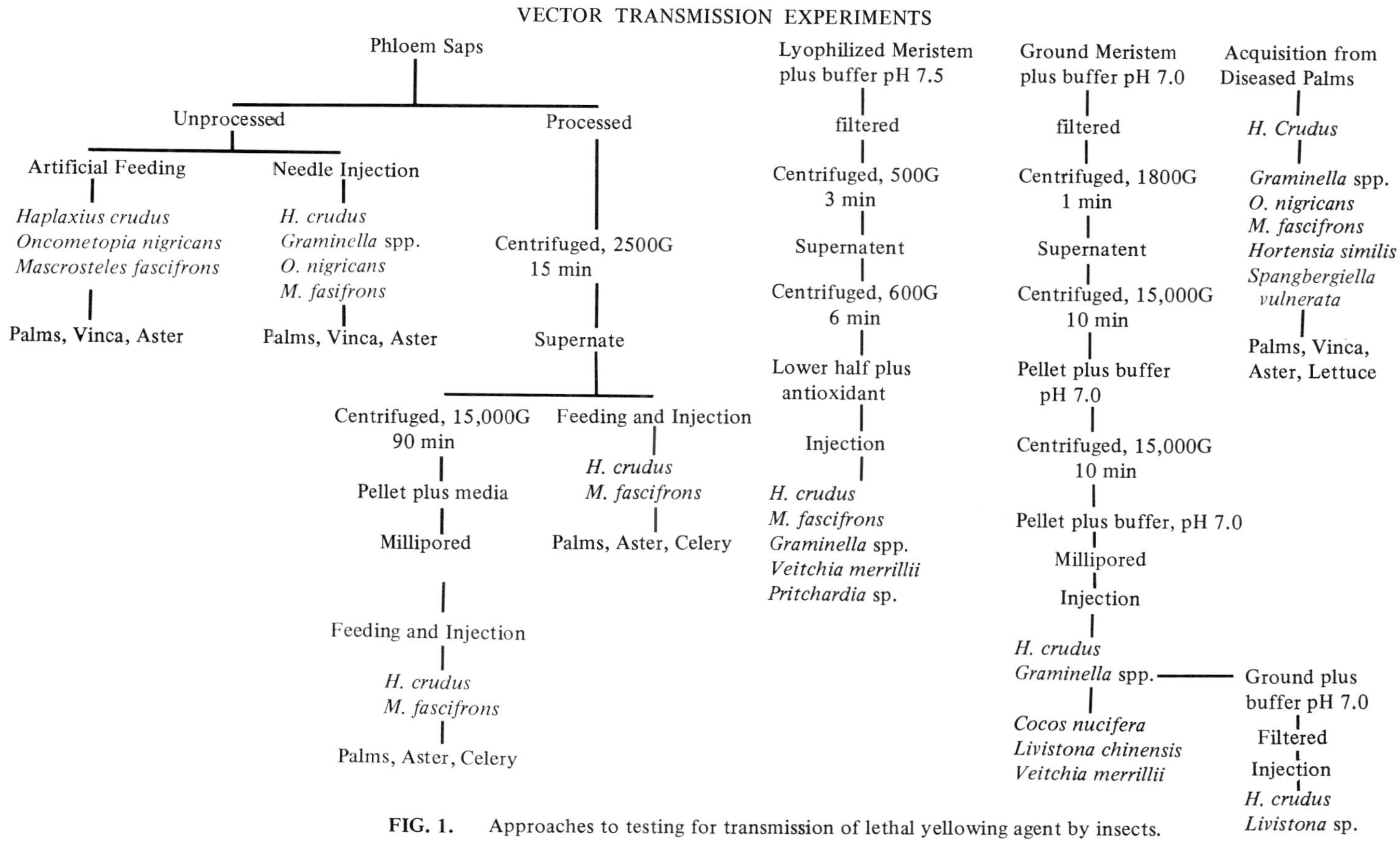

FIG. 1. Approaches to testing for transmission of lethal yellowing agent by insects.

Derbidae–*Cedusa inflata* (Ball);

Membracidae–*Idioderma virescens* (Van Duzee), *Spissitilos festinus* (Say), *Micrutalis* sp.;

Cercopidae–*Clastoptera undulata* (Uhler), *C. Xanthocephala* (Germar), *Prosapia bicincta* (Say);

Glatidae–*Ormenaria rufifiascia* (Walker);

Psyllidae–Unidentified sp.

Tests with Haplaxius crudus. In one experiment, an 11-m tall coconut palm that showed early LY symptoms of premature nutfall and terminal necrosis on newly opened inflorescences was selected for a transmission trial. About 2,300 *H. crudus* adults were caged over seven of the palm fronds of different ages for an average of 82 hr acquisition access period (AAP). At the end of the AAP, 631 adults were recovered and divided into 11 groups. Each group was caged on a 2.5-m tall coconut palm. Very few test insects survived more than 3 weeks on the test palms. None of the 11 palms tested developed LY symptoms after a year in a screenhouse (Tsai, 1977a).

One experiment was designed to test whether *H. crudus* adults could acquire the LY agent from palm heart tissues containing relatively high concentrations of MLO (Thomas, 1974, 1978). A total of 508 *H. crudus* adults were tested on four palm hearts containing 2 to 3 ten-cm primordial leaves. The bases of the palm hearts were immersed in citrate buffer solution. At the end of a 24-hr AAP, 195 adults remained alive and were caged on four 2.5-m tall coconut palms. Another 600 *H. crudus* adults were tested on five palm hearts immersed in sterile distilled water. At the end of a 24-hr AAP, 346 adults were placed on five 2.5-m tall coconut palms. No LY symptoms developed on the test plants after 12 months (Tsai, 1977a).

The failure of *Haplaxius* adults to transmit the LY agent could be attributed to the short life span of the adults under our experimental conditions. Possibly there was insufficient time for the pathogen to incubate inside test insects that lived for such brief periods. Thus, another experiment was designed to use *H. crudus* nymphs in an attempt to acquire the pathogen from roots of a diseased coconut palm as shown by Waters (1976). One hundred eighty-two of 345 nymphs were recovered after a 5-day AAP. They were subsequently reared on St. Augustinegrass (*Stenotaphrum secundatum* (Walt.) Kuntze) as described by Tsai *et al.* (1976). Ninety-four of the emerged adults were divided into two groups on two healthy coconut palms, but the test plants remained healthy (Tsai, 1977a).

In another study, 174 nymphs and 111 adults of *H. crudus* were allowed to feed on phloem sap from diseased coconut palms through a membrane. At the end of a 16-hr AAP, a total of 117 *H. crudus* nymphs survived and were placed in the rearing chambers containing St. Augustinegrass. Ninety-three adults of *H. crudus* emerged from the chambers and were placed on a healthy coconut palm. At the end of AAP, 82 adult *H. crudus* were placed on a healthy coconut palm. However, no transmission resulted (Tsai, 1977a).

Another experiment was conducted by collecting adult *H. crudus* directly from 20 diseased coconut palms in a nursery near Miami, Florida. Three hundred and fifty-six adults were randomly collected from the underside of palm leaflets in one week and released in one cage containing a healthy coconut palm. Four hundred twenty adults were released on another caged coconut palm the following week. This experiment was repeated by collecting 242 *H. crudus* adults from the diseased palms and caging them with three 2.2-m-tall coconut palms. Also, 260 *H. crudus* adults from other source palms were caged on six 1-m-tall *Veitchia* palms (*Veitchia merrillii* (Becc) Moore). No transmission resulted from the above tests (Tsai, 1977a). Still another extensive experiment was carried out to test *H. crudus* as a vector. Confinement of a large number of the *Haplaxius* adults collected from LY epidemic areas on different susceptible hosts also failed to reproduce the symptoms of LY on the test plants (Table I).

The crude phloem exudate collected from the diseased *Veitchia* palms was injected into 218 adult *Haplaxius*. These adults were later released in a screenhouse containing a 2.5-m-tall coconut palm and potted St. Augustinegrasses. Crude sap squeezed from the infected *Veitchia* palms was also utilized for injecting 184 *H. crudus* adults. After injection, those insects were caged on a *Pritchardia affinis*.

Another experiment was conducted by feeding or injecting *H. crudus* with the supernate of phloem sap exudate from diseased palms centrifuged at 2,500 G for 15 min to obtain a clear sap. At the end of 8-hr AAP, 84 of 102 *H. crudus* were recovered, and placed on one coconut palm. The same inoculum was injected into 41 *H. crudus*. Fourteen surviving adults were kept on a test palm. Negative results were obtained from the above tests during 10 months of symptom observation. The next experiment was designed to obtain a higher concentration of pathogen by centrifuging the supernate at 15,000 G for 90 min. The pellet was mixed with healthy phloem exudate which passed through a 0.45 μm Millipore filter. The filtrate was

TABLE I. Unsuccessful Attempts to Transmit Lethal Yellowing (LY) Agent Using *Haplaxius crudus* from LY Epidemic Areas

Test Plant	Number of Test Insects per plant	Number of Trials
Pritchardia remota	152	2
Pritchardia thurstonii	205	2
Livistona chinensis	212	2
Arikuryroba schizophylla	156	2
Arenga engleri	247	2
Chrysalidocarpus cabadae	316	1
Phoenix reclinata	248	3
Phoenix dactylifera	253	2
Latania sp.	175	2
Pandanus utilis[a]	43	2
Carloduvica palmata[b]	17	1

[a]Pandanaceae; [b]Cyclanthaceae

fed and injected into the adults of *H. crudus*. After 7-hr AAP, 44 *H. crudus* were caged on one healthy coconut palm. The same inoculum was injected into 62 adult *H. crudus*. Thirty-seven survivors were caged on a test palm. No positive transmission occurred (Tsai, 1977a).

A similar experiment was conducted by injecting *H. crudus* with the inoculum processed from ground meristem (Fig. 1). Four groups of *H. crudus* consisting of 37, 15, 13, and 216 adults were injected and placed on a *Pritchardia* sp., *Veitchia* palm, coconut palm and *Livistona chinensis* respectively. No transmission resulted.

Other inoculum obtained from the lyophilized meristem of an infected *Veitchia* palm (Fig. 1) was injected into 82 nymphs and 35 adults of *H. crudus*. Subsequently, they were confined in a screenroom containing one *Veitchia* palm and St. Augustine grasses. No transmission resulted.

Other species of cicadellids were selected as test insects, based on the knowledge that either they transmit other prokaryote disease agents (Maramorosch, 1974; Maramorosch *et al*., 1970; Tsai, 1979; Whitcomb and Davis, 1970) or that they are able to feed on palms. *Hortensia similis* (Walker), *Oncometopia nigricans* (Walker), *Spangbergiella vulnerata* (Wolcott), *Graminella* spp. and *Macrosteles fascifrons* (Stal) were selected for various transmission trials. After AAP or injection, the insects were transferred to their preferred hosts for an incubation period of 2-3 weeks, after which time they were transferred to a palm or spear leaf in a cage on a test palm. Again, transmission trials proved negative.*

Tests with Oncometopia nigricans. This leafhopper has been found in coconut groves (Tsai and Mead, 1979), and was found to feed and live three or more months on *Veitchia* and coconut palms (Fisher and Tsai, 1978a; Tsai and Anwar, 1977). A total of 206 *O. nigricans* were given at least a 114-hr AAP on diseased coconut palms. The test insects were caged on periwinkle (*Catharanthus roseus* Don.), lettuce (*Lactuca sativa* L.) coconut and *Veitchia* palms following the AAP until the death of the insects. Twenty-five *O. nigricans* nymphs were injected with crude phloem exudate collected from a diseased coconut palm. Twelve insects which incubated on a periwinkle plant for 14 days were then transferred to a coconut palm. Subsequently, another 35 *O. nigricans* were injected and held on 5 periwinkle plants where they remained until death of the insects. No transmission has resulted (Tsai, 1975, 1977b). A group of 74 *O. nigricans* nymphs were fed with crude phloem exudate through a membrane. At the end of a 16-hr AAP, 60 nymphs were distributed on two periwinkle plants. No symptoms were noted after 6 months (Tsai, 1977b).

Tests with Hortensia similis. Attempts were made to test *H. similis* as a LY vector. A total of 1,484 insects were given 72 to 96-hr AAPs on detached spear leaves of infected coconut palms. None of the 161 test plants including periwinkle, lettuce,

*After this manuscript had gone to press, Tsai (personal communication to K. F. Harris) obtained an apparent transmission of LY agent by *H. crudus*. Confirmatory transmission trials are now in progress.

St. Augustinegrass, bahiagrass (*Paspalum notatum* Flugge), aster (*Callistephus chinensis* N.), pepper (*Capsicum frutescens* L.), tomato (*Lycopersicon exculentum* Mill), carrot (*Dacus carota* L.), marigold (*Tagetes patula* L.), coconut palm and *Veitchia* palm developed symptoms after a year (Tsai, 1975).

Tests with Spangbergiella vulnerata. *Spangbergiella vulnerata* is abundant on St. Augustinegrass and other grasses under coconut plantings both in Jamaica and Florida and has been proved to feed on palms (Fisher and Tsai, 1978). Therefore, one experiment was conducted to test for LY transmission by this insect. Five hundred *Spangbergiella* leafhoppers were collected from stock colonies and caged on a spear leaf of lethal yellowing infected coconut palm for a 48-hr AAP. At the end of AAP, about 400 leafhoppers were recovered and released in a screenroom containing two 2.2 m-tall healthy coconut and two 1.5 m-tall healthy *Veitchia* palms. Fourteen 15-cm pots of St. Augustinegrass were also provided as breeding host for the test insect. There was no detectable transmission (Tsai, 1977b).

Tests with Macrosteles fascifrons. Three groups of *M. fascifrons* consisting of 772 adults were allowed to feed on detached fronds of *Veitchia* palm for 72 to 168-hr AAP's. They were then transferred to three young *Veitchia* palms. Four hundred and ninety *M. fascifrons* nymphs were fed with crude phloem exudate from diseased coconut palms through a membrane. At the end of a 16-hr AAP, 233 nymphs were divided into 11 groups, and each group was kept on an aster plant. The same inoculum was also injected into 42 *M. fascifrons* adults. Twenty-three survivors were placed on an aster plant.

Another experiment was conducted by feeding and injecting *M. fascifrons* with the supernatant fluids from the phloem sap from diseased palms centrifuged at 2,500 G for 15 min. At the end of an 8-hr AAP, 78 of 130 *M. fascifrons* were recovered. The *M. fascifrons* adults were placed on three celery (*Apium graveolens* L.) and two aster plants. The same inoculum was injected into 36 *M. fascifrons* adults. Twenty-one surviving *M. fascifrons* were kept on single aster and celery plants. The next experiment was designed to obtain a higher concentration of pathogen by centrifuging the diseased palm sap at 15,000 G for 90 min. The pellet was mixed with healthy phloem sap which had passed through a 0.45 μm Millipore filter. The filtrate was fed and injected into the adults of *M. fascifrons*. After 7-hr AAP, 52 *M. fascifrons* were divided between aster and two celery plants. The same inoculum was injected into adult insects. Twenty-eight of 44 injected *M. fascifrons* were divided on two aster and two celery plants. No positive transmission occurred before the death of the test plants (Tsai, 1977b).

The inoculum processed from the lyophilized meristem of a diseased *Veitchia* palm (Fig. 1) was injected into 27 *M. fascifrons* nymphs. The insects were kept on oat plants (*Avena sativa* L.) for 7 days prior to transferring to a *Pritchardia* palm. No transmission resulted.

Tests with Graminella nigrifrons and G. cognita. Three *Graminella* species (*G. nigrifrons* Forbes, *G. cognita* Caldwell, and *G. floridana* (Delong and Mohr)) are found abundantly on bermudagrass (*Cynodon dactylon* (L.) Pers.) grown on the golf courses in south Florida, and they are often found in the samples of rotary flight traps (Tsai and Mead, 1979). Only *G. nigrifrons* and *G. cognita* were used in the trials. A total of 1,050 *G. nigrifrons* and *G. cognita* were given 48 to 120-hr AAP's on infected coconut palms. At the end of AAP, 664 leafhoppers were retrieved and kept in a screenroom containing three potted corn plants (*Zea mays* L.), two 2.5-m tall *Veitchia* and one coconut palm.

Crude phloem exudate collected from the cut stem of a diseased *Veitchia* palm was injected into a mixture of 69 *Graminella* adults, and they were caged on a 2.5-m tall coconut palm. Other crude sap squeezed from the diseased *Veitchia* palms was injected into a mixture of 180 *Graminella* adults. One hundred-fifty insects were caged on one *Pritchardia affinis*, the remaining 30 insects were kept on a *Pritchardia* palm.

Other inoculum processed from ground meristem (Fig. 1) was injected into two groups of a mixture of 48 and 73 *Graminella* adults. The injected insects were kept on corn plants for 10-19 days, after which time the first group of 48 insects was caged on a young *Pritchardia affinis*. The other group of 73 insects was ground in a 600 milliosmol buffer solution consisting of 100 ml distilled water, 2.25 g glycine, 0.6 g $MgCl_2$, 5.0 g sucrose, and cysteine at 0.075 g per 100 ml. The inoculum was filtered through 1.2 and 0.45 μm Millipore filters before injection into 88 *H. crudus*. These injected *H. crudus* were caged on a *Livistona* palm.

The inoculum processed from the lyophilized meristem of a diseased *Veitchia* palm was also injected into 38 *Graminella* nymphs. After injection, the insects were kept on oats for 10 days before transferring to a healthy *Veitchia* palm. No symptom has yet developed on any of these test plants.

Tests with Tetranynchus Mite. We wished to determine if a mite (*Tetranynchus tumidus* Banks), commonly found on the coconut palm, could produce a phytotoxic substance causing LY symptoms. About 2,000 *T. tumidus* collected from an LY area were established on 42 container-grown coconut palms 2-m in height and 66 potted *Veitchia* palms 1-m in height in two rooms (8.4 x 6.0 m) covered with a 52 mesh saran screen. The average number of mites per leaflet was 12.5 over a period of 18 months. The damage on the palm leaf as a result of mite feeding was noticeable, but there were no LY symptoms on the 198 palms tested. Although this mite is frequently found in palm plantings, it is evident that LY is not caused by a toxin produced by *T. tumidus* (Tsai, 1977b).

Tests with Inflorescence-Visiting Insects. To test the palm inflorescence as a vector feeding and LY inoculation site, a block of 3 to 4-m tall *Veitchia* palms was selected in a LY epidemic area. The unopened inflorescences of palm were periodically removed. Neighboring blocks of palms served as controls. After 14 months, 20% of 75 palms with inflorescence removal succumbed to LY as compared to 60% of control blocks. The results suggest that inflorescence-visiting insects may be involved in transmitting the LY agent. The frequent inflorescence visitors of *Veitchia* palms are *Apis mellifera* L. (Apidae), *Polistes fuscatus* (Fabricius) (Vespidae), *Agapostemon splendens* (Lepeletier) (Halictidae), *Pachodynerus erynnis* (Lepeletier) (Eumenidae), *Augochlora pura* (Say) (Halictidae), *Dorymyrmex pryamicus* (Roger) (Formicidae), *Allograpta cubana* (Curran) (Syrphidae), and *Idioderma virescens* (Van Duzee) (Membracidae) (Tsai, 1978).

Vector Phototropic Test. One experiment was designed to test whether the suspected vector of LY is positively phototropic. One 52 mesh screen light trap (2.2 x 0.9 x 0.9 m) was constructed of a wooden frame to house four potted coconut palms approximately 0.9 m tall. On top of the trap, a funnel and a 15-watt ultra violet lamp were mounted. The trap was placed in LY epidemic areas for a period of 12 months. Each of three sets of palms was exposed to light-trapped insects for four months and then sprayed with malathion and kept in a screenhouse for a year. None of the 12 test plants developed LY symptoms as a result of exposure to the light-trapped insects (Tsai, 1977b).

Tests with Palm Aphid. Palm aphids, *Cerataphis variabilis* (HRL) have been found on coconut palms in Florida, Jamaica, and West Africa. This species feeds on the phloem tissue where MLO have been visualized in diseased coconut palms. The insect is highly host specific, particularly on Malayan Dwarf coconut palms. Only on rare occasions is it found on LY susceptible 'Jamaican Tall' coconut, *Pritchardia* and *Veitchia* palms. For this reason, and because aphids rarely, if ever, transmit MLOs (Tsai, 1979), the insect is only considered to be a secondary suspect.

Several attempts were made to transmit the LY agent with this aphid. Many technical difficulties were encountered during the course of the transmission trials. These were mainly due to high mortality rate in transferring the insects from one palm to another. Colonies were reared on 'Malayan Dwarf' palms. The early instar nymphs were transferred by a camel hair brush and alatae were transferred by means of an aspirator or by attaching the cut spear leaves which were colonized by the aphids to LY-infected leaves for 5 to 7-day AAPs. Only about 5% of the several thousands of transferred aphids survived after AAP. Furthermore, only about 1% of the remaining test aphids could be successfully transferred to two of each coconut and *Veitchia* palms. No transmission was noted during 10 months observation in a screenroom.

Tests with Cuscuta campestris. Other workers have been able to transmit MLO s from diseased plants to other plant species via several species of dodder (Dale and Kim, 1969; Dijkstra and Lee, 1972; Hibben and Wolanski, 1971). A similar technique was tried on coconut palms. In a series of experiments, healthy dodder (*Cuscuta campestris* Yuncker) seedlings were established on 12 healthy potted periwinkles (*Catharanthus roseus* Don.). The periwinkles with dodder were then tied on 12 apparently healthy coconut palms grown in an active LY area. Strands of dodder were constantly trained onto the leaflets of the coconut palms. Eight of 12 source palms eventually showed striking LY symptoms. The strands of dodder from those trees expressing LY symptoms were bridged to healthy periwinkles, and the connections between the hosts were maintained until the death of the palms. Then the periwinkles were transported back to the greenhouse for observation of symptom development. The tested periwinkles showed no sign of yellowing or other disorder after eight months in the greenhouse (Tsai, 1977b).

11.7 ATTEMPTS TO MECHANICALLY TRANSMIT LY AGENT

Early success in transmission of the LY agent by mechanical inoculation was reported in Florida (Price *et al*., 1967, 1968). Later, several attempts were made in both Jamaica and Florida to transmit LY agent in this manner without any success (Grylls and Hunt 1971b; Heinze, 1972). In one experiment, the unprocessed phloem exudates from the hearts of diseased coconut and *Veitchia* palms were separately rubbed on the unopened spear leaves of 1.5-m tall coconut and *Veitchia* palms. The spear leaves were dusted with 600-mesh carborundum and then rubbed with the coarse end of the pestle. Three coconut and three *Veitchia* palms were tested, but without positive results. In another experiment, unprocessed phloem exudate was injected into a petiole base of 1-m tall *Veitchia* palm. The leaflets of the same palm were slashed with a razor blade and dosed with the same inoculum. The other cut end of a petiole of this palm was tied to a vial containing the inoculum in an attempt to introduce the sap into the vascular tissue. Another experiment was performed using inoculum processed from the ground meristem (Fig. 1). The spear leaf of a 1-m tall *Veitchia* palm was dusted with carborundum and then mechanically inoculated with this inoculum. None of the above test plants showed LY symptoms.

11.8 DETERMINATION OF INCUBATION PERIOD IN PALM

In Jamaica experiments with caged palms and removal of cages at certain intervals indicated that airborne insects were responsible for the spread of the LY agent (Heinze, 1972). From such experiments, the incubation period of the LY agent in young coconut palms was estimated to be 3-11 months (Heinze and Schuiling, 1973). A similar experiment was initiated in July 1976 in Florida to determine the length of the LY incubation period in *Veitchia* palm. A series of ten container-grown *Veitchia* palms at 2-3 m height were transferred to LY endemic area and exposed to the potential vector(s) of LY. At the end of one month exposure, the test palms were treated with insecticide and transferred back to a screenhouse. Weekly spray schedule was carried out in the screenhouse. A total of 100 palms were tested in 10 months. The results showed that 5% of the test palms succumbed to LY. The length of incubation period ranged from 3-11 months.

11.9 ATTEMPTS TO CULTURE COCONUT EMBRYO AND TISSUE IN VITRO

In an effort to improve LY transmission studies, an *in vitro* coconut system was sought. A modified Murashige and Skoog (1962) medium was developed for optimal growth of embryos of 'Jamaican Tall' and 'Malayan Dwarf' coconut palms (Fisher and Tsai, 1978, 1979). A single callus line was induced from solid endosperm and subcultured on modified Schenk and Hildebrandt (1972) medium. Attempts to induce organogenesis in the callus were not successful. However, coconut seedlings from embryo culture could eventually provide: (a) material to study the pathogen-host relationship with the electron-microscope; (b) large number of uniform test plants in a relatively small space for experiments; and (c) an *in vitro* means of culturing LY agent in living tissues (Fisher and Tsai, 1978b).

11.10 DISCUSSION

Lethal yellowing research has proceeded very slowly, due mainly to such innate problems as the slow growth of palms, a long incubation period of pathogen, the unknown biology, ecology and mass rearing techniques of the vector, and difficulties in cultivation or isolation of the causal agent. In addition, coconut palm is a woody monocotyledon, lacking any mechanism for secondary growth of vascular tissues. It does not produce bark, cambium, and wood in definite rings, nor is it able to heal a wound as do many trees. Therefore, it is impossible to graft a palm of this nature which further hinders the opportunity to search for the etiologic agent.

In light of failures to isolate an etiological agent or to find an insect vector, perhaps we are entitled to ask: is this disease an exception to all known principles of plant disease transmission? Probably it is not, but we must remember that unusual answers are not excluded by present-day knowledge. For example, the possibility of mixed infection by MLO and viroid exists. Such an infection would require two components to induce the lethal effect and the syndrome. Should this theory hold true, more than one species of vector would be required to transmit the causal agents. Moreover, there may be more than one species of alternate weed host, which could be a monocotyledon or dicotyledon, for the causal agents. It would then be erroneous to cultivate MLO as a sole agent or to search for a single vector. This possibility, although very remote, ought not to be ignored.

Disease etiology: Evidence obtained from electron microscopy and the remission of symptoms after antibiotic treatment convincingly supports the concept that an MLO is associated with LY of palms in Florida and Jamaica. However, conclusive evidence is still lacking that an MLO causes the disease or that an MLO is the only pathogen involved in LY syndrome. It has been speculated that the causal agent of LY belongs to the genus *Spiroplasma* or to an entirely new genus (Sherman and Maramorosch, 1977).

The cultivation of an MLO from coconut tissue of *Kaincope*-diseased plants in West Africa has been reported (Giannotti *et al.*, 1975). However, the culture has not been maintained, nor has it been deposited in the international type culture collection. Future insect transmission experiments may depend on our ability to extract viable MLO from LY-infected palms and to maintain such extracts *in vitro*. Thus the cultivation of LY agent *in vitro* remains one of the major goals at this time. Another possible approach is the cultivation of insect vector cells *in vitro*. Cell lines from various organs of suspected vectors, such as the salivary glands, could be grown in monolayers for possible cultivation of the presumed MLO.

Vector search: Although the discovery of MLO in palm tissues infected with LY has directed attention to phloem feeding insects, several thousands of the frequent and infrequent arthropod visitors of palms have been tested in both Jamaica and Florida without positive results. The failure to transmit LY agent by suspected vectors does not prove that LY is not a vectored disease, but does suggest that it may not be transmitted by existing methodology. We have felt for many years that the vector must be a common insect. However, it is possible that the vector may be one that is infrequently found on palms. If coconut palm were not a preferred host of such a vector, it might only feed on palms when preferred host plants became scarce. For this reason, studies with infrequent visitors of coconut palms should be continued.

Of all homopterans common to Florida and Jamaica, *Haplaxius crudus* is the only one frequently associated with the palms in both regions, and sufficiently common to be a good suspect. Despite the strong circumstantial evidence suggesting *H. crudus* as a vector of LY, a large number of these insects has been tested by a variety of techniques. This insect did not transmit LY agent under the conditions of those trials at least. However, we must recognize that certain technical difficulties in the tests performed in Florida might influence the transmission results.

The evidence of reduction of LY incidence by excising the unopened inflorescences from the trees suggests that flower-visiting insects are also involved in transmitting the LY agent. Clark (1977) suggested that a new spiroplasma associated with honey bee disease might be transmitted through flowers. It is often observed both in Florida and Jamaica that sexually mature and flowering 'Jamaica Tall' palms are most susceptible to LY infection. This variety is solely dependent upon flower-visiting insects for pollination, whereas 'Malayan Dwarf' variety, which is resistant to LY, is self-pollinated. The bees, wasps and other pollinating insects are worldwide in distribution, their foraging range is usually several miles and they could be carried to a long distance by air currents. Research has been initiated to pursue the study of inflorescence visitors as possible LY vectors. At present, the biology of most potential vectors, with the exception of *H. crudus* (Tsai *et al*., 1976; Tsai and Kirsch, 1978; Tsai *et al*., 1979), is not well understood. Most of the species of Cicadellidae tested were polyphagous, and none of them are able to complete their life cycle on coconut palms. They are capable of living on palms for only short periods of time, thus making it difficult, if not impossible, to test for all possible modes of transmission.

Epidemiology: The various replanting programs in Florida and Jamaica prove that LY susceptible coconut palms will grow in soils in which old coconut palms affected by LY had been growing for many years. The young palms are not affected by planting in soils where older palms developed striking LY symptoms. Observations were also made in Jamaica and Florida on LY affected coconut palms growing in various soil conditions such as sandy beaches, loamy soil and swampy muck in Florida's Everglades. Such evidence indicates that soil conditions or nematodes are unlikely to induce or be involved in any significant way in the LY syndrome.

The relative slowness with which LY moves through a focus suggests a vector which normally lives and breeds on other alternate hosts and feeds only occasionally on coconut palms. On the other hand, the quiescence of the disease in the Key West area after an original epidemic suggests that either the vector or alternate hosts are not now present, or that the spread is from palm to palm. This question is not yet resolved. The spread of LY in the Key West area was in the direction of the southeast prevailing winds. It is possible that the vector was carried by such winds from Cuba across the Straits of Florida. The geographical relation of Florida Keys and Cuba could explain the distribution of LY. In Jamaica, the seashore is more exposed to winds carrying insects into the coconut plantation, so that high incidence occurs along the northeast areas of the island. Of course, human importation of diseased plants is another possible means of inter-island spread.

Palms found to be susceptible to LY disease are usually at least two years old. Rare instances are known where palms under 18 months in age have become infected. The reasons for the different susceptibilities in age could be attributed to the physical or biochemical differences in physiology between the young and old palms. This could however also be due to the height of flight activity of arthropod vectors. In any case, the need to use older palms as test plants presents a serious limitation to the study of the transmission of this disease, since such palms are difficult to grow and the eventual hulks of palm have many special requirements during the prolonged incubation period.

It would be desirable to have more standard and more easily manipulated experimental plants; hence, our interest lies in palm tissue culture. To date, there has been successful plantlet regeneration in African oil palm (*Elaeis guineensis* Jacq.) by tissue culture techniques (Smith and Thomas, 1973). However, vegetative propagation of coconut palms has not been successful (Fisher and Tsai, 1978b; Schwabe, 1977). Successful tissue culture techniques would greatly improve coconut research by allowing for (a) production of homozygous lines for hybrid plants, (b) testing the plantlets produced by tissue culture for LY and insect resistance, (c) collection and exchange of germ plasm material, reducing the risks of inadvertent spread of LY agent, and (d) investigating the nature of transmission of LY agent.

It is especially important to transmit LY agent from palms to an herbaceous host. Such an advance would greatly accelerate LY research programs. Many herbaceous plants that are hosts of other mycoplasmal disease agents are fast growing and would provide a ready source of infected material for research purposes. Fast-growing, small-sized herbaceous plants would also reduce space requirements and the durations of experiments. Also, suspected vectors could be much more readily tested and maintained on the herbaceous hosts than on coconut palms.

In contrast to areas in which coconut palms are an important food crop, the impact of LY in Florida has been on the environmental quality of urban life (Fisher, 1975). The estimated one million coconut palms in south Florida provide about 6,400 hectares of shade canopy. The elimination of coconut palms means not only to expose the city streets, parks and beaches to the intensive tropical sun, but also profound changes in urban environments and the tourist industry. South Florida depends heavily on the tropical landmark of the graceful sway of coconut palms.

The disappearance of this factor can be detrimental to the area's economy.

Host range: Evidence indicates that LY attacks not only coconut palms but also 26 other species representing at least 7 of the 15 major groups of palms (Moore, 1978). MLO were found in these species of diseased palms in the LY active areas (Parthasarathy, 1974; Thomas, 1974, 1976, 1978). The findings have prompted the Florida Department of Agriculture to place the following species in the LY quarantine: Seashore Palm, *Allagoptera arenaria*; Sugar Palm, *Arenga engleri*; Arikury Palm, *Arikuryroba schizophylla*; Palmyra Palm, *Borassus flabellifer*; Cluster Fishtail Palm, *Caryota mitis*; Cabada Palm, *Chrysalidocarpus cabadae*; Coconut Palm, *Cocos nucifera*; Buri Palm, *Corypha elata*; Hurricane Palm, *Dictyosperma album*; Puerto Rican Gaussia Palm, *Gaussia attenuata*; Spindle Palm, *Hyophorbe vershaffeltii*; Latan Palm, *Latania* sp.; Chinese Fan Palm, *Livistona chinensis*; Mazari Palm, *Nannorrhops ritchiana*; Canary Island Date Palm, *Phoenix canariensis*; Date Palm, *Phoenix dactylifera*; Senegal Date Palm, *Phoenix reclinata*; Wild Date Palm, *Phoenix sylvestris*; Kona Palm, *Pritchardia affinis*; Fiji Island Fan Palm, *Pritchardia pacifica*; Thurston Fan Palm, *Pritchardia thurstonii*; Chinese Windmill Palm, *Trachycarpus fortunei*; Christmas Palm, *Veitchia merrillii*; Montgomery's Palm, *Veitchia montgomeryana*; Hildebrandt's Palm, *Ravenea hildebrandtii*; and Sentry Palm, *Howea belmoreana*. This has imposed costly restrictions on the nursery industry. Other states such as California, Hawaii, Mississippi, Texas and Louisiana have similarly enacted quarantines that forbid importation of palm materials. Even though there is no indication of seed transmission of LY agent, the Fairchild Tropical Garden in Miami, Florida, the world's largest palm collection, and the members of the Palm Society have stopped their worldwide palm seed exchanges to minimize the chance of accidental introduction of LY disease into other parts of the world. The majority of the coconut varieties from many parts of the world that were tested in Jamaica have shown high susceptibility to LY (Harries, 1971); therefore, the potential of LY to cause damage in every coconut growing country is very great (Harries, 1978).

There are many examples of important food crops threatened by LY. For example, the date palm, *Phoenix dactylifera*, is on the LY quarantine list, and the potential threat of LY to date production in California, Middle East, and Africa is very real (McCoy *et al*., 1976). Another economically important palm, the palmyra palm (*Borassus flabellifer*) is susceptible to LY. This palm provides superior quality lumber and sugar for people throughout tropical Africa and Asia. Finally, it is possible that LY agent might infect other monocotyledonous plants such as cereal crops and pasture grasses. Thus, this disease has considerable potential for causing environmental and human catastrophe throughout the world; Thurston (1973) has classified it as a disease of intermediate international threat potential.

Resistance: At present the only known control for the disease is the use of resistant varieties. The current method of determination of coconut palms resistance to LY is by exposing the palms to LY infection in the field for an indefinite number of years. Until techniques for transmitting LY agent are developed, this means of selecting for resistant varieties will remain most practical. Any palms that possess factors for resistance to LY agent and/or vector will survive. One of the major contributions of the Coconut Industry Board's Research Department in Jamaica is the selection and testing of 'Malayan Dwarf' coconut palms. Although the majority of the coconut varieties tested in Jamaica have shown very high susceptibility to the LY, this variety is shown to be 96% resistant. All hybrids showing a promising degree of resistance have 'Malayan Dwarf' parentage. More recently, a hybrid between 'Malayan Dwarf' and 'Panama Tall' has been developed, the 'Maypan' in Jamaica (Harries and Romney, 1974). This variety has certain horticultural characteristics and is better suited to Florida growing conditions than the 'Malayan Dwarf.' But 'Maypan' is more susceptible to LY than 'Malayan Dwarf.' 'Malayan Dwarf' appears to have other undesirable characteristics, such as special susceptibility to palm aphid attack, and intolerance to salt spray and poor drainage. It requires very high maintenance and nutrient levels. Also, it lacks much of the graceful appearance of the familiar 'Jamaican Tall' coconut palms.

Although 'Malayan Dwarf' has been extensively used for replanting in Florida and Jamaica with apparent success, we should be aware of the possiblity that a mutant strain of LY agent might arise and become a major limiting factor. It is dangerous to replant 'Malayan Dwarf' as a monocrop. Sole dependence on 'Jamaican Tall' variety of coconut palm in the past left us extremely vulnerable to LY. Thus, the rationale for replanting with a number of different species of palms is quite clear. Meanwhile, the search for other sources of resistance should be continued. There are still other varieties in the world which have not been introduced and tested for resistance to LY. For example, other varieties with poor fruit quality but aesthetic appearance may be resistant to LY, which would make them suitable for ornamental purposes in Florida.

Although 'Malayan Dwarf' coconut palms are resistant to LY, the infected tissues of 'Malayan Dwarf' have been found to contain MLO (Waters, 1975; Thomas personal communications). They may serve as a symptomless carrier and share a common vector of LY agent. Therefore, they should not be shipped from the infected areas to the other parts of the world. Any introductions of 'Malayan Dwarf' coconut palm to any area should come from their native land—Malaysia—where the LY disease does not occur.

Similar diseases: Although the symptomatology described for LY in the Western Hemisphere is quite similar to that of St. Paul Wilt, Kribi and Kaincope disease in West Africa (Addison, 1978; Dollet, 1978; Johnson and Harries, 1976; Maramorosch, 1964; Steiner, 1976), LY also has symptoms in common with other diseases of coconut palms, such as bud rot and red ring disease. One must realize and consider the fact that diseases with similar symptoms may be caused by different organisms, and diseases with different symptoms may be caused by the same organism. Nevertheless, epidemiological studies, together with etiological information, indicate that LY is identical to *Kaincope* disease in West Africa (McCoy, 1976). A cixiid planthopper, *Eumyndus* sp. (Homoptera: Cixiidae), was reported on coconut palms in West Africa (Anonymous, 1976). It is important to learn more about the ecology of this insect and its role in the spread of *Kaincope* disease. Information about any of the coconut diseases of uncertain etiology (Maramorosch, 1964) would be of great value to LY research.

The international LY effort: Study and comparison of palm diseases throughout the world required international cooperation. Two teams of scientists, consisting of plant pathologist, plant physiologist, biochemist, electron microscopist, entomologist, botanist, horticulturist and plant breeder, are actively fighting LY from a necessarily worldwide view in Jamaica and Florida. Even so, for a problem of this magnitude, the manpower engaged in research work is minimal compared to other cash crops. Tomlinson (1972) has pointed out that for each scientist who works on a tropical crop, there are a hundred scientists working on temperate crops. It is difficult to transpose information from temperate plants to a tropical plant, such as coconut palms, with a very complex structure (Parthasarathy and Klotz, 1976a,b; Tomlinson, 1961). It is for this reason that, in 1973, the International Council on Lethal Yellowing of Coconut Palms (ICLY) was established at a symposium held at the Fairchild Tropical Garden, Miami, Florida. The main objectives of the ICLY are to promote international collaborative efforts among researchers working with LY and diseases of similar etiology, to attract expertise in related fields, to identify and integrate priority research areas, to procure financial support for specific projects, and to organize future conferences. The second meeting of the ICLY was held concurrently with the 4th session of the FAO Technical Working Party in Jamaica in 1975. The third meeting of the ICLY was held in Jupiter, Florida in 1977. The meeting brought together about 60 scientists from Colombia, France, Ghana, Jamaica, Surinam, United Kingdom, the United States, and West Germany. The participants have identified important future LY research activities which include: (a) culturing of MLO from host plants and suspected vectors, (b) transmission of LY agent by suspected vectors, (c) search for possible alternate hosts of LY agent, (d) breeding and evaluation of LY resistance in palms, (e) further studies on the physiology of palms, (f) development of techniques for rapid detection of LY, (g) investigation of LY transmission by seed. The fourth meeting of the ICLY was held in Fort Lauderdale, Florida in 1979. It was attended by over 60 scientists representing 18 countries. Participants included researchers on LY and related diseases of coconut palms and also experts working on other mycoplasmal diseases. The research priorities were set for: (a) expansion of transmission trials with *H. crudus* and continuation of tests with other suspected vectors, (b) intensification of MLO cultivation from LY-infected palms, (c) development of techniques for rapid detection of LY, and (d) selection and breeding of LY resistance in ornamental palms in Florida. The purpose of these ICLY meetings is to provide a forum for theoretical and practical discussions on current research and future research needs in the context of the world's coconut economy.

Summary: The lethal yellowing disease of coconut palms poses a difficult and complex problem, one with a worldwide impact on the economy and the environment. Unequivocal proof of the theory that its causal agent is a vector-transmitted MLO requires a concerted effort by scientists from several disciplines. The problem is more difficult than might be imagined because of the lack of knowledge concerning the vector, the difficulty of culturing the causal agent, and the basic deficiencies in our understanding of the morphology, structure and physiology of palms. However, we are hopeful that the LY problem will be solved and that these graceful palms will not disappear from the tropical landscape.

11.11 ACKNOWLEDGMENT

I gratefully acknowledge Dr. R. F. Whitcomb's valuable advice throughout this study. Appreciation is extended to Drs. D. M. DeLong, K. F. Harris, and G. H. Kaloostian for reviewing the manuscript and to Mr. Bill Brooks for supplying the palms used in my research.

11.12 REFERENCES

Addison, E. A. (1978). The Cape St. Paul wilt desease in Ghana. *Proc. 3rd Mtg. Int'l Council on Lethal Yellowing, Univ. of Florida.* FL-78-2: 7.

Anonymous (1976). Yearly Progress Rpt. on Coconut Breeding. FAO. Rome. 49 pp.

Beakbane, A. B., Slater, C. H. W., Posnette, A. F. (1972). Mycoplasmas in the phloem of coconut, *Cocos nucifera* L., with lethal yellowing disease. *J. Hort. Sci.* 47: 265.

Carter, W. (1966). Report to the Government of Jamaica on lethal yellowing disease of coconut. FAO. United Nations, Rome. TA2158 24 pp.

Chen, R. A. (1966). Nutritional aspects of lethal yellowing in coconuts. *Trop. Agric., Trin.* 43: 211-218.

Clark, T. B. (1977). *Spiroplasma* sp., a new pathogen in honey bees. *J. Invertebrate Pathol.*, 29: 112-113.

Dabek, A. J. (1978). Current attempts to transmit lethal yellowing disease in Jamaica. *Proc. 3rd Mtg. Int'l Council on Lethal Yellowing, Univ. of Florida.* FL-78-2: 22.

Dale, J. L., and Kim, K. S. (1969). Mycoplasma-like bodies in dodder parasitizing aster yellows-infected plants. *Phytopathology.* 59: 1765-1766.

De La Torre, C. (1906). La enfermedad de los cocoteros. *Revta de Letras y Ciencias, Univ. la Habana.* 2: 269-281.

Dijkstra, J., and Lee, P. E. (1972). Transmission by dodder of sandal spike desease and the accompanying mycoplasma-like organisms via *Vinca rosea. Neth. J. Pl. Path.* 78: 218-224.

Dollet, M. (1978). Kaincope disease and Kribi disease, two West African coconut yellows associated with the presence of mycoplasma-like organisms. *Proc. 3rd Mtg. Int'l Council on Lethal Yellowing, Univ. of Florida.* FL-78-2: 12.

Dring, B. M. (1963). Rapport au gouvernement du Togo sur la maladie de Kaincope du cocotier. Rome, FAO. 29 pp.

Eden-Green, S. J. (1978). Lethal yellowing transmission tests in Jamaica, 1975-1977. *Proc. 3rd Mtg. Int'l Council on Lethal Yellowing, Univ. of Florida.* FL-78-2: 22.

Eden-Green, S. J. and Schuiling, M. (1976). Root acquisition transmission tests with *Haplaxius crudus* (Homoptera: Cixiidae) and *Proarna hilaris* (Homoptera: Cicadidae). *Principes. J. Palm Soc.* 20: 66.

Fawcett, W. (1891). Report on the coconut disease at Montego Bay. *Bull. Bot. Dept. Jamaica.* 23: 2.

Fisher, J. B. (1975). Environmental impact of lethal yellowing disease of coconut palms. *Environ. Conservation.* 2. 299-304.

Fisher, J. B., and Tsai, J. H. (1978a). Feeding sites of leafhoppers and planthoppers on plant tissues. *Proc. 3rd Mtg. Int'l Council on Lethal Yellowing, Univ. of Florida.* FL-78-2: 23.

Fisher, J. B., and Tsai, J. H. (1978b). In vitro growth of embryos and callus of coconut palm. *In Vitro* 14: 307-311.

Fisher, J. B., and Tsai, J. H. (1979). A branched coconut seedling in tissue culture. *Principes. J. Palm Soc.* 23: 128-131.

Fiskell, J. G. A., Martinez, A. P., and Van Weerdt, L. G. (1959). Chemical studies on the roots and leaves of coconut palms affected by lethal yellowing. *Proc. Fla. St. Hort. Soc.* 72: 408-413.

Giannotti, J., Arnaud, F., Dollet, M., Delattre, R., and deTaffin, G. (1975). Culture of mycoplasma from the roots and inflorescences of coconut palms attacked by Kaincope disease. *Oleagineux* 30: 13-18.

Grant, T. J. (1967). Lethal yellowing disease of coconut. FAO Rep. No. TA2367 to the Government of Jamaica.

Grylls, N. C., and Bor , N. A. (1968). Investigations on the etiology of lethal yellowing of coconuts in Jamaica. 2, the role of insects as possible vectors. FAO Technical Working Party on Coconut Production, Protection, and Processing. 3rd Session, Indonesia.

Grylls, N. E., and Hunt, P. (1971a). A review of the study of the aetiology of coconut lethal yellowing disease. *Oleagineux* 26: 311-315.

Grylls, N. E., and Hunt, P. (1971b). Studies on the aetiology of coconut lethal yellowing in Jamaica, by mechanical and bacteria inoculations and by insect vectors. *Oleagineux* 26: 543-549.

Grylls, N. E., Hunt, P., and Bor, N. A. (1968). An approach to the study of the aetiology of coconut lethal yellowing disease. Paper presented at the First International Congress of Plant Pathology. London. July, 1968.

Harries, H. C. (1971). *Eleventh Rep. Cocon. Ind. Bd. Res. Dept., Jamaica,* 13-35.

Harries, H. C. (1978). Susceptibility to lethal yellowing disease in global perspective. *Proc. 3rd Mtg. Int'l Council on Lethal Yellowing, Univ. of Florida*. FL-78-2: 29.

Harries, H. C., and Romney, D. H. (1974). Maypan: an F_1 hybrid coconut variety for commercial production in Jamaica. *World Crops*, 26 pp.

Heinze, K. G. (1972). Lethal yellowing disease of coconut. FAO Rep. No. TA3152 to the Government of Jamaica.

Heinze, K., and Schuiling, M. (1970). Tenth Rep. Res. Dept. Coconut Industry Board, Jamaica, W. I. 84 pp.

Heinze, K. G., and Schuiling, M. (1973). Versuche zur Ermittlung der Inkubationszeit der tödlichen Vergilbung der Kokospalme (Lethal Yellowing disease of coconut palm) in der Wirtspflanze. *Anz. Schädlingskde. Pflanzen–Umweltschutz* 46: 70-73.

Hibben, C. R., and Wolanski, B. (1971). Dodder transmission of a mycoplasma from ash witches' broom. *Phytopathology.* 61: 151-156.

Howard, F. W. (1978). Survey of Homoptera associated with palms in Florida. *Proc. 3rd Mtg. Int'l Council on Lethal Yellowing, Univ. of Florida.* FL-78-2: 24.

Hunt, P., Dabek, A. J., and Schuiling, M. (1974). Remission of symptoms following tetracycline treatment of lethal yellow-infected coconut palms. *Phytopathology*. 64: 307-312.

Innes, R. F. (1949). The manganese content of leaf and inflorescence tissue in relation to the "unknown disease" of the coconut palm in Jamaica. *Trop. Agric., Trin.,* 26: 57.

Johnson, C. G. (1976). A review of the vector experiments in Jamaica from 1962 to 1971. *Principes. J. Palm Soc*. 20: 65.

Johnson, C. G., and Harries, H. C. (1976). Results of a recent survey of Cape St. Paul wilt of coconut in West Africa. *Principes. J. Palm Soc.* 20: 69.

Latta, R. (1966). Attempts to relate nematodes to lethal yellowing disease of coconut palms in Jamaica. *Trop. Agric., Trin.* 43: 59-68.

Leach, R. (1946). The unknown disease of the coconut palm in Jamaica. *Trop. Agric., Trin.* 23: 50-60.

Leather, R. L. (1959). Further investigations into the Cape St. Paul wilt of coconuts at Keta, Ghana. *Emp. J. Exp. Agric.*, 27: 105.

Maramorosch, K. (1964). A survey of coconut diseases of uncertain etiology. FAO, Rome, 39 pp.

Maramorosch, K. (1974). Mycoplasmas and Rickettsiae in relation to plant diseases. *Ann. Rev. Microb.* 282: 301-324.

Maramorosch, K., Granados, R. R., and Hirumi, H. (1970). Mycoplasma diseases of plants and insects. *Advan. Virus Res.* 16: 135-193.

Martyn, E. B. (1945). Coconut diseases in Jamaica. 1. Bronze leaf wilt and other diseases affecting the bud of coconuts. 2. Diseases affecting the leaves, crown and stem of coconuts. *Trop. Agric. Trin.*, 22: 51-59; 69-76.

Martyn, E. B. (1949). Further observations on the "unknown disease" of coconuts. *Trop. Agric. Trin.*, 26: 110-112.

Martinez, A. F. (1965). Lethal yellowing of coconut in Florida. *FAO Plant Prot. Bull.* 13: 25-29.

McCoy, R. E. (1972). Remission of lethal yellowing in coconut palm treated with tetracycline antibiotics. *Plant Dis. Rep.* 56: 1019-1021.

McCoy, R. E. (1976). Comparative epidemiology of the lethal yellowing Kaincope and cadang-cadang diseases of coconut palm. *Plant Dis. Rep.* 60: 498-502.

McCoy, R. E., Thomas, D. L., and Tsai, J. H. (1976). Lethal Yellowing: A potential danger to date production. *Ann. Rep. Date Growers' Inst.* 53: 4-8.

Mijailova, P. T. (1967). Informe sobre la investigacion de la enfermedad "Pudricion del cogollo del cocotero." *Revta Agric. Habana* 1: 74-110.

Moore, H. E. (1978). Comments on the host family of lethal yellowing and suggestions for replacing susceptible species. *Proc. 3rd Mtg. Int'l. Council on Lethal Yellowing, Univ. of Florida.* FL-78-2: 8.

Murashige, T., and Skoog, F. (1962). A revised medium for rapid growth and bioassay with tobacco tissue culture. *Physiol. Plant* 15: 485-497.

Nutman, F. J., and Roberts, F. M. (1955) Lethal Yellowing: The "Unknown Disease" of coconuts in Jamaica. *Emp. J. Exp. Agric.* 23: 257-267.

Ollagnier, M., and Weststeijn, G. (1961). Coconut diseases in the islands of Caribbean: Comparison with Kaincope disease in Togo. *Oleagineaux* 16: 729-736.

Parthasarathy, M. V. (1973). Mycoplasma-like organisms in the phloem of palms in Florida affected by lethal yellowing. *Plant Dis. Rep.* 57: 861-862.

Parthasarathy, M. V. (1974). Mycoplasma-like organisms associated with lethal yellowing disease of palms. *Phytopathology* 64: 667-674.

Parthasarathy, M. V., and Klotz, L. H. (1976a). Palm "Wood." I. Anatomical aspects. *Wood Science and Technology* 10: 215-229.

Parthasarathy, M. V., and Klotz, L. H. (1976b). Palm "Wood." II. Ultrastructural aspects of sieve elements, tracheary elements and fibers. *Wood Science and Technology* 10: 247-271.

Plavsic-Banjac, B., Hunt, P., and Maramorosch, K. (1972). Mycoplasma-like bodies associated with lethal yellowing disease of coconut palms. *Phytopathology* 62: 298-299.

Price, W. C., Martinez, A. P., and Roberts, D. A. (1967). Mechanical transmission of coconut lethal yellowing. *FAO Plant Prot. Bull.* 15: 105-108.

Price, W. C., Martinez, A. P., and Roberts, D. A. (1968). Reproduction of the coconut lethal yellowing syndrome by mechanical inoculation of young seedlings. *Phytopathology* 58: 593-596.

Schenk, R. U., and Hildebrandt, A. C. (1972). Medium and techniques for induction and growth of monocotyledonous and dicotyledonous plant cell cultures. *Can J. Bot.* 50: 199-204.

Schieber, E., and Hichez-Frias, E. (1970). Lethal yellowing disease of coconut palms in the Dominican Republic. *Phytopathology* 60: 1542.

Schuiling, M., Johnson, C. G., Eden-Green, S. J., and Waters, H. (1976). Recent attempts to find a vector associated with lethal yellowing of coconut. *Principes. J. Palm Soc.* 20: 65.

Schwabe, W. W. (1977). Attempts at vegetative propagation of coconut palm. *Proc. Int'l Symp. on Coconut Research and Development, 1976, India.*

Seal, J. L. (1928). Coconut bud rot in Florida. Univ. of Florida, *Florida Agric. Exp. Sta. Bull.*, No. 199. 27 pp.

Sherman, K. E., and Maramorosch, K. (1977). Present status of lethal yellowing disease of the coconut palm. *J. Plantation Crops* 5:75-83.

Smith, K. W. and Thomas, J. A. (1973). The isolation and in vitro cultivation of cells of *Elaeis guineensis. Oleagineux* 28: 123-127.

Steiner, K. G. (1976). Studies on the Kaincope disease of coconut palms in Togo. *Principes. J. Palm Soc.* 20: 68-69.

Thomas, D. L. (1974). Possible link between declining palm species and lethal yellowing of coconut palms. *Proc. Fla. State Hort. Soc.* 87: 502-504.

Thomas, D. L. (1976). Possible hosts of lethal yellowing. *Principes. J. Palm Soc.* 20: 59.
Thomas, D. L. (1978). Electron microscopy of palm species with suspected lethal yellowing. *Proc. 3rd Mtg. Int'l Council on Lethal Yellowing, Univ. of Florida.* FL-78-2: 15.
Thurston, H. D. (1973). Threatening plant diseases. *Ann. Rev. Phytopathol.* 11: 27-52.
Tomlinson, P. B. (1961). Anatomy of the monocotyledons. II Palmae. Oxford Univ. Press. London.
Tomlinson, P. B. (1972). Lethal yellowing of coconut—The importance of basic research. *Bull. Fairchild Tropical Garden* 27: 7-12.
Tsai, J. H. (1975). Transmission studies of three suspected insect vectors of lethal yellowing of coconut palm. *FAO Plant Prot. Bull.* 23: 140-145.
Tsai, J. H. (1977a). Attempts to transmit lethal yellowing of coconut palms by the planthopper, *Haplaxius crudus. Plant Dis. Rep.* 61: 304-307.
Tsai, J. H. (1977b). Attempt to transmit lethal yellowing of coconut palms by suspected vectors. *Proc. Int'l Symp. on Coconut Research and Development, 1976, India.*
Tsai, J. H. (1978). Vector studies in Florida. *Proc. 3rd Mtg. Int'l Council on Lethal Yellowing, Univ of Florida.* FL-78-2: 24.
Tsai, J. H. (1979). Vector transmission of mycoplasmal agents of plant diseases. *In* "The Mycoplasmas" (M. F. Barile, S. Razin, J. G. Tully and R. F. Whitcomb, eds.), Vol. 3. pp. 265-307. Academic Press, New York.
Tsai, J. H., and Anwar, M. (1977). Molting and longevity of *Oncometopia nigricans* (Homoptera: Cicadellidae) on various host plants. *Florida Entomologist* 60: 105-108.
Tsai, J. H., and Kirsch, O. H. (1978). Bionomics of *Haplaxius crudus* (Van Duzee) (Homoptera: Cixiidae). *Environ. Ent.* 7: 306-308.
Tsai, J. H., and Mead, F. (1979). A comparative study of Auchenorryncha insects associated with coconut and sabal groves. Submitted to *Florida Entomologist.*
Tsai, J. H., Woodiel, N. L., and Kirsch, O. H. (1976). Rearing techniques for *Haplaxius crudus* (Homoptera: Cixiidae). *Florida Entomologist* 59: 41-43.
Tsai, J. H., Houha, B., and Donselman, H. (1979). Longevity and survivorship of *Haplaxius crudus* (Homoptera: Cixiidae) on various palms susceptible to lethal yellowing disease. Submitted to *Florida Entomologist.*
Waters, H. (1975). *Fifteenth Rep. Cocon. Ind. Bd. Res. Dept., Jamaica.* p 51.
Waters, H. (1976). Feeding behavior of a putative vector of lethal yellowing, *Haplaxius crudus* Van Duzee (Homoptera: Cixiidae). *Principes J. Palm Soc.* 20: 65-66.
Waters, H., and Osborne, I. (1978). Preliminary studies upon lethal yellowing and the distribution of MLO in coconut palms. *Proc. 3rd Mtg. Int'l Council on Lethal Yellowing, Univ. of Florida.* FL-78-2: 15.
Westwood, D. (1953). Coconut diseases, Kets. Ho. Gold Coast, Dept. of Agric. 25 pp.
Whitcomb, R. F., and Davis, R. E. (1970). Mycoplasma and phytarboviruses as plant pathogens persistently transmitted by insects. *Ann. Rev. Entomol.* 15: 405-464.
Woodiel, N. L., and Tsai, J. H. (1978). A rotary flight trap used for sampling *Haplaxius crudus* (Homoptera: Cixiidae) in coconut groves. *J. N. Y. Ent. Soc.* 86: 37-44.
Woodroof, J. G. (1970). Coconuts: production, processing, products. AVI Publ. Co., Inc., Westport, Conn. 241 pp.

Chapter 12

INSECT INVOLVEMENT IN THE TRANSMISSION OF BACTERIAL PATHOGENS

Monty D. Harrison
Department of Botany and Plant Pathology

J. Wayne Brewer and Laura D. Merrill
Department of Zoology and Entomology

Colorado State University
Fort Collins, Colorado

12.1 INTRODUCTION 201
12.2 INSECT TRANSMISSION OF BACTERIAL PATHOGENS 202
  12.2.1 Bacterial Soft Rots 203
  12.2.2 Cucurbit Wilt 208
  12.2.3 Bacterial Wilt of Corn 211
  12.2.4 Firc Blight 214
  12.2.5 Olive Knot 220
  12.2.6 Bacterial Rot of Apple 222
  12.2.7 Southern Bacterial Wilt 224
12.3 SUMMARY 225
12.4 ACKNOWLEDGMENTS 276
12.5 REFERENCES 276

12.1 INTRODUCTION

Bacterial plant pathogens cause serious economic losses throughout the world and attack members of virtually all major families of higher plants. Bacterial taxonomy is in a transitional state, but most workers agree that there are 180-200 species of plant pathogenic bacteria.

Bacterial pathogens are remarkedly successful despite numerous handicaps. Most species do not form spores and thus lack the resistant structures commonly produced by fungi. These bacteria depend instead on the protection provided by the host plant during unfavorable periods. The bacterial cell is thin walled, fragile and easily

ISBN 0-12-326450-2

damaged by dessication, heat and irradiation. This limits survival outside the host to relatively short periods and renders them susceptible to adverse environmental conditions. Furthermore, plant pathogenic bacteria lack an efficient mechanism, independent of animal vectors, for long distance dispersal. Unlike windblown fungal spores bacterial cells are sticky and not easily airborne. Instead, they depend mainly on windblown or splashing water for local dissemination and upon the transport of infected host tissues for long distance movement. Finally, bacterial pathogens are unable to directly penetrate plant cuticles. Some enter natural openings, but many require wounds created by other agents to gain entrance into the host plant.

Insects aid bacterial plant pathogens to overcome some of these deficiencies. Assistance in survival, dissemination and penetration of host tissue are the most important ways that insects are involved with plant pathogenic bacteria. When this assistance is combined with the natural advantages of bacteria over other pathogens in the areas of rapid reproduction, rapid entry into infection courts and independent motility, the pathogens become extremely destructive.

12.2 INSECT TRANSMISSION OF BACTERIAL PATHOGENS

With few exceptions insect-bacterial plant pathogen relationships are non-obligate and in many cases strictly accidental. Since bacterial inoculum is often presented as a sticky exudate, insects which contact it are readily contaminated and thus become potential vectors. In a few cases close biological relationships appear to have developed between insects and plant pathogenic bacteria which are highly beneficial to one or both of the biological entities.

Insects aid bacterial pathogens in several distinct ways which are listed below with selected examples of these relationships: (1) They may aid survival of the pathogens. *Erwinia tracheiphila* (Smith) Bergey *et al.*, the cause of cucurbit wilt, *Erwinia stewartii* (Smith) Dye, which incites Stewart's wilt of corn and *Erwinia carotovora* (Jones) Bergey *et al.*, which causes soft rot of onion and blackleg of potato, have been reported to survive within the bodies of their insect vectors. (2) They may disseminate both primary and secondary inoculum of the pathogen from plant to plant. Whiteflies are reported to disseminate *Xanthomonas phaseoli* (E. F. Smith) Dowson from infected to healthy plants (Sabet and Ishag, 1969) where infection may occur via natural openings. Similar relationships exist between *Xanthomonas campestris* (Pammel) Dowson and *Erwinia amylovora* (Burrill) Winslow *et al.* and their insect vectors. (3) They may provide wounds necessary for the entry of bacterial pathogens into healthy hosts. This relationship involves wounding of the host by the insect followed by invasion by bacteria from another source. This situation is illustrated by the relationship between *Agrobacterium tumefaciens* (Smith and Townsend) Conn and soil insects discussed by Agrios (1969). Insects may provide wounds and also deposit bacteria into wounds which they have created. This is true in the case of cucurbit wilt and bacterial rot of iris caused by *Erwinia tracheiphila* and *E. carotovora*, respectively. Feeding wounds are the most common avenues of invasion but oviposition wounds are also effective in the case of olive knot caused by *Pseudomonas savastanoi* (Smith) Stevens (Petri, 1909, 1910), bacterial rot of apple caused by *P. melophthora* Allen and Riker (Allen and Riker, 1932; Allen, 1931) and other diseases. (4) They may help bacterial pathogens circumvent unfavorable environmental conditions as in the case of heart rot of celery caused by *Erwinia carotovora*, where the insect vector deposits the bacterium in a location where it is protected from drying, irradiation and heat (Leach, 1927). (5) In some cases insect vectors may provide substrates for the growth of bacteria which in turn adversely affect the host plant. A saprophytic species of *Xanthomonas* sometimes grows on substrates provided by fruit fly larvae in coffee berries and causes an undesirable flavor in the coffee bean (Stolp, 1960)

J. G. Leach (1940) and Walter Carter (1962, 1973) have reviewed the subject of insect transmission of bacterial plant pathogens. More recent work, however, has cast new light on some insect-bacterial plant pathogen relationships which bear further attention here.

Selected examples of insect-bacterial plant pathogen relationships for which considerable evidence exists are discussed below. Table I lists all of the insect-bacterial disease relationships which we have found in the literature. Many of the relationships reported lack adequate supporting data or the data are conflicting; for example, (1) bacterial gall of Douglas fir [*Agrobacterium pseudotsugae* (Hansen and Smith) Savulescu], (2) black rot of Brassicaceae [*Xanthomonas campestris* (Pammel) Dowson], (3) blade blight of oats (*Pseudomonas avenae* Manns), (4) spot disease of cauliflower [*Pseudomonas maculicola* (McCulloch) Stevens], (5) *Erwinia espinosa* F. M. Brown associated with an aphid gall on witch hazel, all of which were discussed by Leach (1940) and (6) walnut blight [*Xanthomonas juglandis* (Pierce) Dowson], (7) pod twist of French beans (*Pseudomonas flectens* Johnson), (8) ring rot of potatoes [*Corynebacterium sepedonicum* (Spieckermann and Kotthoff) Skaptason and Burkholder] and (9) citrus canker [*Xanthomonas citri* (Hasse) Dowson] which were discussed by Carter (1973) (see Table I for references on these and other diseases).

12.2.1 Bacterial Soft Rots

Bacterial soft rots, which occur in all parts of the world, are some of the most destructive diseases of many vegetables and some ornamental plants. Soft rot usually affects fleshy storage tissues where it appears as a soft watery or slimy decay, often accompanied by an offensive odor. When potato stem tissues are invaded they develop a characteristic black discoloration known as blackleg.

Extensive research has been done on bacterial soft rot and potato blackleg for many years. The role of insect vectors in the epidemiology of soft rot in several crops has received considerable attention.

12.2.1.1 Blackleg and Bacterial Soft Rot of Potato. Potato blackleg, caused by *Erwinia carotovora* var. *atroseptica* (van Hall) Dye and perhaps also be *E. carotovora* var. *carotovora* (Jones) Dye, is a soft rot disease that occurs wherever potatoes are grown. The bacteria decay seedpieces, produce blackleg symptoms in stems and decay tubers in storage. Annual losses frequently reach five percent but crops from individual fields may sustain greater losses.

Early studies by Leach (1930b), Davidson (1948) and Bonde (1950) indicated that the pathogen was commonly present in the soil and needed only favorable conditions, mainly wet soils and low oxygen levels, to rot potato seedpieces and invade the stem to produce blackleg. In the past invasion by the soil borne bacteria through the cut surface of the seedpiece, which failed to heal properly under adverse soil conditions, was considered the major factor in disease production (Leach, 1930b). Recently, however, several workers including Graham (1958), Graham and Harper (1967), Graham and Hardie (1971), Lazar and Bucur (1964), Logan (1969) and Voronkovich (1960) have demonstrated that the pathogen is probably not soil borne but rather survives from season to season in contaminated tubers.

Leach (1940) summarized work done by himself (Leach, 1925, 1926, 1930a, 1931, 1933) and others (Bonde, 1928, 1930b, 1939a and b; Huff, 1928) which suggested a mutualistic symbiotic relationship between the seedcorn magot, *Hylemya platura* (Meigen) (Diptera: Anthomyiidae), and *Erwinia carotovora*. Leach believed that the bacterial pathogen survived in the soil and that the seedcorn maggot larvae were important in wounding the potato seed pieces and providing entry points for the bacteria. He also demonstrated that several bacterial species, sometimes including *E. carotovora*, survived inside the pupae and emerged with the adults of this insect. Eggs laid by contaminated adults were shown to be externally contaminated at the time of oviposition. These bacteria were reported to be normal inhabitants of the insect gut and according to Leach (1926, 1931) and Huff (1928) aided the insect nutritionally by reducing the plant tissues into a form more usable by the larvae. Thus, according to Leach (1940) both flies and bacteria benefited from the relationship, although Leach (1931) believed that this insect was merely an "incidental" vector of the pathogen.

The epidemiological significance of this mutualistic relationship is unknown. Bonde (1939b) isolated soft rot bacteria from only 7 percent of the eggs, 15 percent of the puparia, and 10 percent of the adults of *Hylemya platura*. (Bonde also reported that he was unable to isolate soft rot bacteria from *H. platura* larvae; however, he considered these tests too limited to be conclusive.) Since Bonde (1939b) isolated *E. carotovora* from seedcorn maggot puparia that had overwintered in the field, the insects could provide an important overwintering site for the pathogen, particularly where only disease-free seed potatoes are planted and sanitation is practiced.

Bonde (1930b) reported that a related fly, *Hylemya florilega* (Zetterstedt), was also an important vector of *E. carotovora* in potatoes and probably had a relationship to the bacterium similar to that of *H. platura*, and Tamimi and Banfield (1969) reported like results with *Drosophila busckii* Coquillett (Diptera: Drosophilidae). An analogous relationship apparently exists with *Hylemya brassicae* (Weidemann), the cabbage maggot, and soft rot in the Brassicaceae (Johnson, 1930).

Investigations during the past 10 years have focused on the transmission of

E. carotovora by insects which are not normally pests of potatoes. Such insects acquire the pathogen from waste dumps or potato cull piles (Kloepper *et al.*, 1976; Graham *et al.*, 1976; Kloepper, 1977) or even from infected plants (Molina *et al.*, 1974) and transmit it to fresh injuries on aerial portions of healthy potato plants (Molina *et al.*, 1974; Kloepper *et al.*, 1976; Harrison *et al.*, 1977; Kloepper, 1977). In experiments with adult *Drosophila melanogaster* Meigen, Kloepper *et al.* (1976) found that injured tissues became less susceptible to transmission by insects with time, indicating that the condition of the wound, that is, fresh and moist as opposed to old and dry, is important in transmission. Recently several authors have isolated *E. carotovora* from many miscellaneous insects, primarily Diptera (such as *Drosophila* spp.), collected in the field (Tamimi and Banfield, 1969; Graham and Hardie, 1971; Kloepper *et al.*, 1976; Graham *et al.*, 1976; Harrison *et al.*, 1977; Kloepper, 1977; Phillips, 1977). Graham *et al.* (1976) showed that insects collected at an isolated vegetable dump were contaminated with the same serotypes of *E. carotovora* var. *carotovora* that were later isolated from potato plants 120 to 150 meters away. These plants had been grown from seed potatoes derived from *Erwinia*-free stem cutting stocks, and these authors believed that contamination of the stock occurred via the insects. Molina *et al.* (1974) reported that transmission occurred most often to plants nearest to the inoculum source, although Kloepper *et al.* (1976) recovered *E. carotovora* from wounded potato plants located 183 meters downwind from a potato cull pile infested with *Erwinia*-contaminated Diptera. Phillips and Kelman (1978) used a fluorescent antibody serum technique and a unique serotype of *E. carotovora* var. *carotovora* to test dissemination of blackleg bacteria by insects. They inoculated potato tubers in a cull pile with the organism and demonstrated that this serotype was present in insects and on potato leaves and tubers in the field up to 26 meters from the source.

The current evidence strongly suggests that insects play a role in the dissemination of *Erwinia carotovora* in potatoes. Insect vectors are probably often responsible for the initial reintroduction of the bacteria into fields planted with *Erwinia*-free potato stocks. The likelihood of gross recontamination from this source is remote, but after introduction the bacteria can be spread rapidly in the crop during harvesting and handling operations and perhaps also by insects such as *Hylemya* spp. which harbor the bacteria in their bodies.

12.2.1.2 Soft Rot of Onion. Taylor (1952) reported that larvae of the black onion fly, *Tritoxa flexa* (Wiedemann) (Diptera: Otitidae), were always found associated with bacterial decay in onion bulbs. Other species also found commonly included *Hylemya antiqua* (Meigen) (Diptera: Anthomyiidae), *H. platura* (Meigen) and *Eumeris strigatus* Fallén (Diptera: Syrphidae). Taylor reported that unidentified bacteria on the eggs of *T. flexa* produced a soft rot in onions. When eggs were surface sterilized no soft rot occurred and larvae failed to develop past the first instar. Taylor concluded that one to several species of microorganisms (he did not identify the pathogen) were necessary for completion of larval development in the onion bulb.

Doane (1953) demonstrated that the onion maggot, *Hylemya antiqua*, was important to the survival of *Erwinia carotovora*, the cause of soft rot of onion. *H. antiqua* has several generations a year and consequently attacks onions of various sizes. Small bulbs may be rapidly and completely mined by the larvae so that soft rot is not important in killing these plants. However, when larger bulbs are attacked, soft rot frequently follows and completes the destruction of the bulb. Doane (1953) found that soft rotted onions in storage usually had evidence of maggot injury. He also showed that eggs and larvae of the onion maggot are frequently surface-contaminated with *E. carotovora*, but that the bacterium was not carried internally in these stages. However, he demonstrated that the bacterium was present inside the pupa of the insect and reported that it could be isolated from puparia stored under refrigeration for 10 months. Consequently, it is possible that *E. carotovora* overwinters inside the puparia and that the insect protects the bacterium during this period. The feeding activities of the onion maggot also aid the bacterium by providing wounds for entry into the plant. The bacterium aids the insects nutritionally, as suggested by Friend *et al.* (1959) who reported that larvae of *H. antiqua* could not be reared past the second instar on sterilized onions. These workers could rear the species axenically on a chemically defined diet but growth was accelerated when microorganisms (particularly an unidentified species of *Bacillus*) were added. Apparently live microorganisms are not required since washed, dead *Bacillus* sp. cells produced the same effect. Brooks (1963) concluded from the work of Friend *et al.* (1959) that *H. antiqua* requires microbial breakdown of onion tissue for adequate nutrition. Thus it appears that a close relationship exists between *H. antiqua* and bacterial rot of onions caused by *E. carotovora*. The insect aids the survival of the bacterium through a period of unfavorable environmental conditions and also produces wounds by which the bacterium enters the plant. Attacks by the bacterial pathogen apparently reduce the onion tissue to a form more usable by the fly larvae and perhaps the bacteria themselves provide some of the nutritional requirements of the insects. The fact that *H. antiqua* and other insects are attracted to rotting onions (Doane, 1953) also suggests that tissue in that condition might be important to the survival of the insects.

12.2.1.3 Soft Rot of Iris. Injuries caused by the iris borer, *Macronoctua onusta* Grote (Lepidoptera: Noctuidae), are apparently entry points for the soft rot bacterium, *Erwinia carotovora* (Jones) Bergey *et al.* into iris. Ries (1929) noted that the insect "paves the way" for introduction of the bacterium in what appears to be the first reference to this insect-bacterial association. Later McDaniel (1935) reported that the large larval galleries of this insect in the iris rhizomes "invariably" became infected by *E. carotovora*. He reported that soft rot was more destructive to the plant than was the borer and indicated that the association was beneficial to the borer by suggesting that the bacterium "reduces the interior of the rhizome to a slimy rotten mass in which the borer thrives."

Thanos (1948) reported that wounding was necessary for entrance of *E. carotovora* into the iris plant but presented no experimental data to support this idea.

He offered circumstantial evidence for this: he never found soft rot infected iris plants without one or more borers, and he never found borers in healthy plants. Thanos (1948) isolated *E. carotovora* from the intestinal tract and associated structures of surface sterilized iris borer larvae as well as from fecal pellets. He transferred iris borers from caged diseased plants to healthy plants and demonstrated that the larvae could transmit the bacterium, whether or not they were surface sterilized prior to transfer.

Howard and Leach (1962, 1963) also reported that the iris borer and bacterial soft rot were consistently associated. They noted that control of the disease was dependent upon effective insect control, although they stated that the insects were probably not important in secondary spread since the larvae do not move readily from plant to plant. Howard and Leach (1963) believed that the principal role of the insects involved wounding of plants, inoculation of the bacteria from the soil into the wounds and, through continued wounding, the prevention of wound-cork formation which otherwise would seal off the pathogen and stop infection.

12.2.1.4 Heart Rot of Celery. No research has been published on insect relationships with *Erwinia carotovora* infections of celery since Leach (1940) summarized the literature. Carter (1973), however, commented briefly on Leach's work. *E. carotovora* (Jones) Bergey *et al.* attacks various parts of the celery plant, but the disease is most destructive when the "heart leaves" (the young leaves near the center of the plant) are attacked. This "heart rot" of celery does not kill the plant, but the destruction of the central bud and the subsequent elongation of the stem make the plants unmarketable. Leach (1927) reared several insect species from rotting celery leaves; *Scaptomyza graminum* (Fallén) (Diptera: Drosophilidae) and *Elachiptera costata* (Loew) (Diptera: Chloropidae) were most commonly obtained. He collected eggs of both of these species and maintained them on moist celery leaves. A brown rot occurred on the leaves around the larval entry points, and typical soft rot developed. *E. carotovora* was isolated from the decaying tissue. Leach (1927) reported that leaves maintained under similar conditions in the absence of the insects did not develop soft rot, but apparently he did not wound the leaves. Thus it is not possible to determine with certainty if the insects were acting as vectors or merely providing infection courts for bacteria already present on the leaf surfaces. In his book Leach (1940) suggested that there was a close association between the soft rot organism and these two insect species but noted that experimental evidence for such a relationship was lacking.

Leach (1927) reported that heart rot of celery was most destructive during hot, dry weather in contrast to infections by the same organism in other crops which were most damaging under moist conditions. He believed that this was due to changes in oviposition behavior of the insect vectors during dry weather. The eggs are laid on celery leaves, normally in protected areas where relative humidity is high. When the weather is damp, these insects oviposit almost anywhere on the plant, normally on the most accessible external parts. The resulting soft rot thus occurs on the exterior leaves which are normally discarded prior to marketing;

thus little crop damage occurs. During dry weather the eggs are laid in the more protected, inner portions of the plant and the soft rot bacterium destroys the central bud, resulting in an unmarketable plant as described above. Ogilvie *et al.* (1934) observed that soft rot in celery was associated with high populations of a "carrot fly" in Great Britain and that the rot spread from the larval tunnels of these flies. These authors also reported that soft rot was similarly associated with *Scaptomyza graminum*, and supported Leach's (1927) observations that the spread of the disease occurred during warm dry periods.

Richardson (1938) reported that *E. carotovora* was also spread by the feeding of the tarnished plant bug, *Lygus lineolaris* (Palisot de Beauvois) (Heteroptera: Miridae), but Leach (1940), commenting on Richardson's work, indicated that this insect is probably not an important vector under natural conditions.

12.2.2 Cucurbit Wilt

Bacterial wilt of cucurbits, incited by *Erwinia tracheiphila* (Smith) Bergey *et al.*, is a serious disease of cucumber and muskmelon and, to a lesser extent, pumpkin and squash. It occurs throughout the range of the host plants. The pathogen is strictly a vascular parasite and produces wilting by blocking the vascular system. After infection wilting occurs, beginning with a single leaf, but eventually spreading to the entire plant and killing it. A stringy, viscous, white bacterial ooze emerging from freshly cut stems is considered diagnostic for the disease in the field.

According to Carter (1973) the pathogen is entirely dependent on two insect vectors, the striped cucumber beetle, *Acalymma vittata* (Fabricius) and the spotted cucumber beetle, *Diabrotica undecimpunctata howardi* Barber, (both Coleoptera: Chrysomelidae) for overwintering, inoculation and dissemination. Although the concept of the total dependence of *E. tracheiphila* on these two chrysomelid beetles is commonly accepted, a careful reveiw of the literature offers little experimental evidence to support this conclusion.

Erwin F. Smith (1893) first described cucurbit wilt and suggested that the "blight" was associated with punctures or insect feeding wounds. This appears to be the first suggestion of insect transmission of the bacterium. In a later paper, Smith (1894) concluded that this bacterial disease of cucumber, cantaloupe and squash was almost certainly transmitted by insects although he offered no experimental evidence to support this contention. Smith later (1895) infected plants by wounding them with needles and applying pure cultures of the bacterium. In this paper he stated that it was a simple supposition that insects transmit *Erwinia tracheiphila*.

Rand (1915) apparently conducted the earliest field experiments on the transmission of cucurbit wilt by cucumber beetles. He placed 50 screen cages, each containing three to four plants, in a cucumber field. He excluded cucumber beetles from 46 of these cages although he stated that flea beetles and aphids could move in and out of the cages readily. Only one of the caged plants developed wilt and that cage was found to contain a striped cucumber beetle "which was summarily disposed of." Rand introduced five or six striped cucumber beetles (*Acalymma vittata*) into the remaining four cages. These insects were collected from a cucurbit field where they were assumed to have overwintered. Rand stated that the circumstances of their capture precluded their having fed on diseased plants prior to their being caged with the healthy plants. Apparently no wilt had yet appeared in the field, but Rand seems not to have considered the possibility that the beetles may have fed on debris or other infected plant species. Plants in one of the cages containing cucumber beetles developed wilt symptoms and Rand concluded that cucumber beetles were major vectors of the pathogen.

Later, in a more extensive study, Rand and Enlows (1916) reported no cucurbit wilt in 58 beetle proof cages, although it was prevalent in plants around the cages. They also demonstrated that wilt was not transmitted from a diseased plant to a healthy one when the two were confined in the same pot without insects. Although adding a bacterial suspension to the soil around potted plants produced wilt in a few cases, it seemed to have been associated with root injury.

In this same study Rand and Enlows attempted to show that the pathogen overwintered in the striped cucumber beetle, *A. vittata*. The insects were allowed to feed on wilted plants, then were stored at 10°C for two months, however, they did not transmit the pathogen. Also, when specimens of *A. vittata* and *Diabrotica undecimpunctata howardi* were collected from the field in December, transmission did not occur. Thus these workers demonstrated that *A. vittata* was able to transmit the wilt bacterium but no evidence for winter survival in the beetles was presented.

Rand and Enlows (1920) later demonstrated that both *A. vittata* and *D. undecimpunctata howardi* were effective vectors of *E. tracheiphila* and that *A. vittata* could infect up to four plants successively after one exposure to a diseased plant. However, their continued efforts to demonstrate overwintering of the bacterium in beetles were still not successful. They could not keep beetles alive for more than one week in cold storage in this study, and in the spring they could not recover any of numerous beetles that were placed in the fall in bottomless cages extending 15 inches into the soil in the field. Because of their observation that wilt frequently occurred on the earliest planted cucurbitaceous crops, Rand and Enlows (1920) concluded that the bacterium overwintered in *A. vittata*. Additional circumstantial evidence presented by these workers was that the number of wilting plants peaked about one month after peak beetle populations. Rand and Cash (1920) extended these experiments by dissecting viscera and mouth parts of *A. vittata* and occasionally *D. undecimpunctata howardi*, using sterile technique, then using this material to inoculate cucumber plants. About 0.5 percent of the plants inoculated with viscera and only a few inoculated with mouth parts became infected. Rand and Cash also collected beetles from the field 15 to 18 days prior to the appearance of wilt in the area and caged them on cucumber plants which subsequently developed wilt symptoms. In another experiment, Rand and Cash (1920) isolated *Erwinia tracheiphila* from the viscera of surface sterilized beetles collected in the fall. Also, when feces of infected beetles were allowed to drop into wounds of cucumber leaves, the plants developed wilt symptoms. Again the authors concluded that the pathogen overwintered in a "small percentage" of *A. vittata*. However, no conclusive data were presented to support this hypothesis.

Doolittle (1921) studied overwintering of *E. tracheiphila* by collecting *Acalymma vittata* when it first appeared (May 25-June 6, 1917) and before cultivated cucurbits were planted. The beetles were caged with healthy cucurbit plants which subsequently wilted. Doolittle did not report the condition of control plants, if any. He noted that *A. vittata* fed on *Echinocystis (Micrampeles) lobata* (Michx.) Torr. and Gray (a wild cucumber) before other cucurbits appeared but reported that he had never observed wilt symptoms on *E. lobata*. Apparently no bacterial isolations were attempted from the wilted plants, from the overwintered beetles or from *E. lobata*.

Leach (1964) conducted a series of experiments on the role of *Acalymma vittata* and *Diabrotica undecimpunctata howardi* as overwintering host of *E. tracheiphila*. He collected the two species in early spring and in mid- and late summer. He also allowed beetles to feed on wilted cucumber plants. Leach sectioned and stained the beetles, but found very few bacteria in the intestines of the insects, even after they had fed on wilted cucumber plants. He also noted that no intestinal caeca were present where bacteria might accumulate. Isolations from 10 beetles (five of each species) yielded only one culture of *Erwinia tracheiphila*, and that was from a beetle which had just fed on a wilted plant. In addition most of the stained bacteria observed were in the midgut mixed with food. Leach concluded that survival of the pathogen in beetles over winter was questionable and noted that the beetles feed for 40-50 days in early spring before cultivated cucurbits are available.

With the exception of Leach (1964) researchers who have studied the problem agree that the pathogen is dependent on cucumber beetles, especially *A. vittata*, for winter survival. Although this concept does not appear to be supported by the data presented, it has been perpetuated in the literature, culminating in statements such as, "The disease is entirely dependent on two insect vectors–the striped cucumber beetle [*Diabrotica* (=Acalymma) *vittata* Fabr.] and the twelve-spotted cucumber beetle [*D. duodecimpunctata* Oliv. (=*D. undecimpunctata howardi* Barber)]–for inoculation, dissemination and overwintering of the pathogen" (Carter, 1973). The research done thus far simply does not support this type of association in regard to overwintering. If the bacterium overwintered in the gut of the beetles, and Leach (1964) has shown this to be highly improbable, feeding on plants prior to the presence of cultivated cucurbits would likely "clean out" the gut. There is, in addition, Leach's evidence (1964) that bacteria contaminated beetles do not survive as well as healthy ones during overwintering. Leach's point (1964) that the striped cucumber beetle feeds for a considerable time on plants, mainly in the family Rosaceae, before cultivated cucumber plants are available is a convincing one. His suggestion that these plants be given consideration as sources of primary inoculum appears not to have been pursued. Carter (1973) did not comments on Leach's 1964 paper or cite it at all but acknowledged that the beetles' feeding on these early plants would have a cleansing effect on the intestinal tract.

Carter suggested, without offering evidence, that the low levels of inoculum overwintering in the beetles were increased by the infection of wild cucumbers or other plants in early spring, so that the insects were carrying fresh inoculum when cultivated cucumbers were attacked later. A more reasonable explanation is that the bacterium overwinters in the plants or debris that the emerging beetles feed upon in early spring. Certainly this problem needs additional research before many questions are resolved but the available evidence as opposed to supposition and conjecture suggests that the overwintering site for the bacterium is not known and that the main contribution of the cucumber beetles is in the transmission of the pathogen. Certainly the evidence does not prove that *E. tracheiphila* is "totally dependent" on these beetles.

12.2.3 Bacterial Wilt of Corn

Stewart's bacterial wilt of corn, incited by *Erwinia stewartii* (Erw. Smith) Bergey *et al.*, was first discovered on Long Island in 1895 by F. C. Stewart, who observed losses of 10 to 100 percent on sweet corn (Pepper, 1967). The pathogen is primarily a vascular parasite although other tissues are sometimes affected. The pathogen attacks all types of corn, *Zea mays* L., but varieties differ in their susceptibility. Robert (1955) called the extensive streaking that develops when leaves are attacked Stewart's leaf blight, separating it from bacterial wilt which occurs when the water conducting tissues of the stalk are clogged. Yellow bacterial ooze from cut stems is considered an important diagnostic symptom of bacterial wilt. The leaf infection (Stewart's leaf blight) rarely spreads to the stalks, however, and thus cut stalks do not generally display the ooze symptoms (Carter, 1973).

Stewart's wilt is distributed throughout most of the U. S. and is also found in Canada, Mexico, Costa Rica, Puerto Rico, China, the U.S.S.R., Italy, Yugoslavia, Switzerland and Poland (Pepper, 1967), although in some of these areas more than one bacterium may be involved (Ivanoff, 1933b, Nemlienko, 1951). Elliott (1935) discussed the history of the disease and presented an excellent review of the research on corn wilt dissemination to that time.

Stewart's wilt was thought to be seed borne for many years. Various lines of evidence introduced by different workers including Smith (1909, 1920), Rand and Cash (1921) and Thomas (1924) indicated that the bacterium was transmitted with the seed. However, experiments by Rand (1923) in which he caged corn plants to exclude insects suggested that some factor other than seed or soil transmission was involved. In 1923 Rand demonstrated that the corn flea beetle, *Chaetocnema pulicaria* Melsh (Coleoptera: Chrysomelidae), was responsible for secondary spread of *E. stewartii.* Later Rand and Cash (1924) showed that the toothed flea beetle, *C. denticulata* (Ill.) was also a vector. These workers also suggested that the spotted cucumber beetle (=southern corn rootworm), *Diabrotica undecimpunctata howardi* Barber (Coleoptera: Chrysomelidae), was also involved in the early season spread of inoculum by introducing bacteria into larval feeding wounds on corn roots. Much of the evidence presented in this paper was circumstantial. Rand and Cash later

(1933) presented data which, although not conclusive, tended to verify their earlier hypothesis. Ivanoff (1933b) showed that the bacteria were located in the vascular system of the chalazal region of contaminated kernels. However, his experiments demonstrated that a root tip emerging from the kernel was not infected, even though surrounded by bacteria, unless it was wounded. Infection of root tips quickly followed wounding by artificial means. Injury to root tips caused by white grubs, *Phyllophaga* sp. (Coleoptera: Scarabaeidae), growing in artificially contaminated soil also resulted in infection. Ivanoff (1933a and b) found that larvae of the northern corn rootworm, *Diabrotica longicornis* Say (Coleoptera: Chrysomelidae), transmitted the pathogen from infected to healthy plants.

Elliott and Poos (1934) reported that the Stewart's wilt bacterium overwintered in 19% (of 175 individuals) of *Chaetocnema pulicaria* in Virginia. These authors isolated the pathogen (identified in culture and by inoculation tests) from surface-sterilized adult flea beetles collected from orchard grass and alfalfa in April, 1934. Other individuals from this collection were caged in a greenhouse on healthy corn plants which subsequently developed Stewart's wilt. *E. stewartii* was later isolated from the diseased plants. These workers concluded that it is probable that these beetles, and possibly other insects, were responsible for overwintering as well as dissemination of the pathogen. Later Poos (1955) provided additional support for this hypothesis when he demonstrated that the bacteria could survive very low temperatures inside *C. pulicaria* and could pass intact through the beetle's alimentary tract.

Poos and Elliott (1935) attempted to isolate the wilt bacterium from about 40 other insect species but failed to do so. They reported that as many as 75 percent of *C. pulicaria* were found to contain the bacterium when collections were taken from different host plants and localities under varying conditions in the field.

Frutchey's extensive work (1936) showed that the seedcorn maggot, *Hylemya platura* (Meig.) (Diptera: Anthomyiidae), was also important in development of corn wilt. *Erwinia stewartii* was isolated from external parts and digestive tracts of the larvae. *H. platura* larvae were also frequently found in wilting corn plants. When adult *H. platura* were caged on corn plants in the field, 72% of 93 plants developed wilt symptoms, and all of the diseased plants had larval injury. However, Leach (1940) pointed out that this work did not show whether the insects were acting as vectors or merely as wounding agents that allowed the bacterium to enter the roots. *E. stewartii* was also isolated from wheat wireworms, *Agriotes mancus* (Say) (Coleoptera: Elateridae), and the identity of the isolates was confirmed by inoculation tests (Frutchey, 1936). Attempts to isolate bacteria from the intestinal tract of *A. mancus* were not successful. Corn plants in the greenhouse fed upon by *A. mancus* developed wilt symptoms, but if the wireworms were kept in sterilized soil for two weeks the pathogen could no longer be isolated from them.

In a study comparing the effectiveness of various insects as vectors of corn wilt, Poos and Elliott (1936) found that the pathogen was most easily transmitted by *Chaetocnema pulicaria*, less frequently by *C. denticulata* and least often by *Diabrotica undecimpunctata howardi*, the spotted cucumber beetle. Limited efforts to transmit the disease by larvae of *D. undecimpunctata howardi* were not successful. Adults of numerous other insects were tested as vectors by first caging them on diseased plants and then transferring them to healthy plants. None of the 13 species tested transmitted the pathogen under these conditions.

Poos and Elliott (1936) isolated the *E. stewartii* from several insect species collected "from or near corn" during the growing season. In studies on retention of the bacterium, Poos and Elliott (1936) allowed adult *C. pulicaria* a five day acquisition period on infected plants then transferred them to healthy corn plants every five days. The beetles remained infectious for essentially the rest of their lives. Poos and Elliott believed that the wilt bacterium developed too slowly for the beetles to be re-infested from the plants on which they were feeding. In other experiments, *C. pulicaria* adults retained the bacterium for 22-49 days after acquisition, adult *C. denticulata* for 37 days and adult *D. undecimpunctata howardi* for 5-9 days. The authors suggested that the bacterium overwintered in bodies of the vector. However, better evidence for this hypothesis was presented later by Elliott and Poos (1940) when they reported isolating *E. stewartii* from *C. pulicaria* at least once during every month of the year except February. Collections of the beetle were made whenever weather conditions allowed them to become active (the authors did not state whether the beetles were feeding during these active periods). *E. stewartii* was isolated less consistently from *C. denticulata* collected under similar conditions and the authors stated that they did not believe the pathogen overwintered in that species. Later the range of host plants of *E. stewartii* was extended by Elliott (1935), Elliott and Robert (1939) and Poos (1939) to include Kentucky blue grass (*Poa pratensis* L.), orchard grass (*Dactylis glomerata* L.), wheat (*Triticum aestivum* L), eastern gama grass [*Tripsacum dactyloides* (L.)] and other species. It seems quite probable that wild plants could be intermediate hosts between the insects and corn, as suggested by Carter (1973). It is also possible that the pathogen overwinters in plant debris or other symptomless plants not yet determined to be hosts for the bacterium. The evidence seems clear that certain insects, especially *C. pulicaria*, are effective vectors of the wilt bacterium and may protect the pathogen for periods extending to over 40 days, depending on the vector species involved. However, clear evidence for survival through the winter is still lacking.

Elliott and Poos (1940) showed that the percentage of infested beetles in the spring was lower than in the previous fall. They also found that the percentage of infested beetles increased as the season progressed. However, Heichel *et al.* (1977) found that the incidence of *E. stewartii* in the vector *C. pulicaria* was unrelated to the seasonal patterns of disease incidence. They reported that the first disease cycle apparently resulted from bacteria transmitted by overwintering flea beetles and the second cycle by the new beetle generation.

The importance of insect vectors in the epidemiology of bacterial wilt is indicated by the circumstantial evidence presented by Stevens (1933, 1934, 1937) and summarized by Leach (1940) which indicated that warm winter temperatures are directly related to disease incidence the following summer. Stevens (1934) devised a temperature index based on the sum of the mean temperatures for December, January and February. Stevens' index, as modified by Boewe (1956), is a valuable tool in predicting disease severity; however, because of the success of resistant varieties the temperature index has been ignored for several years (Pepper, 1967). Recently Castor *et al.* (1975b), using the Stevens-Boewe system, have devised a computer program to accurately predict disease severity on local and regional bases in Pennsylvania.

In summary, *Erwinia stewartii* is primarily vectored by the corn flea beetle, *Chaetocnema pulicaria*. Several other insects are also involved in transmitting the pathogen to above- and below-ground parts of the plants. The disease is controlled through the use of resistant varieties (Pepper, 1967) and insecticides (Heichel *et al.*, 1977) when warm winters allow the survival of large numbers of *C. pulicaria*. Overwintering is generally thought to occur within this beetle, although there is no conclusive evidence that this is so. Several native grasses on which the beetles feed have been found to harbor the bacterium. *E. stewartii* also overwinters in corn seed.

12.2.4 Fire Blight

Fire blight of apple, pear and related species, caused by *Erwinia amylovora* (Burrill) Winslow *et al.*, was the first of all bacterial diseases shown to be vectored by insects (Waite, 1891). Since then as many as 100 or more insect species have been shown to be involved in some way in the transmission, survival and inoculation of this bacterium into host plants. Virtually all of the mechanisms by which insects aid bacterial pathogens are represented by the fire blight disease.

12.2.3.1 Dissemination of Inoculum. Although it is well established that *Erwinia amylovora* may be easily disseminated by splashing rain (Gossard and Walton, 1917, 1922; Stevens *et al.*, 1918; Tullis, 1929) and perhaps by wind as aerosols (Southey and Harper, 1971), there is ample evidence that insect vectors are also extremely important. In areas where rainfall is uncommon during blossoming periods and severe epidemics occur, insects are almost certainly the major vector of the bacterium in these circumstances (Miller and Schroth, 1972; Thomson *et al.*, 1975a). Even in areas where rainfall is common insects are often considered to be the major means of dissemination (Pierstorff, 1931; Chester, 1901).

Transmission of inoculum from flower to flower by bees and other flower visiting insects. Numerous experiments have been made in which honey bees caged with infected and healthy flowers or exposed to cultures of the pathogen have transmitted the bacterium to the healthy flowers (Emmett and Baker, 1971; Waite, 1906; Pierstorff and Lamb, 1934; Martin, 1921). Conversely, protecting healthy flowers with cloth bags has prevented infection in numerous studies (Waite, 1892, 1906; O'Gara, 1908; Pierstorff, 1931). These experiments clearly indicate the importance of insects in transmission of fire blight inoculum. Miller (1929) demonstrated the efficiency with which honey bees may transmit the bacterium. He dipped a bee's proboscis into an infected flower, then into the receptacle cups of 15 consecutive pear flowers, all of which subsequently blighted. Although bees are considered to be the most active vector of *Erwinia amylovora* during the blossoming period (Waite, 1906) many other insects have been implicated as probable vectors. Waite (1896, 1906) suggested that many floral visitors, including honey bees, wild bees, wasps, flies and others were probably able to transmit the bacterium. Thomas and Parker (1933) found over 200 species of insects visiting open apple flowers and O'Gara (1910) reported that 50 species of insects in addition to honey bees, wild bees, wasps and flies visited blighted blossoms in the orchard. Thomas and Ark (1934a) cultured *Erwinia amylovora* from ants, aphids, beetles, flies and thrips collected from pear blossoms. Phillips (1933) collected 3,555 insects in 51 genera (mostly Diptera) from apple flowers in New York. Emmett (1971) reported that in England many flies visited pear flowers in early spring but were outnumbered by bees. He found that bees were rare in summer but that syrphid flies and a few calliphorid flies were present at that time. Parker (1936) and Thygesen *et al.* (1973) offered experimental evidence that flies and insects other than bees were capable of transmitting the fire blight pathogen. Parker found that four of eight flies collected from blossoms or cankers transmitted the bacterium to healthy blossoms and that two of 11 ants similarly collected were also able to transmit. Thygesen *et al.* (1973) tested 60 spp. of insects and found that less than 50% of them transmitted the bacteria. They found that the larger insects were the most effective vectors.

The available evidence suggests that most insects which visit blighted and healthy flowers can probably serve as vectors of fire blight. The fact that epiphytic populations of *Erwinia amylovora* appear on flowers and other plant parts in relatively high populations before symptoms appear (Miller and Schroth, 1972; Thomson *et al.*, 1975b) suggests that insects may carry small numbers of bacteria to those sites where they multiply epiphytically and cause infections when conditions are favorable. The transmission of secondary inoculum from flower to flower by insects is adequately proven but the role of insects in transmission to other plant parts, i.e. young shoots and leaves, is still a matter of debate. Several researchers (Stewart, 1913a and b; Burrill, 1915; Stewart and Leonard, 1915; Merrill, 1917; Lathrop, 1918; Tullis, 1929; Emmett and Baker, 1971; Wilde *et al.*, 1971) demonstrated that a number of sucking insects, especially aphids (Homoptera: Aphididae), *Lygus* bugs (Heteroptera: Miridae) and leafhoppers (Homoptera: Cicadellidae) could transmit *Erwinia amylovora* to shoots, leaves and fruits when contaminated insects were caged on healthy tissues. Lathrop (1918) studied three species of leafhoppers which attack apple in New York and found that only one, *Empoasca fabae* (Harris), transmitted the bacterium when the insects were transferred from infected to healthy trees. He reported that the other two species, *Empoasca canariensis* Metcalf and *Edwarsiana rosae* (Linnaeus) fed on older tissues while *E. fabae* fed on tender terminal growth and suggested that this explained the lack of transmission by the former species. Waters (1922) isolated *E. amylovora* from the mouthparts of the leafhopper, *Sephena cinerea* Kirkaldy, which was abundant in one district in New Zealand, and suggested that it served as a vector of the pathogen. Tullis (1929), in a study involving 4,500 *Aphis pomi* DeGeer and *Dysaphis sorbi* (Kaltenbach) (both Homoptera: Aphididae), showed that artifically contaminated insects caged on healthy twigs transmitted the bacteria with low frequency (three of 800), but aphids which had fed on naturally infected shoots failed to transmit the bacterium successfully. Burrill (1915) and Brooks (1926), however, reported successful infection of high percentages of healthy shoots with *Sitobion avenae* (Fabricius) (Homoptera: Aphididae), *Aphis pomi* and *Empoasca fabae* which were transferred from infected leaves. Wilde *et al.* (1971) demonstrated that the pear psylla, *Psylla pyricola* Foerster (Homoptera: Psyllidae), successfully transmitted *E. amylovora* in seven of 596 tests performed in Canada from 1964-1969. They used a relatively new method to determine if the bacterium had been transmitted, i.e. determining the increase in *E. amylovora* specific bacteriophage in pear tissue upon which the psylla had fed. This technique may be helpful in detecting lower numbers of bacterial cells than was previously possible. There is insufficient evidence at present to determine the actual importance of insects in disseminating *E. amylovora* to host shoots under natural conditions. However, peaks in aphid and *Lygus* bug activity were reported to coincide with or immediately precede peaks in new fire blight infections (Stewart, 1913a; Thomas and Parker, 1933), and new blight infections continued to occur later in the season in orchards where aphids were not controlled (Merrill, 1915). This circumstantial evidence suggests that insects are important in the shoot phase of fire blight.

Transmission of primary inoculum. Erwinia amylovora survives from season to season in cankers on trunks and larger branches of infected trees. It is generally agreed that splashing rain can disseminate the pathogen from cankers to blossoms, especially if the cankers are located high in the tree, but insects also appear to play a major role. In arid regions where severe blight outbreaks often occur during dry periods (Miller and Schroth, 1972) insects are almost certainly responsible for bacterial dissemination. Large numbers of insects have been observed visiting oozing cankers (Waite, 1896; Jones, 1909. Parker, 1936), and it has been suggested that they carry the bacteria from the cankers to the blossoms. Numerous insect groups have been reported to visit oozing cankers including aphids (Jones, 1909; Parker, 1936), beetles (Jones, 1909; Parker, 1936), flies (Parker, 1936; Parker *et al.*, 1956) and various hemipterans (Jones 1909). Authors disagree on the role of bees and wasps as vectors of inoculum from cankers to blossoms or shoots. Parker (1936) reported never having observed honey bees near oozing cankers even when cankered tissue was placed near beehives, and Jones (1909), although observing numerous insects on oozing cankers, never saw bees there. Waite (1896), however, reported that bees and wasps were attracted to bacterial exudate on cankers and suggested that they carried bacteria from cankers to blossoms. Although a wide variety of insects have been shown to visit oozing cankers and to become contaminated by *E. amylovora*, proof of their role in transmitting primary inoculum in nature is lacking. However, studies have been made which demonstrate that some insects are capable of transmitting *E. amylovora* from cankers to blossoms. Parker (1936) collected flies and ants from cankers and found that four of eight individual flies, *Bibio albipennis albipennis* Say (Diptera: Bibionidae) and *Cynomyopsis cadaverina* (Robineau-Desvoidy) (Diptera: Calliphoridae) and two of 11 ants transmitted the bacterium to healthy blossoms. Tullis (1929) allowed aphids to crawl on bark taken from active cankers then transferred the insects to flower styles. He reported that 28 of 174 blossoms blighted and concluded that aphids probably transmitted primary inoculum. It is unknown how often this occurs in nature, however. Miller and Schroth (1972) and Thomson *et al.* (1975a) demonstrated that epiphytic populations of *E. amylovora* occurred on flowers long before fire blight symptoms appeared, even during rainless periods. They believed that the bacterium was disseminated by insects from cankers to blossoms and then further disseminated by pollinating insects. In the absence of rain insects are the only logical vectors of primary inoculum. Thomson *et al.* (1975b) noted that flowers in a single cluster were randomly colonized by *E. amylovora* and suggested that this supported the concept of insect dissemination.

12.2.4.2 Insects Provide Wounds for Infection. Conflicting evidence has been presented regarding the importance of wounds in infection by *E. amylovora*. Some workers (Stewart, 1913b; Hildebrand and MacDaniels, 1935; Crosse *et al.*, 1972) reported that infection does not occur when bacteria are applied to uninjured leaves or stems but that injured tissues are readily infected. However, several authors have presented evidence which shows that infection of leaf and stem tissue often occurs through natural openings such as stomata, hydathodes, lenticels and glandular trichomes (Miller, 1929; Tullis, 1929; Erskine, 1975). These reports raise questions about the role of insect wounds as infection courts; however, sufficient evidence exists to show that insects may contribute to fire blight epidemiology by producing wounds and, in some cases, by inoculating bacterial cells into the wounds they have created.

Wounding and inoculating. The most efficient way for insects to transmit pathogens is to deposit the cells directly into the wounds they have created. This eliminates problems of wound dessication and exposure of inoculum to adverse environments prior to invasion. We know that insects transmit *E. amylovora* in this way from previous research. Stewart (1913b) reported that *Lygus lineolaris* (Palisot de Beauvois) (Heteroptera: Miridae) inoculated healthy shoots when contaminated bugs fed on healthy tissues. Non-contaminated insects did not cause infection, indicating that bacteria from other sources were not merely invading injuries created by the bugs. Stahl and Luepschen (1977) also found that *Lygus* spp. exposed to infected pear fruits then transferred to healthy fruits infected 11.4-23.7% of them while fruits exposed to non-contaminated insects remained healthy. Similar results have been reported for aphids (Jones, 1909; Burrill, 1915; Merrill, 1917; Brooks, 1926). Plurad *et al.* (1967) questioned the role of aphids, particularly *Aphis pomi*, the species most frequently associated with fire blight infection, as vectors of the bacterium. They found that *A. pomi* introduced from one to 130 bacterial cells into healthy tissues during 30 minutes feeding but was unable to consistently introduce the 70 cells into the healthy tissues which the authors reported were necessary for successful infection. Hildebrand (1937), however, reported producing a successful infection in an apple blossom with a single cell. Perhaps large numbers of insects could cause significant levels of infection since at least some of them would deposit the minimum inoculum for successful infection (Plurad *et al.,* 1967). The experiments which have included adequate controls indicate that piercing sucking insects do injure and inoculate healthy shoots, leaves and fruits. Other studies suggest that this phenomenon occurs with other insects but these have not included adequate controls to insure that epiphytic bacteria have not merely invaded the injuries created by the insects. Such studies have involved *Scolytus rugulosus* (Ratzeburg) (Coleoptera: Scolytidae) (Jones, 1909, 1910, 1911a and b), two species of thrips (Thysanoptera: Thripidae) (Bailey, 1935) and *Pegomya rubivora* (Coquillet) maggots (Diptera: Anthomyiidae) in raspberry canes (Folsom, 1954).

Wounding without inoculation. Cassidy (1888) suggested that trunk boring insects created wounds in tree trunks which served to admit *E. amylovora* cells, resulting in cankers on the trees. Since then several reports and some experimental evidence have been presented which support Cassidy's work. Jones (1910, 1911a and b) reported fire blight infection developing around fresh punctures made by *Scolytus rugulosus.* Tullis (1929) noted that curculio (Coleoptera: Curculionidae) injuries on fruits were infected with blight and hypothesized that codling moth larvae (Lepidoptera: Olethreutidae) also provided avenues for infection. Stewart (1913a) also suggested that bacteria enter via curculio injury. Whetzel (1906) reported that unspecified boring insects provided wounds to which other insects, such as flies, transmitted the bacterium, resulting in cankers. These reports–that insects provided the wounds through which *E. amylovora* disseminated from some other source gained entry into healthy tissues–were usually without experimental proof. Several studies, however, have provided stronger evidence that this does occur. Stewart (1913b) smeared shoots with *E. amylovora* then caged non-contaminated *Lygus lineolaris* on the "inoculated" tissues. Fire blight infection occurred on the tissues contaminated with the bacteria where the insects were allowed to feed while none was observed on "inoculated" shoots without the insects or on non-contaminated shoots exposed to the bugs in the absence of the bacterium. Stahl and Luepschen (1977) sprayed a water suspension of *E. amylovora* on healthy pear fruit and allowed it to dry. When non-contaminated *Lygus* spp. were allowed to feed on the fruit 25.5% of the fruit became infected compared to only 11.3% of the controls. This suggests that the bacterium is able to infect healthy fruits either through natural openings or other wounds but that the presence of insects may increase disease severity. Thygesen *et al.* (1973) reported that *E. amylovora* enters intact tissue through wounds made by gall midge larvae and suggested that "insect injury may be of greater importance than formerly presumed." The potential importance of insect injuries as infection courts assumes greater importance since the discovery of epiphytic populations of *E. amylovora* on flowers, shoots and leaves (Miller and

Schroth, 1972; Thomson *et al.,* 1975a). If inoculum is present on plant surfaces infection courts created by hail (Miller and Schroth, 1972), insects or other agents could result in widespread outbreaks of disease. The role of insect injuries in the development of cankers on trunks and larger limbs where one would expect infection through natural openings to be rare seems to be a significant possibility which is unproven but which should be investigated more thoroughly.

12.2.4.3 Survival of Inoculum. The role of insects in the survival of *E. amylovora* inoculum has attracted the attention of researchers for nearly 50 years. Rosen (1930, 1931, 1933), working in Arkansas, reported isolating *E. amylovora* from beehives and hive material throughout the summer, winter and early spring before the appearance of fire blight infections. Rosen (1933) injected a water suspension of material from beehives into healthy pear shoots and found that only 21 of 3,534 injections produced infection, suggesting very low populations of bacteria in the inoculum. Gossard (1916a) had reported earlier that he was unable to isolate the bacteria from honey in the spring and many other workers have reported similar results. Thomas (1930), Pierstorff and Lamb (1934), Hildebrand (1935), Hildebrand and Phillips (1936), Parker (1936) and Beyme *et al.* (1975) found that *E. amylovora* could be recovered from wood frames, waxy surfaces, honey combs and hive scrapings for only short periods of time (usually from one to 21 days with a maximum of 55 days after inoculation of the hives) and concluded that there was no possibility of the bacterium overwintering in the hive. Pierstorff and Lamb (1933a and b, 1934) artificially contaminated hives with *E. amylovora*, transported them to clean orchards and enclosed them under cloth cages with healthy trees. They found that no infection occurred.

The work of Rosen (1930, 1931, 1933) appears to have been carefully done and the fact that he achieved only a low level of transmission suggests that only a few cells may survive the winter in the hive. Perhaps the techniques employed by others were not sensitive enough to detect low bacterial numbers. Newer media and methods could be used to resolve the question of winter survival in the hive. This work is needed since this aspect of the life cycle of *E. amylovora* would be of great epidemiological importance. Lelliott (1959) reported that the fire blight disease order in Great Britain provides for the restriction of the movement of beehives from farms affected by fire blight, suggesting that concern about beehive contamination still exists.

The survival of the *E. amylovora* in honey has also been investigated. Results of most studies have shown that *E. amylovora* survives only a short time in honey (Thomas, 1930; Pierstorff and Lamb, 1934; Beyme *et al.,* 1975). Survival time has been reported to range from 47 hours (Gossard and Walton, 1916a) to 48 days (McLarty, 1923).

Erwinia amylovora appears capable of surviving only a relatively short time in association with the bodies of insect vectors. Beyme *et al.* (1975) found the pathogen in the digestive tract of honey bees but Hildebrand and Phillips (1936) and Pierstorff and Lamb (1934) reported that it could be recovered from bees for only one or two days after contaminated food sources were removed. They concluded that it was not possible for the organism to overwinter on or in the honeybee.

Ark and Thomas (1936) found that *E. amylovora* survived in three species of adult Diptera for three to six days. Their methods did not eliminate the possibility of recontamination of the insects during the study. Emmett and Baker (1971), however, showed that the bacterium could remain alive on *Sarcophaga carnaria* Linnaeus (Diptera: Sarcophagidae) for more than nine days. Plurad *et al.* (1965) reported that *E. amylovora* persisted at least 72 hours in *Aphis pomi* adults after they had fed for five minutes on apple sap contaminated with the organism. Neal *et al.* (1964) collected and surface-sterilized *Aphis pomi* eggs during the winter and obtained bacterial isolates which reacted positively with an *Erwinia amylovora* bacteriophage. All but one of the isolates, however, were avirulent on apple tissue suggesting that they were not *E. amylovora*. The fact that one isolate was apparently *Erwinia amylovora* suggests that studies on the survival of the organism with different stages of insect vectors should be pursued using more sophisticated techniques available today.

Little recent work has been done on survival; however, even the older work suggests that insects, if able to harbor *Erwinia amylovora* for only three to nine days, will extend its normal life and increase its potential to produce infection.

12.2.5 Olive Knot

Olive knot, caused by *Pseudomonas savastanoi* (Smith) Stevens, is found in California (Wilson, 1935), the Mediterranean region and other areas of the world. The pathogen is commonly disseminated by splashing rain and enters trees through injuries in the bark. In California the pathogen is almost entirely dependent on splashing water for dissemination (Horne *et al.*, 1912a and b; Wilson, 1935), although these authors speculated that insects and perhaps birds may accidentally carry the bacterium from tree to tree. Petri (1909, 1910), however, working in Italy, reported that the olive fruit fly, *Dacus oleae* (Gmelin) (Diptera: Tephritidae), was consistently contaminated with *P. savastanoi* and that the adult fly inoculated olive trees by depositing contaminated eggs into the tissues during oviposition. Petri described a close biological relationship between *P. savastanoi* and *D. oleae* in which all stages of the insect were constantly contaminated by the bacterium. Petri found specialized structures in the insect, including four blind pockets (ceca) located in the anterior portion of the mid intestine, several sack-like evaginations off the rectum just anterior to where the rectum and vagina unite to form the cloaca and an esophageal diverticulum or "pharynx gland" in the head. He reported that all of these structures were filled with bacteria. Petri noted that the bacteria disappeared from the intestinal pouches and the rectal evaginations before pupation. The bacteria appeared to survive in the esophageal diverticulum and from that reservoir to recontaminate the entire adult intestinal tract, including the four pockets in the intestine and the anal evaginations, thus assuring constant association between the insect and the intestinal bacteria. Petri also described a unique mechanism for insuring contamination of the egg before oviposition. Directly opposite the rectal evaginations a slit was present in the membrane separating the oviduct from the rectum, thus allowing the eggs to be pressed against the bacterial filled evaginations as they passed down the oviduct. The bacteria reportedly penetrated the egg through the micropyle, contaminating it internally and insuring contamination of the larvae at hatching.

Stammer (1929) confirmed the presence of the bacteria-filled structures, including the esophageal diverticulum, in *D. oleae* and found similar structures in numerous related insect species in the family Tephritidae. He agreed with Petri that the rectal evaginations constituted the transmission mechanism by which contamination of the eggs occurred to perpetuate the bacterial contamination of the insect.

Petri believed that *P. savastanoi* was the primary organism constantly associated with the intestinal tract of the olive fly. He was not often able to culture the bacterium from the insects but its presence was assumed because the inoculation of olive branches with mixed cultures of bacteria isolated from the esophageal diverticulum produced typical olive know symptoms. Petri also reported that another bacterial symbiote, *Ascobacterium luteum*, was nearly always present with *P. savastanoi* in the intestinal tract of *D. oleae.*

Yamvrias *et al.* (1970) reported that examinations of the esophageal diverticulum and the eggs of *D. oleae* revealed the presence of several gram negative bacteria, some of which belonged to the genus *Pseudomonas*. Comparison of the isolates with *P. savastanoi* using various morphological and biochemical tests revealed that all of the isolates differed from *P. savastanoi* in at least three characteristics. They successfully isolated bacteria from only 20% of the 200 adults tested and only a small percentage of the eggs sampled yielded bacteria.

Hellmuth (1956), however, isoalted *P. savastanoi* from larvae, pupae and adults of *D. oleae*, thus confirming Petri's work. She also isolated an undescribed *Pseudomonas* species, which she named *Pseudomonas mutabilis*, from larvae, pupae and adults of several species of Tephridae by using a special medium. *Ascobacterium luteum* was also isolated and extensively studied. Based upon her work Hellmuth concluded that *Ascobacterium* should be placed in the genus *Agrobacterium* and named it *A. luteum*. This report raises interesting questions regarding the role of *Ascobacterium* (*Agrobacterium*) *luteum* in the production of the olive knots, since other species in the genus *Agrobacterium* are known to produce similar symptoms on other hosts.

It seems probable from the more recent studies that the intestinal tract of *D. oleae* and related species are constantly contaminated with symbiotic bacteria and that the specialized structures described by Petri and Stammer do indeed serve to perpetuate the organisms from generation to generation. It seems doubtful, however, that *Pseudomonas savastanoi* is the primary organism present in the intestinal tract, as Petri believed, or even a significant part of the intestinal microflora in many cases. The data which support its presence, namely the inoculation of olive branches with mixed bacterial populations isolated from insect organs, fail to take into account the possibility that other bacteria, such as *Agrobacterium luteum* (Hellmuth,

1956), may be able to produce the symptoms or that *P. savastanoi* may be present on the plant surface as an epiphyte. These questions and others need further investigation. The data, however, indicate that *P. savastanoi* can be present in the intestinal tract of *D. oleae* at least in some instances (Petri, 1909, 1910; Hellmuth, 1956) and under these circumstances may be transmitted by the fly as Petri has described.

The symbiotic relationships between the intestinal bacteria, including *Pseudomonas savastanoi*, and *Dacus oleae* have been the subject of intensive investigation in recent years. Studies in which the bacterial microflora of the intestine have been eliminated by incorporating streptomycin in the diet of adult flies have consistently shown that the adult females oviposit normally but that the larvae are unable to survive in olive fruits (Hagen *et al.,* 1963; Hagen, 1966; Fytizas and Tzanakakis, 1966a, b, c; Fytizas, 1969a, b; Rey, 1969). Hagen *et al.* (1963) and Hagen (1966) found that the larvae from streptomycin-fed adults, although unable to develop in olive tissue, developed normally into adult flies which in turn produced viable eggs when these larvae were grown on an artificial medium containing protein hydrolysate. Hagen (1966) suggested the *P. savastanoi* hydrolyzes protein in the olive tissue and makes it available to *D. oleae* larvae. He suggested that the bacterium may also synthesize methionine and threonine, essential amino acids not present in the olive fruit. Fytizas and Tzanakakis (1966b) reported that *D. oleae* larvae did not develop in freshly harvested olive fruits but that the inhibition of larval development and pupation was less as the fruit ripened and that no inhibition of development occurred in fruit picked in October and stored for two or more months at four to five degrees C. They concluded that changes in the mesocarp of the olive as the fruit ripened in the field or was held in cold storage made it possible for larval development to be completed.

Rey (1969) attributed the inability of larvae produced by streptomycin-treated adults to develop in olive fruits to the absence of *Pseudomonas savastanoi* due to aseptic rearing methods and Hagen (1966) and Hagen *et al.* (1963) implied that the absence of *P. savastanoi* in the adult flies explained the failure of the larvae to survive and pupate in olive fruits. These workers apparently did not consider the fact that the streptomycin treatment probably also eliminated other bacteria including *Ascobacterium* (*Agrobacterium*) *luteum* which has been reported to be lipolytic (Petri, 1910). This view is partially reinforced by the fact that Fytizas and Tzanakakis (1966c) found that larvae produced by bacteria-free flies which had been recontaminated by *Pseudomonas savastanoi* were able to develop in olive tissue but the level of pupation was still lower than with larvae from non-streptomycin-treated flies. Considerably more study is needed to fully elucidate the role of *Pseudomonas savastanoi* and other symbiotes in the life history of the olive fly and the role of the olive fly in the epidemiology of olive knot.

12.2.6 Bacterial Rot of Apple

Bacterial rot of apple, caused by *Pseudomonas melophthora* Allen and Riker, was reported in 1931 to be associated with the apple maggot, *Rhagoletis pomonella*

(Walsh) (Diptera: Tephritidae). Allen (1931) isolated the pathogen from rotting areas around larval burrows and from adult apple maggot flies. He inoculated apple fruits with the bacterial isolates and produced decay, thus establishing that the bacterium, and not the insect alone, was responsible for the rot. Based upon the isolation of *P. melophthora* from adult flies he suggested that the insects may serve as vectors of the pathogen. Subsequent work by Allen and Riker (1932) and Allen *et al.* (1934) showed that there was indeed a relationship between the apple maggot and *P. melophthora*. The adult female fly punctures the surface of apple fruits during oviposition, and the bacterium is apparently introduced as a contaminant on the surface of the egg. Allen *et al.* (1934) isolated *P. melophthora* from the surfaces of puparia and young and mature larvae, and from adult flies. They found that the bacterium was probably also present inside the larvae and adults since surface sterilized individuals yielded *P. melophthora* as readily as non-surface sterilized insects. They failed to recover the bacterium from surface sterilized eggs and puparia and concluded that it was associated with those stages only as a surface contaminant. Two larvae which hatched from surface sterilized eggs were not contaminated with *P. melophthora*. When these individuals were placed in apple tissue they died within a few days, no rot developed in the fruit and no bacteria were isolated from the tissue. These data suggested that the bacterium may be necessary for the normal development of the larvae, since sterile larvae died without increasing in size. Little further work has been done on the relationship between *Rhagoletis pomonella* and *P. melophthora* but it has been generally accepted that the bacterium serves as an obligate extracellular symbiote which renders the apple habitable for the apple maggot larvae (Miyazaki *et al.,* 1968). These latter authors suggested that the bacterium may provide essential amino acids which are absent in apple tissue. They showed that *P. melophthora* synthesizes methionine and cystine which are not present in apple tissue and suggested that these amino acids may be the factors which allow *Rhagoletis* larvae to develop in rotting tissue. They recognized, however, that since the nutritional requirements of *Rhagoletis pomonella* are not known it was not possible to draw final conclusions regarding the significance of this work.

Considerable speculation has been made regarding the physical relationship between *Rhagoletis* adults and *P. melophthora*. The bacterium has been isolated from the alimentary canal of a female fly (Allen and Riker, 1932) and from surface sterilized adults (Allen *et al.,* 1934). These results, coupled with the reports that eggs are contaminated on the surface, suggest that some mechanism to insure that the eggs are contaminated when inserted into healthy apple tissue exists. Work by Dean (1933, 1935) has shown that the adult *Rhagoletis pomonella* possesses an esophageal diverticulum in the head region which he named the esophageal bulb. Numerous diverticula are also present in the epithelium of the rectum and the rectum and vagina unite in the ovipositor to form a common cloacal passage through which the eggs pass before deposition. Dean offered no evidence for the role of these anatomical structures in the survival and perpetuation of *P. melophthora* in the flies. Leach (1940), however, called attention to the close similarities between the anatomy of *R. pomonella* and the olive fly, *Dacus oleae*, where these structures are believed to function in the perpetuation of *Pseudomonas savastanoi* and other symbiotic bacteria in the insect. He suggested that work should be done to determine if the structures in *R. pomonella* functioned as they are believed to function in *Dacus oleae*—to perpetuate the pathogen between molts and to recontaminate the system. No studies have been made to clarify this point, although a recent study using a fluorescent antibody staining technique (Baerwald and Boush, 1968), which demonstrated *P. melophthora* cells in adult flies between the interthoracic longitudinal muscles and in the abdominal hemocoel, failed to mention the esophageal bulb and the rectal diverticula. Thus the possibility that these structures serve to perpetuate the bacterium and insure egg contamination remains unproven. If this is true, however, the flies may significantly aid the survival of the bacterium. This area needs much more research before the questions of the epidemiological significance of internal contamination by *P. melophthora* can be resolved.

12.2.7 Southern Bacterial Wilt

Southern bacterial wilt, also known as Granville wilt of tobacco, brown rot of potato, tomato and related plants and the Moko disease of banana, caused by *Pseudomonas solanacearum* (Smith) Smith, is widespread throughout the world. It is especially prevalent in tropical and subtropical regions. The disease has been associated with potential insect vectors virtually since its discovery. Insects have been associated with the disease primarily in the dissemination of inoculum to natural infection courts and in providing wounds for bacterial entry.

12.2.7.1 Dissemination of Inoculum to Natural Infection Courts. There are several strains of *Pseudomonas solanacearum* which cause bacterial wilt (Moko disease) of bananas but one, the SFR strain (Buddenhagen and Elsasser, 1962) and the tomato race (Vakili and Baldwin, 1966), which also attacks *Musa* spp., are commonly spread to natural infection courts by insects. Buddenhagen and Elsasser (1962) hypothesized that the bacterium probably spread from indigenous *Heliconia* spp. to bluggoe banana by non-insect means and was then readily spread among the bluggoe population and occasionally to other bananas by insects. They believed that the most common avenue of infection was through fresh scars left by abscissed male flowers, and later Buddenhagen and Kelman (1964) reported that abscission of both male inflorescences and flower bracts exposed open vessels through which the bacterium invaded the plants. These fresh scars are present for only about three months per year. When inflorescences were enclosed in muslin bags to prevent insect visitation, no infection occurred (Buddenhagen and Elsasser, 1962; Vakili and Baldwin, 1966), while non-protected inflorescences became diseased, thus providing strong circumstantial evidence that insects serve as vectors of the bacterium. Stover (1972) reported that pruning wounds and other injuries also served as avenues of invasion for insect disseminated bacteria.

Buddenhagen and Elsasser (1962) isolated *P. solanacearum* from five percent of the bees (*Trigona corvina* Cockerell, Hymenoptera: Apidae) and wasps collected in an affected area. Other authors also agreed that bees, wasps and flies which visit flowers are probably common vectors of the disease. Stover (1972), Buddenhagen and Kelman (1964) and Revilla (1967), suggested that bees (*Trigona* spp.), wasps (*Polybia* spp., Hymenoptera: Vespidae) and flies (*Drosophila* spp., Diptera: Drosophilidae) regularly visited natural abscission sites and in doing so acquired the bacterium and carried it to healthy plants. Vakili and Baldwin (1966) reported that bees and fruit flies made "continuous" visits to abscission sites of staminate flowers and bracts. Buddenhagen and Kelman (1964) found that infection by the SFR strain of *P. solanacearum* normally occurs through these natural infection courts. Much of the evidence for insect transmission is circumstantial, nevertheless it indicates that insects do serve as vectors of this disease.

12.2.7.2 Provide Wounds for Infection. Since the earliest work on the infection of solanaceous plants, as well as those in other families, by *Pseudomonas solanacearum*, insects have been implicated as potential vectors of the bacterium. Smith (1896) transmitted the pathogen from diseased to healthy potato plants in the greenhouse using the Colorado potato beetle, *Leptinotarsa decemlineata* (Say) (Coleoptera: Chrysomelidae), as the vector. He hypothesized that blister beetles, flea beetles and other chrysomelids as well as other leaf eating insects could vector the pathogen. Smith first believed that insects were largely responsible for the spread of the disease but later (1914) found that the bacterium, when present in the soil, could invade wounds on roots and other parts of the plants. He concluded that insects may be less important (except in certain circumstances) than he had originally believed. Nolla (1931) later demonstrated that the beetle, *Diabrotica graminea* Baly (Coleoptera: Chrysomelidae), was able to transmit the pathogen to eggplants when fed on diseased tissue and caged on healthy plants. Plants caged without beetles remained healthy.

12.2.7.3 Wounding Without Inoculation. Wounding is necessary for infection to occur and infection normally occurs through below ground plant parts (Kelman, 1953). Some workers have suggested that root injury caused by root feeding insects provides avenues through which soil bacteria invade healthy plants. Evidence for this phenomenon is sparce but it is likely that it is a factor in the epidemiology of this tissue.

12.3 SUMMARY

The involvement of insects in the transmission of bacterial plant pathogens is indisputable. At least 60 diseases have been reported to be transmitted by insects (Table 1). The exact nature of the insect involvement and its epidemiological significance is, however, not clear in many cases.

Insects generally are not absolutely required for the success of bacterial diseases, but their involvement often supplements other mechanisms of survival, dissemina-

TABLE I. Bacterial Plant Pathogens and Arthropod Associates

Pathogen[a]	Disease[b]	Arthropod Associate[c]	Reference[d]
Aerobacter aerogenes (Kruse) Beijerink (name no longer recognized) see *Bacillus gossypina* Stedman	cotton boll rot		
Aerobacter sp.	soft rot of Brassicaceae [Cruciferae]	Diptera: Anthomyiidae	
		Hylemya brassicae (Weidemann) (cabbage maggot) [*Hylemyia brassicae* (Bouché)]	Doane and Chapman, 1964b
Agrobacterium pseudotsugae (Hansen and Smith) Savulescu [*Bacterium pseudotsugae* Hansen and Smith]	gall of Douglas fir, *Pseudotsuga* sp.	Hemiptera (Homoptera): Cicadidae	
		Platypedia areolata (Uhler)	Hansen and Smith, 1937
		Hemiptera (Homoptera): Phylloxeridae	
		Adelges cooleyi (Gillette) (Cooley spruce gall adelgid) [*Cheres cooleyi* Gillette]	Hansen and Smith, 1937
Agrobacterium tumefaciens (Smith and Townsend) Conn [*Pseudomonas tumefaciens* (Smith and Townsend) Duggar; *Bacterium tumefaciens* Smith and Townsend]	crown gall of grape (*Vitis* spp.), willow (*Salix* spp.), apple (*Malus* spp.), raspberry (*Rubus* spp.), etc.	Coleoptera: Curculionidae	
		Cryptorhynchus lapathi (Linnaeus) (poplar-and-williow borer) [*Cryptorrhyncus lapathi*]	Magerstein, 1931
		Coleoptera: Elateridae	
		"wireworms"	Granovsky, 1940[g]
		Coleoptera: Scarabaeidae	
		"white grubs"	Granovsky, 1940[g]

	crown gall	Coleoptera: ?	
		“beetle larvae”	Granovsky, 1940[g]
		Diptera: ?	
		“maggots”	Granovsky, 1940[g]
		Hemiptera (Homoptera): Aphididae	
		Eriosoma lanigerum (Hausmann) (woolly apple aphid)	Waite, 1912[g]
		Hymenoptera:?	
		“bees”	Hedgcock, 1910
Bacillus avenae Manns see *Pseudomonas avenae* Manns		unidentified insects	Granovsky, 1931[g]
Bacillus gossypina Stedman [*Bacillus gossypinus* Stedman] (not listed in Buchanan and Gibbons, 1974); *Aerobacter aerogenes* (*Kruse*) Beijerink (name no longer recognized); *Bacillus pumilus* Meyer and Gottheil; other unidentified *Bacillus* spp.: *Erwinia carotovora* (Jones) Bergey *et al.* [*Erwinia aroidea* (Townsend) Bergey *et al.*]; (*Erwinia chrysanthemi* Burkholder *et al.*; *Erwinia herbicola* (Löhnis) Dye; other unidentified *Erwinia*	cotton (*Gossypium* spp.) boll rot; necrosis of immature cotton bolls; internal disease of cotton bolls	Coleoptera: Nitidulidae	
		Carpophilus dimidiatus (Fabricius) (corn sap beetle) [*Carpophilus mutilatus*]	Stedman, 1894
		Epuraea depressa (Illiger) [*Epuroea aestiva*]	Stedman, 1894
		Hemiptera (Heteroptera): Coreidae	
		Hypselonotus fulvus (De Geer)	Mendes, 1956
		Leptoglossus oppositus (Say)	Morrill, 1910
		Leptoglossus phyllopus (Linnaeus) (leaffooted bug)	Morrill, 1910
		miscellaneous coreids	Morrill, 1910
		Hemiptera (Heteroptera): Lygaeidae	
		Lygaeus turcicus Fabricius	Morrill, 1910
		Oncopeltus fasciatus (Dallas) (large milkweed bug)	Morrill, 1910
		Oxycarenus gossypinus Distant	Hayward, 1967
		Hemiptera (Heteroptera): Miridae	
		Adelphocoris rapidus (Say)	Morrill, 1910

Pathogen[a]	Disease[b]	Arthropod Associate[c]	Reference[d]
Bacillus gossypina		(rapid plant bug) [*Calocoris rapidus* Say]	
		Gerhardiella delicatus (Uhler) [*Psallus delicatus* Uhl.]	Morrill, 1910
		Hemiptera (Heteroptera): Pentatomidae	
		Acrosternum hilare (Say) (green stink bug) [*Nezara hilaris* Say]	Morrill, 1910
		Chlorochroa ligata (Say) (conchuela) [*Pentatoma ligata* Say]	Morrill, 1910
		Chlorochroa sayi Stål (Say stink bug) [*Pentatoma sayi* Stål]	Morrill, 1910
		Edessa meditabunda (Fabricius)	Mendes, 1956
		Euchistus conspersus Uhler	Ashworth *et al.*, 1970
		Euchistus impictiventris Stal [brown stinkbug]	Ashworth *et al.*, 1969 Ashworth *et al.*, 1970
		Euchistus servus (Say) (brown stink bug)	Morrill, 1910
		Nezara viridula (Linnaeus) (southern green stinkbug)	Morrill, 1910 Nowell, 1918 Mendes, 1956
		Nezara spp.	Wickens, 1942[f]
		Piezodorus purus Stal[e]	Pearson, 1934
		Thyanta custator (Fabricius)	Morrill, 1910
		miscellaneous or unidentified pentatomids	Morrill, 1910 Pearson, 1934 Wickens, 1942[f]
		Hemiptera (Heteroptera): Pyrrhocoridae	
		Dysdercus fasciatus[e]	Pearson, 1934
		Dysdercus honestus Blöte[e]	Mendes, 1956
		Dysdercus intermedius[e]	Pearson, 1934

Dysdercus mendesi Blöte[e]	Mendes, 1956
Dysdercus nigrofasciatus Stål[e]	Pearson, 1934, Wickens, 1942[f]
Dysdercus ruficollis (Linnaeus)[e]	Mendes, 1956
Dysdercus suturellus (Herrich-Schäffer) (cotton stainer)	Morrill, 1910
Dysdercus spp. (not enumerated)	Nowell, 1918 Cauquil and Mildner, 1965 Cognee and Frinking, 1966 Cauquil, 1973
Euryophthalmus succinctus (Linnaeus) [*Largus succinctus* Linnaeus]	Morrill, 1910
stainers	Wickens, 1942[f]
Hemiptera: ?	
unidentified species	Pearson, 1934
Lepidoptera: Cosmopterigidae	
Sathrobrota simplex (Walsingham) [*Pyroderces simplex* Wlsm.]	Hayward, 1967
Lepidoptera: Gelechiidae	
Pectinophora gossypiella (Saunders) (pink bollworm) [*Platyedra gossypiella*]	Hayward, 1967 Cognee and Frinking, 1966
Lepidoptera: Noctuidae	
Diparopsis watersi (Rothschild)	Cognee and Frinking, 1966 Hayward, 1967
Earias biblaga Walker	Hayward, 1967
Earias insulana (Boisduval)	Hayward, 1967
Heliothis zea (Boddie) (bollworm, corn earworm, tomato fruitworm) [*Heliothis armigera* (Hübner)]	Hayward, 1967
Lepidoptera: Olethreutidae	
Cryptophlebia leucotreta (Meyrick) [*Argyroploce leucotreta* Meyr.]	Cognee and Frinking, 1966 Hayward, 1967

Pathogen[a]	Disease[b]	Arthropod Associate[c]	Reference[d]
Bacillus gossypina		Lepidoptera: ?	
		larvae of nonspecified lepidopterans	Cognee and Frinking, 1966[g]
		unidentified insects	Edgerton, 1912[g] Cauquil, 1963[g] Ashworth *et al.*, 1970[g]
Bacillus pini (not listed in Buchanan and Gibbons, 1974)	galls on zirbelkiefer or stone pine, *Pinus cembra*	Hemiptera (Homoptera): Phylloxeridae	
		Adelges sp. or spp. ["Chermes lice"]	von Tubeuf, 1911
		unidentified insects	von Tubeuf, 1910[g], 1911[g]
Bacillus pumilus Meyer and Gottheil see *Bacillus gossypina* Stedman	cotton boll rot		
Bacterium dianthi Chester [*Bacillus dianthi* Arch. and Boll.] (not listed in Buchanan and Gibbons, 1974)	a bacterial disease of carnation, *Dianthus caryophyllus*	Thysanoptera: ?	
		unidentified species	Pussard-Radulesco, 1930[g]
Corynebacterium isidiosum (McCulloch) Jensen	wilt of alfalfa, *Medicago sativa*	Coleoptera: Curculionidae	
		Sitona hispidula (Fabricius) (clover root curculio)	Hill *et al.*, 1971
Corynebacterium michiganense (Smith) Jensen [*Phytomonas michiganensis*]	canker of tomato, *Lycopersicon* spp.	Coleoptera: Chrysomelidae	
		Diabrotica undecimpunctata howardi Barber (spotted cucumber beetle) [*Diabrotica duodecimpunctata* Fabr.]	Ark, 1944
		Hemiptera (Heteroptera): Miridae	
		Lygus lineolaris (Palisot de	Ark, 1944

		Beauvois) (tarnished plant bug) [misidentified as *Lygus pratensis* L.]	
		Hemiptera (Homoptera): Aphididae	
		Myzus persicae (Sulzer) (green peach aphid)	Ark, 1944
		Thysanoptera: Thripidae	
		Caliothrips fasciatus (Pergande) (bean thrips) [*Heliothrips fasciatus* Perg.]	Ark, 1944
		Thrips tabaci Lindeman (onion thrips)	Ark, 1944
Corynebacterium sepedonicum (Spieckermann and Kotthoff) Skaptason and Burkholder [*Phytomonas sepedonica* Spieckermann and Kotthoff]	ring rot of potato, *Solanum tuberosum*	Coleoptera: Chrysomelidae	
		Epitrix cucumeris (Harris) (potato flea beetle)	Duncan and Généreux, 1960
		Leptinotarsa decemlineata (Say) (Colorado potato beetle)	List and Kreutzer, 1942 Duncan and Généreux, 1960
		"flea beetle"	Metzger and Binkley, 1940[f,g]
		Coleoptera: Meloidae	
		Epicauta pennsylvanica (De Geer) (black blister beetle)	List and Kreutzer, 1942
		Hemiptera (Heteroptera): Miridae	
		Lygus lineolaris (Palisot de Beauvois) (tarnished plant bug)	Duncan and Généreux, 1960
		Hemiptera (Homoptera): Aphididae	
		Myzus persicae (Sulzer) (green peach aphid)	Duncan and Généreux, 1960
		unidentified species	Duncan and Généreux, 1960[g]
		Hemiptera (Homoptera): Cercopidae	
		unidentified species	Duncan and Généreux, 1960[g]
		Hemiptera (Homoptera): Cicadellidae	
		Macrosteles fascifrons (Stål) (aster leafhopper)	Duncan and Généreux, 1960

Pathogen[a]	Disease[b]	Arthropod Associate[c]	Reference[d]
Corynebacterium sepedonicum		unidentified species	Duncan and Généreux, 1960
		Orthoptera: Acrididae	
		Melanoplus bivittatus (Say) (twostriped grasshopper)	Brentzel and Munro, 1940
		Melanoplus differentialis (Thomas) (differential grasshopper)	Brentzel and Munro, 1940 List and Kreutzer, 1942
		Melanoplus packardii Scudder (Packard grasshopper)	Brentzel and Munro, 1940
		Melanoplus sanguinipes (Fabricius) (migratory grasshopper) [*Melanoplus mexicanus*]	Brentzel and Munro, 1940
		Orthoptera: ?	
		"grasshoppers"	List and Kreutzer, 1942[g]
		unidentified insects	Perrault, 1948[g]
Erwinia amylovora (Burrill) Winslow *et al.* [*Micrococcus amylovorus* Burrill; *Bacillus amylovorus* (Burrill) Trevisan]	fire blight (bacterial blight; pear blight; blight canker; blossom-blight; apple twig blight; collar blight; sour-sap; insect blight) of pomaceous fruits, esp. *Malus* spp., *Pyrus* spp., and *Crataegus* spp.	Acari: ?	
		unidentified mite species	Gossard and Walton, 1922[g]
		Arachnida: Araneidae	
		unidentified species	Thygesen *et al.*, 1973
		Coleoptera: Cantharidae	
		Cantharis andersoni Frost [*Cantaris livida*]	Thygesen *et al.*, 1973
		Coleoptera: Curculionidae	
		Rhynchaenus canus (Horn) [*Orchestes canus*]	Gossard, 1916b
		Rhynchaenus pallicornis (Say) (apple flea weevil) [*Orchestes pallicornis*]	Gossard and Walton, 1922
		unidentified species	Whetzel and Stewart, 1909[g] Stewart, 1913a[g] Tullis, 1929[g]

Coleoptera: Elateridae	
Melanotus oregonensis LeConte	Thomas and Ark, 1934a
unidentified species	Cate, 1918[g]; Jones, 1914
Coleoptera: Nitidulidae	
Glischrochilus fasciatus (Olivier)	Parker, 1936
Meligethes sp.	Emmett and Baker, 1971
Coleoptera: Scolytidae	
Scolytus pyri Beck[e]	Barry, 1847 (insects thought to cause fire blight)
Scolytus rugulosus (Ratzeburg) (shothole borer) [*Eccoptogaster rugulosus*]	Jones, 1909, 1910, 1911a, Gossard, 1916b
Coleoptera: ?	
"apple tree borer"	Orton and Adams, 1914[g], 1915[g]
"cucumber beetle"	Jones, 1914[g]
"fruit bark-beetle"	Orton and Adams, 1915[g]
unidentified species	Jones, 1909[g] Thomas and Ark, 1934[g] Thygesen *et al.*, 1973[g]
Dermaptera: Forficulidae	
unidentified species	Thygesen *et al.*, 1973
Diptera: Anthomyiidae	
Hylemya antiqua (Meigen) (onion maggot) [*Hylemia antiqua*]	Thomas and Ark, 1934a Parker, 1936
Pegomya lipsia (Walker) [*Hylemia lipsia* Walker]	Thomas and Ark, 1934a Parker, 1936
Pegomya rumicifoliae Huckett [*Pegomyia calyptrata* Zett.]	Thomas and Ark, 1934a Parker, 1936
Pegomya sp. [*Pegamya* sp.]	Miller and Schroth, 1972
Scatophaga stercoraria (Linnaeus) [*Scopeuma stercorarium*]	Emmett and Baker, 1971

Pathogen[a]	Disease[b]	Arthropod Associate[c]	Reference[d]
Erwinia amylovora		Diptera: Bibionidae	
		Bibio albipennis albipennis Say [*Bibio albipennis* Say]	Parker, 1936
		Diptera: Bombyliidae	
		unidentified species	Jones, 1914[g]
		Diptera: Calliphoridae *Cynomyopsis cadaverina* (Robineau-Desvoidy) [*Cynomyia cadaverina* Desv.]	Parker, 1936
		Lucilia sp.	Emmett and Baker, 1971
		Phaenicia sericata (Meigen) [*Lucilia seriata*]	Ark and Thomas, 1936
		Pollenia rudis (Fabricus) (cluster fly)	Stewart and Leonard, 1916
		Diptera: Cecidomyiidae	
		Dasineura crataegi (Winnertz) [*Dasyneura crataegi*]	Thygesen *et al.*, 1973
		Diptera: Drosophilidae	
		Drosophila funebris (Fabricius)	Thomas and Ark, 1934a, Parker, 1936
		Drosophila melanogaster Meigen	Ark and Thomas, 1936
		Diptera: Lauxaniidae	
		Homoneura bispina (Loew) [*Sapromyza bispina* Loew]	Stewart and Leonard, 1916
		Minettia sp.	Miller and Schroth, 1972
		Diptera: Muscidae	
		Musca domestica Linnaeus (house fly)	Jones, 1909 , Ark and Thomas, 1936
		Muscina assimilis (Fallen) [squash root maggot]	Thomas and Ark, 1934a, Parker, 1936[g]
		Muscina stabulans (Fallén) (false stable fly)	Thomas and Ark, 1934a, Parker, 1936[g]

Phaonia variegata (Meigen)[e]	Emmett, 1971
Diptera: Sarcophagidae	
Sarcophaga carnaria Linnaeus	Emmett and Baker, 1971
Diptera: Syrphidae	
Eristalis tenax (Linnaeus) (drone fly)	Emmett and Baker, 1971
Neoascia sp.	Emmett and Baker, 1971
Syrphus sp.	Thygesen *et al.*, 1973
unidentified species	Jones, 1914[g] Gossard and Walton, 1922[g] Miller and Schroth, 1972[g]
Diptera: ?	
unidentified species	Waite, 1896[g]; 1906[g]; Whetzel, 1906[g]; Jones, 1909[g]; O'Gara, 1910[g]; Stewart, 1915[g]; Thomas and Ark, 1934a[g]; Thygesen *et al.*, 1973[g]; Stahl and Luepschen, 1977[g]
Hemiptera (Heteroptera): Miridae	
Adelphocoris rapidus (Say) (rapid plant bug)	Stewart and Leonard, 1915
Campylomma verbasci (Meyer-Duer)	Stewart and Leonard, 1915
Heterocordylus malinus Reuter	Stewart and Leonard, 1915
Ilnacora malina (Uhler)	Gossard and Walton, 1922
Lygidea mendax Reuter (apple red bug)	Stewart and Leonard, 1915
Lygocoris pabulinus[e]	Emmett and Baker, 1971
Lygus elisus Van Duzee (pale legume bug)	Stahl and Luepschen, 1977
Lygus invitus (Say)	Stewart and Leonard, 1915
Lygus lineolaris (Palisot de Beauvois) (tarnished plant bug) [frequently misnamed *Lygus pratensis* Say]	Forbes, 1884a; 1884b[f]; Stewart, 1913a; 1913b; Burrill, 1915; Thomas and Ark, 1934a; Stahl and Luepschen, 1977

Pathogen[a]	Disease[b]	Arthropod Associate[c]	Reference[d]
Erwinia amylovora		*Orthotylus flavosparsus* (Sahlberg)	Stewart and Leonard, 1915
		Orthotylus marginalis[e]	Emmett and Baker, 1971
		Phytocoris ulmi (Linnaeus)[e]	Thygesen *et al.*, 1973
		Plagiognathus politus Uhler	Stewart and Leonard, 1916
		Poeciloscytus basalis Reuter	Stewart and Leonard, 1915
		Hemiptera (Heteroptera): Pentatomidae	
		unidentified species	Jones, 1914
		Hemiptera (Heteroptera): ?	
		"red bugs"	Orton and Adams, 1915[g]
		unidentified species	Jones, 1914[g]
			Gossard and Walton, 1922[g]
			Thygesen *et al.*, 1973[g]
		Hemiptera (Homoptera): Aphididae	
		Aphis pomi De Geer (apple aphid)	Jones, 1909, 1910
		[*Aphis mali* Fabricius]	Stewart, 1913a
			Burrill, 1915
			Brooks, 1926
			Tullis, 1929
			Thomas and Parker, 1933
			Neal *et al.*, 1964
			Plurad *et al.*, 1965, 1967
		Dysaphis plantaginea (Passerini) (rosy apple aphid) [*Anuraphis roseus*]	Gossard and Walton, 1922
		Dysaphis sorbi (Kaltenbach) [*Aphis sorbi*]	Tullis, 1929
		Eriosoma lanigerum (Hausmann)	Jones, 1909, 1910
		(woolly apple aphid) [*Schizoneura*	Orton and Fowler, 1915
		lanigera; Eriosoma lanigera]	Rosen, 1933
		Myzus persicae (Sulzer) (green peach aphid)	Burrill, 1915

Rhopalosiphum fitchii (Sanderson) (apple grain aphid) [*Rhopalosiphym prunifolium*]	Gossard and Walton, 1922
Siphocoryne avenae[e]	Gossard and Walton, 1916b
Sitobion avenae (Fabricius) [*Aphis avenae* Fabricius]	Burrill, 1915; Brooks, 1926; Thomas and Ark, 1934a
"green aphid"	Merrill, 1917[g]
"green apple aphis"	Swingle, 1911[g]
"woolly aphids"	Gossard, 1916b[g]
unidentified species	Jones, 1909[g],1910; Whetzel and Stewart, 1909[g],Jones,1914[g], Burrill, 1915[g],Merrill,1915[g];Gossard, 1916a[g], Tullis, 1929[g], Rosen, 1933[g];Thomas and Ark,1934a[g];Parker, 1936[g]Lelliott, 1968[g];Emmett and Baker, 1971[g]; Thygesen *et al.*, 1973.
Hemiptera (Homoptera): Cercopidae	
Philaenus spumarius (now several species)	Emmett and Baker, 1971
Hemiptera (Homoptera): Cicadellidae	
Edwardsiana rosae (Linnaeus) (rose leafhopper) [*Empoa rosae* (Linnaeus)]	Lathrop, 1918
Empoasca canariensis Metcalf [*Empoasca unicolor* Gillette]	Lathrop, 1918
Empoasca fabae (Harris) (potato leafhopper) [*Empoasca mali*]	Burrill, 1915; Stewart and Leonard, 1916; Lathrop, 1918
Erythroneura sp.	Marshall *et al.*, 1941
Sephena cinerea Kirkaldy[e]	Waters, 1922
Typhlocyba pomaria McAtee (white apple leafhopper)	Marshall *et al.*, 1941
unidentified species	Whetzel and Stewart, 1909[g]; Brooks, 1926[g]

Pathogen[a]	Disease[b]	Arthropod Associate[c]	Reference[d]
Erwinia amylovora		Hemiptera (Homoptera): Membracidae	
		Stictocephala bubalus (Fabricius) (buffalo treehopper)	Jones, 1911a[g]; Gossard and Walton, 1922[g]
		Hemiptera (Homoptera): Psyllidae	
		Psylla pyricola Foerster (pear psylla)	Stewart and Leonard, 1916; Wilde *et al.*, 1971
		Psylla simulans[e]	Emmett and Baker, 1971
		unidentified species	Thygesen *et al.*, 1973
		Hemiptera (Homoptera): ?	
		"mealybugs"	Rosen, 1933[g]
		Hemiptera (?): ?	
		unidentified species	Jones, 1909[g]
		Hymenoptera: Andrenidae	
		Andrena varians[e]	Emmett and Baker, 1971
		Andrena sp.	Thygesen *et al.*, 1973
		Hymenoptera: Apidae	
		Apis mellifera Linnaeus (honey bee) [*Apis mellifica* Linnaeus] (also see "bees," below)	Waite, 1901[g], 1906[g]; O'Gara, 1910[g]; Gossard, 1916a[g], 1916b[g]; Gossard and Walton, 1916b[g], 1922[g]; Phillips, 1916[g]; Martin, 1921[g]; Milbrath, 1922[g]; McLarty, 1923[g]; Thomas, 1930[g]; Rosen, 1930[g], 1933[g]; Pierstorff and Lamb, 1933a[g], 1934[g]; Thomas and Ark, 1934a[g], 1934b[g]; Thomas and Vansell, 1934[g]; Hildebrand, 1935[g]; Hildebrand and Phillips, 1936[g]; Parker, 1936[g]; Ark and

	Thomas, 1936[g]; Keitt and Ivanoff, 1941[g]; Lelliott, 1968[g]; Emmett and Baker, 1971; Thygesen *et al.*, 1973; Beyme *et al.*, 1975
Bombus hortorum (Linnaeus)	Emmett and Baker, 1971
Bombus terrestris (Linnaeus)	Emmett and Baker, 1971
Hymenoptera: Formicidae	
Formica fusca Linnaeus (silky ant) [*Formica fusca* var. *subsericea*]	Thomas and Ark, 1934a; Parker, 1936
Formica pallide-fulva Latreille (now several subspecies)	Thomas and Ark, 1934a
Formica pallide-fulva nitidiventris Emery [*Formica pallidefulva* Latr. subsp. *schaufussi* Mayr var. *incerta* Emery]	Parker, 1936
Formica sp.	Parker, 1936
Lasius alienus (Foerster) (cornfield ant) [*Lasius niger* L. var. *americanus* Emery; synonymy unclear]	Parker, 1936
Lasius niger (Linnaeus) (now 3 or more species)	Thomas and Ark, 1934a
Prenolepis imparis imparis (Say) [*Prenolepis imparis*]	Thomas and Ark, 1934a
unidentified species	Swingle, 1911[g]; Jones, 1914[g]; Hotson, 1916[g]; Gossard, 1916b[g]; Gossard and Walton, 1922[g]; Thomas and Ark, 1934a[g]; Thygesen *et al.*, 1922[g]

Pathogen[a]	Disease[b]	Arthropod Associate[c]	Reference[d]
Erwinia amylovora		Hymenoptera: Tenthredinidae	
		Hoplocampa testudinea (Klug) (European apple sawfly)	Emmett and Baker, 1971
		Hymenoptera: Vespidae	
		Polistes sp.	Thomas and Ark, 1934a
		Vespula sp.	Thomas and Ark, 1934a
		Hymenoptera: ?	
		"bees" [probably indicates *Apis mellifera* L. but may include other bee species]	Waite, 1892[g], 1896[g], 1898[g]; Cook, 1901[g]; Reed, 1914[g]; Swingle, 1921[g]; Tullis, 1929[g];
		"wasps"	Thygesen *et al.*, 1973[g]; O'Gara, 1910[g]; Miller, 1929[g]
		"wasps"	Waite, 1896[g], 1906[g]; O'Gara, 1910[g]; Swingle, 1921[g]; Thygesen *et al.*, 1973[g]
		"wild bees"	Waite, 1906[g]; O'Gara, 1910[g]; Gossard, 1916b[g]; Thomas and Ark, 1934a[g]
		Lepidoptera: Geometridae	
		"canker worm"	Jones, 1914[g]
		Lepidoptera: Lasiocampidae	
		Malacosoma neustria (Linnaeus)	Thygesen *et al.*, 1973
		Lepidoptera: Lymantriidae	
		"tussock moth"	Jones, 1914[g]
		Lepidoptera: Noctuidae	
		Lithophane antennata (Walker) (green fruitworm)	Jones, 1914[g]
		Lepidoptera: Olethreutidae	
		Laspeyresia pomonella (Linnaeus) (codling moth) [*Carpocapsa pomonella*]	Jones, 1914[g]; Thomas and Ark, 1934a

Lepidoptera: ?	
Diloba caeruleocephala (Linnaeus) [*Episema coeruleocephala*]	Thygesen *et al.*, 1973
"butterfly"	Thygesen *et al.*, 1973[g]
"moths"	Thomas and Ark, 1934a[g]
Orthoptera: Acrididae [*Locustidae*]	
unidentified species	Thygesen *et al.*, 1973
Thysanoptera: Thripidae	
Frankliniella occidentalis (Pergande) [*Frankliniella moultoni* Hood]	Bailey, 1935
Taeniothrips inconsequens (Uzel) (pear thrips)	Bailey, 1935; Jones, 1914[g]
Thysanoptera: ?	
"common grain or weed thrips"	Cate, 1918[g]
unidentified species	Thomas and Ark, 1934a[g]
nectar and pollen-feeding insects in general (includes references dealing with the survival of the bacterium in nectar)	Waite, 1891[g]; O'Gara, 1908[g], 1910[g]; Gossard and Walton, 1917[g], 1922[g]; Stevens *et al.*, 1918[g]; Tullis, 1929[g]; Miller, 1929[g]; Pierstorff, 1931[g]; Thomas and Ark, 1934b[g]; Ivanoff and Keitt, 1941[g]; Lelliott, 1968[g]; Thomson *et al.*, 1975a[g]
"sucking insects"	Chester, 1901[g]; O'Gara, 1910[g]; Gossard and Walton, 1922[g]
unidentified insects	Barry, 1847[g] (who stated that fire blight is "solely produced by some minute venomous insect" that "operates mainly in the night"); Arthur, 1885[g]; Cassidy, 1888[g]; Waite, 1896[g];

Pathogen[a]	Disease[b]	Arthropod Associate[c]	Reference[d]
Erwinia amylovora			Whetzel, 1906[g]; O'Gara, 1908[g], 1910[g]; Stewart, 1915[g]; Cate, 1918[g]; Gossard and Walton, 1922[g]; Miller, 1929[g]; Miller and Schroth, 1972[g]; Thomson *et al.*, 1975b[g]
Erwinia amylovora f. sp. *rubi* Starr *et al.*	fire blight of raspberry, *Rubus idaeus*	Diptera: Anthomyidae	
		Pegomya rubivora (Coquillet) (raspberry cane maggot) [*Phorbia rubivora*]	Folsom, 1954
Erwinia aroideae (Townsend) Bergey *et al.* See *Erwinia carotovora* (Jones) Bergey *et al.*			
Erwinia cacticida (Johnston and Hitchcock) Magrou [*B. cacticidus* Johnston and Hitchcock]	a bacteriosis of prickly pear, *Opuntia* spp.	Coleoptera: Cerambycidae	
		Moneilema spp.	Johnston and Hitchcock, 1923
		Diptera: Drosophilidae	
		unidentified species	Johnston and Hitchcock, 1923
		Hemiptera (Heteroptera): Coreidae	
		Chelinidea canyona [*Chelinides canyona*][e]	Johnston and Hitchcock, 1923
		Chelinidea tabulata (Burmeister) [*Chelinides tabulata*]	Johnston and Hitchcock, 1923
		Chelinidea vittiger Uhler [*Chelinides vittigera*; 2 varieties tested]	Johnston and Hitchcock, 1923
		Lepidoptera: Pyralidae	
		Cactoblastis cactorum (Berg)	Johnston and Hitchcock, 1923

		Melitara junctolineella Hulst (now at least 3 spp. in the genera *Melitara* Walker and *Olycella* Dyar)	Johnston and Hitchcock, 1923
		Melitara prodenialis Walker	Johnston and Hitchcock, 1923
		Mimorista flavidissimalis Grote	Johnston and Hitchcock, 1923
		unidentified species	Johnston and Hitchcock, 1923
Erwinia carnegieana Standring	bacterial necrosis of giant cactus, *Carnegiea gigantea* and *Cereus giganteus*	Coleoptera: Histeridae	
		Hololepta yucateca Marseul	Boyle, 1949
		Coleoptera: ?	
		unidentified species	Boyle, 1949[g]
		Diptera: Syrphidae	
		Volucella sp.	Boyle, 1949
		Lepidoptera: Pyralidae	
		Cactobrosis fernaldialis Hulst	Boyle, 1949
		Lepidoptera: ?	
		unidentified moth species	Boyle, 1949[g]
Erwinia carotovora (Jones) Bergey *et al.*	banana (*Musa* spp.) tip over disease and rhizome rot	Coleoptera: Curculionidae	
		Cosmopolites sordiduf (Germar) (banana root borer)	Stover, 1972
		Odoiporus longicollis (Olivier)	Edward *et al.*, 1973
		Hymenoptera: Formicidae	
		unidentified species	Stover, 1972[g]
Erwinia carotovora (Jones) Bergey *et al.* [*Bacillus carotovorus; Bacillus phytophthorus; Erwinia atroseptica; Erwinia aroideae*]	Brassicaceae [Cruciferae] soft rot	Coleoptera: Curculionidae	
		Ceutorhynchus quadridens (Panzer) (cabbage seedstalk curculio)	Skowroński, 1976
		Coleoptera: Staphylinidae	
		unidentified species	Johnson, 1930;
		Collembola: ?	
		unidentified species	Johnson, 1930[g]

Pathogen[a]	Disease[b]	Arthropod Associate[c]	Reference[d]
Erwinia carotovora		Diptera: Anthomyiidae	
		Hylemya brassicae (Weidemann) (cabbage maggot) [*Hylemyia brassicae* (Bouché)]	Johnson, 1930; Bonde, 1930a, 1939a; Doane and Chapman, 1964a, 1964b[f]
		Hylemya platura (Meigen) (seedcorn maggot) [*Hylemia cilicrura* Rond.]	Al-Tamimi, 1969
		unidentified species	Johnson, 1930
		Diptera: Drosophilidae	
		Drosophila busckii Coquillett [*Drosophila buskeii*]	Tamimi and Banfield 1969; Al-Tamimi, 1969
		Diptera: probably Drosophilidae	
		"fruit flies"	Tamimi and Banfield, 1969[g]
		Diptera: Muscidae	Robbs, 1946
		Atherigona sp.	Chiu *et al.*,1958
		Fannia sp.	
		Diptera: Sarcophagidae	
		Sarcophaga sp.	Chiu *et al.*, 1958
		Diptera: ?	
		unidentified species	Walker,1927[g];Robbs,1946[g]
		Hemiptera (Homoptera):Aphididae	
		Myzuz persicae(Sulzer) (green peach aphid)	Muller and Skiebe, 1962
		Hymenoptera: Apidae	
		Apis melliflora Linnaeus (honeybee)	Chiu *et al.*,1958
		Hymenoptera: Cynipidae	
		unidentified species	Johnson, 1930
		Hymenoptera: Tenthredinidae	
		Athalia rosae (Linnaeus)	Chiu *et al.*,1958

		Lepidoptera: Pyralidae *Hellula phidilealis*	Robbs, 1946
		Lepidoptera: Yponomeutidae *Plutella xylostella* (Linnaeus) (diamondback moth) [*Plutella maculipennis* (Curtis)]	Robbs, 1946; Chiu *et al.*, 1958
Erwinia carotovora (Jones) Bergey *et al.* [*Bacillus carotovorus*]	celery (*Apium graveolens*) heart rot	Diptera: Chloropidae *Elachiptera costata* (Leow)	Leach, 1927
		Diptera: Drosophilidae *Scaptomyza graminum* (Fallén)	Leach, 1927; Ogilvie *et al.*, 1934
		Diptera: ? "carrot flies"	Ogilvie *et al.*, 1934[g]
		several unidentified species	Leach, 1927[g]
		Hemiptera (Heteroptera): Miridae *Lygus lineolaris* (Palisot de Beauvois) (tarnished plant bug) [misidentified as *Lygus pratensis*]	Richardson, 1938
Erwinia carotovora (Jones) Bergey *et al.* [*Bacillus carotovorus* L. R. Jones]	*Iris* spp. soft rot	Diptera: Syrphidae *Eumerus strigatus* (Fallén) (onion bulb fly) [Lesser bulb fly]	McDaniel, 1935
		Lepidoptera: Noctuidae *Macronoctua onusta* Grote (iris borer)	Ries, 1929; McDaniel, 1935; Thanos, 1948; Howard and Leach, 1962, 1963
Erwinia carotovora (Jones) Bergey *et al.* [*Bacillus phytophthorus* Appel; *Bacillus carotovorus* Jones]	soft rot of miscellaneous vegetables (*Capsicum* [pepper], *Daucus carota sativa* [carrot], *Lactuca sativa* [lettuce], and *Spinacia oleracea* [spinach])	Diptera: Anthomyiidae *Hylemya platura* (Meigen) (seedcorn maggot); *Hylemyia cilicrura* Rondani]	Bonde, 1939a
		Diptera: Drosophilidae *Drosophila busckii* Coquillett [*Drosophila busckeii*]	Tamimi and Banfield, 1969

Pathogen[a]	Disease[b]	Arthropod Associate[c]	Reference[d]
Erwinia carotovora		unidentified species	Robbs, 1946
		Diptera: probably Drosophilidae "fruit flies"	Tamimi and Banfield, 1969[g]
		Diptera: Lonchaeidae *Lonchaea pendula*[e]	Robbs, 1946
		Diptera: Psilidae *Psila rosae* (Fabricius) (carrot rust fly)	Bonde, 1939a
		Lepidoptera: Gelechiidae *Gnorimoschema* sp.	Robbs, 1946
		Lepidoptera: Pyralidae *Ostrinia nubilalis* (Hübner) (European corn borer) [*Pyrausta nubilalis* Hübner]	McKeen, 1948
Erwinia carotovora (Jones) Bergey *et al.*	onion (*Allium cepa*) soft rot	Diptera: Anthomyiidae *Hylemya antiqua* (Meigen) (onion maggot) [*Hylemyia antiqua* (Meigen)]	Taylor, 1952; Doane, 1953; Gorlenko *et al.*, 1956; Friend *et al*, 1959[f]
		Hylemya platura (Meigen) (seedcorn maggot) [*Hylemyia cilicura* (Rondani)]	Taylor, 1952; Doane, 1953
		Diptera: Otitidae *Euxesta notata* (Wiedemann)	Doane, 1953
		Tritoxa flexa (Wiedemann)	Taylor, 1952; Doane, 1953
		Diptera: Syrphidae *Eunerus strigatus* (Fallén) (onion bulb fly) [lesser bulb fly]	Taylor, 1952 Doane, 1953
		Thysanoptera: Thripidae unidentified species	Robbs, 1946
Erwinia carotovora (Jones) Bergey *et al.* (including *E. carotovora* var. *atroseptica* and	potato (*Solanum tuberosum*) blackleg, soft rot and seed piece decay	Coleoptera: Chrysomelidae *Epitrix cucumeris* (Harris) (potato flea beetle)	Phillips, 1977[g]

E. carotovora var. *carotovora*) [*Bacillus phytophthorus* Appel; *Bacillus carotovorus; Bacillus solanisaprus; Bacillus aeroidea; Erwinia phytophthora* (Appel) Holland]	Coleoptera: Dermestidae	
	Dermestes lardarius Linnaeus (larder beetle)	Bonde, 1939b
	Coleoptera: Elateridae	
	Cryptohypnus abbreviatus (Say)	Bonde, 1939b
	unidentified species	von Hegyi, 1910
	Coleoptera: Staphylinidae	
	unidentified species	Bonde, 1939a, 1939b Kloepper, *et al.* ,1979
	Coleoptera: ?	
	unidentified species	Harrison *et al.*, 1977[g]
	Diplopoda: ?	
	unidentified species	Lapwood, 1976[g]
	Diptera: Anthomyiidae	
	Hylemya florilega (Zettersted) [*Hylemyia trichodactyla*, seed-potato maggot; *Delia florilega*]	Bonde, 1930b, 1939b, 1950 Harrison *et al.*, 1977
	Hylemya platura (Meigen) (seedcorn maggot) [*Pegomya fusciceps; Phorbia fusciceps* Zett; *Hylemyia cilicrura* Rond]	Sellards, 1905[f]; Leach, 1925, 1926, 1930a[g], 1933[f]; Huff, 1928[f]; Bonde, 1928, 1930b, 1939a, 1939b, 1950; Landis, *et al.*, 1971; Kloepper, 1977; Phillips, 1977; Kloepper *et al.*, 1979; Al-Tamimi, 1969
	Hylemya sp. or spp. [*Delia* spp.]	Bonde, 1939b; Kloepper *et al.*, 1976; Kloepper, 1977; Harrison *et al.*,1977
	Nupedia infirma[e]	Harrison *et al.*, 1977
	Scatophaga furcata (Say) [*Scatophaga squalida*]	Harrison *et al.*, 1977
	Scatophaga stercoraria (Linnaeus)	Harrison *et al.*, 1977
	Diptera: Calliphoridae	
	Phormia sp.	Kloepper, 1977; Kloepper *et al.*, 1979

Pathogen[a]	Disease[b]	Arthropod Associate[c]	Reference[d]
Erwinia carotovora		*Protophormia terraenovae* (Robineau-Desvoidy)	Kloepper, 1977; Kloepper *et al.*, 1979
		unidentified species	Kloepper *et al.*, 1979
		Diptera: Ceratopogonidae	
		unidentified species	Kloepper *et al.*, 1979
		Diptera: Chironomidae	
		unidentified species	Kloepper, 1977; Kloepper *et al.*, 1979
		Diptera: Drosophilidae	
		Drosophila busckii Coquillett [*Drosophila buskeii* Coq.]	Bonde, 1939b; Kloepper, 1977
		Drosophila funebris (Fabricius)	Bonde, 1939b
		Drosophila melanogaster Meigen	Molina *et al.*, 1974; Allan and Kelman, 1975, 1977; Kloepper *et al.*, 1976; Kloepper, 1977, Kloepper *et al.*, 1979, in manuscript a, in manuscript b
		Drosophila sp. or spp.	Bonde 1930b; Graham *et al.*, 1976 Kloepper *et al.*, 1976; Kloepper, 1977; Kloepper, *et al.*, 1979; Harrison *et al.*, 1977
		Scaptomyza pallida (Zett.) [*Parascaptomyza pallida*]	Harrison *et al.*, 1977 Kloepper *et al.*, 1979
		Diptera: probably Drosophilidae "fruit flies"	Al-Tamimi, 1969 Tamimi and Banfield, 1969[g] Al-Tamimi, 1969[g]
		Diptera: Ephydridae	
		Allotrichoma sp. or spp.	Kloepper *et al.*, 1976; Kloepper, 1977; Kloepper *et al.*, 1979
		unidentified species	Kloepper *et al.*, 1979
		Diptera: Milichiidae	
		unidentified species	Kloepper *et al.*, 1979

Diptera: Muscidae	
Azelia sp.	Harrison *et al.*, 1977
Coenosia tricolor[e] (Zett.)	Harrison *et al.*, 1977
Fannia sp.	Graham *et al.*, 1976
Musca domestica Linnaeus (house fly)	Bonde, 1930b, 1939b
unidentified species	Kloepper *et al.*, 1979
Diptera: Scatopsidae	
Scatopse fuscipes Meigen	Kloepper, 1977; Kloepper *et al.*, 1979
Scatopse sp. or spp.	Kloepper *et al.*, 1976; Harrison *et al.*, 1977
Diptera: Sciaridae	
Bradysia sp.	Kloepper, 1977; Kloepper *et al.*, 1979
Sciara tridentata Rübsaamen [*Sciara tribentata*]	Bonde, 1930b, 1939b
Diptera: Sepsidae	
Nemopoda nitidula (Fallén)	Harrison *et al.*, 1977
Sepsis spp.	Harrison *et al.*, 1977
Themira minor (Haliday)	Harrison *et al.*, 1977
Themira putris (Linnaeus)	Harrison *et al.*, 1977
Diptera: Sphaeroceridae	
Leptocera sp. or spp.	Graham *et al.*, 1976; Kloepper *et al.*, 1976; Harrison *et al.*, 1977; Kloepper, 1977; Kloepper *et al.*, 1979
Diptera: Syrphidae	
Eumerus strigatus (Fallén) (onion bulb fly)	Bean, 1947[g]
Diptera: Tethinidae	
Pelomyia cruciata Hendel	Kloepper, 1977; Kloepper *et al.*, 1979
Diptera: Trichoceridae	
Trichocera sp. [false crane fly]	Bonde, 1930b, 1939b

Pathogen[a]	Disease[b]	Arthropod Associate[c]	Reference[d]
		Diptera: ?	
		unidentified species	Leach, 1926[g]; Harrison *et al.*, 1977[g]; Phillips, 1977[g]; Phillips and Kelman, 1978[g]; Kloepper *et al.*, 1979
		Hemiptera (Homoptera): Aphididae	
		unidentified species	Harrison *et al.*, 1977[g]
		Hemiptera (Homoptera): Cicadellidae	
		Empoasca fabae (Harris) (potato leafhopper)	Phillips, 1977[g]
		Macrosteles fascifrons (Stål) (aster leafhopper)	Phillips, 1977[g]; Kloepper *et al.*, 1979
		Hymenoptera: Braconidae	
		Pentapleura sp. or spp.	Kloepper *et al.*, 1979
		Hymenoptera: Formicidae	
		Solenopsis sp.	Robbs, 1946
		Hymenoptera: Vespidae	
		Vespa sp.	Graham *et al.*, 1976
		Isoptera: ?	
		unidentified species	Robbs, 1946[g]
		Lepidoptera: ?	
		"moths"	
		unidentified insects	Harrison *et al.*, 1977[g]; Robbs, 1946[g]; Lapwood, 1976[g]; Allan and Kelman 1977[g]; Kloepper, 1977[g]; Phillips, 1977[g]; Phillips and Kelman, 1978[g]; Kloepper *et al.*, 1979 in manuscript a
Erwinia carotovora (Jones) Bergey *et al.* [including *Erwinia aroideae*, which these authors treat as a separate species]	(pathogens of beech, cauliflower, and cucumbers)	Coleoptera: probably Chrysomelidae	
		"Colorado beetle" (probably *Leptinotarsa decemlineata* (Say))	Hvozdyak *et al.*, 1973[g]

Erwinia chrysanthemi Burkholder *et al.* [*Erwinia carotovora* var. *zeae*]	stalk rot of maize, *Zea mays*	Lepidoptera: Pyralidae	
		Chilo partellus (Swinhoe)	Thind and Singh, 1976
Erwinia espinosa F.M. Brown (not listed in Buchanan and Gibbons, 1974)	a gall of witch-hazel, *Hamamelis* spp.	Hemiptera (Homoptera): Aphididae	
		Hamamelistes spinosus Shimer	Lutz and Brown, 1928
Erwinia herbicola (Löhnis) Dye and possibly other species	black rot of rice (*Oryza sativa*) grains	Hemiptera (Heteroptera): Pentatomidae	
		unidentified species	Tanii *et al.*, 1974
		unidentified insects	Tanii *et al.*, 1974[g]
Erwinia salicis (Day) Chester [*Pseudomonas saliciperda* Lindeijer; *Bacterium salicis* Day]	watermark disease and canker of willow, *Salix* spp.	Coleoptera: Curculionidae	
		Cryptorhynchus lapathi (Linnaeus) (poplar-and-willow borer) [*Cryptorrhynchus lapathi* L.]	Lindeijer, 1932a, 1932b, 1933 Callan, 1939 Pelkwijk, 1946[f]
		Diptera: Cecidomyiidae	
		Helicomyia saliciperda (Dufour) [*Rhabdophaga saliciperda* Duf.]	Callan, 1940
		Rhabdophaga sp.	Callan, 1940
		Hymenoptera: Tenthredinidae	
		Euura atra Jur.[e]	Callan, 1940
		Hymenoptera: ?	
		"willow wood wasp"	Gray, 1940[g]
		unidentified insects	Day, 1924[g] Lindeijer, 1933[g] Callan, 1939[g]
Erwinia stewartii (Smith) Dye [*Aplanobacter stewarti* (Smith) McCulloch; *Bacterium stewarti* (Smith) Smith; *Phytomonas stewarti* (Smith) Bergey *et al.*]	Stewart's wilt of corn, *Zea mays*	Coleoptera: Chrysomelidae	
		Chaetocnema confinis Crotch (sweetpotato flea beetle)	Poos and Elliott, 1936 Elliott and Poos, 1940
		Chaetocnema denticulata (Illiger) (toothed flea beetle)	Rand and Cash, 1924, 1933 Poos and Elliott, 1935, 1936 Elliott and Poos, 1940 Poos, 1955

Pathogen[a]	Disease[b]	Arthropod Associate[c]	Reference[d]
Erwinia stewartii		*Chaetocnema ectypa* Horn (desert corn flea beetle)	Poos and Elliott, 1936
		Chaetocnema pulicaria Melsheimer (corn flea beetle) [brassy flea beetle]	Forbes, 1894[f], 1905[f] (Forbes described damage which he thought was caused by this insect but which appears in retrospect to have been due to Stewart's wilt.) Rand, 1923[g]; Rand and Cash, 1924, 1933; Elliott and Poos, 1934, 1940; Poos and Elliott, 1935, 1936; Wood, 1937; Elliott, 1938; Poos, 1939, 1945, 1955; Elliott and Robert, 1939; Voronkevich, 1958; Castor *et al.*, 1975a, 1975b; Heichel *et al.*, 1977
		Diabrotica longicornis (Say) (northern corn rootworm)	Ivanoff, 1933a, 1933b; Ivanoff *et al.*, 1933; Elliott and Poos, 1940
		Diabrotica undecimpunctata howardi Barber (spotted cucumber beetle; sourthern corn rootworm) [*Diabrotica duodecimpunctata* (Fab.)]	Rand and Cash, 1924, 1933; Poos and Elliott, 1935, 1936; Elliott and Poos, 1940
		Diabrotica virgifera LeConte (western corn rootworm)	Pepper, 1967
		Disonycha glabrata (Fab.)	Poos and Elliott, 1936; Elliott and Poos, 1940

Epitrix cucumeris (Harris) (potato flea beetle)	Poos and Elliott, 1936; Elliott and Poos, 1940
Systena blanda Melsheimer (palestriped flea beetle) [*Systena taeniata blanda* (Melsh.)]	Poos and Elliott, 1936; Elliott and Poos, 1940
"flea beetles"	Rand, 1923[g]; Rand and Cash, 1933[g]; Wood, 1937[g]
Coleoptera: Coccinellidae	
Ceratomegilla fuscilabris (Mulsant)	Poos and Elliott, 1936; Elliott and Poos, 1940
Coleoptera: Elateridae	
Agriotes mancus (Say) (wheat wireworm)	Frutchey, 1936
Coleoptera: Phalacridae	
Phalacrus sp.	Elliott and Poos, 1940
Stilbus apicalis (Melsheimer)	Poos and Elliott, 1936; Elliott and Poos, 1940
Stilbus viduus Casey	Poos and Elliott, 1936; Elliott and Poos, 1940
Coleoptera: Scarabaeidae	
Phyllophaga sp.	Ivanoff, 1932, 1933b
Diptera: Anthomyiidae	
Hylemya platura (Meigen) (seedcorn maggot) [*Hylemyia cilicrura* Rondani]	Frutchey, 1936
Hemiptera (Heteroptera): Miridae	
Halticus citri (Ashmead)	Poos and Elliott, 1936; Elliott and Poos, 1940
Lygus lineolaris (Palisot de Beauvois) (tarnished plant bug) [misidentified as *Lygus pratensis* (L.)]	Poos and Elliott, 1936 Elliott and Poos, 1940
Hemiptera (Heteroptera): Nabidae	
Nabis ferus (L.)	Poos and Elliott, 1936 Elliott and Poos, 1940
Hemiptera (Heteroptera): Pentatomidae	
Thyanta custator (Fabricius)	Poos and Elliott, 1936

Pathogen[a]	Disease[b]	Arthropod Associate[c]	Reference[d]
Erwinia stewartii		Hemiptera (Homoptera): Aleyrodidae	
		unidentified species	Thomas, 1924[g]
		Hemiptera (Homoptera): Aphididae	
		Macrosiphum euphorbiae (Thomas) (potato aphid) [*Illinoia solanifolii* (Ashmead)]	Poos and Elliott, 1936; Elliott and Poos, 1940
		unidentified species	Thomas, 1924[g]
		Hemiptera (Homoptera): Cicadellidae	
		Chlorotettix viridius Van Duzee	Poos and Elliott, 1936
		Cicadula divisa (Uhler)	Poos and Elliott, 1936
		Empoasca fabae (Harris) (potato leafhopper)	Poos and Elliott, 1936
		Euscelis bicolor (Van Duzee) [*Stirellus bicolor* (Van Duzee)]; *Phlepsius irroratus* (Say); *Thamnotettix nigrifrons* (Forbes)	Poos and Elliott, 1936; Elliott and Poos, 1940
		unidentified species	Rand, 1923[g]
		Lepidoptera: Pyralidae	
		Ostrinia nubilalis (Hübner) (European corn borer) [*Pyrausta nubilalis* Hbn.]	Nemlienko, 1951[f]; Pepper, 1967
		Orthoptera: Acrididae	
		Melanoplus differentialis (Thomas) (differential grasshopper)	Poos and Elliott, 1936
		Thysanoptera: Thripidae	
		Anaphothrips obscurus (Müller) (grass thrips)	Poos and Elliott, 1936; Elliott and Poos, 1940
		Hercinothrips femoralis (O. M. Reuter) (banded greenhouse thrips) [*Heliothrips femoralis* Reuter]	Poos and Elliott, 1936

unidentified insects	Smith, 1920[g]; Rand, 1923[g]; Elliott and Poos, 1934[g]

In addition to the species listed above, Poos and Elliott (1936) and Elliott and Poos (1940) attempted to isolate the pathogen from more than 77 species in the following groups but were unable to do so:

Coleoptera: Anthicidae: 1 sp.; Anthribidae: 1 sp.; Cantharidae: 1 sp.; Cerambycidae: 1 sp.; Chrysomelidae: 16 spp.; Coccinellidae: 4 spp.; Curculionidae: 6 spp.; Elateridae: 2 spp.; Languriidae: 1 sp.; Lathridiidae: 1 sp.; Meloidae: 1 sp.; Melyridae: 1 sp.; Nitidulidae: 2 spp.; Phalacridae: 2 spp.; Scarabaeidae: 3 spp.

Collembola: Sminthuridae: 1 sp.

Hemiptera (Heteroptera): Anthocoridae: 1 sp.; Coreidae: 1 sp.; Corimelaenidae: 1 sp.; Lygaeidae: 2 spp.; Miridae: 2 spp.; Pentatomidae: 1 sp.

Hemiptera (Homoptera): Aphididae: 3 spp.; Cercopidae: 1 sp.; Cicadellidae: more than 13 spp.; Delphacidae: 1 sp.; Fulgoridae: 2 spp.; Membracidae: 1 sp.

Lepidoptera: Noctuidae: 1 sp.

Orthoptera: Tettigoniidae: 1 sp.

Pathogen[a]	Disease[b]	Arthropod Associate[c]	Reference[d]
		Thysanoptera: Aeolothripidae: 1 sp. Thripidae: 1 sp.	
Erwinia tracheiphila (Smith) Bergey *et al.* [*Bacillus tracheiphilus* Smith]	wilt of Cucurbitaceae	Coleoptera: Chrysomelidae	
		Acalymma vittata (Fabricius) (striped cucumber beetle) [*Diabrotica vittata* (Fabr.)]	Smith, 1895; Rand, 1915; Rand and Enlows, 1916, 1920; Rand and Cash, 1920; Doolittle, 1921; Mills, 1939; Brooks and Allen, 1941; Gould, 1944; Hervey and Schroeder, 1946; Alban and Keirns, 1948[g]; Fronk and Peterson, 1954; Williams and Lockwood, 1956; Young and Ditman, 1959; Leach, 1964; Staples *et al.*, 1967; Da Costa and Jones, 1971; Haynes and Jones, 1975
		Crepidodera cucumeris[e]	Rand and Enlows, 1916
		Diabrotica balteata LeConte (banded cucumber beetle)	Da Costa and Jones, 1971
		Diabrotica undecimpunctata howardi Barber (spotted cucumber beetle) [*Diabrotica duodecimpuntata* Oliv.]	Rand and Enlows, 1916, 1920; Rand and Cash, 1920; Alban and Keirns, 1948; Fronk and Peterson, 1954; Williams and Lockwood, 1956; Young and Ditman, 1959; Leach, 1964; Staples *et al.*, 1967; Da Costa and Jones, 1971; Haynes and Jones, 1975

Diabrotica undecimpunctata undecimpunctata Mannerheim (western spotted cucumber beetle) [*Diabrotica undecimpunctata* Mann.]	Da Costa and Jones, 1971
Diabrotica spp.	Rand and Enlows, 1916
Epitrix cucumeris (Harris) (potato flea beetle)	Rand and Enlows, 1920
"cucumber beetles"	Burkholder, 1960[g]
"flea beetles"	Rand, 1915[g]; Rand and Enlows, 1916[g]
Coleoptera: Coccinellidae	
Epilachna borealis (Fabricius) (squash beetle)	Rand and Enlows, 1916, 1920
unidentified species	Rand and Enlows, 1920[g]
Coleoptera: Elateridae	
unidentified species	Rand and Enlows, 1920[g]
Diptera: ?	
unidentified species	Rand and Enlows, 1920[g]
Hemiptera (Heteroptera): Coreidae	
Anasa tristis (De Geer) (squash bug) [*Coreus tristis* De Geer]	Smith, 1895; Rand and Enlows, 1916; 1920; Alban and Keirns, 1948[g]
Hemiptera (Homoptera): Aphididae	
Aphis gossypii Glover (melon aphid; cotton aphid)	Rand and Enlows, 1916 1920
unidentified species	Rand, 1915[g] Rand and Enlows, 1916[g]
Hymenoptera: Apidae	
Apis mellifera Linnaeus (honey bee)	Rand and Enlows, 1920
Lepidoptera: Sesiidae	
Melittia satyriniformis (Hübner) (squash vine borer)	Alban and Keirns, 1948[g]

Pathogen[a]	Disease[b]	Arthropod Associate[c]	Reference[d]
		unidentified insects	Smith, 1893[g]; Rand and Enlows, 1920[g]
Erwinia sp.	*Gardenia* spp. bud drop	Hemiptera (Homoptera): ? "mealy bugs"	Wilson, 1927[g]
		Hymenoptera: Formicidae unidentified species	Wilson, 1927[g]
Erwinia sp. or spp.	pathogens of onion (*Allium cepa*) and tomato (*Lycopersicon* spp.)	"beet pest larvae"	Hvozdyak *et al.*, 1973[f,g]
Erwinia spp., including *E. carotovora* (Jones) Bergey *et al.* [*E. aroideae* (Townsend) Bergey *et al.*], *E. chrysanthemi* Burkholder *et al.*, and *E. herbicola* (Löhnis) Dye see *Bacillus gossypina* Stedman	cotton boll rot		
Escherichia coli (Migula) Castellani and Chalmers	coconut (*Cocos* sp.) bud rot	Dermaptera: ?, unidentified species	Johnston, 1912[g]
		unidentified insects	Johnston, 1912[g]
Pseudomonas angulatum (Fromme and and Murray) Holland [*Bacterium angulatum* Fromme and Murray] (not listed in Buchanan and Gibbons, 1974)	angular leaf spot of tobacco, *Nicotiana* spp.	Lepidoptera: Sphingidae *Manduca sexta*, Linneaus (tobacco hornworm) *Phlegethontius sexta* Joh.	Johnson, 1934
Pseudomonas aptata (Brown and Jamieson) Stevens [*Bacterium aptatum* Brown and Jamieson]	a bacterial disease of sugar beet (*Beta vulgaris*) and nasturtium (*Tropaeolum*) leaves	unidentified insects	Brown and Jamieson, 1913[g]

Pseudomonas avenae Manns and *Bacillus avenae* Manns (*B. avenae* is not listed in Buchanan and Gibbons, 1974)	blade blight of oats, *Avena* spp.	Hemiptera (Homoptera): Aphididae "oat aphides"	Manns, 1909[g]
Pseudomonas cerasi Griffin	gummosis of cherries, *Prunus* spp.	unidentified insects	Rees, 1915[g]
Pseudomonas flectens Johnson	pod twist of beans, *Phaseolis* spp.	Thysanoptera: Phlaeothripidae	
		Haplothrips gowdeyi Franklin (black flower thrips)	Johnson, 1956
		Thysanoptera: Thripidae	
		Frankliniella sp.	Johnson, 1956
		Megalurothrips usitatus (Bagnall) [*Taeniothrips nigricornis* Schmutz]	Johnson, 1956
		Pseudanaphothrips achaetus (Bagnall)	Johnson, 1956
		Thrips tabaci Lindeman (onion thrips)	Johnson, 1956
Pseudomonas lachrymans (Smith and Bryan) Carsner	angular leaf spot of cucumber, *Cucumis sativus*	Coleoptera: Chrysomelidae	
		Acalymma vittata (Fabricius) (striped cucumber beetle) [*Diabrotica vittata* Fab.]	Carsner, 1918; Beecher and Doolittle, 1950
		Diabrotica undecimpunctata howardi Barber (spotted cucumber beetle; southern corn rootworm) [*Diabrotica duodecimpunctata* L.]	Carsner, 1918; Beecher and Doolittle, 1950
		unidentified insects	Carsner, 1918
Pseudomonas maculicola (McCulloch) Stevens [*Bacterium maculicolum* McCulloch]	peppery leaf spot and spot disease of cauliflower, *Brassica oleracea* var. *botrytis*	Hemiptera (Heteroptera): Pyrrhocoridae	
		Euryophthalmus convivus (Stål)	Goldsworthy, 1926
		unidentified insects	Clayton, 1924[g]
Pseudomonas marginata (McCulloch) Stapp	scab of *Gladiolus* spp.	Acari: Acaridae	
		Rhizoglyphus echinopus[e]	Forsberg, 1959, 1965
		Rhizoglyphus solani Oudemans group	Engelhard, 1969

Pathogen[a]	Disease[b]	Arthropod Associate[c]	Reference[d]
		Rhizoglyphus spp.	Jefferson *et al.*, 1956
Pseudomonas melophthora Allen and Riker [*Phytomonas melophthora; Bacterium melophthora*]	Apple (*Malus* spp.) rot	Diptera: Tephritidae	
		Rhagoletis pomonella (Walsh) (apple maggot)	Illingworth, 1912[g](does not mention rot, but it is apparent in the photos of Maggot damage); Allen, 1931; Allen and Riker, 1932; Allen *et al.*, 1934; Boush and Matsumura, 1967; Baerwald and Boush, 1968; Miyazaki *et al.*, 1968
Pseudomonas phaseolicola (Burkholder) Dowson	halo blight of bean, *Phaseolus* spp.	Thysanoptera: Thripidae	
		Hercinothrips femoralis (O. M. Reuter) (banded greenhouse thrips) [*Heliothrips femoralis* Reuter]	Buchanan, 1932[f] Leach, 1940
Pseudomonas saliciperda Lindeijer see *Erwinia salicis* (Day) Chester	watermark disease and canker of willow		
Pseudomonas savastanoi (Smith) Stevens [*Bacterium savastanoi* Smith]	olive (*Oleae* spp.) knot	Diptera: Tephritidae	
		Dacus oleae (Gmelin) (olive fruit fly)	Petri, 1909, 1910; Stammer, 1929; Hellmuth, 1956; Hagen *et al.*, 1963; Hagen, 1966; F Fytizas and Tzanakakis, 1966a, 1966b, and 1966c
			Rey, 1969; Fytizas, 1969a[f], 1969b[f]; Yamvrias *et al.*, 1970
		unidentified leaf miner	Martelli, 1958[g]
		unidentified insects	Smith, 1908[g]; Horne *et al.*, 1912a[g], 1912b[g]

Pseudomonas savastanoi var. *fraxini* (Brown) Dowson	canker of ash, *Fraxinus* spp.	Coleoptera: Scolytidae	
		Leperisinus fraxini (Panzer)	Oganova, 1957
Pseudomonas solanacearum (Smith) Smith [*Phytomonas solanacearum*]	Moko disease or wilt of banana, plantain and related plants (*Musa* spp.)	Coleoptera: Curculionidae	
		Cosmopolites sordidus (Germar) (banana root borer)	Robbs, 1946
		Odoiporus paganus[e]	Wardlaw, 1961
		Diptera: Drosophilidae	
		Drosophila spp.	Buddenhagen and Kelman, 1964
		Diptera: ?	
		"fruit flies"	Vakili and Baldwin, 1966[g]
		unidentified species	Buddenhagen and Kelman, 1964[g]
		Hymenoptera: Apidae	
		Trigona corvina Cockerell[e]	Buddenhagen and Elsasser, 1962
		Trigona spp.	Buddenhagen and Kelman, 1964
		Hymenoptera: Vespidae	Buddenhagen and Elsasser, 1964
		Polybia spp.[e]	Buddenhagen and Kelman, 1964
		Hymenoptera: ?	
		"bees"	Vakili and Baldwin, 1966[g]
		"wasps"	Buddenhagen and Elsasser, 1962[g]
		unidentified insects	Robbs, 1946[g]; Roth, 1961[g]; Wardlaw, 1961[g]; Buddenhagen and Elsasser, 1962[g]; Vakili and Baldwin, 1966[g]; Berg, 1971[g]
Pseudomonas solanacearum (Smith) Smith [*Bacillis solanacearum* Smith; *Bacterium solanacearum* (Smith) Smith; *Phytomonas solanacearum*]	wilt and brown rot of Solanaceae (tomato, potato, eggplant, etc.); Granville wilt of tobacco, *Nicotiana* spp.	Coleoptera: Chrysomelidae	
		Diabrotica graminea Baly[e]	Nolla, 1931
		Leptinotarsa decemlineata (Say) (Colorado potato beetle) [*Doryphera decemlineata*]	Smith, 1896, 1914
		"flea beetles"	Smith, 1896[g]
		other chrysomelids	Smith, 1896

Pathogen[a]	Disease[b]	Arthropod Associate[c]	Reference[d]
Pseudomonas solanacearum		Coleoptera: Meloidae	
		Epicauta spp.	Robbs, 1946
		unidentified species	Smith, 1896[g]
		Hymenoptera: Formicidae	
		Solenopsis sp. (foot washing ant)	Robbs, 1946
		Isoptera: ?	
		unidentified species	Robbs, 1946[g]
		Spirostreptida: Spirostreptidae	
		"millipedes"	Robbs, 1946[g]
		unidentified insects	Smith, 1896[g]; Smith, 1914[g]; Miller and Harvey, 1932[g]; Robbs, 1946[g]; Kelman, 1953[g]
Pseudomonas tabaci (Wolf and Foster) Stevens [*Phytomonas tabaci*]	wildfire of tobacco, *Nicotiana* spp.	Coleoptera: Chrysomelidae	
		Epitrix cucumeris (Harris) (potato flea beetle)	Robbs, 1946
		Epitrix parvula Fabricius	Robbs, 1946
		Lepidoptera: Sphingidae	
		Manduca sexta (Linnaeus) (tobacco hornworm) [*Protoparce sexta*]	Robbs, 1946
Pseudomonas tolaasii Paine	brown blotch of cultivated mushrooms, *Agaricus* spp.	Diptera: Phoridae	
		unidentified species	Gandy, 1967, 1968
also see "unidentified bacteria," below		Diptera: Sciaridae	
		unidentified species	Gandy, 1967, 1968
Pseudomonas sp.	soft rot of Brassicacea [Cruciferae]	Diptera: Anthomyiidae	
		Hylemya brassicae (Weidemann) (cabbage maggot) [*Hylemyia brassicae* (Bouché)]	Doane and Chapman, 1964b

Pseudomonas sp.	a disease of coriander, *Coriandrum sativum*	Hemiptera (Heteroptera): Miridae	
		Lygus rugulipensis Popp.	Ferenc, 1971
		unidentified insects	Ferenc, 1971[g]
Pseudomonas sp. in the fluorescent group	decay of the capitulum of sunflower, *Helianthus annuus*	Diptera: Tephritidae	
		Neotephritis finalis (Loew)	Fucikovsky, 1972
Pseudomonas spp.	a disease of grain crops	Diptera: Chloropidae	
		Oscinella frit (Linnaeus) (frit fly) [*Oscinella pusilla* (Meigen)]	Ryzhkova, 1962
Salmonella montevideo	(a contaminant in stored wheat)	Coleoptera: Curculionidae	
		Sitophilus oryzae (Linnaeus) (rice weevil)	Husted *et al.*, 1969
Streptomyces sp. or spp. and unidentified bacteria (acting with fungi and arthropods) also see "unidentified bacteria," below.	deterioration of stored grain	Acarina: Acaridae	
		Acarus siro Linnaeus	Sinha *et al.*, 1962
		Acarina: Ascidae	
		Melichares keegani Fox.	Sinha *et al.*, 1962
		Acarina: Cheyletidae	
		Cheyletus eruditus (Schrank)	Sinha *et al.*, 1962
		Cheyletus trouessarti Oudemans	Sinha *et al.*, 1962
		Acarina: Glycyphagidae	
		Aeroglyphus robustus (Banks) (warty grain mite)	Sinha *et al.*, 1962
		Glycyphagus destructor (Schrank)	Sinha *et al.*, 1962
		Acarina: Haemogamasidae	
		Haemogamasus pontiger (Berlese)	Sinha *et al.*, 1962
		Acarina: Laelaptidae	
		Haemolaelaps casalis Berlese	Sinha *et al.*, 1962
		Acarina: Tarsonemidae	
		Parsonemus sp.	Sinha *et al.*, 1962
		Acarina: Tydeidae	
		Tydeus interruptus Sig Thor.	Sinha *et al.*, 1962
		Coleoptera: Cucujidae	

Pathogen[a]	Disease[b]	Arthropod Associate[c]	Reference[d]
Streptomyces sp. or spp.		*Ahasverus advena* (Waltl) (foreign grain beetle)	Sinha *et al.*, 1962
		Coleoptera: Curculionidae	
		Sitophilus granarius (Linnaeus) (granary weevil)	van Bronswijk and Sinha, 1971
		Coleoptera: Lathridiidae	
		Lathridius minutus (Linnaeus) (squarenosed fungus beetle)	Sinha *et al.*, 1962
		Coleoptera: Tenebrionidae	
		Tribolium castaneum (Herbst) (red flour beetle)	van Bronswijk and Sinha, 1971
		Lepidoptera: Pyralidae	
		Pyralis farinalis Linnaeus (meal moth)	Sinha *et al.*, 1962
		Psocoptera: Liposcelidae	
		Liposcelis divinatorius (Müller)	Sinha *et al.*, 1962
Xanthomonas campestris (Pammel) Dowson [*Pseudomonas campestris* (Pammel); *Bacillus campestris* Pammel; *Phytomonas campestris*]	brown or black rot, leaf blight and stump rot of Brassicaceae [Cruciferae]	Coleoptera: ?	
		unidentified species	Smith, 1897[g]
		Hemiptera (Heteroptera): ?	
		"bugs"	Smith, 1897[g]
		"harlequin cabbage bug"	Smith, 1898a[g]
		Hemiptera (Homoptera): Aphididae	
		Brevicoryne brassicae (Linnaeus) (cabbage aphid)	Robbs, 1946; Stapleton and McDermott, 1950
		Myzus persicae (Sulzer) (green peach aphid)	Stapleton and McDermott, 1950
		unidentified species	Brenner, 1904[g] Clayton, 1924[g], 1929[g]

		Lepidoptera: Noctuidae *Trichoplusia ni* (Hübner) (cabbage looper) [*Autographa brassicae* Riley; *Plusia brassicae*, southern cabbage worm]	Smith, 1898b[f]; Clayton, 1929
		Lepidoptera: Pieridae	
		Ascia monusta Linnaeus	Robbs, 1946
		Pieris rapae (Linnaeus) (imported cabbageworm)	Smith, 1897; Clayton, 1929
		Lepidoptera: Yponomeutidae	
		Plutella xylostella (Linnaeus) (diamondback moth) [*Plutella maculipennis* Curtis]	Clayton, 1929
		Orthoptera: ?	
		"grasshoppers"	Russell, 1898[g]
		Thysanoptera: ?	
		unidentified species	Clayton, 1929[g]
		unidentified insects	Smith, 1897[g]; Russell, 1898[g]; Clayton, 1924[g]
Xanthomonas citri (Hasse) Dowson [*Pseudomonas citri* Hasse]	*Citrus* spp. canker	Lepidoptera: Gracillariidae	
		Phyllocnistis citrella Stainton	Hill, 1918; Hutson and Pinto, 1934; Voute, 1934; Bokura, 1936; Latif and Yunis, 1950; Sohi and Sandhu, 1968; Sinha *et al.*, 1972
		Phyllocnistis saligna (Zeller)	Kurisaki, 1920
		Lepidoptera: ?	
		Psorosticha zizyphi Stainton[e]	Hutson and Pinto, 1934
		unidentified leaf miners	Nirvan, 1962[g]
		unidentified insects	Stevens, 1915[g]
Xanthomonas juglandis (Pierce) Dowson [*Pseudomonas juglandis* Pierce; *Phytomonas juglandis*	walnut (*Juglans* spp.) blight	Acari: Eriophyidae	
		Eriophyes tristriatus var. *erinae* Nalepa	Rudolph, 1943a, 1943b

Pathogen[a]	Disease[b]	Arthropod Associate[c]	Reference[d]
(Pierce) Bergey *et al.*; *Bacterium juglandis* Pierce]		Acari: ?	
		"mites"	Miller and Bollen, 1946[g]
		Coleoptera: ?	
		unidentified species	Smith *et al.*, 1912[g]
		Diptera: Tephritidae	
		Rhagoletis completa Cressan (walnut husk fly)	Tsiropoulos, 1976
		Diptera: ?	
		"husk maggot"	Rudolph, 1933[g]
		unidentified species	Smith *et al.*, 1912[g]
		Hemiptera (Homoptera): Aphididae	
		Chromaphis juglandicola (Kaltenbach) (walnut aphid)	Rudolph, 1933; Miller and Bollen, 1946[g]
		unidentified species	Smith *et al.*, 1912[g] Miller and Bollen, 1946[g]
		Lepidoptera: Olethreutidae	
		Laspeyresia pomonella (Linnaeus) (codling moth)	Rudolph, 1933[g]; Michelbacher, 1945
		Lepidoptera: Pyralidae	
		Amyelois transitella (Walker) (navel orangeworm) [*Paramyelois transitella* (Walker)]	Michelbacher and Davis, 1961
		unidentified insects	Smith *et al.*, 1912[g] Miller and Bollen. 1946[g]
Xanthomonas malvacearum (Erw. Smith) Dowson [*Bacterium malvacearum*]	blackarm and angular leaf spot of cotton, *Gossypium* spp.	Coleoptera: Chrysomelidae	
		"cucumber beetles"	Faulwetter, 1917[g]
		"flea beetles"	Faulwetter, 1917[g]
		"leaf-eating beetles"	Faulwetter, 1917[g]
		Hemiptera (Homoptera): Cicadeliidae	
		"jassids"	Faulwetter, 1917

		Order and family unknown *Helopeltis* spp.[e]	Hayward, 1967
Xanthomonas malvacearum (Erw. Smith) Dowson see *Bacillus gossypina* Stedman	cotton boll rot		
Xanthomonas manihotis (Arthaud-Berthet and Bondar) Starr [*Phytomonas manihotis*]	bacteriosis of cassava, *Manihot* spp.	Diptera: Lonchaeidae *Silba pendula*[e] Bezz [*Lonchaea pendula*]	Robbs, 1946
Xanthomonas oryzae (Uyeda and Ishiyama) Dowson	blight of rice, *Oryza sativa*	Hemiptera (Heteroptera); Coreidae *Leptocorisa acuta* (Thunberg)	Mohiuddin *et al.*, 1976
		unidentified insects	Mohiuddin *et al.*, 1976[g]
Xanthomonas pelargonii (Brown) Starr and Burkholder	leaf spot of geranium, *Pelargonium hortorum*	Hemiptera (Homoptera): Aleyrodidae *Trialeurodes vaporariorum* (Westwood) (greenhouse whitefly)	Bugbee, 1962; Bugbee and Anderson, 1963
Xanthomonas phaseoli **(Erw. Smith) Dowson**, *X. phaseoli* var. *fuscans* (Burkholder) Starr and Burkholder and *X. phaseoli* f. sp. *vignicola* (Burkholder) Sabet	**bacterial blight of beans** (*Phaseolus* spp.) and cowpeas (*Vigna* spp.)	Coleoptera: Chrysomelidae *Cerotoma ruficornis* (Oliver)	Kaiser and Vakili, 1978
		Coleoptera: Curculionidae *Chalcodermus ebeninus* Boheman[e]	Kaiser and Vakili, 1978
		Diaprepes abbreviatus Linnaeus [*Diaprepes abbreviata* L.]	Kaiser and Vakili, 1978
		Hemiptera (Heteroptera): Pentatomidae *Nezara viridula* (Linnaeus) (southern green stink bug)	Kaiser and Vakili, 1978
		Hemiptera (Homoptera): Aleyrodidae *Bemisia tabaci* (Gennadius) (sweetpotato whitefly)	Sabet and Ishag, 1969
		Hemiptera (Homoptera): Cicadellidae *Empoasca* sp.	Kaiser and Vakili, 1978

Pathogen[a]	Disease[b]	Arthropod Associate[c]	Reference[d]
Xanthomonas ricini (Yoshi and Takimoto) Dowson	leaf blight of castor bean, *Ricinus communis*	Hemiptera (Homoptera): Aleyrodidae	
		Bemisia tabaci (Gennadius) (sweetpotato whitefly)	Khalil and Sabet, 1971
		unidentified species	Khalil and Sabet, 1971[g]
Xanthomonas rubrilineans (Lee *et al.*) Starr and Burkholder [*Phytomonas rubrilineans*] (not listed in Buchanan and Gibbons, 1974)	top rot of sugarcane, *Saccharum officinarum*	Coleoptera: Curculionidae	
		Rhabdoscelus obscurus (Boisduval) (New Guinea sugarcane weevil)	Mungomery, 1944
Xanthomonas translucens (Jones, Johnson and Reddy) Dowson (including *X. translucens* f. spp. *undulosa* Hagborg, *hordei* Hagborg [*hordei-avenae*] and *cerealis* Hagborg)	blight of barley, *Hordeum* spp.; *Xanthomonas* streak of cereals and grasses	Hemiptera (Homoptera): Aphididae	
		unidentified species	Jones *et al.*, 1917[g] Boosalis, 1952[g]
		Thysanoptera: ?	
		unidentified species	Jones *et al.*, 1917[g]
Xanthomonas vasculorum (Cobb) Dowson [*Bacterium vascularum* (Cobb) Grieg-Smith]	gumming disease of sugarcane, *Saccharum officinarum*	Coleoptera: Chrysomelidae	
		Monochirus multispinosus Germar[e]	North, 1935
		Diptera: Anthomyiidae	
		Atherigona tibiseta Malloch[e]	North, 1935
		Diptera: Calliphoridae	
		Calliphora augur Fabricius[e]	North, 1935
		Phaenicia sericata (Meigen) [*Lucilia sericata* (Meigen)]	North, 1935
		Diptera: Chloropidae	
		Oscinisoma sp. [*Oscinosoma* sp.]	North, 1935
		Scoliophthalmus sp.	North, 1935
		Diptera: Empididae	
		Drapetis ciliatocosta Bezzi[e]	North, 1935
		Diptera: Lauxaniidae	
		Homoneura indesisa Malloch[e]	North, 1935

unidentified species	North, 1935
Diptera: Muscidae	
unidentified species	North, 1935
Diptera: Sarcophagidae	
Helicobia australis J. and T.[e]	North, 1935
Sarcophaga aurifrons Macquart[e]	North, 1935
Sarcophaga spp.	North, 1935
Diptera: ?	
unidentified species	North, 1935[g]
Hemiptera (Heteroptera): Lygaeidae	
Nysius vinitor Bergr.	North, 1935
Hemiptera (Heteroptera): Miridae	
"capsid"	North, 1935
Hemiptera (Heteroptera): Nabidae	
Nabis capsiformis (Germar) (pale damsel bug)	North, 1935
Hemiptera (Heteroptera): ?	
"acanthiid"	North, 1935[g]
Hemiptera (Homoptera): Aphididae	
Melanaphis sacchari (Zehntner) (sugarcane aphid) [*Aphis sacchari* Zehnter]	North, 1935
Hemiptera (Homoptera): Delphacidae	
Perkinsiella saccharicida Kirkaldy (sugarcane delphacid)	North, 1935
Hemiptera (?): ?	
Cuspicona simplex[e]	North, 1935
Hemiptera (?): ?	
Megacoelum modestum[e]	North, 1935
Hemiptera (?): ?	
Mirperus ? *scutellaris* (sic)[e]	North, 1935

Pathogen[a]	Disease[b]	Arthropod Associate[c]	Reference[d]
Xanthomonas vasculorum		unidentified insects	North, 1935[g]
		In addition to the species listed above, North (1935) tested insects in the following groups for transmission with negative results:	
		Coleoptera: Anthribidae: 1 sp.; Carabidae: 1 sp.; Chrysomelidae: 10 spp.; Coccinellidae: 1 sp.; Cucujidae: 1 sp.; Curculionidae: 4 spp.; Melyridae: 1 sp.; Mycetophagidae: 1 sp.; Scarabaeidae: 2 spp.; Staphylinidae: 1 sp.	
		Hemiptera (Heteroptera): Coreidae: 3 spp.; Lygaeidae: 6 spp.; Miridae: 1 sp.; Pentatomidae: 9 spp.; Reduviidae: 1 sp.	
		Hemiptera (Homoptera): Cicadellidae [Jassidae]: 9 spp.; Fulgoridae: 6 spp.; Membracidae: 1 sp.	
		Neuroptera: Hemerobiidae: 1 sp.	
		Orthoptera: Acrididae: 1 sp.; Tettigoniidae: 5 spp.	
		Orthoptera (Dictyoptera): Blattidae: 1 sp.	
		Psocoptera: Psocidae: 1 sp.	
		Thysanoptera: Thripidae: 1 sp.	
Xanthomonas sp.	leaf streak of corn, *Zea mays*	Lepidoptera: Noctuidae	
		Spodoptera frugiperda (J. E. Smith) (fall armyworm)	Camino *et al.*, 1974
Xanthomonas sp. or spp. (saprophytic)	bad flavor (potato flavor or pea-like flavor) in Kivu	Diptera: Tephritidae	
		Ceratitis annonae[e]	Stolp, 1960

	coffee, *Coffea* sp.	*Trihithrum inscripta*[e]	Stolp, 1960
		Trirhithrum coffeae Bezzi[e]	Ingram, 1965
unidentified bacteria	soft rot of green banana (*Musa* spp.) fruits	unidentified insects	Stover, 1972[g]
unidentified bacteria	soft rot of Brassicaceae [Cruciferae]	Diptera: Anthomyiidae	
		Hylemya brassicae (Weidemann) (cabbage maggot) *Hylemyia brassicae* (Bouché)	Doane and Chapman, 1964b
unidentified bacterium or bacteria	black heart of celery, *Apium graveolens*	Hemiptera (Heteroptera): Miridae	
		Lygus lineolaris (Palisot de Beauvois)	Ontario Dept. of Agriculture, 1918
unidentified bacteria	unspecified *Citrus* diseases	Diptera: Tephritidae	
		Dacus ferrugineus tryoni Froggatt[e]	Queensland Dept. of Agriculture, and Stock, 1945[f]
unidentified bacteria	damage to coconut, *Cocos nucifera*	Coleoptera: Lymexylonidae	
		Melittomma insulare Fairmaire[e]	Brown, 1954
unidentified bacterium or bacteria	rot of lettuce, *Lactuca sativa*	Hemiptera (Homoptera): Aphididae	
		unidentified species	Burger, 1914[g]
unidentified bacterium or bacteria	bacteriosis of maize, *Zea mays*	Diptera: Chloropidae	
		Elachiptera cornuta Fallén	Vilkova, 1962[f]
		Lepidoptera: Pyralidae	
		Ostrinia nubilalis (Hübner) (European corn borer) [*Pyrausta nubilalis* Hbn.]	Nemlienko, 1951
		unidentified insects	Nemlienko, 1951[g]
unidentified bacteria	pit, decomposition of the gills and cap surface and unnamed diseases of cultivated mushrooms, *Agaricus* spp.	Diptera: Cecidomyiidae	
		Heteropeza pygmaea Winnertz	Hussey and Wyatt, 1958, 1960, 1962; Hussey *et al.*, 1960[f]
		Mycophila barnesi Edwards[e]	Hussey and Wyatt, 1958
		Mycophila speyeri (Barnes)	Hussey and Wyatt, 1958
unidentified bacterium	canker or bacteriosis of	Acari: Tetranychidae	

Pathogen[a]	Disease[b]	Arthropod Associate[c]	Reference[d]
	oleander, *Nerium* sp. or spp.	*Tetranychus telarius* Linnaeus	Tonelli, 1913
		Acari: ?	
		"mites"	Tonelli, 1913[g]
		Diptera: ?	
		unidentified species	Tonelli, 1913[g]
		Hemiptera (Homoptera): Coccidae	
		Lecanium sp.	Tonelli, 1913
		Hemiptera (Homoptera): Diaspididae	
		Aspidiotus nerii Bouché (oleander scale)	Tonelli, 1913
		unidentified insects	Tonelli, 1913[g]
unidentified bacterium or bacteria	bacterial canker of poplar, *Populus* spp.	Coleoptera: Cerambycidae	
		Saperda carcharias Linnaeus[e]	Régnier, 1933[f]
		Saperda populnea (Linnaeus)	Régnier, 1933[f]
		Diptera: Anisopodidae	
		Mycetobia pallipes Meigen	Régnier, 1933[f]
		Sylvicola fenestralis (Scopoli) [*Anisopus fenestralis* Scop.]	Régnier, 1933[f]
		Diptera: Mycetophilidae	
		unidentified species	Régnier, 1933[f]
		Hemiptera (Homoptera): Aphididae	
		Pterocomma populeum (Kaltenbach)	Régnier, 1933[f]
		unidentified species	Régnier, 1933[f,g]
		Hemiptera (Homoptera): Cercopidae	
		Aphrophora salicis De Geer[e]	Régnier, 1933[f]
		Aphrophora spumaria (Linnaeus)	Régnier, 1933[f]
		unidentified species	Régnier, 1933[f]
		Hemiptera (Homoptera): Cicadellidae	
		Idiocerus populi Linnaeus[e]	Régnier, 1933[f]
		Idiocerus scurra (Germar)	Régnier, 1933[f]
		"jassids"	Régnier, 1933[f,g]

		Lepidoptera: Cossidae	
		Cossus cossus Linnaeus	Régnier, 1933[f]
		Zeuzera pyrina (Linnaeus) (leopard moth)	Régnier, 1933[f]
		Lepidoptera: Olethreutidae	
		Gypsonoma aceriana Dup.[e]	Régnier, 1933[f]
		Lepidoptera: Sesiidae	
		Aegeria vespiformis Linnaeus[e]	Régnier, 1933[f]
		Paranthrene tabaniformis var. *intermedia* Le Cerf[e]	Bleton and Fieuzet, 1941
		Sciapteron tabaniforme Rott.[e]	Régnier, 1933[f]
		Trochilium apiforme Clerck[e]	Régnier, 1933[f]
		unidentified insects	Régnier, 1933[f,g]
unidentified bacterium or bacteria	necrosis of young tomato (*Lycopersicon* sp. or spp.) fruits	Thysanoptera: ?	
		unidentified species	Pussard-Radulesco, 1930[g]
unidentified bacteria	damage to shoots of sugar beet, *Beta vulgaris*	"root borer"	Petrenko and Karaseva, 1969[f,g]
unidentified bacteria and/or fungi	decay of Brassicaceae [Cruciferae]	Diptera: Anthomyiidae	
		Hylemya brassicae (Weidemann) (cabbage maggot) [*Phorbia brassicae* Bouché; *Chortophila brassicae* Bouché]	Schoene, 1916[f]; Herrick and Colman, 1922[f]; Miles, 1931[f]
unidentified bacteria and/or fungi	root rot, crown rot, and basal stem rot of cereals and grasses	Coleoptera: Curculionidae	
		Limnobaris deplanata Casey [*Anacentrinus deplanatus* (Casey)]	Hanson and Milliron, 1941
		Sphenophorus parvulus Gyllenhal (bluegrass billbug) [*Calendra parvula* (Gyllenhal)]	Hanson and Milliron, 1942; Hanson *et al.*, 1950
		Diptera: Cecidomyiidae	
		Mayetiola destructor (Say) (Hessian fly) [*Phytophaga destructor* (Say)]	Boosalis, 1954

Pathogen[a]	Disease[b]	Arthropod Associate[c]	Reference[d]
unidentified bacteria and several species of fungi (Other authors, e.g. Davey and Smith, 1933, indicated that the fungi alone were responsible for the disease.)	souring and decay of figs, *Ficus* spp.	Acari: Cheyletidae	
		Cheyletus sp.[e]	Hansen and Davey, 1932
		Acari: Eriophyidae	
		Eriophyes fici Cotte (fig mite)	Hansen and Davey, 1932
		Acari: ?	
		Sejus pomi Parrott[e]	Hansen and Davey, 1932
		Acari: ?	
		unidentified predaceous mites	Hansen and Davey, 1932[g]
		Thysanoptera: Phlaeothripidae	
		Leptothrips mali (Fitch)	Hansen and Davey, 1932
		Liothrips ilex (Moulton)[e]	Hansen and Davey, 1932
		Thysanoptera: Thripidae	
		Caliothrips fasciatus (Pergande) (bean thrips) [*Heliothrips fasciatus* (Perg.)]	Hansen and Davey, 1932
		Frankliniella cephali (Crawford) [*Frankliniella tritici* var. *californicus* (Moulton)]	Hansen and Davey, 1932
		Frankliniella tritica (Fitch) (flower thrips)	Hansen and Davey, 1932
		Frankliniella sp.	Hansen, 1929
		Thrips tabaci Lindeman (onion thrips)	Hansen, 1929
		[*Thrips bremneri* (Moulton)]	Hansen and Davey, 1932
		unidentified insects	Condit, 1919[g]
unidentified bacteria and fungi	damage to wheat, *Triticum* spp., by this insect	Diptera: Cecidomyiidae	
		Sitodiplosis mosellana (Géhin) (wheat midge)	Miller and Halton, 1961
unidentified bacteria (acting with fungi and arthropods) also see *Streptomyces* sp. or spp., above	deterioration of stored grain	Coleoptera: Curculionidae	
		Sitophilus oryzae (Linnaeus) (rice weevil) [*Sitophilus oryza* (L.)]	Van Wyk *et al.*, 1959
		Coleoptera: Tenebrionidae	

		Tribolium confusum Jacquelin duVal (confused flour beetle)	Van Wyk *et al.*, 1959
unidentified bacteria and fungi	rot of fruit of grapefruit, *Citrus paradisi*	Diptera: Tephritidae	
		Anastrepha ludens (Loew) (Mexican fruit fly)	McPhail *et al.*, 1969
unidentified bacteria and/or fungi	soft rot and decay of prickly pear, *Opuntia* spp.	Coleoptera: Curculionidae	
		Cactophagus spinolae (Gyllenhal)[e]	Sellers, 1951; Pettey, 1953
		Lepidoptera: Pyralidae	
		Cactoblastis cactorum (Berg) (cactus moth)	Dodd, 1929, 1936, 1940
unidentified bacteria and fungi	a disease of red clover (*Trifolium pratense*) roots	Coleoptera: Scolytidae	
		Hylastinus obscurus (Marsham) (clover root borer)	Leath and Byers, 1973
		Coleoptera: Curculionidae	
		Phrydiuchus topiarius Germar[e]	Nesterenko and Zaritsky, 1938

[a]Except as noted, bacterial names are those listed in Buchanan and Gibbons (1974). Names in brackets are synonyms used in the references cited in this table.

[b]Host plants listed are those which are mentioned in the references cited in this table.

[c]Arthropod names were updated when possible. Common names in parentheses are those approved by the Entomological Society of America (Sutherland, 1978). Names in brackets are synonyms used in the references cited in this table. Question marks indicate data not available to us. Both arthropods shown to be and shown not to be associated with particular diseases are listed, as well as those for which relationships have been suggested but not proven. Arthropods were listed as specifically as possible.

[d]All pertinent references for the arthropod-disease relationships listed were included. Some minor papers that briefly mentioned work done by others and popularized accounts of previously reported work were not included. Those listed here report original research, personal observation on the first suggestion of a possible arthropod-bacterial association. All borderline references were included. Research dealing indirectly with particular arthropod-disease associations was included where appropriate.

[e]We were unable to verify this arthropod name.

[f]Author did not adequately identify pathogen by scientific or common name, but disease appeared to be that which was listed.

[g]Arthropod or arthropod group was identified by common name only.

tion and inoculation of bacteria and strengthens the inherent weaknesses of the pathogens.

Many significant epidemiological concepts developed prior to 1940 were based upon incomplete or inconclusive evidence and still remain unresolved. These include the concept involving the role of insects in the overwintering of bacterial pathogens, their long distance dissemination and their perpetuation from generation to generation in their vectors.

The questions of the actual significance of insect involvement compared to rain splash and other means of inoculum dissemination, and the establishment of random infection foci in healthy crops from which spread by other agents can occur are of great practical significance but also remain unresolved.

A renewed research effort utilizing new techniques for bacterial identification, detection and isolation is badly needed to determine the answers to these and other questions.

12.4 ACKNOWLEDGMENTS

The authors wish to express their deep appreciation to Mrs. Jane Wilkins for typing the manuscript; Ms. Mary Lou Hanagan for proof reading the entire chapter; Ms. Linda Lahman, Mr. Thomas Nehrkorn, Mr. Charles McVean and Mr. Jose Antonio Saraiva for translating some papers into English; Ms. Frances Barney, Ms. Dorothy Leising, Ms. Loretta Saracino and Ms. Julie Wessling for library assistance and to Dr. John Capinera, Dr. Howard Evans and Dr. W. Don Fronk for their critical review of the manuscript.

12.5 REFERENCES

Agrios, G. N. (1969). Plant Pathology. Academic Press, N. Y. pp. 342-349.

Al-Tamimi, K. M. (1969). Transmission of bacterial soft rot by fruit flies. M. S. thesis, Univ. of Massachusetts, Amherst, Mass.

Alban, E. K., and Keirns, V. E. (1948). The effect of rotenone, commercial, and aerosol grade DDT dusts on the total yield, grade, and maturity of seven cucumber varieties. *Proc. Am. Soc. Hortic. Sci.* 51: 448-452.

Allan, E., and Kelman, A. (1975). Detection of *Erwinia carotovora* var. *atroseptica* in mixed cultures, potato tubers, soil, insects and on potato leaves by immunofluorescent staining procedures (abstr.). *Proc. Am. Phytopathol. Soc.* 2: 68.

Allan, E. and Kelman, A. (1977). Immunofluorescent stain procedures for detection and identification of *Erwinia carotovora* var. *atroseptica*. *Phytopathology* 67: 1305-1312.

Allen, T. C. (1931). Bacteria producing rot of apple in association with the apple maggot, *Rhagoletis pomonella*. *Phytopathology* 21: 338.

Allen, T. C., Pinckard, J. A., and Riker, A. J. (1934). Frequent association of *Phytomonas melophthora* with various stages in the life cycle of the apple maggot, *Rhagoletis pomonella*. *Phytopathology* 24: 228-238.

Allen, T. C., and Riker, A. J. (1932). A rot of apple fruit caused by *Phytomonas melophthora*, n. sp., following invasion by the apple maggot. *Phytopathology* 22: 557-571.

Ark, P. A. (1944). Studies on bacterial canker of tomato. *Phytopathology* 34: 394-400.

Ark, P. A., and Thomas, H. E. (1936). Persistence of *Erwinia amylovora* in certain insects. *Phytopathology* 26: 375-381.

Arthur, J. C. (1885). Report of the botanist. Diseases of the pear. Pear blight. *N. Y. Agric. Exp. Stn. Geneva Annu. Rep.* 3: 357-367.

Ashworth, L. J., Jr., Hildebrand, D. C., and Schroth, M. N. (1970). *Erwinia*- induced necrosis of immature cotton bolls. *Phytopathology* 60: 602-607.

Ashworth, L. J., Jr., Hunter, R. E., and Leigh, T. F. (1969). Bacterial boll rot of cotton: transmission of an *Erwinia* sp. by the brown stinkbug. *Proc. 29th Cotton Dis. Counc.*, New Orleans. p. 25.

Baerwald, R. J., and Boush, G. M. (1968). Demonstration of the bacterial symbiote *Pseudomonas melophthora* in the apple maggot, *Rhagoletis pomonella*, by fluorescent-antibody technique. *J. Invertebr. Pathol.* 11: 251-259.

Bailey, S. F. (1935). Thrips as vectors of plant disease. *J. Econ. Entomol.* 28: 856-863.

Barry, P. (1847). Insect blight. *Genesee Farmer* 8 (2nd ser.): 218-219.

Bean, J. L. (1947). *Eumerus strigatus* reared from decayed potatoes. *J. Econ. Entomol.* 40: 452-454.

Beecher, F.S., and Doolittle, S.P. (1959). Dissemination of angular leaf spot of cucumber. *Plant Dis. Rep*. 34: 382-383.

Berg, L. A. (1971). Weed hosts of the SFR strain of *Pseudomonas solanacearum*, causal organism of bacterial wilt of bananas. *Phytopathology* 61: 1314-1315.

Beyme, D., Ficke, W., Kleinhempel, H., Schaefer, H.J., and Bremer, R. (1975). Modellversuche zur Überlebensdauer von *Erwinia amylovora* im Verdauungstrakt der Honigbiene, im Honig und an Teilen des Bienenstocks. [Model experiments on the longevity of *Erwinia amylovora* in the digestive tract of the honey bee, in honey, and on parts of the beehive.] *Arch. Phytopathol. Pflanzenschutz* 11: 203-211 (saw English summary only).

Bleton, C. A. and Fieuzet, L. (1941). Les parasites du peuplier au Maroc. 2e note. Sur la biologie de *Paranthrene tabaniformis*, Rott. var. *intermedia*, Le Cerf (Lépidoptère: Sesiidae). [The parasites of poplar in Morocco. 2nd note. On the biology of *Parathene tabaniformis*, Rott. var. *intermedia*, Le Cerf (Lepidoptera: Sesiidae).] *Bull. Soc. Sci. Nat. Maroc*. 20: 57-59. (not seen). Abtsr. in *Rev. Appl. Entomol.* A 32: 232 (1944).

Boewe, G. H. (1956). Stewart's disease prospects for 1956. *Plant Dis. Rep.* 40: 367-368.

Bokura, U. (1936). [Cares in preventing citrus canker.] *J. Plant. Prot., Tokyo* 23: 431-435 (not seen).

Bonde, R. (1928). The transmission of potato black-leg by the seedcorn maggot in Maine. *Phytopathology* 18: 459.

Bonde, R. (1930a). The cabbage maggot as a disseminating agent of bacterial rots in the Cruciferae (abstr.). *Phytopahtology* 20: 128.

Bonde, R. (1930b). Some conditions determining potato-seed-piece decay and blackleg induced by maggots (abstr.). *Phytopathology* 20: 128.

Bonde, R. (1939a). Comparative studies of the bacteria associated with potato blackleg and seed-piece decay. *Phytopathology* 29: 831-851.

Bonde, R. (1939b). The role of insects in the dissemination of potato blackleg and seed-piece decay. *J. Agric. Res.* 59: 889-917.

Bonde, R. (1950). Factors affecting potato blackleg and seed-piece decay. *Maine Agric. Exp. Stn. Bull.* 482. 31p.

Boosalis, M. G. (1952). The epidemiology of *Xanthomonas translucens* (J. J. and R.) Dowson on cereals and grasses. *Phytopathology* 42: 387-395.

Boosalis, M.G. (1954). Hessian Fly in relation to the development of crown and basal stem rot of wheat. *Phytopathology* 44: 224-229.

Bournier, A. (1970). Principaux types de dégats de Thysanptères sur les plantes cultivées. [The principal types of damage by Thysanoptera on cultivated plants.] *Ann. Zool. Ecol.*

Anim. 2: 237-259. Abstr. in *Rev. Appl. Entom.* -A 60: 3278 (1972).

Boush, G. M. and Matsumura, F. (1967). Insecticidal degradation by *Pseudomonas melophthora*, the bacterial symbiote of the apple maggot. *J. Econ. Entomol.* 60: 918-920.

Boyle, A. M. (1949). Further studies of the bacterial necrosis of the giant cactus. *Phytopathology* 39: 1029-1052.

Brenner, W. (1904). Die Schwarzfäule des Kohls. [Blackrot of cabbages.] *Zentralbl. Bakteriol.* II 12: 725-735.

Brentzel, W. E. and Munro, J. A. (1940). Bacterial ring rot of the potato. Investigation on possible dissemination by grasshoppers. *N. Dak. Agric. Exp. Stn. Bull.* 295. 8 p.

Brooks, A. N. (1926). Studies of the epidemiology and control of fireblight of apple. *Phytopathology* 16: 665-695.

Brooks, J. W. and Allen, T. C. (1941). Tests of certain insecticidal dusts against the striped cucumber beetle. *J. Econ. Entomol.* 34: 295-297.

Brooks, M. A. (1963). Microorganisms of healthy insects. *In* "Insect Pathology. An Advanced Treatise" (E. A. Steinhaus, ed.), Vol. 1, pp. 215-250. Academic Press, N. Y.

Brown, E. S. (1954). The biology of the coconut pest *Melittomma insulare* (Col., Lymexylonidae), and its control in the Seychelles. *Bull. Entom. Res.* 45: 1-66.

Brown, N. A., and Jamieson, C. O. (1913). A bacterium causing a disease of sugar-beet and nasturium leaves. *J. Agric. Res.* 1: 189-210.

Buchanan, D. (1932). A bacterial disease of beans transmitted by *Heliothrips femoralis* Reut. *J. Econ. Entomol.* 25: 49-53.

Buchanan, R. E. and Gibbons, N. E. Eds. (1974). Bergey's Manual of determinative bacteriology, 8th ed. Williams and Wilkens Co., Baltimore. 1268 p.

Buddenhagen, I.W. and Elsasser, T.A. (1962). An insect-spread bacterial wilt epiphytotic of bluggoe banana. *Nature (Lond.)* 194: 164-165.

Buddenhagen, I. W. and Kelman, I. W. (1964). Biological and physiological aspects of bacterial wilt caused by *Pseudomonas solanacearum*. *Annu. Rev. Phytopathol.* 2: 203-230.

Bugbee, W. M. (1962) Whitefly transmission of *Xanthomonas pelargonii* (abstr.). *Phytopathology* 52: 5.

Bugbee, W. M., and Anderson, N. A. (1963). Whitefly transmission of *Xanthomonas pelargonii* and histological examination of leafspots of *Pelargonium hortorum*. *Phytopathology* 53: 177-178.

Burger, O. F. (1914). Report of assistant plant pathologist. Lettuce diseases-bacterial rot. *Fla. Agric. Exp. Stn. Annu. Rep.* 1913: 87-88.

Burkholder, W. H. (1960). Some observations on *Erwinia tracheiphila*, the causal agent of the cucurbit wilt. *Phytopathology* 50: 179-180.

Burrill, A. C. (1915). Insect control important in checking fire blight. *Phytopathology* 5: 343-347.

Callan, E. McC. (1939). *Cryptorrhynchus lapathi* L. in relation to the watermark disease of the cricket-bat willow. *Ann. Appl. Biol.* 26: 135-137.

Callan, E. McC. (1940). Hymenopterous parasites of willow insects. *Bull. Entomol. Res.* 31: 35-44.

Camino, A., de Leon, J.M., and de Leon, C. (1974). Bacterial leaf streak in corn, *Zea mays*. *Agrociencia* (18): 63-70 (saw English summary only). Abstr. in *Biol. Abstr.* 61:2882 (1976).

Carsner, E. (1918). Angular-leafspot of cucumber: dissemination, overwintering, and control. *J. Agric. Res.* 15: 201-220 and plates 13-16.

Carter, W. (1962). Insects in Relation to Plant Disease. John Wiley and Sons, N.Y. 705 p.

Carter, W. (1973). Insects in Relation to Plant Disease, 2nd ed. John Wiley and Sons, N.Y. 759 p.

Cassidy, J. (1888). Apple twig blight. *Colo. Agric. Exp. Stn. Annu. Rep.* 1: 64-70.

Castor, L.L., Ayers, J.E., McNab, A.A., and Krause, R.A. (1975a). Computerized forecasting system for Stewart's bacterial desease on corn (abstr.). *Phytopathology* 65: 835.

Castor, L.L., Ayers, J.E., McNab, A.A., and Krause, R.A. (1975b). Computerized forecasting system for Stewart's bacterial desease on corn. *Plant Dis. Rep.* 59: 533-536.

Cate, C.C. (1918). Pear-blight control in Rogue River Valley, Oregon. *Better Fruit* 13(3): 5-6.

Cauquil, J. (1963). Premières observations sur les pourritures de capsules en République Centrafricaine. [First observations on boll rots in the Central African Republic.] *Coton Fibres Trop.* 18(2): 243-250.

Cauquil, J. (1973). La pourriture des capsules du cotonnier: essai de mise en place d'une methode de lutte. [Cotton boll rot: attempt to establish a method of control.] *Coton Fibres Trop.* 28(3): 413-448. Abstr. in *Rev. Appl. Entomol.* -A 63: 595 (1975).

Cauquil, J., and Mildner, P. (1964). Première étude sur le comportement varietal de cotonnier en présence des pourritures de capsules. [Preliminary study of varietal reaction of cotton in the presence of boll rots.] *Coton Fibres Trop.* 20(4): 539-548. Abstr. in *Rev. Appl. Mycol.* 45: 259 (1966).

Chester, F.D. (1901). Pear blight and pear canker. *Del. Coll. Agric. Exp. Stn. Bull.* 52. 8 p.

Chiu, W.-F., Yuen, C.-S, and Wu, C. A. (1958). [On the overwintering and dissemination of the soft rot organism *Erwinia aroideae* (Towns.) Holland. *Chih Wu Ping Li Hsueh Pao* 4: 8-15 (saw English summary only). Abstr. in *Rev. Appl. Mycol.* 38: 44 (1965).

Clayton, E.E. (1924). Investigations of cauliflower diseases on Long Island. *N. Y. Agric. Exp. Stn. Geneva Bull.* 506. 15 p.

Clayton, E.E. (1929). Studies on the black rot or blight disease of cauliflower. *N. Y. Agric. Exp. Stn. Geneva Bull.* 576. 44 p.

Cognée, M., and Frinking, H. D. (1966). Rôle de quelques bacteries dans le développement des pourritures secondaires des capsules du cotonnier. [Role of some bacteria in the development of secondary cotton boll rots.] *Coton Fibres Trop.* 21(3): 249-261. Abstr. in *Biol. Abstr.* 48: 7406 (1967).

Condit, I.J. (1919). Some causes of the souring and splitting of figs. *Fig. and Olive J.* 3(9): 13-16.

Cook, A.J. (1901). Pear blight and bees. *Calif. Cultiv.* 17: 83-84.

Crosse, J.E., Goodman, R.N., and Shaffer, W.H., Jr. (1972). Leaf damage as a predisposing factor in the infection of apple shoots by *Erwinia amylovora*. *Phytopathology* 62: 176-182.

Da Costa, C.P., and Jones, C.M. (1971). Resistance in cucumber *Cucumis sativus* L. to 3 species of cucumber beetles. *Hortscience* 6: 340-342.

Davey, A.E., and Smith, R.E. (1933). The epidemiology of fig spoilage. *Hilgardia* 7: 523-551.

Davidson, R.S. (1948). Factors affecting the development of bacterial soft rot of potato tuber initials. *Phytopathology* 38: 673-687.

Day, W.R. (1924). The watermark disease of the cricket bat willow (*Salix caerulea*). Oxf. For. Mem. no. 3. 30 p. (not seen). Abstr. in *Rev. Appl. Mycol.* 4: 321-323 (1925).

Dean, R.W. (1933). Morphology of the digestive tract of the apple maggot fly, *Rhagoletis pomonella* Walsh. *N.Y. Agric. Exp. Stn. Geneva Tech. Bull.* 215: 3-17.

Dean, R.W. (1935). Anatomy and postpupal development of the female reproductive system in the apple maggot fly, *Rhagoletis pomonella* Walsh. *N.Y. Agric. Exp. Stn. Geneva Tech. Bull.* 229. 21 p.

Doane, C.C. (1953). The onion maggot in Wisconsin and its relation to rot in onions. Ph.D. dissertation, Univ. of Wisconsin.

Doane, J.F., and Chapman, R.K. (1964a). Development of the cabbage maggot, *Hylemyia brassicae* (Bouché) on aseptic and decaying rutabaga tissue. *Entomol. Exp. Appl.* 7: 115-119.

Doane, J.F., and Chapman, R.K. (1964b). The relation of the cabbage maggot, *Hylemyia brassicae* (Bouché) to decay in some cruciferous crops. *Entomol. Exp. Appl.* 7: 1-8.

Dodd, A.P. (1929). The Progress of Biological Control of Prickly Pear in Australia. Commonw. Prickly Pear Board, Brisbane, Australia. 44 p. Abstr. in *Rev. Appl. Entomol.* -A 18: 287-288.

Dodd, A.P. (1936). The control and eradication of prickly-pear in Australia. *Bull. Entomol. Res.* 27: 503-522.

Dodd, A.P. (1940). The Biological Campaign Against Prickly Pear. Commonw. Prickly-Pear Board, Brisbane, Australia. 177 p. Abstr. in *Rev. Appl. Entomol.* -A 29: 587 (1941).

Doolittle, S.P. (1921). Overwintering of the bacterial wilt of cucurbits. *Phytopathology* 11: 299-300.

Duncan, J., and Généreus, H. (1960). La transmission par les insectes de *Corynebacterium sepedonicum* (Spieck & Kott.) Skaptason et Burkholder. [The transmission by insects of

Corynebacterium sepedonicum (Spieck & Kott.) Skaptason and Burkholder.] *Can. J. Plant Sci.* 40: 110-116.

Edgerton, C.W. (1912). The rots of the cotton boll. *La. Agric. Exp. Stn. Bull.* 137. 113 p.

Edward J.C., Tripathi, S.C. and Singh, K.P. (1973). Observations on a 'tip-over' disease of banana in Allahabad. *Curr. Sci.* 42: 696-697. Abstr. in *Rev. Appl. Entomol.* -A 63: 565 (1975).

Elliott, C. (1935). Dissemination of bacterial wilt of corn. *Iowa State Coll. J. Sci.* 9: 461-480.

Elliott, C. (1938). Bacterial wilt of sweet corn in 1938. *Plant Dis. Rep.* 22: 401-402.

Elliott, C. , and Poos, F.W. (1934). Overwintering of *Aplanobacter stewartii*. *Science* n.s. 80-289-290.

Elliott, C., and Poos, F.W. (1940). Seasonal development, insect vectors, and host range of bacterial wilt of sweet corn. *J. Agric. Res.* 60: 645-686.

Elliott, C., and Robert, A.L. (1939). *Tripsacum dactyloides*, another host of *Aplanobacter stewarti*. *Phytopathology* 29: 284-285.

Emmett, B.J. (1971). Insect visitors to pear blossom. *Plant Pathol.* 20: 36-40.

Emmett, B.J., and Baker, L.A.E. (1971). Insect transmission of fireblight. *Plant Pathol.* 20: 41-45.

Engelhard, A.W. (1969). Bulb mites associated with deseases of gladioli and other crops in Florida (abstr.). *Phytopathology* 59: 1025.

Erskine, J.M. (1975). Preventing the entry of dangerous plant diseases into South Africa - the fire blight threat. *S. Afr. J. Sci.* 71: 265-269. Abstr. in *Biol. Abstr.* 62: 1624 (1976).

Faulwetter, R.C. (1917). Dissemination of the angular leaf-spot of cotton. *J. Agric. Res.* 8: 457-475.

Ferenc, N. (1971). A koriander (*Coriandrum savtivum* L.) károsodásával kapcsolatos phytophylakologiai vizagálatok eredeményei. [Results of phyto-phylacological investigations related to the injury of coriander (*Coriandrum sativum* L.).] *Herba. Hung.* 10(1): 37-46 (saw English summary only). Abstr. in *Biol. Abstr.* 54: 4379 (1972).

Folsom, D. (1954). Bacterial fire blight of raspberry associated with raspberry cane maggot. *Plant Dis. Rep.* 38: 338-339.

Forbes, S. A. (1884a). The tarnished plant bug. *Farmers' Rev.*, Feb. 28, p. 150.

Forbes, S. A. (1884b). The tarnished plant bug. *In* Noxious and Beneficial Insects of the State of Illinois. *Ill. State Entomol. Rep.* (1883) 13: 115-135.

Forbes, S. A. (1894). [Corn flea beetle, *Chaetocnema pulicaria* Mels.] Noxious and Beneficial Insects of the State of Illinois. *Ill. State Entomol Rep.* (1891-1892) 18. 171 p.

Forbes, S. A. (1905). [Corn flea beetle, *Chaetocnema pulicaria* Mels.] Noxious and beneficial insects of the state of Illinois. *Ill. State Entomol. Rep.* 23. p. 109-110.

Forsberg, J. L. (1959). Relationship of the bulb mite *Rhizoglyphus echinopus* to bacterial scab of gladiolus (abstr.). *Phytopathology* 49: 538.

Forsberg, J. L. (1965). The relationship of *Pseudomonas marginata, Stromatina gladiola*, bulb mites, and chemical control treatments to the occurrence of scab and *Stromatinia* rot of gladiolus (abstr.). *Phytopathology* 55: 1058.

Friend, W. G., Salkeld, E. H., and Stevenson, I. L. (1959). Nutrition of onion maggots, larvae of *Hylemya antiqua* (Meig.), with reference to other members of the genus *Hylemya*. *Ann. N. Y. Acad. Sci.* 77: 384-393.

Fronk, W. D., and Peterson, L. E. (1954). Some effects of treating muskmelons with insecticides. *J. Econ. Entomol.* 47: 807-811.

Frutchey, C. W. (1936). A study of Stewart's disease of sweet corn caused by *Phytomonas stewarti*. *Mich. Agric. Exp. Stn. Tech. Bull.* 152. 25 p.

Fucikovsky, L. (1972). Pudricion bacteriana del capitulo del girasol (*Helianthus annus*). [Bacterial decay of the capitulum of the sunflower *Helianthus annuus* (abstr.).] *Phytopathology* 62: 758.

Fytizas, E. (1969a). Inhibition du développement larvaire de la descendance de *Dacus oleae* (Gmel.) (Diptera: Tephritidae) par la streptomycine, incorporée à la nourriture d'adultes. I. Doses de streptomycine efficaces pour l'inhibition. [Inhibition of larval development of the progeny of *D. oleae* by streptomycin incorporated in the food of the adults. I. Doses of streptomycin efficacious for inhibiton.] *Z. Angew. Entomol.* 63(3): 268-271. Abstr. in *Rev. Appl. Entomol.*-A 59: 768 (1971).

Fytizas, E. (1969b). Inhibition du développement larvaire de la descendance de *Dacus oleae* Gmel. (Diptera, Tephritidae) par la streptomycine, incorporeé à la nourriture des adultes. II. Administrations de la streptomycine répétées, en présence des symbiotes. [Inhibition of the larval development of the progeny of *Dacus oleae* Gmel. by streptomycin incorporated into the adult food. II. Repeated administration of streptomycin in the presence of the symbionts.] *Z. Angew. Entomol.* 64: 35-38. Abstr. in *Rev. Appl. Entomol.*-A 60: 40 (1972).

Fytizas, E., and Tzanakakis, M. E. (1966a). Développement des larves de *Dacus oleae* dans des olives, lorsque leurs parents ont reçu la streptomycine, incorporée à leur nourriture. [Development of larvae of *D. oleae* in olives, when their parents have received streptomycin incorporated in their food.] *Ann. Epiphyt.* 17(1): 53-59. Abstr. in *Rev. Appl. Entomol.*-A 55: 612-613 (1967).

Fytizas, E., and Tzanakakis, M. E. (1966b). Some effects of streptomycin, when added to the adult food, on the adults of *Dacus oleae* (Diptera: Tephritidae) and their progeny. *Ann. Entomol. Soc. Am.* 59: 269-273.

Fytizas, E., and Tzanakakis, M. E. (1966c). Sur la possibilité d'une lutte contre le *Dacus oleae* par l'utilization des antibiotiques. [On the possibility of controlling *D. oleae* by the use of antiobiotics.] *In* Achtiende International Symposium over Fytofarmacie en Fytiatrie, 3 Mei 1966. *Meded. Rijksfac. Landbwet. Gent* 31(3): 343-1170. p. 782-788. Abstr. in *Rev. Appl. Entomol.*-A 56: 254 (1968).

Gandy, D. G. (1967). The epidemiology of bacterial blotch of the cultivated mushroom. *Glasshouse Crops Res. Inst. Annu. Rep.* 1966: 150-154.

Gandy, D. G. (1968). A technique for screening bacteria causing brown blotch of cultivated mushrooms. *Glasshouse Crops Res. Inst. Annu. Rep.* 1967: 150-154.

Glorlenko, M. V., Voronkevich, I. V., and Maksimova, T. S. (1956). [Mutual relations between the onion fly and onion syrphid and the bacteria that cause damp rot in plants.] *Zool. Zhr.* 35: 116-120 (not seen).

Goldsworthy, M. C. (1926). Studies on the spot disease of cauliflower; a use of serum diagnosis. *Phytopathology* 16: 877-884.

Gossard, H. A. (1916a). Is the hive a center for distributing fireblight: Is aphid honedew a medium for spreading blight? *J. Econ. Entomol.* 9: 59-62.

Gossard, H. A. (1916b). The role of insects as carriers of fireblight. *Rep. Proc. Mont. Hortic. Soc.* 19: 84-90.

Gossard, H. A., and Walton, R. C. (1916a). *Bacillus amylovorus* in honey and in honeydew (abstr.). *Phytopathology* 6: 113.

Gossard, H. A., and Walton, R. C. (1916b). Fire-blight investigations. *Ohio Agric. Exp. Stn. Mon. Bull.* 1: 274-276.

Gossard, H. A., and Walton, R. C. (1917). Fire-blight infection. Rain proves an important carrier of disease in open orchards. *Ohio Agric. Exp. Stn. Mon. Bull.* 2: 357-364.

Gossard, H. A., and Walton, R. C. (1922). Dissemination of fire blight. *Ohio Agric. Exp. Stn. Bull.* 357: 79-126.

Gould, G. E. (1944). The biology and control of the striped cucumber beetle. *Ind. Agric. Exp. Stn. Bull.* 490. 28 p.

Graham, D. C. (1958). Occurrence of soft rot bacteria in Scottish soils. *Nature (Lond.)* 181: 61.

Graham, D. C., and Hardie, J. L. (1971). Prospects for the control of potato blackleg disease by the use of stem cuttings. *Proc. 6th Br. Insectic. Fungic. Conf.* 1: 219-224.

Graham, D. C., and Harper, P. C. (1967). Potato blackleg and tuber soft rot. *Scottish Agric.* 46: 1-7.

Graham, D. C., Quinn, C. E., and Harrison, M. D. (1976). Recurrence of soft rot coliform bacterial infections in potato stem cuttings: An epidemiological study on the central nuclear stock production farm in Scotland, 1967-1974. *Potato Res.* 19(1): 3-20.

Granovsky, A. A. (1931). Insects in relation to fruit injuries. *Minn. Hortic.* 59: 60-61.

Granovsky, A. A. (1940). The relation of subterranean insects to the raspberry crown gall. *Hoosier Hortic.* 22: 67-69.

Gray, E. (1940). The willow wood wasp and watermark disease of willows. *Br. Vet. J.* 96: 370-373.

Hagen, K. S. (1966). Dependence of the olive fly, *Dacus oleae*, larvae on symbiosis with *Pseuomonas savastanoi* for the utilization of olive. *Nature (Lond.)* 209: 423-424.

Hagen, K. S., Santas, L., and Tsecouras, A. (1963). A technique of culturing the olive fly *Dacus oleae* Gmel. on synthetic media under xenic conditions. *In* Radiation and Radioisotopes Applied to Insects of Agricultural Importance. Proc. Symp. Athens, 22-26 April 1963, Int. At. Energy Agency, Vienna. p. 333-356.

Hansen, H. N. (1929). Thrips as carriers of fig-decaying organisms. *Science n. s.* 69: 356-357.

Hansen, H. N., and Davey, A. E. (1932). Transmission of smut and mold in figs. *Phytopathology* 22: 247-252.

Hansen, H. N., and Smith, R. E. (1937). A bacterial gall disease of Douglas fir, *Pseudotsuga taxifolia. Hilgardia* 10: 569-577.

Hanson, E. W., and Milliron, H. E. (1941). The relation of the curculionid, *Anacentrinus deplanatus*, to root rot and basal stem rot of barnyard grass, *Echinochloa crusgalli. Phytopathology* 31: 832-837.

Hanson, E. W., and Milliron, H. E. (1942). The relation of certain weevils to root rot and basal stem rot of cereals and grasses (abstr.). *Phytopathology* 32: 22.

Hanson, E. W., and Christensen, J. J. (1950). The relation of the bluegrass billbug, *Calendra parvula* (Gyllenhal), to the development of basal stem rot and root rot of cereals and grasses in North Central United States. *Phytopathology* 40: 527-543.

Harrison, M. D., Quinn, C. E., Sells, I. A., and Graham, D. C. (1977). Waste potato dumps as sources of insects contaminated with soft rot coliform bacteria in relation to re-contamination of pathogen-free potato stocks. *Potato Res.* 20(1): 37-52.

Haynes, R. L., and Jones, C. M. (1975). Wilting and damage to cucumber by spotted and striped cucumber beetles. *Hortscience* 10: 265-266.

Hayward, J. A. (1967). Cotton in Western Nigeria. 2. Entomological problems. *Cotton Grow. Rev.* 44: 117-135.

Hedgcock, G. G. (1910). Field Studies of the crown gall of the grape. *U. S. Dept. Agric. Bur. Plant Ind. Bull.* 183. 40 p.

Heichel, G. H., Sands, D. C., and Kring, J. B. (1977). Seasonal patterns and reduction by carbofuran of Stewart's bacterial wilt of sweet corn. *Plant Dis. Rep.* 61: 149-153.

Hellmuth, H. (1956). Untersuchungen zur Bakteriensymbiose der Trypetiden (Diptera). [Studies on the bacterial symbionts of Trypetidae.] *Z. Morphol. Oekol. Tiere* 44: 483-517.

Herrick, G. W., and Colman, W. (1922). The cabbage maggot, with special reference to its control. *N. Y. Agric. Exp. Stn. Cornell Bull.* 413. 15 p.

Hervey, G. E. R., and Schroeder, W. T. (1946). The varietal response of cucumbers to DDT control. *J. Econ. Entomol.* 39: 403-404.

Hildebrand, E. M. (1935). Longevity of the fire blight organism in the honeybee environment (abstr.) *Phytopathology* 25: 20.

Hildebrand, E. M. (1937). Infectivity of the fire-blight organism. *Phytopathology* 27: 850-852.

Hildebrand, E. M., and MacDaniels, L. H. (1935). Modes of entry of *Erwinia amylovora* into the flowers of the principal pome fruits (abstr.). *Phytopathology* 25: 20.

Hildebrand, E. M., and Phillips, E. F. (1936). The honeybee and the beehive in relation to fire blight. *J. Agric. Res.* 52: 789-810.

Hill, G. F. (1918). History of citrus canker in the Northern Territory. *North. Ter. Aust. Bull.* 18. 8 p. Abstr. in *Rev. Appl. Entomol.* -A 7: 201 (1919).

Hill, R. R., Jr., Murray, J. J., and Zeiders, K. E. (1971). Relationships between clover root curculio injury and severity of bacterial wilt in alfalfa. *Crop Sci.* 11: 306-307.

Horne, W. T., Parker, W. B., and Daines, L. L. (1912a). The method of spreading of the olive knot disease (abstr.). *Phytopathology* 2: 96.

Horne, W. T., Parker, W. B., and Daines, L. L. (1912b). The method of spreading of the olive knot disease. *Phytopathology* 2: 101-105.

Hotson, J. W. (1916). Observations on fire blight in the Yakima Valley, Washington. *Phytopathology* 6: 288-292.

Howard, C. M., and Leach, J. G. (1962). Relationship of the iris borer (*Macronoctua onusta*) to the spread and development of soft rot (abstr.). *Phytopathology* 52: 1219.

Howard, C. M., and Leach, J. G. (1963). Relation of the iris borer to bacterial soft rot of iris. *Phytopathology* 53: 1190-1193.

Huff, C. G. (1928). Nutritional studies on the seed-corn maggot *Hylemyia cilicrura* Rondani. *J. Agric. Res.* 36: 625-630.

Hussey, N. W., and Wyatt, I. J. (1958). Preliminary observations on the cecid problem. *Mushr. Grow. Assoc. Bull.* 100: 150-158.

Hussey, N. W., and Wyatt, I. J. (1960). Insecticidal control of paedogenetic cecid larvae in mushroom beds. II. Incorporation of insecticides in the casing layer. *Ann. Appl. Biol.* 48: 347-351.

Hussey, N. W., and Wyatt, I. J. (1962). Insecticidal control of paedogenetic cecid larvae in mushroom beds. III. Further experiments on control of *Heteropeza pygmaea* Winn. *Ann. Appl. Biol.* 50: 423-429.

Hussey, N. W., Wyatt, I. J., and Hughes, J. T. (1960). Insecticidal control of paedogenetic cecid larvae in mushroom beds. I. Incorporation of insecticides into compost. *Ann. Appl. Biol.* 48: 336-346.

Husted, S. R., Mills, R. B., Foltz, V. D., and Crumrine, M. H. (1969). Transmission of *Salmonella montevideo* from contaminated to clean wheat by the rice weevil. *J. Econ. Entomol.* 62: 1489.

Hutson, J. C., and Pinto, M. P. D. (1934). Two caterpillar pests of citrus. *Trop. Agric.* 83: 188-193.

Hvozdyak, R. I., Horal, V. M., Sydorenko, S. S., Lappa, N. V., Korobko, O. P., and Bichuk, Yu. P. (1973). Entomopatohenni vlastyrosti fitopatohennykh bakterii rodu *Erwinia*. [Entomopathogenic properties of phytopathogenic bacteria of the genus *Erwinia*.] *Dopov akad nauk ukr rsr ser B heol heofiz khim biol* 35: 80-82 (saw English summary only). Abstr. in *Biol. Abstr.* 59: 5925 (1975).

Illingworth, J. F. (1912). A study of the biology of the apple maggot (*Rhagoletis pomonella*), together with an investigation of methods of control. *N. Y. Agric. Exp. Stn. Cornell Bull.* 324: 125-187.

Ingram, W. R. (1965). An evaluation of several insecticides against berry borer and fruit fly in Uganda robusta coffee. *E. Afr. Agric. For. J.* 30: 259-262.

Ivanoff, S. S. (1932). Stewart's disease of corn (abstr.). *Phytopathology* 22: 13-14.

Ivanoff, S. S. (1933a). Bacterial wilt of corn (abstr.). *Phytopathology* 23: 18.

Ivanoff, S. S. (1933b). Stewart's wilt disease of corn, with emphasis on the life history of *Phytomonas stewarti* in relation to pathogenesis. *J. Agric. Res.* 47: 749-770.

Ivanoff, S. S., and Keitt, G. W. (1941). Relations of nectar concentration to growth of *Erwinia amylovora* and fire blight infection of apple and pear blossoms. *J. Agric. Res.* 62: 733-743.

Ivanoff, S. S., Riker, A. J., and Dickson, J. G. (1933). Bacterial wilt of corn caused by more than one strain of bacteria. *Wis. Agric. Exp. Stn. Bull.* 425: 107-108.

Jefferson, R. N., Bald, J. G., Morishita, F. S., and Close, D. H. (1956). Effect of vapam on *Rhizoglyphus* mites and gladiolus diseases. *J. Econ. Entomol.* 49: 584-589.

Johnson, D. E. (1930). The relation of the cabbage maggot and other insects to the spread and development of soft rot of Cruciferae. *Phytopathology* 20: 857-872.

Johnson, E. M. (1934). Dissemination of angular leaf spot of tobacco by the southern tobacco worm. *Phytopathology* 24: 1381-1383.

Johnson, J. C. (1956). Pod twist: a previously unrecorded bacterial disease of French bean (*Phaeolis vulgaris* L.). *Queensl. J. Agric. Sci.* 13: 127-158.

Johnston, J. R. (1912). The history and cause of the coconut bud-rot. *U. S. Dep. Agric. Bur. Plant Ind. Bull.* 228. 175 p.

Johnston, T. H., and Hitchcock, L. (1923). A bacteriosis of prickly pear plants (*Opuntia* spp.). *Trans. Proc. Soc. S. Aust.* 47: 162-164.

Jones, B. J. (1914). The natural modes of distribution of fire blight in Caifornia. *Calif. State Comm. Hortic. Mon. Bull.* 3: 505-511.

Jones, D. H. (1909). Bacterial blight of apple, pear and quince trees. *Ont. Dep. Agric. Bull.* 176. 64 p.

Jones, D. H. (1910). Plant diseases. *Ont. Agric. Coll. and Exp. Farm Annu. Rep.* 35: 128-130.

Jones, D. H. (1911a). Further research on fire blight. *Ont. Agric. Coll. and Exp. Farm Annu. Rep.* 36: 163-167.

Jones, D. H. (1911b). *Scolytus rugulosus* as an agent in the spread of bacterial blight in pear trees. *Phytopathology* 1: 155-158.

Jones, L. R., Johnson, A. G., and Reddy, C. S. (1917). Bacterial-blight of barley. *J. Agric. Res.* 11: 625-644.

Kaiser, W. J., and Vakili, N. G. (1978). Insect transmission of pathogenic xanthomonads to bean and cowpea in Puerto Rico. *Phytopathology* 68: 1057-1063.

Keitt, G. W., and Ivanhoff, S. S. (1941). Transmission of fire blight by bees and its relations to nectar concentration of apple and pear blossoms. *J. Agric. Res.* 62: 745-753.

Kelman, A. (1953). The bacterial wilt caused by *Pseudomonas solanacearum*. A literature review and bibliography. *N. C. Agric. Exp. Stn. Tech. Bull.* 99. 194 p.

Khalil, O., and Sabet, K. A. (1971). A study of certain aetiological aspects of bacterial leaf-blight disease of castor bean. *Sudan Agric. J.* 6: 26-33.

Kloepper, J. W. (1977). The role of insects in the epidemiology of potato blackleg. M. S. thesis, Colorado State Univ., Ft. Collins, Co.

Kloepper, J. W., Harrison, M. D., and Brewer, J. W. (1976). Contamination of insects by two varieties of *Erwinia carotovora* in nature and their ability to transmit the pathogens to potato plants (abstr.). *Proc. Am. Phytopathol. Soc.* 3: 317.

Kloepper, J. W., Harrison, M. D., and Brewer, J. W. (1979). The association of *Erwinia carotovora* var. *atroseptica* and *Erwinia carotovora* var. *carotovora* with insects in Colorado. *Am. Potato J.* 56: 351-361.

Kloepper, J. W., Harrison, M. D., and Brewer, J. W. (1980). Insect transmission of *Erwinia carotovora* var. *carotovora* and *Erwinia carotovora* var. *atroseptica* to potato plants in the field. In manuscript (a).

Kloepper, J. W., Harrison, M. D., and Brewer, J. W. (1980). The effect of temperature on the transmission of *Erwinia carotovora* var. *atroseptica* and *Erwinia carotovora* var. *carotovora* by insects. In manuscript (b).

Kurisaki, M. (1920). [On the life history of citrus leaf-miner, *Phyllocnistis saligna* Zell. and the relation between this miner and citrus canker, *Pseudomonas citri* Hasse.] *Konchu Sekai* 24(2): 39-44 (not seen). Abstr. in *Rev. Appl. Entomol.* -A 8: 194 (1920).

Landis, B. J., Onsager, J. A., Fox, L., and Foiles, L. L. (1971). Chemical control of the seed-corn maggot *Hylemya platura* (Meigen), and seed piece decay in potato seed pieces. *Am. Potato J.* 48: 374-380.

Lapwood, D. H. (1976). Field observations on blackleg in England. *Bull. Organ. Eur. Mediterr. Prot. Plant* 6: 237-239.

Lathrop, F. H. (1918). Leaf-hoppers injurious to apple trees. *N. Y. Agric. Exp. Stn. Geneva Bull.* 451: 183-200.

Latif, A., and Yunis, M. (1950). Observations on the relationship between citrus leaf-miner (*Phyllocnistis citrella* Stn.) and citrus canker (*Pseudomonas citri* Hasse). *Indian J. Entomol.* 12: 27-29.

Lazar, I., and Bucur, El. (1964). Recent research in Roumania on blackleg and bacterial soft rot of potato. *Eur. Potato J.* 7: 102-111.

Leach, J. G. (1925). The seed-corn maggot and potato blackleg. *Science n. s.* 61: 120.

Leach, J. G. (1926). The relation of the seed-corn maggot (*Phorbia fusciceps* Zett.) to the spread and development of potato blackleg in Minnesota. *Phytopathology* 16: 149-176.

Leach, J. G. (1927). The relation of insects and weather to the development of heart rot of celery. *Phytopathology* 17: 663-667.

Leach, J. G. (1930a). Further studies on the seed-corn maggot and potato black leg (abstr.). *Phytopathology* 20: 127.

Leach, J. G. (1930b). Survival of the potato-black-leg pathogen in the soil and some factors influencing the infection (abstr.). *Phytopathology* 20: 127.

Leach, J. G. (1931). Further studies on the seed-corn maggot and bacteria with special reference to potato blackleg. *Phytopathology* 21: 387-406.

Leach, J. G. (1933). The method of survival of bacteria in the puparia of the seed-corn maggot (*Hylemia cilicrura* Rond.). *Z. Angew. Entomol.* 20: 150-161.

Leach, J. G. (1940). Insect transmission of plant diseases. McGraw-Hill Book Co., Inc., N. Y. 615 p.

Leach, J. G. (1964). Observations on cucumber beetles as vectors of cucurbit wilt. *Phytopathology* 54: 606-607.

Leath, K. T. and Byers, R. A. (1973). Attractiveness of diseased red clover roots to the clover root borer. *Phytopathology* 63: 428-431.

Lelliott, R. A. (1959). Fire blight of pears in England. *Agriculture (Lond.)* 65: 564-568.

Lelliott, R. A. (1968). Fire blight in England. *In* Rep. Int. Conf. on Fire Blight (*Erwinia amylovora* (Burr.) Winslow *et al.*), Canterbury, 1967. *Europ. and Mediterr. Plant Prot. Organ. Publ. Ser. A.*, No. 45-E: 10-14.

Lindeijer, E. J. (1932a). De bacterienziekte van de wilg veroorzaakt door *Pseudomonas saliciperda* n. sp., [The bacterial disease of the willow caused by *Pseudomonas saliciperda* n. sp.] Thesis, Univ. Amsterdam (Hollandia-Drukkerij) Baarn. 82 p. (Transl. by E. W. J. Reyers, 1937, available from the Natl. Agric. Libr.).

Lindeijer, E. J. (1932b). Een bacterie-ziekte van de wilg (II.). [A bacterial disease of the willow (II.).] *Tijdschr. Plantenziekten*. 38(1): 9-11 (not seen).

Lindeijer, E. J. (1933). Die Bakterienkrankheit der Weide (verursacht durch *Pseudomonas saliciperda* n. sp.). [The bacterial disease of the willows (caused by *Pseudomonas saliciperde* n. sp.).] *Phytopathol. Z.* 6: 371-374. (Transl. by E. W. J. Reyers, 1939, available from the Natl. Agric. Libr.).

List, G. M., and Kreutzer, W. A. (1942). Transmission of the causal agent of the ringrot disease of potatoes by insects. *J. Econ. Entomol.* 35: 455-456.

Logan, C. (1969). The survival of the potato blackleg pathogen overwinter. *Rec. Agric. Res. Min. Agric. N. Ireland* 17: 115-121.

Lutz, F. E., and Brown, F. M. (1928). A new species of bacteria and the gall of an aphid. *Am. Mus. Novit.* No. 305: 1-4.

Magerstein, V. (1931). Rakovina vrby. [Crown gall of the willow.] *Ochrana Rostlin* 11: 135-137 (not seen). Abstr. in *Rev. Appl. Mycol.* 11: 274 (1932).

Manns, T. F. (1909). The blade blight of oats, a bacterial disease. *Ohio Agric. Exp. Stn. Bull.* 210: 89-167.

Marshall, G. E., Childers, N. F., and Brody, H. W. (1941). Leafhoppers can weaken apple trees and reduce the crop. *Proc. Ohio. Hortic. Soc.* 74: 61-66.

Martelli, G. P. (1958). Infezioni fogliari di rogua dell' Olivo in gallerie di insetti minatori. [Leaf infections of olive knot disease in galleries of mining insects.] *Inf. Fitopatol.* 4: 59-60 (not seen). Abstr. in *Rev. Appl. Mycol.* 38: 759 (1959).

Martin, J. P. (1921). The honey bee and pear blight. *Univ. Calif. J. Agric.* 7(6): 4, 17, 22.

McDaniel, E. I. (1935). Control of iris borer (*Macronoctua onusta* Grote). *Mich. Agric. Exp. Stn. Q. Bull.* 18(2): 92-94

McKeen, C. D. (1948). An occurrence of soft rot in peppers and its relation to the corn borer. *Sci. Agric.* 28: 142-143.

McLarty, H. R. (1923). Fire blight [*Bacillus amylovorus* (Burr.) Trev.] Report of the Dominion Laboratory of Plant Pathology, Summerland, B. C. Can. Dep. Agric. Div. Bot. Rep. Dom. Bot., Can. Exp. Farms: 64-66.

McPhail, M., Benschoter, C. A., and Lopez D, F. (1969). Fumigation with ethylene dibromide and ethylene chlorobromide to reduce the development of rot in grapefruit infested with Mexican fruit fly larvae. *J. Econ. Entomol.* 62: 1238-1239.

Mendes, L. O. T. (1956). Podridao interna does capulhos do algodoeiro obtida por meio de insetos. [Internal boll-rot of cotton, obtained by means of insects.] *Bragantia* 15: 9-11 (not seen). Abstr. in *Rev. Appl. Entomol.* -A 45: 72 (1957).

Merrill, J. H. (1915). Notes on an apparent relation between aphids and fire blight (*Bacillus amylovorus*) *J. Econ. Entomol.* 8: 402-403.

Merrill, J. H. (1917). Further data on the relation between aphids and fire blight (*Bacillus amylovorus* Bur. Trev.). *J. Econ. Entomol.* 10: 45-46.

Metzger, C. H., and Binkley, A. M. (1940). Some evidence on the spread of bacterial wilt. *Am. Potato J.* 17: 198-201.

Michelbacher, A. E. (1945). Control of codling moth on walnuts: progress report. *J. Econ. Entomol.* 38: 347-355.

Michelbacher, A. E., and Davis, C. S. (1961). The navel orangeworm in Northern California. *J. Econ. Entomol.* 54: 559-562.

Milbrath, D. G. (1922). Resume of pear blight history and methods of control. *Calif. Dep. Agric. Mon. Bull.* 11: 760-765.

Miles, H. W. (1931). The control of the cabbage root fly. *J. Minist. Agric.* 37: 1227-1231.

Miller, B. S., and Halton, P. (1961). The damage to wheat kernels caused by the wheat blossom midge (*Sitodiplosis mosellana*). *J. Sci. Food Agric.* 12: 391-398.

Miller, J. H., and Harvey, H. W. (1932). Peanut wilt in Georgia. *Phytopathology* 22: 371-383.

Miller, P. W. (1929). Studies of fire blight of apple in Wisconsin. *J. Agric. Res.* 39: 579-621.

Miller, P. W., and Bollen, W. B. (1946). Walnut bacteriosis and its control. *Oreg. Agric. Exp. Stn. Bull.* 9. 107 p.

Miller, T. D., and Schroth, M. N. (1972). Monitoring the epiphytic population of *Erwinia amylovora* on pear with a selective medium. *Phytopathology* 62: 1175-1182.

Mills, H. B. (1939). Montana insect pests for 1937 and 1938. 27th Rep. State Entomol. Mont. *Mont. Agric. Exp. Stn. Bull.* 366. 32 p.

Miyazaki, S., Boush, G. M., and Baerwald, R. J. (1968). Amino acid synthesis by *Pseudomonas melophthora*, bacterial symbiote of *Rhagoletis pomonella* (Diptera). *J. Insect Physiol.* 14: 513-518.

Mohiuddin, M. S., Rao, Y. P., Mohan, S. K., and Verma, J. P. (1976). Role of *Leptocorisa acuta* Thun. in the spread of bacterial blight of rice. *Curr. Sci.* 45: 426-427.

Molina, J. J., Harrison, M. D., and Brewer, J. W. (1974). Transmission of *Erwinia carotovora* var. *Atroseptica* by *Drosophila melanogaster* Meig. I. Acquisition and transmission of the bacterium. *Am. Potato J.* 51: 245-250.

Morrill, A. W. (1910). Plant-bugs injurious to cotton bolls. *U. S. Dep. Agric. Bur. Entomol. Bull.* 86. 110 p.

Müller, H. J., and Skiebe, K. (1962). [On relationships between bacterial rot and aphid infestation of Chinese cabbage, *Brassica pekinensis* Rupr., and their significance for the breeder.] *Züchter* 32: 210-215 (not seen). Abstr. in *Rev. Appl. Mycol.* 42: 285 (1963).

Mungomery, R. W. (1944). Report of the division of entomology and pathology. *44th Rep. Bur. Sugar Exp. Stn. Queensl.*: 21-24 (not seen). Abstr. in *Rev. Appl. Entomol.* -A 34: 1-2 (1946).

Neal, J. W., Jr., Enns, W. R., and Goodman, R. N. (1964). Sensitivity of bacterial isolates from eggs of *Aphis pomi* to *Erwinia amylovora* bacteriophages. *J. Econ. Entomol.* 57: 831-832.

Nemlienko, F. E. (1951). [A study of bacteriosis of maize cobs.] *Rep. Lenin Acad. Agric. Sci.* 1951 (2): 35-40 (not seen). Abstr. in *Rev. Appl. Mycol.* 30: 564 (1951).

Nesterenko, P. A., and Zaritsky, S. N. (1938). ["Hollowness" in clary sage (*Salvia sclarea* L.).] *C. R. Acad. Sci. USSR (N. S.)* 20: 741-744 (not seen). Abst. in *Rev. Appl. Entomol.* -A 29: 15 (1941).

Nirvan, R. S. (1962). Citrus canker and its control. *Hortic. Adv.* 5: 171-175. Abstr. in *Biol. Abstr.* 42: 1224 (1963).

Nolla, J. A. B. (1931). Studies on the bacterial wilt of the Solanaceae in Puerto Rico. *J. Dep. Agric. P. R.* 15: 287-308.

North, D. S. (1935). The gumming disease of sugar-cane, its dissemination and control. Agric. Rep. Colon. Sugar Refining Co., Sydney, 10 (tech.). 149 p.

Nowell, W. (1918). Internal disease of cotton bolls in West Indies. *W. Indian Bull.* 17: 203-235.

Oganova, E. A. (1957). [On bacterial canker of ash.] *Soobshch. Inst. Lesa, Akad. Nauk U. S. S. R.* 8: 81-89 (not seen). Abstr. in *Rev. of Appl. Mycol.* 38: 281 (1959).

O'Gara, P. J. (1908). Pear blight and its control on the Pacific coast. *Wash. State Hortic. Assoc. Rep.* 5: 36-55.

O'Gara, P. J. (1910). Control of pear blight on the Pacific coast. *Better Fruit* 5(2): 49-51, 54-56; 5(5): 30-43, 52-57.

Ogilvie, L., Mulligan, B. O., and Brian, P. W. (1934). Progress report on vegetable diseases, VI. *Annu. Rep. Agric. Hortic. Res. Stn. Univ. Bristol, 1934*: 175-190.

Ontario Department of Agriculture. (1918). Black heart of celery. *In* Report of the Minister of Agriculture, Vegetable Specialist. *Ont. Dep. Agric. Annu. Rep.* 1917. p. 61.

Orton, C. R., and Adams, J. F. (1914). Collar-blight of apple trees in Pennsylvania (abstr.). *Phytopathology* 4: 43.

Orton, C. R., and Adams, J. F. (1915). Collar-blight and related forms of fire blight. *Pa. Agric. Exp. Stn. Bull.* 136. 23 p.

Parker, K. G. (1936). Fire blight: overwintering, dissemination, and control of the pathogen. *N. Y. Agric. Exp. Stn. Cornell Memo.* 193. 42 p.

Parker, K. G., Fisher, E. G., and Mills, W. D. (1956). Fire blight on pome fruits . . . and its control. *N. Y. Agric. Exp. Stn. Cornell Ext. Bull.* 966. 23 p.

Pearson, E. O. (1934). Investigations on cotton stainers and internal boll disease. *In* Rep. 2nd Conf. Cotton Grow. Probl., 1934. Empire Cotton Growing Corp., Lond. p. 146-152.

Pelkwijk, A. J. ter. (1946). Onderzoek naar de biologie van den elzensnuitkever *Cryptorrhynchus lapathi* L. [The biology of the poplar and willow borer, *C. lapathi*.]. *Tijdschr. Plantenziekten.* 52: 22-25 (not seen). Abstr. in *Biol. Abstr.* 21: 1535 (1947).

Pepper, E. H. (1967). Stewart's bacterial wilt of corn. *Am. Phytopathol. Soc. Monogr.* 4. 36 p.

Perrault, C. (1948). Etudes sur la pourriture du cerne des pommes de terre causée par *Corynebacterium sependonicum* (Spieck & Kott.) Skaptason & Burkholder. I. Les agents de dissemination. [Studies on ringrot of potatoes caused by *C. Sependonicum*. I. Agents of dissemination.] *Sci. Agric.* 28: 244-260.

Petrenko, M. B., and Karaseva, L. V. (1969). Izmenenie soderzhaniya uglevodov v sakharnoi svekle pod vilyaniem rizosfernykh bakterii. [Changes in the carbohydrate content in the sugar beet due to the action of rhizosphere bacteria.] *Biol. Nauki* 12(8); 89-93 (not seen). Abstr. in *Biol. Abstr.* 51: 13074 (1970).

Petri, L. (1909). Ricerche sopra i batteri intestinali della Mosca olearia. [Research on the intestinal bacteria of the olive fly.] *Mem. R. Stn. Patol. Veg. Roma.* 130 p. and 1 plate.

Petri, L. (1910). Untersuchungen über die Darmbakterien der Olivenfliege. [Research on the intestinal bacteria of the olive fly.] *Zentralbl. Bakteriol.* II 26: 357-367. (A summary of Petri, 1909. English translation. *In* Brewer, J. W., and Harrison, M. D., ed. 1973. Readings in Insect-Plant Disease Relationships. MMS Information Corp., N. Y. p. 192-221.

Pettey, F. W. (1953). The boring beetles of prickly pear in South Africa and their importance in the control of *Opuntia megacantha. Union S. Afr. Dep. Agric. Sci. Bull.* 340: 1-36.

Phillips, E. F. (1916). Professor Gossard's theory on fire blight transmission. *J. Econ. Entomol.* 9: 362-363.

Phillips, E. F. (1933). Insects collected on apple blossoms in western New York. *J. Agric. Res.* 46: 851-862.

Phillips, J. A. (1977). The ecology and epidemiology of soft rot coliform bacteria with special emphasis on the role of insect transmission of the pathogen. M. S. thesis, Univ. of Wisconsin. 75 p.

Phillips, J. A., and Kelman, A. (1978). Tracing the dissemination of *Erwinia carotovora* from an inoculum source with a direct fluorescent antibody technique (abstr.). *Am. Potato J.* 55: 389.

Pierstorff, A. L. (1931). Studies on the fire blight organism, *Bacillus amylovorus. N. Y. Agric. Exp. Stn. Cornell Memo.* 136. 53 p.

Pierstorff, A. L., and Lamb, H. N. (1933a). Bees not cause of epidemics of fire blight. *Glean. Bee Cult.* 61: 216-217.

Pierstorff, A. L., and Lamb, H. N. (1933b). Dissemination of fire blight bacteria by honeybees (abstr.). *Phytopathology* 23: 27.

Pierstorff, A. L., and Lamb, H. N. (1934). The honeybee in relation to the overwintering and primary spread of the fire blight organism. *Phytopathology* 24: 1347-1357.

Plurad, S. B., Goodman, R. N., and Enns, W. R. (1965). Persistance of *Erwinia amylovora* in the apple aphid (*Aphis pomi* De Geer), a probable vector. *Nature (Lond.)* 205: 206.

Plurad, S. B., Goodman, R. N., and Enns, W. R. (1967). Factors influencing the efficacy of *Aphis pomi* as a potential vector for *Erwinia amylovora. Phytopathology* 57: 1060-1063.

Poos, F. W. (1939). Host plants harboring *Aplanobacter stewarti* without showing external symptoms after inoculation by *Chaetocnema pulicaria. J. Econ. Entomol.* 32: 881-882.

Poos, F. W. (1945). DDT to control corn flea beetle on sweet corn and potato leafhopper on alfalfa and peanuts. *J. Econ. Entomol.* 38: 197-199.

Poos, F. W. (1955). Studies of certain species of *Chaetocnema. J. Econ. Entomol.* 48: 555-563.

Poos, F. W., and Elliott, C. (1935). Bacterial wilt of corn and its insect vectors (abstr.). *Phytopathology* 25: 32.

Poos, F. W., and Elliott, C. (1936). Certain insect vectors of *Aplanobacter stewarti. J. Agric. Res.* 52: 585-608.

Pussard-Radulesco, E. (1930). Recherches giologiques et cytologiques sur quelques Thysanoptères. [Biological and cytological studies of some thysanopterans.] *Ann. Epiphyt.* 16(3-4): 103-189.

Queensland Department of Agriculture and Stock. (1945). Annu. Rep. for 1944-1945 (not 1945. Abstr. in *Rev. Appl. Entomol.* 35: 241-247 (1947).

Rand, F. V. (1915). Dissemination of bacterial wilt of cucurbits (preliminary note). *J. Agric. Res.* 5: 257-260.

Rand, F. V. (1923). Bacterial wilt or Stewart's disease of corn. *The Canner* 56: 164-165.

Rand, F. V., and Cash. L. C. (1920). Some insect relations of *Bacillus tracheiphilus* Erw. Sm. *Phytopathology* 10: 133-140.

Rand, F. V., and Cash, L. C. (1921). Stewart's disease of corn. *J. Agric. Res.* 21: 263-264.

Rand, F. V., and Cash, L. C. (1924). Further evidence of insect dissemination of bacterial wilt of corn. *Science n. s.* 59: 67-69.

Rand, F. V., and Cash, L. C. (1933). Bacterial wilt of corn. *U. S. Dep. Agric. Tech. Bull.* 362. 31 p.

Rand, F. V., and Enlows, E. M. A. (1916). Transmission and control of bacterial wilt of cucurbits. *J. Agric. Res.* 6: 417-434.

Rand, F. V., and Enlows, E. M. A. (1920). Bacterial wilt of cucurbits. *U. S. Dep. Agric. Bull.* 828. 43 p.

Reed, G. M. (1914). An unusual outbreak of apple blossom blight. *Phytopathology* 4: 27-30.

Rees, H. L. (1915). Bacterial gummosis of cherries. *West Wash. Agric. Exp. Stn. Puyallup Bull.* 3(2): 12-16.

Régnier, R. (1933). Contribution à l'étude du rôle des insectes dans la propagation du cancre des peupliers. [Contribution on the study of the role of insects in the propagation of canker of poplars.] Congr. Int. Entomol. Paris, 1932 5(2): 645-650 (not seen). Abstr. in *Rev. Appl. Entomol.* -A 22: 193 (1934).

Revilla, M. V. (1967). El Moko o casa del platono [Bacterial blight of banana.] *Agric, Ganad, Trop., Peru* 1(2): 36-37. Abstr. in *Rev. Plant Pathol.* 48: 99 (1969).

Rey, J. M. (1969). Desarrollo de una dieta larvaria para el cultivo de *Dacus oleae* Gmel. (Dipt. Tripetidae). [Development of a larval diet for the rearing of *D. oleae*.] *Graellsia* 24: 253-259 (not seen). Abstr. in *Rev. Appl. Entomol.* -A 59: 794 (1971).

Richardson, J. K. (1938). Studies on blackheart, soft-rot, and tarnished plant bug injury of celery. *Can. J. Res.* sect. C 16: 182-193.

Ries, D. T. (1929). Two iris insects. *Bull. Am. Iris Soc.* July, 1929: 27-46.

Robbs, C. F. (1946). Contribuicao ao estudo das bacterias que atacam plantas no Brasil. I. Contribuicao ao estudo da "podridao negra" das Cruciferas no Distrito Federal. II. Consideracoes sobre insetos possivelmente transmissores de doencas bacterianas em plantas do Brasil. III. Lista preliminar das bacterias fitopathogenicas observados no Brasil. [Contribution to the study of bacteria that attack plants in Brazil. I. Contribution to the study of black rot of Cruciferae in the Federal District. II. Considerations of insects which may be vectors of bacterial diseases of plants in Brazil. III. Preliminary list of phytopathogenic viruses observed in Brazil.] *Bol. Soc. Brasil. Agron.* (Rio de Janeiro) 9: 195-221. Abstr. in *Biol. Abstr.* 22: 972 (1948).

Robert, A. L. (1955). Bacterial wilt and Stewart's leaf blight of corn. *U. S. Dep. Agric. Farmers' Bull.* 2092. 13 p.

Rosen, H. R. (1930). Overwintering of the fire blight pathogen, *Bacillus amylovorus*, within the beehive. *Science* 72: 301-302.

Rosen, H. R. (1931). Fire blight from infected beehives. *Am. Fruit Grow. Mag.* 51(3): 5, 31-32.

Rosen, H. R. (1933). Further studies on the overwintering and dissemination of the fire blight pathogen. *Arkansas Agric. Exp. Stn. Bull.* 283. 102 p.

Roth, L. M. (1961). Some observations on insects attacking *Heliconia*. *Ann. Entomol. Soc. Am.* 54: 613-616.

Rudolph, B. A. (1933). Bacteriosis (blight) of the English walnut in California and its control. *Calif. Agric. Exp. Stn. Bull.* 564. 88 p.

Rudolph, B. A. (1943a). A possible relationship between the walnut erinose mite and walnut blight. *Science* 98: 430-431.

Rudolph, B. A. (1943b). The walnut erinose mite a carrier of walnut blight. *Diamond Walnut News* 25: 4-5.

Russell, H. L. (1898). A bacterial rot of cabbage and allied plants. *Wis. Agric. Exp. Stn. Bul.* 65. 39 p.

Ryzhkova, Ye. V. (1962). Phytopathogenic symbionts of *Oscinella frit* L. and *O. pusilla* Mg. (Diptera, Chloropidae) and their practical use. *Entomol. Rev.* 41(4): 490-494.

Sabet, K. A., and Ishag, F. (1969). Studies on the bacterial diseases of Sudan crops. VIII. Survival and dissemination of *Xanthomonas phaseoli*. *Ann. Appl. Biol.* 64: 65-74.

Schoene, W. J. (1916). The cabbage maggot: its biology and control. *N. Y. Agric. Exp. Stn. Geneva Bull.* 419: 99-160.

Sellards, E. H. (1905). A potato maggot. *In* Report of the Entomologist. Fla. Agric. Exp. Stn. Rep. p. 19-20.

Sellers, W. F. (1952). The collection of the cactus weevil, *Cactophagus spinolae* (Gylh.), in Mexico and its dispatch to South Africa. *Bull. Entomol. Res.* 43: 43-50.

Sinha, M. K., Batra, R. C., and Uppal, D. K. (1972). Role of citrus leaf-miner (*Phyllocnistis citrella* Stainton) on the prevalence and severity of citrus canker (*Xanthomonas citri* (Hasse) Dowson). *Madras. Agric. J.* 59: 240-245.

Sinha, R. N., Liscombe, E. A. R., and Wallace, H. A. H. (1962). Infestation of mites, insects and microorganisms in a large wheat bulk after prolonged storage. *Can. Entomol.* 94: 542-555.

Skowroński, Z. (1976). Etiologia i możliwości zwalczania zgnilizny róż kalafiorów nasiennych. [Aetiology and the possibilities of control of soft rot of curds of seed cauliflowers.] *Rocz. Nauk Roln., ser. E* 5(2): 35-54 (saw English summary only). Abstr. in *Rev. Plant Pathol.* 56: 102-103 (1977).

Smith, E. F. (1893). Two new and destructive diseases of cucurbits (abstr.). *Bot. Gaz.* 18: 339.

Smith, E. F. (1894). Two new and destructive diseases of cucurbits: (1) the muskmelon *Alternaria*; (2) a bacterial disease of cucumbers, cantaloups and squashes (abstr.). *Proc. Am. Assoc. Adv. Sci.* 42: 258-259.

Smith, E. F. (1895). *Bacillus tracheiphilus* sp. nov., die Ursache des Verwelkens verschiedener Cucurbitaceen. [*B. tracheiphilus* sp. nov., the cause of wilt of several Cucurbitaceae.] *Zentralbl. Bakteriol.* II 1: 364-373.

Smith, E. F. (1896). A bacterial disease of the tomato, eggplant and Irish potato (*Bacillus solanacearum* nov. sp.). *U. S. Dep. Agric. Div. Veg. Physiol. Pathol., Bull.* 12. 28 p.

Smith, E. F. (1897). *Pseudomonas campestris* (Pammel), the cause of a brown rot in cruciferous plants. *Zentralbl. Bakteriol.* II 3: 284-291, 408-415, 478-486.

Smith, E. F. (1898a). The black-rot of the cabbage. *U. S. Dep. Agric. Farmers' Bull.* 68. 22 p.

Smith, E. F. (1898b). The spread of plant diseases—a consideration of some of the ways in which parasitic organisms are disseminated. *Trans. Mass. Hortic. Soc.* 1: 117-133.

Smith, E. F. (1908). Recent studies of the olive-tubercle organism. *U. S. Dep. Agric. Bur. Plant Ind. Bull.* 131: 25-43.

Smith, E. F. (1909). Seed corn as a means of disseminating *Bacterium stewarti. Science n. s.* 30: 223-224.

Smith, E. F. (1914). Bacteria in relation to plant diseases, v. 3. Carnegie Institution, Wash. D. C. 309 p.

Smith, E. F. (1920). An introduction to bacterial diseases of plants. W. B. Saunders Co., Philadelphia. 688 p.

Smith, R. E., Smith, C. O., and Ramsey, H. J. (1912). Walnut culture in California: walnut blight. *Calif. Agric. Exp. Stn. Bull.* 231: 119-398.

Sohi, G. S., and Sandhu, M. S. (1968). Relationship between citrus leaf-miner (*Phyllocnistis citrella* Stainton) injury and citrus canker [*Xanthomonas citri* (Hasse) Dowson] incidence on citrus leaves. *J. Res. Punjab Agric. Univ.* 5: 66-69.

Southey, R. F. W., and Harper, G. J. (1971). The survival of *Erwinia amylovora* in air-borne particles: tests in the laboratory and in the open air. *J. Appl. Bacteriol.* 34: 547-556.

Stahl, F. J., and Luepschen, N. S. (1977). Transmission of *Erwinia amylovora* to pear fruit by *Lygus* spp. *Plant Dis. Rep.* 61: 936-939.

Stammer, H. J. (1929). Die Bakteriensymbiose der Trypetiden (Diptera). [The bacterial symbionts of Trypetidae (Diptera).] *Z. Morphol. Oekol. Tiere* 15: 481-523. Abstr. in *Biol. Abstr.* 6: 1379 (1932).

Staples, R., Ahmad, M., Chawdhry, S., and Diaz, G. (1967). Effectiveness of several insecticides in controlling cucumber beetles in eastern Nebraska. *J. Econ. Entomol.* 60: 463-466.

Stapleton, N. M., and McDermott, L. A. (1950). Role of the aphids, *Brevicoryne brassicae* and *Myzus persicae*, in the spread of *Xanthomonas campestris*, the causal organism of black rot of turnips. *Can J. Res.* 28 (C): 699-705.

Stedman, J. M. (1894). Cotton boll-rot: a new disease of cotton. *Ala. Agric. Exp. Stn. Bull.* 55. 12 p.

Stevens, F. L., Ruth, W. A., and Spooner, C. S. (1918). Pear blight wind borne. *Science n. s.* 48: 449-450.

Stevens, H. E. (1915). Citrus canker–III. *Fla. Agric. Exp. Stn. Bull.* 128. 20 p.

Stevens, N. E. (1933). The winter of 1932-1933. *Plant Dis. Rep.* 17: 14.

Stevens, N. E. (1934). Stewart's disease in relation to winter temperatures. *Plant Dis. Rep.* 18: 141-149.

Stevens, N. E. (1937). Third experimental forecast of the incidence of bacterial wilt of corn. *Plant Dis. Rep.* 21: 102-107.

Stewart, V. B. (1913a). The fire blight disease in nursery stock. *N. Y. Agric. Exp. Stn. Cornell Bull.* 329: 313-371.

Stewart, V. B. (1913b). The importance of the tarnished plant bug in the dissemination of fire blight in nursery stock. *Phytopathology* 3: 273-276.

Stewart, V. B. (1915). Notes on the fire blight disease. *Phytopathology* 5: 327-334.

Stewart, V. B., and Leonard, M. D. (1915). The role of sucking insects in the dissemination of fire blight bacteria. *Phytopathology* 5: 117-123.

Stewart, V. B., and Leonard, M. D. (1916). Further studies in the role of insects in the dissemination of fire blight bacteria. *Phytopathology* 6: 152-158.

Stolp, H. (1960). Über das Zusammenwirken von Bakterien und Insekten bei der Entstehung einer Geschmacksbeeinträchtigung des Kivu-Kaffees und die Rolle von Bakteriophagen bei der Aufklärung der Zusammenhänge. [On the combined action of bacteria and insects in producing bad flavor in Kivu coffee and the role played by bacteriophages in the clarification of the relationship.] *Phytopathol. Z.* 39: 1-15 (saw English summary only).

Stover, R. H. (1972). Banana, Plantain and Abaca Diseases. Commonwealth Mycological Institute, Kew, Surrey England. 316 p.

Sutherland, D. W. S. (1978). Common names of insects and related organisms. Entomol. Soc. Am. special publ. 78-1, College Park, Md. 132 p.

Swingle, D. B. (1911). Pear and apple blight in Montana. *Mont. Agric. Exp. Stn. Circ.* 2 (revised). 14 p.

Swingle, D. B. (1921). Pear and apple blight in Montana. *Mont. Agric. Exp. Stn. Circ.* 98. 10 p.

Tamimi, K. M., and Banfield, W. M. (1969). Transmission of bacterial soft rot by fruit flies (abstr.). *Phytopathology* 59: 403.

Tanii, A., Baba, T., and Haruki, T. (1974). [Bacteria isolated from black rot of rice grains.] *Ann. Phytopathol. Soc. Jap.* 40: 309-318 (saw English summary only). Abstr. in *Biol. Abstr.* 60: 1154 (1975).

Taylor, P. S. (1952). The biology of the black onion fly, *Tritoxa flexa* (Wied.), Ceroxydidae, Diptera, in Minnesota and its relation to onion rot. M. S. thesis, Univ. Minn.

Thanos, A. (1948). Insect transmission of iris soft rot. M. S. thesis, W. Va. Univ., Morgantown, W. Va. 47 p.

Thind, B. S., and Singh, N. (1976). Maize borer [*Chilo partellus* (Swinhoe)] as carrier of *Erwinia carotovora* var. *zeae*, the causal agent of bacterial stalk rot of maize. *Curr. Sci.* 45: 117-118.

Thomas, H. E. (1930). The longevity of *Bacillus amylovorus* (Burr.) Trev. in association with honey. *Science n. s.* 72: 634.

Thomas, H. E., and Ark, P. A. (1934a). Fire blight of pears and related plants. *Calif. Agric. Exp. Stn. Bull.* 586. 43 p.

Thomas, H. E., and Ark, P. A. (1934b). Nectar and rain in relation to fire blight. *Phytopathology* 24: 682-685.

Thomas, H. E., and Parker, K. G. (1933). Fire blight of pear and apple. *N. Y. Agric. Exp. Stn. Cornell Bull.* 557. 24 p.

Thomas, H. E., and Vansell, G. H. (1934). Relation of bees to fire blight. *Am. Bee J.* 74(4): 160. (Also in *Blue Anchor* 11(1): 8.

Thomas, R. C. (1924). Stewart's disease or bacterial wilt of sugar corn. *Ohio Agric. Exp. Stn. Mon. Bull.* 9: 81-84.

Thomson, S. V., Schroth, M. N., Moller, W. J., and Reil, W. O. (1975a). Occurrence of fire blight of pears in relation to weather and epiphytic populations of *Erwinia amylovora. Phytopathology* 65: 353-358.

Thomson, S. V., Schroth, M. N., Reil, W. O., and Moller, W. J. (1975b). Incidence of *Erwinia amylovora* in bartlett pear flowers of different ages (abstr.). *Proc. Am. Phytopathol. Soc.* 2: 67.

Thygesen, T., Esbjerg, P., and Eiberg, H. (1973). Ildsotoverfring med insekter. [Fire blight transmission by insects.] *Tidskr. Planteavl.* 77: 324-336 (saw English summary only). Abstr. in *Rev. Appl. Entomol.* -A 63: 746 (1975).

Tonelli, A. (1913). Una bacteriosi del leandro (rogna, o cancro, o tubercolosi del leandro). [A bacteriosis of oleander (scab, or canker, or tuberculosis of oleander.)] *Ann. R. Accad. Agric. Torino* 55: 383-400.

Tsiropoulos, G. J. (1976). Bacteria associated with the walnut husk fly, *Rhagoletis completa. Environ. Entomol.* 5: 83-86.

Tullis, E. C. (1929). Studies on the overwintering and modes of infection of the fire blight organism. *Mich. Agric. Exp. Stn. Tech. Bull.* 97. 32 p.

Vakili, N. G., and Baldwin, C. H. (1966). Insect dissemination of the tomato race of *Pseudomonas solanacearum*, the cause of bacterial wilt of certain *Musa* species. *Phytopathology* 56: 355-356.

Van Bronswijk, J. E. M. H., and Sinha, R. N. (1971). Interrelations among physical, biological, and chemical variates in stored-grain ecosystems: a descriptive and multivariate study. *Ann. Entomol. Soc. Am.* 64: 789-803.

Van Wyk, J. H., Hodson, A. C., and Christensen, C. M. (1959). Microflora associated with the confused flour beetle, *Tribolium confusum. Ann. Entomol. Soc. Am.* 52: 452-463.

Vilkova, N. A. (1962). [The stem fly–*Elachiptera cornuta* Fall. (Diptera, Chloropidae) and its importance as a pest of maize.] *Zool. Zh.* 41(4): 586-590 (saw English summary only). Abstr. in *Rev. Appl. Entomol.* -A 51: 254-255.

Von Hegyi (1910). Einige Beobachtungen betreffs der Schwarzbeinigkeit der Kartoffeln. *Z. Pflanzenkr.* 20: 79-81 (not seen).

Von Tubeuf, C. (1910). Knospen-Hexenbesen und Zweig-Tuberkulose der Zirbelkiefer. [Bud witchbroom and branch tubercle of stone pine.] *Naturw. Z. Forst. Landw.* 8(1): 1-12.

Von Tubeuf, C. (1911). Zweigtuberkulose am Ölbaum, Oleander und der Zirbelkiefer. [Branch tuberculosis of olive, oleander and stone pine.] *Naturw. Z. Forst. Landw.* 9: 25-44.

Voronkevich, I. V. (1958). [Dangerous bacterial disease of maize.] *Natur.* (Moscow) 5: 84-86 (not seen). Abstr. in *Rev. Appl. Mycol.* 37: 583 (1958).

Voronkevich, I. V. (1960). [On the survival of the soil bacteria of the genus *Erwinia*-causal agents of soft rot in plants.] *Bull. Soc. Nat. Moscow (Biol.)* 65: 95-105. Abstr. in *Rev. Appl. Mycol.* 40: 202 (1961).

Voûte, A. D. (1934). De djeroek-mineerrups (*Phyllocnistis citrella* St.) [The citrus-leafminer (*P. citrella*).] *Koret Meded. Inst. Pflziekt.* no. 19. 38 p (saw English summary only). Abstr. in *Rev. Appl. Entomol.* -A 23: 162-163 (1935).

Waite, M. B. (1891). Results from recent investigations in pear blight (abstr.). *Bot. Gaz.* 16: 259.

Waite, M. B. (1892). Results from recent investigations in pear blight (abstr.). *Proc. Am. Assoc. Adv. Sci.* 40: 315.

Waite, M. B. (1896). The cause and prevention of pear blight. *In* U. S. Dep. Agric. Yearb. 1895. p. 295-300.

Waite, M. B. (1898). Pear blight and its treatment. *East. N. Y. Hortic. Soc.* 2: 779-790.

Waite, M. B. (1901). The relation of bees to the orchard. *Natl. Beekeepers' Assoc. Proc.* 32: 70-73.

Waite, M. B. (1906). Pear-blight work and its control in California. *Proc. Calif. Fruit Grow. Farmers' Conv.* 1905: 137-155.

Waite, M. B. (1912). Collar blight and other collar and root diseases of the apple. *Proc. W. Va. State Hortic. Soc.* 1911: 66-73.

Walker, J. C. (1948). Diseases of cabbage and related plants. *U. S. Dep. Agric. Farmers' Bull.* 1439 (revised). 38 p.

Wardlaw, C. W. (1961). Banana Diseases Including Plantains and Abaca. John Wiley and Sons, Inc. N. Y. 648 p.

Waters, R. (1922). Fireblight: Incidence of the disease in New Zealand [and] long-distance dissemination of the disease. *N. Z. J. Agric.* 24: 350-357 and 25: 209-214.

Whetzel, H. H. (1906). The blight canker of apple trees. N. Y. Agric. Exp. Stn. Cornell Bull. 236: 177-217.

Whetzel, H. H., and Stewart, V. B. (1909). Fire blight of pears, apples, quinces, etc. *N. Y. Agric. Exp. Stn. Cornell Bull.* 272: 147-169.

Wickens, G. M. (1942). [A report on plant pathology at Barberton, South Africa.] *In* Progress Reports from Exp. Stations, Season 1940-1941. Empire Cotton Growing Corp., Lond. p. 43-64 (not seen). Abstr. in *Rev. Appl. Entomol.* -A 31: 30-34 (1943).

Wilde, W. H. A., Carpenter, J., Liberty, J., and Tunnicliffe, J. (1971). *Psylla pyricola* (Hemiptera: Psyllidae) vector relationships with *Erwinia amylovora. Can. Entomol.* 103: 1175-1178.

Williams, L. E., and Lockwood, J. L. (1956). Control of bacterial wilt of cucumber by antibiotic sprays. *Plant Dis. Rep.* 40: 479-482.

Wilson, A. (1927). Preliminary report on the gardenia bud drop. *Phytopathology* 17: 671-672.

Wilson, E. E. (1935). The olive knot disease: its inception, development, and control. *Hilgardia* 9: 233-264.

Wood, J. I. (1937). Bacterial wilt of corn in 1937. *Plant Dis. Rep.* 21: 301-305.

Yamvrias, C., Panagopoulos, C. G., and Psallidas, P. G. (1970). Preliminary study of the internal bacterial flora of the olive fruit fly (*Dacus oleae* Gmelin). *Ann. Inst. Phytopathol. Benaki* 9: 201-206.

Young, J. R., and Ditman, L. P. (1959). The effectiveness of some insecticides on several vegetable crops. *J. Econ. Entomol.* 52: 477-481.

Chapter 13

INSECT INVOLVEMENT IN THE TRANSMISSION OF FUNGAL PATHOGENS

George N. Agrios

Department of Plant Pathology
University of Massachusetts
Amherst, Massachusetts

13.1 INTRODUCTION 293
13.2 ROLE OF INSECTS IN THE TRANSMISSION OF FUNGAL PLANT PATHOGENS IN THE SOIL 295
13.3 ROLE OF INSECTS IN THE TRANSMISSION OF FUNGI AFFECTING STALKS, STEMS, TRUNKS AND BRANCHES 297
13.3.1 Stalk Rots of Corn 297
13.3.2 Tree Canker Diseases 298
13.3.3 Vascular Wilts 303
13.4 ROLE OF INSECTS IN THE TRANSMISSION OF FUNGI CAUSING FOLIAR DISEASES 305
13.5 ROLE OF INSECTS IN THE TRANSMISSION OF FUNGAL DISEASES AFFECTING BUDS AND BLOSSOMS 308
13.6 ROLE OF INSECTS IN THE TRANSMISSION OF FUNGAL DISEASES AFFECTING FRUITS OR SEEDS IN THE FIELD 309
13.6.1 Rots of Fleshy Fruits 309
13.6.2 Molds and Decays of Grains and Legumes 313
13.7 ROLE OF INSECTS IN THE TRANSMISSION OF FUNGAL DISEASES AFFECTING FRUITS OR SEEDS AFTER HARVEST 315
13.8 CONCLUDING REMARKS 316
13.9 REFERENCES 317

13.1 INTRODUCTION

The number, severity, and diversity of plant diseases caused by fungal pathogens by far surpass that of the diseases caused by all the other pathogens, i.e. bacteria, viruses, mycoplasmas, etc., taken together. In descriptions of the epidemiology of most fungal diseases of plants one frequently encounters statements that

ISBN 0-12-326450-2

indicate that the fungal pathogen is disseminated in nature by means of wind, water, "and insects," or "and probably by insects." Yet, in few fungal diseases of plants has the role of insects in the transmission of fungal plant pathogens been investigated with relative thoroughness; and a close association between the insect vector(s) and the fungal plant pathogen has been unequivocally established in only a few cases. Such associations and relationships between vector(s) and pathogen have been studied infinitely more extensively in viral, mycoplasmal and some bacterial diseases of plants; in fungal diseases the association is most frequently assumed or inferred from limited observations, or from similarities with other diseases (Leach, 1940; Carter, 1973). Experimental proof for many statements suggesting transmission of a fungal pathogen by insects is often lacking and the kind of relationship between vector and pathogen is unknown. Furthermore, there is no information on the role of the vector in the survival of the pathogen and on the overall dissemination and severity of the disease in comparison to wind, water, man and other agents that affect the same. It is, of course, understandable that these relationships have been studied more and are better known in viral, mycoplasmal and some bacterial diseases: in all of these the insect vector(s) is the only vector of the disease in nature. In the great majority of fungal diseases in which an insect may also be a vector, the insect is but one, and, as it is often assumed, not necessarily the most important, of the several factors disseminating the pathogen. It may be, however, that the importance of insects as vectors of fungal plant pathogens in nature has been underestimated because it has been overshadowed by the generality of wind and water dissemination of fungal spores. Yet, there are several fungal plant diseases in which insects have already been implicated as the main means of dissemination of the pathogen—and insects also play a role in fungal diseases by making wounds on plants through which fungi enter the plants.

Few reviews have, so far, dealt with insect transmission of fungal plant pathogens (Heald, 1913; Gardner, 1918; Leach, 1935, 1940; Austwick, 1958; Ingold, 1971; Carter, 1973). Carter's (1973) book contains an excellent and extensive chapter on insects in relation to fungal plant pathogens; diseases in which such relationships have been found are classified according to the fungus-insect relationship and include transmission via: (1) pollination; (2) external contamination; (3) internal contamination; (4) feeding wounding; (5) oviposition wounding; (6) symbiotic associations between insects and fungi, and (7) by the fungi feeding on insect exudates. This type of grouping has the advantage that it focuses attention to the type of fungus-insect relationship but, as these authors state, there is really no cohesiveness within the fungal or insect members of each group, nor is there any phylogenetic significance in this sort of classification. For this reason, and because the above classification can be found in the earlier reviews (Carter, 1973), in this chapter an attempt will be made to bring together available information on the role of insects in the transmission of fungal plant pathogens (with the exception of Dutch elm disease which is the topic of another chapter in this volume) by grouping the diseases according to plant parts or organs affected. This classification is, of course, another classification of convenience that has the advantage that it calls attention to some common features of the diseases and some common feeding or breeding habits of the insect vectors involved.

13.2 ROLE OF INSECTS IN THE TRANSMISSION OF FUNGAL PLANT PATHOGENS IN THE SOIL

Plants exist in the soil for various periods of time as seeds, or as roots and other vegetative propagative organs such as tubers, corms, rhizomes, etc. During these periods, plant parts are exposed to attacks by soil insects and by soil fungi, the latter causing seed decays, damping-off, root and lower stem lesions and rots, as well as lesions and rots on tubers, corms, etc., and even vascular wilts.

Although it is easy to imagine transmission of fungal spores or mycelium on or in the bodies of insect larvae or adults from one seed to another, or from the soil to a seed, no proof that such transmission occurs has been reported. It is possible that seed-attacking soil insects consume the seeds or embryos anyway so that transmission of any fungal pathogens to such seeds would be inconsequential; however, insects often consume only small portions of seeds and it is possible that the seed could still germinate if it were not for the fungal, and possibly other, pathogens brought into the seed by the insect and causing the seed to rot. The fact, on the other hand, that treatment of seeds with appropriate fungicides usually secures excellent seed survival and germination in the soil suggests that soil insects by themselves apparently have little effect on seed survival. Further research in this area should improve our understanding of the role insects may play in reduced stands from seed decay.

Root infections by fungi are often assumed to be facilitated by the feeding of insects on plant roots either because the insects create ports of entry for the fungal pathogens or because, in addition, weaken the plants and predispose them to fungal attacks (Christensen, 1953). Thus, root and basal stem rots of cereals and grasses caused by *Fusarium* spp., *Helminthosporium* spp. and other fungi are increased by infestations of the roots and lower stems with several species of billbugs (Satterthwait, 1932) of which the most common and destructive is *Calendra parvula* (Gyllenhal) (Hanson *et al.*, 1950) and by the Hessian fly (*Phytophaga destructor*) (Boosalis, 1954), while another billbug, *Anacentrus deplanatus*, has been reported associated mainly with root and basal stem rot of barnyard grass and sugarcane (Hanson and Milliron, 1942). Also, root rot, lodging and basal stalk rot of corn caused by *Pythium graminicola, Gibberella* and *Fusarium* spp., and other species of *Pythium* are often associated with or increased by injury to roots by the southern corn root worm (*Diabrotica undecimpuctata howardii* Barber) and the northern corn root worm (*Diabrotica longicornis* Say) (Summers, 1952). The southern corn root worm larvae also feed on peanut fruit and the damage caused to the pods by the larvae facilitates the entry of many fungi, including *Pythium myriotylum*, which is a serious pathogen of the crop (Porter and Smith, 1974). Similarly, rot of white clover and ladino clover tap roots by *Fusarium* spp. is apparently facilitated by damage and girdling of the roots by larvae of the clover root curculio *Sitona* spp. and particularly *S. hispidula* F. (Kilpatrick and Dunn, 1958; James *et al.*, 1977). Alfalfa and red clover seedlings, on the roots of which the larvae of fungus gnats (*Bradysia* spp., *Sciaridae* spp.) feed, are predisposed to attacks by *Fusarium* sp. and plants even slightly injured by the larvae are killed at a much higher rate than are uninjured plants (Leath and Newton, 1969). Larvae of *Bradysia coprophila*, feeding on soybean stems and crowns, apparently provide entrance sites for many pathogens, including *Pythium* and *Fusarium*, resulting in root rots, lodging and little seed production (Graham and McNeill, 1972). Girdling of soybean seedling stems by *Spissistilus fistinus* Say predisposed them to infection by *Sclerotium rolfsii*, the causal agent of sclerotial blight. Direct transmission of the fungus by the insect apparently was of little importance since even simulated girdling damage resulted in increased infection (Herzog *et al.*, 1975).

Oil palms in Indonesia and Malaysia are attacked by the root mining larvae of the Lepidopteron *Sufetula sunidesalis* Walker. However, areas favorable to *Sufetula* coincide to a large extent with those favorable to the fungus *Ganoderma* sp. which causes heavy damage to the roots of oil palms in plantations. Although *Ganoderma*-infected palm roots harbor several species of mycophagous insects, they do not harbor *Sufetula* which, instead, attacks apparently healthy roots but leaves wounds that do not heal and thus facilitate entry by the fungus (Chenon, 1975). Conversely, root diseases such as those caused by *Poria, Fomes* and *Armillaria* may be an important factor in predisposing trees such as grand fir (*Abies grandis*) to attack by bark beetles (Coleoptera, Scotylidae) since both the incidence and the number of Scolytid attacks was greater on trees exhibiting root decay (Hertert *et al.*, 1975).

Few reports provide direct evidence of insect transmission of root infecting fungi. In an attempt to obtain such information, four insect species were studied (Nuorteva and Laine, 1968): *Myelophilus (Blastophagus) piniperda* L., *Hylobius abieties* L., *Hylastes brunneus* Erichs and *Rhizophagus ferrugineus* (Payk.), observed or believed to develop in parts of trees attacked by the fungus *Fomes annosus* (Fr.) Cooke, and known to construct winter quarters at the foot or on the roots of healthy trees. Adults were kept with a pure culture of the fungus for one hour and were then transferred immediately or after varying periods to pure substrate and again left for one hour. All insect species were capable of transmitting the fungus three days after contact with it, and the last two species after five days. *Hylastes* sp. appeared to be particularly effective in transmitting the fungus. Even non-phytophagus insects, however, may be involved in transmission of phytopathogenic fungi. For example, up to 3.6 percent of sclerotia of the fungus *Sclerotium rolfsii* retained their viability after passage through the digestive tracts of cockroaches (Akueshi and Oso, 1976).

Mites apparently also transmit soilborne phytopathogenic fungi. Thus, the mite *Rhizoglyphus echinopus* Fum and Rob not only is a primary pest of gladiolus roots and corms but also acts as a vector of rot-producing organisms such as *Fusarium, Stromatinia* and *Scabbia* (Rexrode, 1968). Similarly, the mites *Sancas-*

sania and *Tyrophagus* can feed on *Aspergillus flavus* as well as peanut pods. Mites that have fed on the fungus become heavily infested with spores and so, when they penetrate the peanut pods and feed on the kernels, transmit the spores to the peanuts and aflatoxin production results (Aucamp, 1969).

A different kind of relationship, resembling that of the sooty molds, exists among certain coccid insects, e.g. *Pseudococcus citri, P. deceptor* and *P. comstocki* and fungi of the genera *Polyporus, Bornetina* and *Boletus* on the roots of coffee, grape and citrus. These fungi are not parasitic but feed on the sap that exudes from the feeding punctures made by the coccid insects on the roots and on the honeydew excreted by these insects. The fungus mycelium forms an impermeable sheath around the roots and interferes with their function (Maublanc and Roger, 1935; Goncalves, 1940; Steyaert, 1946).

13.3 ROLE OF INSECTS IN THE TRANSMISSION OF FUNGI AFFECTING STALKS, STEMS, TRUNKS AND BRANCHES

13.3.1 Stalk Rots of Corn

Corn stalk rots and related diseases caused by *Gibberella zeae, Diplodia zeae, Fusarium moniliforme, Fusarium graminearum*, and less frequently by species of *Penicillium, Colletotrichum, Cephalosporium, Helminthosporium* and other fungi, are apparently aided in their development in several ways by a variety of insects (Christensen and Wilcoxson, 1966), particularly the widespread European corn borer (*Pyrausta nubilalis*). The insects may carry inoculum into the tissues or provide openings through which pathogens enter the stalk. As the insects move through the stalk tissue they also distribute pathogens within the stalk (Koehler, 1960). Frass deposited by the insects within tunnels and between the stalk and the leaf sheath provides an excellent medium for rapid growth of saprophytes and many parasites which then invade living tissue (Christensen and Wilcoxson, 1966). Injuries caused by insects weaken the plants and make them more susceptible to attack by pathogens (Christensen and Schneider, 1950; Chez *et al.*, 1977). By fall, 84 to 93% of internodes or shanks with borer injuries contain conspicuous stalk rot and whenever borer injuries are associated with rots, severe damage results. This suggests that part of the damage attributed to the borer is due to rots that follow insect infestation (Christensen and Schneider, 1950). There is no doubt that fungi move with the larvae as well as in the tissue surrounding the tunnel, but whether corn stalk-rotting fungi are transmitted by insects from one plant to another to any appreciable extent has not been determined with certainty. *Fusarium, Gibberella, Diplodia* and other stalk-rotting fungi, however, can be isolated from borer larvae throughout the year, although they were less prevalent in inactive larvae collected from diseased stalks in the winter (Christensen and Schneider, 1950). At least during late summer and early fall the larvae carry these fungi both internally and externally but the fungi can be readily isolated from infected portions of the corn stalks throughout the fall and winter (Christensen and Schneider, 1950).

Two cases similar to the corn stalk rots exist in the red rot of sugarcane and the stem or bole rot of sisal. In the red rot of sugarcane, caused by the fungus *Colletotrichum falcatum* Went., the fungus enters through the holes and follows the tunnels made in the stalk and the underground parts of the stem by the moth borer *Diabrotica saccharalis* (Fab.) and the sugarcane weevil *Anacentrinus subnudus* Buchanan (Abbott, 1953). In the stem or bole rot of sisal (*Agave sisalensis*), which is caused by *Aspergillus niger*, the fungus is picked up by the sisal weevil (*Scyophophorus interstitialis* Glyh.) when the latter feeds on infected plants and its spores are carried in the weevil retaining their infectivity even after passage through the alimentary tract of the insect (Wienk, 1964-1965).

13.3.2 Tree Canker Diseases

13.3.2.1 Nectria Cankers. Several canker-causing fungi affecting the trunks, branches or twigs of fruit and forest trees are associated with insects. In most of these cases the insects seem to play a secondary or incidental role in the development of disease either by providing wounds that are used as entry ports by fungi, or by occasionally carrying on their bodies fungus spores that are deposited on or in the insect wounds and subsequently cause infections.

The *beech bark disease* is caused by the fungus *Nectria coccinea* var. *faginata* on beech trees (*Fagus* sp.) heavily infested with the scale insect *Cryptococcus fagi.* The fungus can be transmitted by the insect and introduced into the bark through the feeding wounds of the insect, but this sort of transmission is probably quite secondary to that effected by water and wind (Ehrlich, 1934; Hawboldt, 1945; Parker, 1974). The fungus, however, does not seem capable of infecting trees until they have been quite heavily infested with scale insects which occurs about three to five years after the initial infestation. Subsequently, the fungus invades and kills the bark, partially or completely girdling the tree and, where the fungus is abundant, it forms canker-like lesions (Ehrlich, 1934; Parker, 1974; Bonneau and Simonin, 1976).

In the *cacao (Theobroma cacao L) dieback*, caused by *Calonectria (Fusarium) rigidiuscula* (Berk. & Br.) Sacc. and/or *Botryodiplodia theobromae* Pat., the initial courts of entry are provided by the capsid species *Sahlbergella singularis* Hazl. and *Distantiella theobroma* Dist. and sometimes by the borings of scolytids (Crowdy, 1947; Owen, 1956). In single capsid lesions of shoots and twigs the fungus becomes quickly isolated by the host and no severe infections take place. When the trees are massively infested with capsids or are otherwise debilitated, however, the fungus develops unchecked in the debilitated tissue and causes a chronic dieback. Control of the capsids with chemicals also results in the halting of the fungal invasion, production of new vigorous growth, and finally the shedding of the fungus contained in the old dead tissues (Carter, 1973). *Calonectria rigidiuscula* is also considered as the cause of *cushion gall of cacao* but in this disease it is associated with the mealybug *Pseudococcus njalensis* Laing. and possibly others (Brunt and Wharton, 1960). Instead of producing cankers, the fungus is carried and is introduced by the mealybug into cacao inflorescences and young beans which develop into galls that in some areas are so numerous as to sterilize the tree (Wellman, 1972).

In the *burn blight of pines, Nectria cucurbitula* enters twigs and branches of jack and red pine through the feeding punctures made by the spittle bug *Aphrophora saratogensis* Fitch and then spreads through the bark causing necrosis that girdles twigs and branches and advances both up and down. The needles of affected trees die and appear scorched. The spittle bugs not only create the openings through which the fungus enters, but also become contaminated with the spores of the fungus which they then carry to new feeding sites and even serve as agents to weaken the twigs and make them more susceptible to fungus attack (Gruenhagen *et al.*, 1947). A similar association exists between the spittle bug *Aphrophora parallela* Say and the fungus *Diplodia pinea*, the latter causing a serious and lethal bark disease of twigs, branches and stems of Scots pine only in the presence of heavy infestations of spittle bugs (Haddow and Newman, 1942).

Recently (Anonymous, 1976), several insects including *Scatophaga stercoraria* L, *Anthocoris nemorum* L, *Archips podanus* and *Humerobates rostrolamellatus* have been reported as potential vectors of *Nectria galligena*, the cause of a canker disease of apples and many other trees.

13.2.2 Ceratocystic Cankers. Ceratocystis fimbriata Ell. & Halst. causes a canker disease on most stone fruits (almond, prune, apricot, peach) which is often associated with bark injuries produced by mallets or other hard instruments used in shaking off the fruit. The bark injuries are often visited by adults and larvae of certain insects several of which are vectors of the fungus. The most important vectors appear to be the nitidulid beetle *Carpophilus freemani* and the drosophilid fly *Chymomyza procnemoides*. Vectors of lesser importance include *Carpophilus hemipterus, Litargus balteatus, Euzophera semifuneralis, Scolytus rugulosus, Drosophila melanogaster*, and a species of *Tarsonemus*. When insects and mites are completely excluded from injured bark, infection by *C. fimbriata* does not occur (Moller and DeVay, 1968). Studies with *C. freemani* showed that this insect can transmit the fungus to healthy trees after it has been caged on a bark injury less than five minutes and that fully viable spores of the fungus can be detected in the excrement of insects less than one hour after they begin feeding on the fungus. Insects feeding on the fungus may carry its spores internally or externally for up to eight days after they stop feeding on the fungus. The fungus is even carried through a proportion of larvae that undergo pupation and into the emerging adults which are then able to transmit it to trees (Moller and DeVay, 1968).

In the *Ceratocystis canker of aspen (Populus tremoloides)*, also caused by the fungus *Ceratocystis fimbriata*, numerous insects which visit fresh aspen bark injuries can transmit not only *C. fimbriata* but also *C. alba, C. crassivaginata, C. moniliformis, C. minor, C. pilifera, C. populina* and *C. tremulo-aurea* (Hinds, 1972). The insects belong to several families including Nitidulidae, Rhizophagidae, Staphylinidae and

Drosophilidae, and some of them spend parts of their life cycles in fungus mats formed in the wounds. The nitidulid beetles *Epuraea* spp. and *Colopterus truncatus*, the staphylinids (rove beetles) *Nudobius corticalis* and *Quedius raevigatus*, and a root-eating beetle *Rhizophagus brunneus* easily and consistently transmit *C. fimbriata* and other *Ceratocystis* species into artificial wounds and result in canker development. The fungi are carried by the insects both externally and internally and they can often be recovered from adults upon emergence in the spring from overwintering pupae (Hinds, 1972). Similar association exists between beetles and *Ceratocystis fimbriata* in the *stem canker of plane trees* (*Platanus* spp.). *Cryptarcha ampla* and *Laemophloeus biguttatus* beetles collected from infected plane trees and placed on fresh wounds of healthy plane trees transmitted the fungus in two of seven trials while *Coleopterus (Colastus) unicolor, C. morio* and *Carpophilus lugubris* beetles, collected from healthy trees and allowed to feed on a culture of *Ceratocystis fimbriata* for 30 minutes, transmitted the fungus in six of eleven trials (Crone and Bachelder, 1961).

13.3.2.3 Other Cankers. The *tree cricket canker of apple* is caused by the fungus *Leptosphaeria coniothyrium* which enters through wounds often made by the tree crickets *Oecanthus niveus* DeG. and *O. angustipennis* Fitch. The tree crickets feed on soft-bodied insects, decaying plant tissue and other organic matter and are generally heavily contaminated both externally and internally with viable spores of many species of fungi. The tree crickets lay their eggs in the bark of the tree in which they first chew a hole, then deposit the egg and then plug the hole with a pellet of fecal matter or chewed wood. When spores of the canker fungus are carried on or in the cricket they are deposited in the hole with the pellet of wood or fecal matter, infection takes place and a canker develops (Gloyer and Fulton, 1916).

A similar canker and blight is caused by the same fungus *Leptosphaeria coniothyrium* and by another fungus *Didymella applanata* in raspberry but in this host the fungi are associated with the raspberry midge *Thomasianna theobaldi* Barns, not crickets. The wounds caused by larval feeding permit the entry of the two fungi into the wood. The infection process is thought to be facilitated by a substance secreted by the larvae during feeding since rotting does not occur when spore suspensions are applied alone, with or without incisions in the canes; but when larvae are also present, rotting develops at the sites of feeding, from which *Leptosphaeria coniothyrium* and to a lesser extent *Didymella applanata* can be isolated (Stoyanov, 1963).

In the *perennial canker of apple*, the woolly apple aphid (*Eriosoma lanigerum* (Hausm.)) feeds on the bark of the base of the stem 5 to 8 cm below the soil line or, rarely, on true roots and causes the formation of warty galls. In the galls, in turn, more young aphids become established and develop. In early winter, the warty galls burst upon freezing and the injured tissue is then attacked by the fungus *Gloeosporium perennans* which also causes a fruit rot and branch canker

(Childs, 1929). The fungus invades the burst gall and also, during the winter, advances into healthy tissue and produces a canker. In the summer, the host produces new callus tissue that seals off the fungus and the canker heals. The aphids, however, establish themselves on the new callus and the process is repeated. This kind of relationship between the fungus and the insect is considered to be a type of mutualism (Leach, 1940; Austwick, 1958).

A canker associated with simultaneous attack by the fungus *Cytospora* and the clearwing moth *Synanthedon myopaeformis* (Bkh) was found on pear trees of Williams' Bon Chretien in Switzerland (Bolay *et al.*, 1976). The fungus generally penetrates the branches through pruning wounds but the insect lays its eggs in necrotic tissues and crevices in the bark and, as the larvae tunnel in the surrounding healthy bark, they facilitate the progression of the mycelium of the fungus and thus help extend the canker. A similar relationship apparently exists in the *Phomopsis* canker of gooseberry and black currant (*Ribes*) bushes between the fungus (*Phomopsis*) and the larval behavior and tunnel construction by the sessid *Synanthedon tipuliformis* Clerk (Perret and Mussillon, 1975).

In the *birch constriction disease*, the apical birch woodwasp (*Pseudoxiphydria betulae* Ensl.) attacks the lower parts of shoots at which point the shoots become characteristically constricted. The leaves above the constriction wither and die but do not always fall in the autumn. Ninety-two percent of the constrictions on shoots were also infected with the fungus *Melanconium bicolor* where the wood showed wart-like protuberances and was withered and calloused beneath. Mycelium was found in the tunnels of *P. betulae*, the spores being carried beneath the bark with the eggs of the insect at the time of oviposition (Sinadskii, 1967). The effect of the symbiotic relationship is evidently to hasten the breakdown of the shoot tissues.

Mites and termites also may serve as vectors of fungal pathogens, as was shown for the eriophyid mite *Aceria mangifera* Sayed which disseminates the fungus *Fusarium moniliforme*, the cause of mango malformation disease (Summanwar and Raychaudhuri, 1969). Another *Aceria* mite seems to be associated with the fungus *Sphaerotheca phytoptophila*, the cause of witches' broom of hackberry (Skuhravy, 1973). No proof of a vector relationship has been offered but cleistothecia of the fungus were found only where the mite was present and hackberry trees free from witches' broom were also free of the mite. On the other hand, fungi belonging to several families and many species were isolated from the digestive tract and feces of workers of *Reticulitermes flavipes* (Kollar) termites (Coyle, 1974). Although some of the fungi associated with the termites were also present in wood cultures, the relationship between the fungi and the termites remains obscure.

Insects are often associated with wood rotting fungi (Ackerman and Shenefelt, 1973), primarily basidiomycetes. The macro-fruiting bodies of 112 species of fungi were frequented by insects and of these 47 fungi were specifically associated with wood and 30 caused important damage. Several groups of insects showed potential as vectors and also as biological control agents for some timber-destroying fungi (Ackerman and Shenefelt, 1973). Insects and mites have also been reported to be associated with white pine blister rust cankers caused by *Cronartium ribicola* (Furniss *et al.*, 1972), and with comandra blister rust cankers on lodgepole pine caused by *C. commandrae* (Powell, 1975), although no specific vector relationships were established.

In *scleroderis canker* of pine and spruce, caused by the fungus *Scleroderis lagerbergii*, the fungus is apparently transmitted from tree to tree on the bodies of at least three of the most common insect visitors of such trees, the scolytid beetles *Pithyophthorus cariniceps* LeC., *P. pulchellus* Eichh., and *Orthotomicus caelatus* Eich. (Frazer, 1944).

In the *white rot of softwoods* caused by the fungi *Amylostereum areolatum* and *Stereum sanguinolentum* there seems to be a true and highly adapted symbiosis between the fungus and the siricid woodwasps *Urocerus gigas* and *Sirex noctilio, S. juvencus* and *S. cyaneus* (Parkin, 1942; Talbot, 1977). The fungus is carried within the ovipositor of adult females and the eggs become contaminated with fungus spores as they pass down the ovipositor. From contaminated eggs deposited in timber, fungal mycelium grows into the wood ahead of the larvae modifying the wood microenvironment for better development of the woodwasps.

13.3.2.4 Wood-stain Diseases of Conifers. Wood stain or wood discoloration diseases of conifers are caused by the so-called blue-stain fungi of which the most common are species of *Ceratocystis* (earlier known as *Ceratostomella, Ophiostoma*, etc.). Wood stain diseases either affect felled timber in which they reduce the timber's value, or they affect living trees which they usually kill, sometimes rapidly, by causing a vascular wilt (Craighead, 1928; Leach, 1940). Several species of bark beetles, e.g. *Dendroctonus ponderosae, Ips pini, I. grandicollis, I. emarginatus, I. integer*, etc., are associated with blue-stain fungi. The bark beetles serve as vectors of the fungi and provide them with a wound for penetration while the fungus, in turn, reduces the water content in the tree and improves the microenvironment for the developing brood of insects (Craighead, 1928; Nelson and Beal, 1929; Leach *et al.*, 1934; Nelson, 1934). This relationship is described as a true mutualistic symbiosis (Craighead, 1928), while other blue stain fungi-bark beetle associations are described as symbiotic ambrosia cultures (Frazer, 1944) as, for example, the ones between the fungi *Trichosporium tingens* and *T. tingens* var. *macrosporum* and the bark beetles *Myelophilus (Blastophagus) minor* (Htg) and *Ips acuminatus* (Gryll), respectively, in which the fungi are constantly associated with and found regularly in the breeding places of the insects in pine stems (Frazer, 1944).

In the southern United States, beetle attacks, primarily by *Dendroctonus frontalis* Zimm., on short-leaf pines are accompanied by heavy fungus infections soon after tunneling of the insect through the bark and outer wood. Infected wood generally contains a complex of fungi including *Ceratocystis pini, Dacryomyces* spp., *Saccharomyces pini* and *Monilia* spp., and the same fungi are carried by the beetles since they can be cultured both from the exterior and the interior of the insects (Bramble and Holst, 1940).

In spruce, wood-staining is associated with attacks by *Dendroctonus engelmani* followed by infections primarily by the *Leptographium* spp. fungus which is the most consistent associate of the beetle and causes a conspicuous gray stain in the sapwood of infected trees (Davidson, 1951, 1954). Often, however, the fungi *Endoconidiophora coerulescens* or *Ophiostoma* sp. are also present in the beetle galleries and on adult beetles and are believed to be part of the wood staining complex.

13.3.3 Vascular Wilts

The most important fungal vascular wilts of plants, mostly trees, in which insects play a role in the dissemination of the fungus, include the Dutch elm disease, oak wilt, persimmon wilt and mango wilt. Of these, the Dutch elm disease is discussed as a separate chapter in this volume and, therefore, will not be considered further here.

13.3.3.1 Oak wilt. Oak wilt is caused by the fungus *Ceratocystis fagacearum* and is one of the most important diseases of forest and shade trees in North America. The fungus enters the xylem vessels of trees through fresh mechanical wounds to which it is carried by air or insects (Craighead *et al.*, 1953; Zuckerman, 1954; Norris, 1955), through natural root grafts (Yount, 1958) and probably through wounds made by insects that also carry the fungus (Rexrode, 1968). Infected trees develop crinkled, bronze and finally brown foliage while current year's wood may also show dark streaks (Henry and Moses, 1943; True *et al.*, 1960).

The role of insects as vectors of *Ceratocystis fagacearum* has been investigated extensively. Nitidulid beetles were the first to be implicated (Dorsey *et al.*, 1953; Norris, 1955) because they were found to breed in the mycelial mats of the fungus between bark and wood. Several species of nitidutids are involved, including *Carpophilus lugubris* Mur., *Colopterus niger* Say, *Colopterus semitectus* Oliv., *Cryptarcha ample* Er., *Eupurea* spp. and several *Glischrochilus* species (Dorsey *et al.*, 1953; Norris, 1955; Yount *et al.*, 1955). Nitidulids carry the fungus both externally and internally in their feces (Ingold, 1971), and the fungus may survive the winter on or in these insects (Stambaugh and Fregus, 1956). Nitidulids have been shown to transmit the oak wilt fungus when allowed to feed on mycelial mats of the fungus produced on wilted trees or on fungus cultures, or on wounds on infected trees and then transferred to wounds on healthy trees (Norris, 1955; Wood and Skelly, 1968; Gillespie, 1971).

In addition to nitidulid beetles, insects from other groups have also been implicated as vectors of *Ceratocystic fagacearum*. Thus, several scolytid beetles (oak bark beetles) including *Monarthrum fasciatum* Say, *Pseudopityophthorus minutissimus* Zimm. and *P. pruinosus* Eic. have been shown to carry the spores of the fungus when they emerge from the tunnels of diseased trees in which they breed and overwinter and carry them to highly susceptible oaks in the spring (Dorsey *et al.*, 1953; Griswold, 1956). Scolytids, too, can carry the fungus externally and internally in their feces (Rexrode and Jones, 1971). A brentid beetle, *Arrhenodes minuta* Drury and the buprestid *Agrillus bilineatus* Web. both bore holes into oak trees and both carry spores of the oak wilt fungus, and probably act as vectors of the fungus (Buchanan, 1957; Parmeter *et al.*, 1956). Also, the flat-headed borer *Chrysobothris femorata* Oliv. and the mite *Germania bulbicola* Oud. could transmit the fungus to artificially wounded oak seedlings in the greenhouse but whether they do so in nature is not known (Himelick and Curl, 1955). Even *Drosophila* spp. and certain collembolans were shown to obtain fungus spores from mycelial mats and to effectively transmit them into fresh wounds on oak seedlings resulting in successful infections (Griswold, 1953; Curl, 1956). Species of several other insect families, particularly Curculionidae, Buprestidae, Elateridae, Histeridae and Ostomidae, are also possible vectors of the oak wilt fungus but no definitive studies of such associations have been made yet.

Insect transmission of *Ceratocystis fagacearum* not only insures the dissemination of the fungus to new trees but also increases the potential for variability in the fungus through cross-spermatization of the two hermaphroditic but self-sterile mating types A and B of the fungus (Hepting *et al.*, 1952). Infected trees or groups of trees often contain only one mating type. The fungus produces mycelial mats under the bark of infected trees with specialized "pressure cushions" which eventually crack the bark (Zuckerman, 1954). Each mycelial mat produces endoconidia and, on perithecial initials, receptive hyphae of the same mating type as the endoconidia. Since endoconidia cannot fertilize perithecial initials of the same mating type, and since the endoconidia are sticky and are not transmitted by wind, insects, particularly those that feed on the mycelial mat, play the most important role in spreading the endoconidia to new fresh wounds, where they start new infections (Wood and Skelly, 1968), or to other mycelial mats some of which are likely to be of the compatible mating type and, upon fertilization, produce perithecia and ascospores. Although both endoconidia and ascospores are carried by the insect vectors and can start new infections, the latter are hardier and more durable than the endoconidia, and can also express any new characters brought together as a result of the preceding fertilization (Bramble and Holst, 1940).

13.3.3.2 Oak decline. At least some of the oak decline syndrome in Texas is caused by the fungus *Cephalosporium (=Phialophora) diospyri*. A large number of insects of many species, belonging to several insect orders, *Coleoptera, Dermaptera, Hymenoptera,* emerge from *Cephalosporium diospyri* infected trees carrying the fungus (Jares and Van Arsdel, 1977). Several insects are also considered to be vectors of *Ophiostoma kubanicum*, the cause of vascular mycosis of oak in the southeastern European parts of the Russian republic (Kryukova, 1976).

13.3.3.3 Persimmon wilt. It is also caused by the fungus *Cephalosporium diospyri* and the symptoms and effect on the tree resemble those observed in other wilt diseases. It is one of the most devastating hardwood tree diseases known. The fungus enters the plant through any kind of wounds but is apparently also transmitted by the powder-post beetle *Xylobiops basilaris* (Say) and the twig girdler beetle *Onciders cingulatus* (Say) (Crandall and Baker, 1950). Generally, only a small percentage of insect attacks of healthy trees results in transmission of the fungus, probably because under favorable conditions of moisture the attacked host produces a mass of gum that fills the tunnel and forces the insect to leave the tree and also inactivates any fungus spores or mycelium that may be present.

13.3.3.4 Mango wilts. A wilt of mango trees caused by the fungus *Diplodia recifensis* causes rapid dying of large numbers of mango trees in Brazil. Infected trees die quickly if infection begins in the summer. A brown, radial discoloration is apparent in cross sections of infected stems. The fungus appears to be transmitted by the beetle *Xyleborus affinis* (Chaves-Batista, 1947).

In the mango wilt caused by *Ceratocystis fimbriata*, the scolytid beetle *Hypocryphalus mangiferae* (Stebbing), which tunnels in the woody parts of mango trees in Brazil, is the vector of the fungus (Abrahão and Wegmüller, 1974).

13.4 ROLE OF INSECTS IN THE TRANSMISSION OF FUNGI CAUSING FOLIAR DISEASES

Fungi that attack primarily the foliage and other green organs of plants often grow on or near the surface of the plant on which they produce their fruiting structures, their spores and, sometimes, their mycelium (Agrios, 1978). These fungal structures, often produce, or are contained in, sticky liquid substances that give off odors attractive to insects and are therefore visited by them. Sometimes, the insects feed on such sticky exudates or on the fungal spores and mycelium, or they feed on the plant sap that oozes out from the plant tissue rotted by the fungus. Whatever the reason, whenever insects visit or feed on plant surfaces bearing fungi, they often ingest or become contaminated with fungal spores or mycelial fragments which they subsequently carry to other plants or plant parts they visit. In many cases, a wide variety of insects may be involved in the transmission of some fungi, while in others, more or less specific associations have developed between a fungus and one or a few insects that act as vectors.

13.4.1 Diploidization of Rust Fungi

The rust, as well as other fungi, are hermaphroditic but self-sterile fungi producing spermatia (pycniospores) and receptive hyphae in the same spermagonium (pycnium) that cannot fertilize each other. Fertilization can take place only when spermatia of one mating type are carried from one spermagonium to another that has receptive hyphae of the opposite, compatible mating type (Craigie, 1927). The transfer of spermatia, which ooze out of the spermagonium in a sticky, sugary exudate, is accomplished in part by rain water, but most frequently it is accomplished by insects which visit the spermagonia for the sugary exudate, become smeared with spermatia, and, subsequently, transport them to other spermagonia in which they fertilize the receptive hyphae, if they happen to be of the compatible mating type. More than a hundred species of insects including flies, bees, ants, etc. that visit spermagonia and, presumably, transmit spermatia, have been observed (Craigie, 1927).

13.4.2 Sooty Molds

Sooty molds are primarily foliar diseases of trees in the tropics and the subtropics. They are caused by many dark-colored fungi that do not parasitize the plant but grow on the honeydew deposited by insects such as white flies, scales, mealybugs, aphids, etc. The presence of the sooty mold fungi on the leaf surface forms a black layer that interferes with photosynthesis and gas exchange and gives the plant an unsightly appearance (Agrios, 1978). The sooty mold fungi are disseminated primarily by wind. However, many insects other than those that secrete honeydew, including flies, wasps, bees and ladybird beetles disseminate spores and hyphal fragments of the sooty mold fungi (Yamamoto, 1951). These insects seek honeydew as a source of food, become smeared with spores or mycelial fragments of sooty mold fungi and carry them externally, or ingest them and carry them internally, to other leaves on which honeydew is present. There the spores and mycelium fragments begin to grow causing more sooty mold symptoms (Yamamoto, 1951; Moller and DeVay, 1968; Nath, 1973).

13.4.3 Powdery Mildews

The superficial growth of the powdery mildew fungi makes them potential candidates for mycophagous insects and for dissemination by insects that either ingest or become smeared with the spores of these fungi. Thus, thrips were observed to feed on the mildew fungi *Sphaerotheca pannosa* and *Uncinula necator* and to destroy established mildew colonies on rose and grape leaves (Yarwood, 1943), while the colembolan *Lepidocyrtinus* sp. behaved similarly towards the tobacco powdery mildew (*Erysiphe cichoracearum*) on tobacco leaves (Yamaguchi *et al.*, 1970). Although in neither case it was shown that the insects transmitted the fungus, it would seem quite probable that the insects would carry on them spores of the fungus and would transmit them to the next plant they visit.

13.4.4 Coffee Rust

Several kinds of insects, including *Drosophila* sp., seem to be involved in the transmission through contamination of the uredospores of the coffee rust fungus *Hemileia vastatrix* (Amante *et al.*, 1971). Also, certain insects, as e.g. the cecidomyid midge *Mycodiplosis hemileiae* (Barnes), consume the uredospores of the fungus so efficiently that they clear coffee leaves of spores of the rust fungus (Milne, 1975). Although midge larvae eat uredospores of the coffee rust fungus, two hymenoptera parasitic on midge larvae are capable of obtaining fungus spores from the larvae and of transfering the disease to new plantations remote from other coffee areas (Crowe, 1963).

13.4.5 Red Pine Needle Blight

Needle blight of red pine affects the lower parts of current season's needles which then become hooked and drop. Needle blight is apparently caused by the fungus *Pullularia pullulans* which is commonly present on the sheath scales at the base of the needles and invades the bases of the shoots in association with injury caused by a cecidomyid midge. The midge adults oviposit in the sheath of scales at the base of the needles and the larvae form cavities at the base of the needles and probably carry the fungus with them into the needles (Haddow, 1941; Sinadskii, 1967).

13.4.6 Other Foliar Diseases

Some other foliar diseases that seem to be associated with insect vectors include:

Cucurbit anthracnose is caused by the fungus *Colletotrichum lagenarium*. Anthracnose lesions often develop at or near the feeding sites of the spotted cucumber beetle (*Diabrotica undecimpunctata howardi*) and it has actually been demonstrated that the beetle transmits the fungus in at least a small number of plants (Libby and Ellis, 1954).

Dieffenbachia leaf spot is caused by the fungus *Cephalosporium dieffenbachia*. The fungus, however, enters primarily through wounds provided by the stylets of the citrus mealybug *Pseudococcus citri* (Risso) which often infests dieffenbachia (Linn, 1942).

Alternaria blight of rubber trees is caused by a species of *Alternaria* that causes defoliation of young leaves and brown, concentric spots in between the veins, and elongate spots on the veins, of mature leaves. Some rubber clones are infected only in the presence of punctures made by the insect *Tomaspie inca* Guer., while others may be infected slightly in the absence of insect or mechanical injury but become severely infected when the insect is present, presumably because the fungus enters through the feeding punctures of the insects (Martin, 1947).

Oil palm leaf spot is caused by the fungi *Pestalotiopsis palmarum* and *P. glandicola* and results in the destruction of fronds and a significant reduction in yield. The fungi, however, are facilitated in their entry into oil palm leaves by the feeding and oviposition punctures made by a Tingid of the genus *Gargaphia*. It actually appears that *Gargaphia* is also a vector of *Pestalotiopsis* spp. (Genty *et al.*, 1975).

Thistle rust is caused by *Puccinia puncti* but there is circumstantial evidence that the fungus is assisted in its spread by the weevil *Centorhynchus litura*, which has been used in some areas for biological control of thistles (Peschken and Beecher, 1973).

Narcissus smoulder is caused by the fungus *Sclerotinia narcissicola*. Many times infections occur in leaf initials colonized by the mite *Steneotarsonemus laticeps* (Halbert) and on lesions of various sizes on the corresponding tissue after emergence (Gray *et al.*, 1975). Some association between the two organisms seems to be inferred, but its importance is not clear.

13.5 ROLE OF INSECTS IN THE TRANSMISSION OF FUNGAL DISEASES AFFECTING BUDS AND BLOSSOMS

Insects often overwinter in the protective scales of buds of perennial plants and feed on blossoms but, even more frequently, insects visit blossoms for the sweet exudate (nectar) contained in them. Buds also harbor, externally or internally, spores and mycelium of numerous pathogenic fungi, and the delicate nature of blossoms often makes them one of the first organs attacked by several fungi. Whenever insects overwinter, feed on, or visit fungus-infected or contaminated buds and blossoms, their bodies are likely to become smeared with the fungus spores and mycelium which they then transmit to plants or parts of plants they subsequently visit. In some cases, insects may ingest spores and mycelial fragments which retain their viability in the insect and may start new infections if they are deposited on susceptible tissues.

Bud infections in which insects have been implicated as vectors are rather rare. The *bud-blast disease of rhododendron* which is caused by the fungus *Pycnostyanus (Sporocybe) azalea* seems to be associated with the leafhopper *Graphocephala coccinea*, since the insect attacks primarily the medium-sized buds which subsequently develop the bud-blast disease (Baillie and Jepson, 1951). In the bud rot disease of carnations, the causal fungus, *Fusarium poae*, seems to be associated with the mite *Siteroptes graminum*, gravid females of which are usually present among the petals of affected flowers that are often knitted together by hyphae of the fungus (Alford *et al.*, 1976). In neither disease is the mechanism of transmission of the fungus by the insect known and, actually, the evidence that any transmission occurs is only circumstantial.

Several diseases of blossoms have been reported to be associated with insect vectors. In most of them transmission of the fungus by the insect is incidental to the activities of the insect during pollination. Thus, in the *anther smut of campions and pinks* (carnations) caused by the fungus *Ustilago violacea*, the pollen is replaced by the teliospores of the fungus but the petals remain unaffected and continue to attract insects. The visiting insects then become smeared with the smut spores and transfer them to previously unaffected flowers (Brefeld and Falck, 1905). Also, in the *blossom blight of red clover* caused by *Botrytis anthophila*, pollinating bees seem to be the principal vector of the fungus (Silow, 1933). In both of these diseases, however, the fungus is systemic and seedborne and the insects play a role mostly in the secondary spread of these diseases.

In *ergot of cereals and grasses*, caused by the fungus *Claviceps purpurea*, the fungus develops in the flowers, and its conidia are contained in a sweet and sticky substance that is attractive to many insects, particularly flies and beetles. The spores of the fungus have been found to adhere externally on insects and to be passed through the intestinal tract in a viable condition (Mercier, 1911; Langdon and Champ, 1954). Insect transmission of ergot is important primarily for secondary infections and for long-distance transport of the fungus since primary infection is through windborne ascospores released from sclerotia on the ground. Some beetles, however, also feed on sclerotia of the fungus and it is possible that they carry mycelial fragments or spores of the fungus on their bodies which may also cause primary infections (Lambert and McIlveen, 1976).

In the *anthracnose disease of Musa balbiniana*, the causal fungus *Gloeosporium musarum* is transmitted by the hymenoptera *Polybia occidentalis, Synoeca surinama,* and *Trigona* spp. The fungus infects the floral parts of the plant and these are dropped while still quite fresh and producing a sweet exudate. The insects are therefore still attracted to the infected flowers and become smeared with spores of the fungus which they then transmit to more flowers (Cardenosa-Barrigal, 1963).

In the *flower spot of azalea*, caused by the fungus *Ovulinia azaleae*, several species of bees, thrips and ants seem to contribute to the secondary spread of the disease and probably to the long-distance spread of the disease as far as two miles from the source of infection (Smith and Weiss, 1942). The insects carry spores of the fungus on their bodies and drop the spores as they move from flower to flower. Moreover, the insects, with their spiny legs, cause scratches and abrasions on the delicate flowers and these serve as ports of entry for the fungus.

13.6 ROLE OF INSECTS IN THE TRANSMISSION OF FUNGAL DISEASES AFFECTING FRUITS OR SEEDS IN THE FIELD

Fruits and seeds are the main source of food and the breeding grounds for many insects. Many of these insects puncture fruits and seeds in order to obtain food from or in order to lay their eggs in the fruits and seeds. Such insects cause direct damage to these organs that is often sufficient to render the fruits and seeds unsalable or to drastically reduce their market value. Frequently, however, the direct damage caused by the insects is increased manyfold by fungi and other pathogens that invade the punctured fruits or seeds and cause them to rot to a smaller or greater extent. Many of these fungi happen to land on the openings made by the insects and use them as ports of entry. Others, however, are carried by the insects externally or internally and enter the fruits and seeds at the time of their puncture or penetration by the insects. In many cases, the fungi are transmitted by insects that do not puncture healthy fruits or seeds but rather feed on the exudates of infected, rotten fruits, become smeared or internally contaminated with spores or mycelium of the rot-causing fungus, and then, as they move on to healthy fruits, they deposit some of the spores on these fruits. Subsequent infection may then take place through other wounds, natural openings or directly.

13.6.1 Rots of Fleshy Fruits

13.6.1.1 Fig Rots. Several types of fungi attack and cause rotting of figs in the field and, in most such cases, certain types of insects play a more or less important role in the transmission and introduction of the fungus into the fig.

(a) *Endosepsis of figs.* It is caused by the fungus *Fusarium moniliforme* f.s. *fici*. It appears as brown streaks along the flower stalks when the fruits begin to ripen. The streaks advance until the entire fruit content is turned into a pulp, and the whole fruit becomes sagging, dripping and deformed (Caldis, 1927). The pathogen is transmitted from fruit to fruit by the fig wasp *Blastophaga psenes* L. between which and the pollination of the fruit a rather complex relationship has developed. Fig trees are dioecious, male trees producing staminate flowers around the opening and gall flowers in the cavity, while female trees produce only pistilate flowers. The fig wasp lays its eggs in the ovules of the gall flowers, thus stimulating gall development. The eggs hatch and the larvae parasitize the galls until pupation. The emerging adult females are fertilized while still in the fig. When they come out of the fig, they brush against the staminate flowers that surround the eye and become smeared with pollen which they carry to other flowers they visit for oviposition, including pistilate flowers on the female fig. In female figs, however, due to the length of the styles, oviposition fails, but pollination is nevertheless accomplished and they develop into edible figs. At the same time, the wasp may become smeared with spores of the fungus, usually in the cavity of male figs before the wasp comes out of the eye, and externally transfers the spores, and during oviposition deposits them into male and female figs alike, causing endosepsis of the latter (Caldis, 1927).

(b) *Souring of figs.* It refers to spoilage of figs by fermentation organisms. At first it appears as discoloration and wateriness of the fig pulp which then exudes from the eye. Finally, the fig shrivels, dries and may cling to the tree. The fermenting organisms are transmitted both externally and internally by the beetle *Carpophilus hemipterus* L. and *Drosophila melanogaster* Low, the two most common visitors of figs still on the tree (Phillips *et al.*, 1925; Caldis, 1930).

(c) *Fig smuts and molds.* They refer to the black, dusty types of smut caused by *Aspergillus niger* and to the variously colored growths of other fungi (Hansen and Davey, 1937). The smut and mold organisms are carried into green figs, even before the eye scales begin to loosen, on the bodies of predaceous mites and, to a lesser extent, of thrips (Hansen, 1929; Smith and Hansen, 1931; Hansen and Davey, 1932; Davey and Smith, 1933).

A rot of Smyrna figs was reported to be transmitted by the fly *Lonchacea aristella* Beck as a result of oviposition in the profici figs by the first generation, and in the young fruits of cultivated fruits by subsequent generations, with fig-rotting fungi being transmitted into the figs during oviposition (Schewket, 1934).

13.6.1.2 Brown Rot of Stone and Pome Fruits. The disease is caused by *Monilinia (Sclerotinia) fructicola* and *M. laxa* in stone fruits and *M. fructigena* in pome fruits. In stone fruits, the disease has been associated with injury of fruit by insects including the plum curculio (*Conotrachelus nenuphar* Herbst.) (Scott and Ayres, 1910; Snapp *et al.*, 1927; Daines, 1942), the oriental fruit moth (*Grapholitha molesta*) (Daines, 1942; Poulos and Heuberger, 1952; Besson and Joly, 1969; Kable, 1969; Bohm, 1970), the dried fruit beetle (*Carpophilus hemipterus*) (Ogawa, 1957), and two nitidulid beetles (*Carpophilus mutilatus* and *Haptonchus luteolus*) (Tate and Ogawa, 1975). The plum curculio and the oriental fruit moth both feed and oviposit on the fruit, and the wounds they make serve as ports of entry for the brown rot fungus. The insects also become contaminated with the spores of the fungus when they visit infected fruit and then transmit the spores to healthy fruit. Thus, when the dried fruit beetle was allowed to feed on diseased peaches it became contaminated with the spores of *M. fructicola* and was able to transmit them to naturally wounded peaches (Ogawa, 1957). The nitidulid beetles also frequently are contaminated in nature with viable conidia of the brown rot fungus. These beetles are naturally attracted to oriental fruit moth exit holes, transmit the spores to such injuries and result in development of decay, particularly in late maturing stone fruit varieties (Tate and Ogawa, 1975).

In pome fruits, penetration by the fungus *Monilinia fructigena* is facilitated by the feeding holes made by the earwig *Forficula auricularia* L. The insects bore circular holes about 4-6 mm in diameter and form a shallow cavity beneath the skin of fruits when they are beginning to ripen and at which time the fruits are susceptible to brown rot (Croxall *et al.*, 1951).

13.6.1.3 Gray Mold of Grapes. It is caused by the fungus *Botrytis cinerea*, the spores of which are typically windborne. Penetration of grape berries and stems by the fungus, however, seems to be facilitated and increased by the lepidopterans *Argyrotaenia pulchellana* Haw. and *Lobesia botrana*. The larvae of these insects cause damage to fruits and stems directly but they also provide openings through which *Botrytis* spores enter the grapes and cause them to rot (Voight, 1972; Velimirovic, 1975).

13.6.1.4 Black Pod of Cocoa. The disease is caused by the fungus *Phytophthora palmivora* and results in devastating losses of yield. Several insects of different orders seem to play a role in the epidemiology of the disease. Thus, at least ten building ant species, particularly *Crematogaster striatula* Emery and, to a lesser extent, *Camponotus acvapimensis* Mayr. and *Pheidole megacephala* (F.) are considered to contribute to vertical spread of the fungus within the tree, especially during the rainy season, by carrying spore containing soil particles up the cacao tree for nesting purposes (Evans, 1973a; Adenuga, 1975). Also, certain plant feeding Coleoptera, e.g. the nitidulid beetle *Brachypeplus pilosellus*, and certain sugar feeding Diptera, e.g. the fly *Chaetonerius latifemur* End., colonize black pods in the field and may carry viable inoculum externally, on their mouth and mouthparts, or internally in their feces (Evans, 1973a, 1973b). These same insects are probably important vectors of the black pod fungus because of their abundance on cocoa trees, their habit of visiting wounded pods, and their proven efficiency in transmission experiments. These vector species quickly colonize the lesions of pod-wounding insects such as mirids and pod borers and are considered to be the main means of local and long distance horizontal spread of the pathogen (Evans, 1973b).

Two other rots of cocoa pods are also transmitted by insects. In Malaysia, the nymphs and adults of *Helopeltis theivora theobromae* attack immature pods causing lesions on them. Fungi such as *Botryodiplodia theobromae* sometimes enter the wounds and cause pods to rot (Tar, 1974). In Columbia, the cacao bug *Antiteuchus tripterus* (F.) has been implicated as a vector of fungi causing "moniliasis" (Eberhard, 1974).

13.6.1.5 Boll Rots of Cotton. Boll rot of cotton can apparently be caused by several fungi, particularly *Fusarium moniliforme, Alternaria tenuis, Aspergillus flavus,* and *Rhizopus nigricans*. In addition, the lint rot of cotton is caused by *Nigrospora oryzae* and the internal boll disease of cotton is caused by the fungus *Nematospora gossypii*. A variety of insects seem to be involved in the transmission of these fungi and, in some cases, the mechanisms of transmission are also different.

In the boll rot of cotton caused by *Fusarium moniliforme* and *Alternaria tenuis,* infection of cotton bolls developed after penetration by the fungi either through feeding and oviposition wounds made by the boll weevil (*Anthonomus grandis*), cotton bollworm (*Heliothis zeae*) and a lygus bug (*Lygus lineolaries*), or through the nectaries, having been brought there by nectar feeding flies such as *Drosophila melanogaster* and *Trichoplusia ni* (Bagga and Laster, 1968). In the boll rot of cotton caused by *Aspergillus flavus* and related species, in which aflatoxins are also produced, the fungus is soilborne and wind disseminated, but it is also carried externally and internally by several insects, such as the lygus bug *Lygus hesperus* and the stink bug *Chlorochroa sayi*, that frequently visit cotton bolls in the field (Stephenson and Russell, 1974). Entrance of the fungus into cotton bolls, however, seems to depend on the presence of wounds, particularly the large exit holes made by the mature larvae of the pink bollworm *Pectinophora gossypiella* (Saunders), while the entrance tunnels of young larvae do not transmit *Aspergillus flavus* to the interior of cotton bolls (Ashworth *et al.,* 1971).

In the boll rot of cotton caused by *Rhizopus stolonifer,* wounds created by the cotton bollworm *Earias insulana* (Boisd.) and the pink bollworm are responsible for the establishment of almost all of the *Rhizopus* infections present. Bolls without larval feeding or exit holes rarely become infected (Nasr and Azab, 1969; Ashworth *et al.,* 1971).

In the *lint rot of cotton* caused by *Nigrospora oryzae,* the fungus is transmitted with great efficiency by the mite *Siteroptes reniformis*. A small percentage (17%) of bolls may become infected if inoculated only with the fungus, but 100% of the bolls become infected when each boll is inoculated with just 1-3 mites carrying two spores each. The mites readily feed on the mycelium of *N. oryzae*. Since the mite is viviparous, the young are mature at birth, and young females on cultures of *N. oryzae* attach 1-2 spores on the ventral surface of their bodies soon after emergence from the mother (Laemmlen, 1969).

Internal boll disease, or stigmatomycosis, of cotton is caused by the fungus *Nematospora gossypii* and some related species (Ashby and Nowell, 1926). The disease, which results in staining of the cotton fibers in the absence of external symptoms, is associated with the feeding of several species of plant bugs, mostly of the genus *Dysdercus,* which are often referred to as cotton stainers. The insects carry spores and mycelium of the fungus externally on the mouthparts and internally in their deep stylet pouches into which the spores apparently escape during feeding. Even though the fungus is shed off with each molting, the insects are usually recontaminated. The spores do not seem to survive in the intestine. Infections take place when the fungus is introduced through the wall of young cotton bolls and this, presumably, is accomplished by the proboscis of the insect. However, the deposition of spores by the insect occurs at random and in a completely passive manner (Frazer, 1944).

13.6.1.6 Coffee Bean Rot. It is caused by *Nematospora corylii* and *N. gossypii* and causes serious losses in coffee berries in Kenya. The fungi are introduced into the berries through the feeding wounds of the insects *Antestia lineaticolis* Stol. and *A. faceta* Germ. The insects feed on both small and large green berries and if they carry with them spores of the fungi the latter are introduced into the berries and infection results. Infected berries at first turn black and later rot (LePelley, 1942). The number of infected berries is generally proportional to the number of insects to a tree, more than 300 insects per tree apparently resulting in almost total berry infection. Even as few as four insects per tree, however, can bring about infection of enough berries to cause considerable economic loss.

13.6.2 Molds and Decays of Grains and Legumes

13.6.2.1 Ear and Kernel Rots of Corn. Corn is susceptible to kernel and ear rots caused by several different fungi, and some of them are widely distributed. The rots are usually much more serious in humid areas and in years when rainfall is above normal from silking to harvest. However, one of the most important factors in the prevalence and severity of corn rots is the frequency and degree of insect damage (Koehler, 1959; Shurtleff, 1973).

The fungi most commonly causing corn rots are *Fusarium moniliforme* and *F. moniliforme* var. *subglutinans, Gibberella roseum* f. sp. *cerealis* (*G. zeae*), *Diplodia maydis, Cephalosporium acremonium, Negrospora oryzae, Physalospora zeae, Cladosporium herbarum, Penicillium* spp., *Aspergillus* spp., *Rhizopus* spp., *Trichoderma viride,* and others. Some of these fungi are transmitted by insects more frequently, or are assisted by insects in causing infection to a greater extent, than are some other fungi. The insects most commonly involved in the development and spread of corn rots are the corn earworm, *Heliothis zeae* Boddie, and the European corn borer, *Pyrausta nubilalis,* but other borers and other insects, e.g. the picnic beetle *Glischrochilus quadrisignatus,* seem to play important roles as vectors of some of these fungi.

All fungi causing corn rots are almost universally distributed and are more or less wind- and rainborne. The role of insects seems to be primarily as providers of wounds that serve as ports of entry for these fungi, but in many cases the insects also serve as vectors of the fungi by carrying their spores on their bodies (Taylor, 1952; Shurtleff, 1973).

In the corn rots caused by *Fusarium moniliforme* and several other *Fusarium* species, the fungus seems to be transmitted and assisted in penetration by the picnic beetle *Glischrochilus quadrisignatus*. The beetles were frequently observed visiting corn ears and wounds produced by other insects and 92% of the beetles obtained from standing-corn ears harbored one or more species of *Fusarium,* including *F. moniliforme, F. solani, F. oxysporum, F. tricinctum* and *F. roseum* 'Acuminatum,' 'Avenaceum,' 'Equiseti,' 'Gibbosum,' and 'Graminearum' (Windels *et al*., 1976). Also, the same beetle or its larvae and one or more of the *Fusarium* species were present in all corn ears buried by tillage operations, and at least three species, *F. moniliforme, F. Solani* and *F. oxysporum*, were isolated from 60 to 96% of the larvae, pupae and new adult beetles emerging from corn ears rotting in the ground (Windels *et al*., 1976).

The *black and yellow molds of corn* caused by *Asperigullus niger* and *A. flavus*, respectively, almost always begin to grow on corn at the point of injury by some insect, especially where the corn earworm *Heliothis zeae* has penetrated the husks into the ear. The fungus grows rapidly in young, tender ears but its spread through the ear increases with insect damage and decreases with maturity of the ear. Regardless of the species of insect involved, the severity of infection and the amount of aflatoxins, especially aflatoxin B_1, produced increase with the level of insect damage of corn kernels (LaPrade and Manwiller, 1977). Even when insect damage, such as that caused by the corn earworm, European corn borer and the rice weevil (*Sitophilus zeamais*), is not exclusively associated with *Aspergillus* infection and aflatoxin B_1 contamination of the preharvest crop, damage by these and other insects is always associated with increased infection and aflatoxin production (Taubenhaus, 1920; Widstrom *et al.*, 1976; LaPrade and Manwiller, 1977). Apparently, aflatoxin production is also influenced, among other things, by the insect species causing the ear infestation in the field (Widstrom *et al.*, 1975).

Regarding the relationship of insect vectors to mycotoxins, it might be worth noting here that mycotoxin F-2, produced by the fungus *Fusarium roseum* var. *graminearum* on various crops, is ingested by and retained in the bodies of insects such as the confused flour beetle *Tribolium confusum* Duv. and the lesser mealworm *Alphitobius diaperinus* (Panz) even through metamorphosis, so these insects not only may transmit the fungus but they may even be potential carriers of mycotoxins to the food of domestic animals and humans (Eugenio *et al.*, 1970).

13.6.2.2 Nematospora Infection of Rice. Rice often becomes infected with the fungus *Nematospora corylii* following infestation by the rice stinkbug *Oebalus pugnax* (F). The stinkbug feeds on rice kernels and causes discoloration and reduced milling quality. The fungus penetrates rice kernels through the feeding injuries made by the stinkbug and causes further deterioration of the rice (Daugherty and Foster, 1966).

13.6.2.3 Yeast Spot in Soybean. Yeast spot affects soybeans, lima beans, mung beans and cowpeas. It causes young seeds to remain small and become discolored, shrivelled and lightweight and young pods to fail to develop and drop prematurely.

Yeast spot is also caused by *Nematospora corylii* in association with several species of stinkbugs (Pentatomidae) such as *Acrosternum hilare* and *Thyanta custator*, and the Lygaeid *Spilostethus* (*Lygaeus*) *pandurus*. The stinkbugs often carry the fungus on the rostrum and other mouthparts and internally as viable spores in their body and throughout their salivary system (Daugherty, 1967; Foster and Daugherty, 1969; Ershad and Barkhordary, 1974). The fungus is often present with *N. gossypii* in the stylet pouches of the cotton stainer (*Dysdercus*) (Frazer, 1944). The insects apparently retain the fungus in their head for about 90 days and in the thorax and abdomen for lesser durations (Clarke and Wilde, 1970a, 1970b, 1971). The fungus is present in the feces but not in salivary secretions. The insects can obtain the fungus after feeding on plants and on soybeans that had been fed on by infested insects. Infested insects transmit the fungus to soybeans in about 60% of their feedings. There was no significant damage to oil or protein content by yeast spot, unless it occurred in early stages of soybean development (Clarke and Wilde, 1970a, 1970b, 1971).

13.6.2.4 Nigrospora Diseases of Grains. They are caused by the fungus *Nigrospora oryzae* in association with the mite *Pediculopsis*. The disease has been reported on wheat and corn in southern Europe (Alfaro, 1946; Peyronel, 1950). Spikelets and ears are killed in the incipient stages of their development. Among the mycelium and conidia of the fungus in infected plants, large numbers of mites were present, feeding on the fungus and the decomposed host tissue. Non-gravid female mites were observed to transport conidia of the fungus in their abdominal sacs.

Similar situations on wheat and corn were reported to be the results of associations of *Pediculopsis* mites and the fungus *Fusarium poae* (Peyronel, 1950).

13.6.2.5 Molds of Broad Bean. They are caused by *Botrytis fabae, Cladosporium cladosporoides,* and *Pullularia* (*Aureobasidium*) *pullulans*. Microscopic examination of insect bodies and isolation of fungi showed that *Botrytis fabae* is spread only by thrips (*Thysanoptera*) while aphids transmit conidia of *Cladosporium cladosporoides* and *Pullularia* (*Aureobasicium*) *pullulans*. Application of insecticides before flowering protected broad bean from these fungus diseases (Ondrej, 1973).

13.7 ROLE OF INSECTS IN THE TRANSMISSION OF FUNGAL DISEASES AFFECTING FRUITS OR SEEDS AFTER HARVEST

Very little is definitively known about the role insects play in the transmission and development of postharvest fungal diseases of fruits and seeds. It is generally accepted that fungal rots of fleshy fruits and vegetables, and molds and decays of grains and legumes, are greatly increased in number and severity when the fruits or seeds are injured by insects or other means (Chupp and Sherf, 1960; Christensen and Kaufmann, 1969; Agrios, 1978). The fruits, vegetables and seeds may be infected with the fungus and infested with some insect at the time of harvest, and if both, fungus and insect, escape detection during sorting they may continue their activities in storage; or one or both may exist in the storage rooms and the other is brought in with the crop. In either case, the insects can move and multiply in the stored produce and, at the same time, create openings through which fungi also enter the produce and cause it to rot. Moreover, as insects move about by crawling or flying they become smeared with spores of the fungi which they then carry with them to new produce they feed on next. Sometimes, insects such as flies do not actually wound the produce but, as they fly about searching for sugary exudates to feed on, they too become smeared with fungal spores which they then disseminate to other fruits they visit (Willison and Dustan, 1956). Since harvested, ripe fruits are generally much more susceptible to infection by several fungi than they were while still on the plant, spore dispersal by insects is often a sufficient condition for new infections to develop, and, of course, it becomes the main means of spread of fungal diseases in storage where neither wind nor rain can have an effect on the dissemination of fungi. Furthermore, presence of insects in stored grains and legumes helps raise the relative humidity, at least locally, through their respiration and excretions, and the higher relative humidity then allows even greater spore germination and mycelium development with subsequent greater molding and rotting of the produce.

The fungi most frequently involved in rots of fleshy fruits and vegetables in storage include *Penicillium, Fusarium, Botrytis, Rhizopus, Alernaria, Sclerotinia, Monilinia,* and *Colletotrichum* (*Gloeosporium*), while those involved in molds and decays of grains and legumes involve primarily *Aspergillus, Fusarium* and *Penicillium*. The insects involved in the transmission of, or infection by these fungi are the larvae and adults of various Lepidoptera, e.g. the oriental fruit moth *Grapholitha molesta*, Diptera such as the apple maggot *Rhagoletis pomonella*, the Mediterranean fruit fly *Ceratitis capitata* and the house fly *Musca domestica,* and others. Among the insects involved in the transmission or development of fungi causing molds and decays of grains and legumes are the larvae and adults of various Coleoptera, e.g. the rice weevil *Sitophilus oryzae*, the granary weevil *Sitophilus granarius* and the confused grain beetle *Tribolium confusum*, Lepidoptera, e.g. the Argoumois grain moth *Sitotroga cerealella*, the European corn borer, *Pyrausta nubilalis* and the corn earworm *Heliothis zeae*, and others.

13.8 CONCLUDING REMARKS

There is no doubt that insects play a very important role in the transmission of, and the establishment of infections by many fungi. The examples discussed in the previous pages give an idea of how widespread this association may be. Several of these diseases would not occur were it not for the association of the fungus with a particular insect vector; most of these diseases, however, simply occur in much wider areas and with much greater severity than they would in the absence of the association with the insect. It would not be at all risky to suggest that in many more of the known fungal diseases it will be shown that the fungus is more or less intimately associated with insect vectors. Such knowledge would be most welcome because it would not only improve our understanding of these diseases but it could help us control some of the diseases by focusing our attention not only on the pathogen but also on the vector of the pathogen.

However, progress in this area will require more cooperation between plant pathologists and entomologists and more interdisciplinary education for each specialty. The tendency for the pathologists to attribute all the damage to the fungus and to overlook the role of the insect appears to be equaled by the tendency of the entomologists to attribute everything to the insect and to overlook the fact that insect invaders of plants are always, and without any exception, accompanied or followed by fungi and bacteria.

13.9 REFERENCES

Abbott, E. V. (1953). Red rot of sugarcane. Yearbook Agr. U. S. Dept. Agr., 536-539.

Abrahão, J., and Wegmüller, O. (1974). Atividades do casal de *Hypocryphalus mangiferae* (Stebbing) na construcao de celula de oviposicão. *Biologico* 40: 57-58.

Ackerman, J. K., and Shenefelt, R. D. (1973). Organisms, especially insects, associated with wood rotting higher fungi (Basidiomycetes) in Wisconsin forests. *Trans. Wisc. Acad. Sci. Arts and Letters* 61: 185-206.

Adenuga, A. O. (1975). Mutualistic association between ants and some Homoptera–its significance in cocoa production. *Psyche* 82: 24-28.

Agrios, G. N. (1978). Plant Pathology. 2nd Edition. Academic Press, New York and London. 703 p.

Akueshi, C. O., and Oso, B. A. (1976). Studies on *Sclerotium rolfsii.* Part III. Observations on a means of sclerotia dispersal. *Nigerian J. of Pl. Protection* 2: 40-42.

Alfaro, A. (1946). El Acaro *Pediculopsis graminum* Reut. y el hongo *Nigrospora oryzae* (Berk. et Br.) Petch en association parasitario sobre Tigos aragoneses. *Bol. Patol. Vegeta y Entomol. Agr.* 16: 321-334.

Alford, D. V., Woodville, H. C., and Yarham, D. J. (1976). Occurrence of bud rot of carnations in Hertfordshire. *Plant Pathol.* 25: 116-117.

Amante, E., Vulcano, M. A., and Abrahão, J. (1971). Observacoes preliminares sobre a influencia da entomofauna na dispersao dos urdosporos da ferrungem do cafeeiro (*Hemileia vastatrix*). *Biologico* 37: 102-105.

Anonymous. (1976). Annual Report of 1974. Report on research and technical work of the Ministry of Agriculture for Northern Ireland. [20+]. 229 pp.

Ashby, S. F., and Nowell, W. (1926). The fungi of stigmatomycosis. *Ann. Botany (London)* 40: 69-83.

Ashworth, L. J. Jr., Rice, R. E., McCleans, J. L., and Brown, C. M. (1971). The relationship of insects to infection of cotton bolls by *Aspergillus flavus. Phytopathology* 61: 488-493.

Aucamp, J. L. (1969). Role of mite vectors in the development of aflatoxin in groundnut. *J. Stored Prod. Res.* 5: 245-249.

Austwick, P. K. C. (1958). Insects and the spread of fungal disease. *In* "Biological Aspects of the Transmission of Disease," pp. 73-79. Oliver and Boyd, London.

Bagga, H. S., and Laster, M. L. (1968). Relation of insects to the initiation and development of boll rot of cotton. *J. Econ. Entomol.* 61: 1141.

Baillie, A. F. H., and Jepson, W. F. (1951). Bud blast disease of the rhododendron in its relation to the leaf-hopper *Graphocephala coccinea* Forst. *J. Roy. Hort. Soc.* 76: 355-365.

Besson, J., and Joly, E. (1969). La tordeuse orientale du pecher en France. *Phytoma* 21: 25-37.

Böhm, H. (1970). Die Pfirsichmotte, ein nicht mehr seltener Schädling in Österreich. *Pflanzenarzt* 23: 108-109.

Bolay, A., Baggiolini, M., Neury, G., and Antonin, P. (1976). Chancres à *Cytospora* et à sésie sur poiriers en Valais. *Rev. Suisse de Viticult., d'Arboricult. et de Horticult.* 8: 7-16.

Bonneau, P., Simonin, A. (1976). La cochenille du hêtre (*Ctytococcus fagi* Bär.) en Normandie. Eléments de biologie et essais de lutte chimique. *Phytiatri-Phytopharmacie* 25: 3-18.

Boosalis, M. G. (1954). Hessian fly in relation to the development of crown and basal stem rot of wheat. *Phytopathology* 44: 224-229.

Boyce, J. S., Jr., Garren, K. H. (1953). Compatibility types of the oak wilt fungus in 23 Appalachian trees. *Phytopathology* 43: 644-645.

Bramble, W. C., and Holst, E. C. (1940). Fungi associated with *Dendroctonus frontalis* in killing short leaf pines and their effect on conduction. *Phytopathology* 30: 881-899.

Brefeld, O., and Falck, R. (1905). Die Blüteninfektion bei den Brandpilzen und die natürliche Verbreitung der Brandkrankheiten. *Untersuchungen aus dem Gesamtgebiete der Mycol.* 13: 1-74.

Brunt, A. A., and Wharton, A. L. (1960). Transmission of a gall disease of cacao. *Nature* 187: 80.

Buchanan, W. D. (1957). Brentids may be vectors of the oak wilt disease. *Plant Disease Reptr.* 41: 707.

Caldis, P. D. (1927). Etiology and transmission of endosepsis (internal rot) of the fruit of fig. *Hilgardia* 2: 287-328.

Caldis, P. D. (1930). Souring of the figs by yeasts and the transmission of the disease by insects. *J. Agr. Research* 40: 1031-1051.

Cardenosa-Barrigal, R. (1963). La antracnosis del Platana "Cachaco" en el Tolema. *Turrailba* 13: 88-95.

Carter, W. (1973). Insects in Relation to Plant Disease. 2nd Ed. Wiley, New York. 759 p.

Chaves-Batista, A. (1947). Mal do Recife (Grave doenca da Manguiera). Escola Super. Agr. Pernambuco, Recife, 1-106.

Chenon, R. Desmier de. (1975). Presence in Indonesia and Malaysia of a Lepidoptera oil palm root miner, *Sufetula sunidesalis* Walker, and its relationship to attacks by *Ganoderma. Oléagineaux* 30: 449-456.

Chez, D. Hudon, M., and Chiang, M. S. (1977). Résistance du mais à la pyrale (*Ostrinia nubilalis* Hubner) et à la verse parasitaire causée par *Gibberella* zeae (Schw.) Petch. *Phytoprotection* 58: 5-17.

Childs, L. (1929). The relation of woolly apple aphis to perennial canker infection with other notes on the disease. *Oregon State College, Agr. Expt. Sta. Bull.*, No. 243.

Christensen, J. J. (1953). Root rots of wheat, oats, rye, barley. Yearbook Agr., U. S. Dept. Agr., 321-328.

Christensen, C. M., and Kaufmann, H. H. (1969). Grain Storage: The Role of Fungi in Quality Loss. 153 p.

Christensen, J. J., and Schneider, C. L. (1950). European corn borer (*Pyrausta nubilalis* Hbn.) in relation to shank, stalk, and ear rots of corn. *Phytopathology* 40: 284-291.

Christensen, J. J., and Wilcoxson, R. D. (1966). Stalk rot of corn. *Amer. Phytopath. Soc. Monograph* No. 3. 59 p.

Chupp, C., and Sherf, A. F. (1960). Vegetable Diseases and Their Control. The Ronald Press Co., New York. 693 p.

Clarke, R. G., and Wilde, G. E. (1970a). Association of the green stink bug and the yeast spot disease organism of soybeans. I. Length of retention, effect of molting, isolation from feces and saliva. *J. Econ. Entomol.* 63: 200-204.

Clarke, R. G., and Wilde, G. E. (1970b). Association of the green stink bug and the yellow spot disease organism of soybeans. II. Frequency of transmission to soybeans, transmission from insect to insect, isolation from field populations. *J. Econ. Entomol.* 63: 355-357.

Clarke, R. G., and Wilde, G. E. (1971). Association of the green stink bug and the yeast-spot organism of soybeans. III. Effect on soybean quality. *J. Econ. Entomol.* 64: 222-224.

Cotton, R. T. (1956). Pests of Stored Grains and Grain Products. Burgess Publishing Co., Minneapolis, 306 p.

Coyle, C. (1974). Transport of fungi by *Reticulitermes flavipes* (Kollar) (Isoptera: Rhinotermitidae). (Abstract). *Proceedings, Indiana Acad. Sciences* 84: 284-285.

Craighead, F. C. (1928). Interrelation of tree-killing bark beetles (*Dendroctonus*) and blue stains. *J. Forestry* 26: 886-887.

Craighead, F. C., Morris, C. L. and Nelson, J. C. (1953). A preliminary note on the susceptibility of wounded oaks to natural infection by the oak wilt fungus. *Plant Disease Reptr.* 37: 483-484.

Craigie, J. H. (1927). Discovery of the function of the pycnia of the rust fungi. *Nature* 120: 756-767.

Crandall, B. S., and Baker, W. L. (1950). The wilt disease of American persimmon caused by *Cephalosporium diospyri. Phytopathology* 40: 307-325.

Crone, L. J., and Bachelder, S. (1961). Insect transmission of canker stain fungus. *Phytopathology* 51: 576. (Abstr.).

Crowdy, S. H. (1947). Observations on the pathogenicity of *Calonectria rigidiuscula* (Berk. and Br.) Sacc. on *Theobroma cacao* L. *Ann. Appl. Biol.* 34: 45-59.

Crowe, R. J. (1963). Possible insect vectors of the uredespores of *Hemeleia vastratrix* in Kenya. *Trans. British Mycol. Soc.* 46: 24-26.

Croxall, H. E., Collingwood, C. A. and Jenkins, J. E. E. (1951). Observations on brown rot (*Sclerotinia fructigena*) of apples in relation to injury caused by earwig (*Forficula auricularia*). *Ann. Appl. Biol.* 38: 833-843.

Curl, E. A. (1956). Experimental transmission of *Endoconidiophora fagacearum* by Collembola. *Plant Disease Reptr.* 40: 455-458.

Daines, R. H. (1942). Brown rot of peach and its control. *N. J. Agr. Exp. Sta. Circ.* 434. 8 p.

Daugherty, D. M. (1967). Pentatomidae as vectors of yeast-spot disease of soybeans. *J. Econ. Entomol.* 60: 147-152.

Daugherty, D. M., and Foster, J. E. (1966). Organism of yeast spot disease isolated from rice damaged by rice stink bug. *J. Econ. Entomol.* 59: 1282-1283.

Davey, A. E., and Smith, R. E. (1933). The epidemiology of fig spoilage. *Hilgardia* 7: 523-551.

Davidson, R. W. (1951). A deterioration problem in beetle-killed spruce in Colorado. *Phytopathology* 41: 560.

Davidson, R. W. (1954). Species of Ophiostomataceae associated with Engelmann spruce bark beetle. *Phytopathology* 44: 485.

Dorsey, C. K. Jewell, F. F., Leach, J. G., and True, R. P. (1953). Experimental transmission of oak wilt by four species of Nitidulidae. *Plant Disease Reptr.* 37: 419-420.

Eberhard, W. (1974). Insectos y hongos gue atac̷an a la chinche del cacao, *Antiteuchus tripterus. Revista, Facultad National de Agronomia Medellin* (1974), 29: 65-68.

Ehrlich, J. (1934). The beech bark disease, a *Nectria* disease of *Fagus* following *Cryptococcus fagi* (Baer). *Can. J. Research* 10: 593-692.

Ershad, D., and Barkhordary, M. (1974). Host range and vectors of *Nematospora coryli* Peglion in Kerman of Iran. *Iranian J. Plant Pathol.* 10: 34-39.

Eugenio, C., de las Casas, E., Harein, P. K., and Mirocha, C. J. (1970). Detection of the mycotoxin F-2 in the confused flour beetle and the lesser mealworm. *J. Econ. Entom.* 63: 412-415.

Evans, H. C. (1973a). New developments in black pod epidemiology. *Cocoa Growers Bull.* 20: 10-16.

Evans, H. C. (1973b). Invertebrate vectors of *Phytophthora palmivora,* causing black pod disease of cocoa in Ghana. *Ann. Appl. Biol.* 75: 331-345.

Foster, J. E., and Daugherty, D. M. (1969). Isolation of the organism causing yeast spot disease from the salivary system of the green stink bug. *J. Econ. Entomol.* 62: 424-427.

Francke-Grosmann, H. (1952). On the ambrosia culture of the two pine bark beetles, *Myelophilus minor* Htg. and *Ips acuminatus* GYLL. *Medd. Skogsforskn Inst. Stockholm* 41: 52.

Frazer, H. L. (1944). Observations on the methods of transmission of internal boll disease of cotton by the cotton-staining bug. *Ann. Appl. Biol.* 31: 271-290.

Frederick, D. J., Sloan, N. F. and Skowron, W. S., Jr. (1976). Potential insect transmission of *Scleroderis lagerbergii* by scolytid beetles. *Plant Dis. Reptr.* 60: 411-413.

Furniss, M. M., Hungerford, R. D. and Wicker, E. F. (1972). Insects and mites associated with western white pine blister rust cankers in Idaho. *Canadian Entomologist* 104: 1713-1715.

Gardner, M. W. (1918). Mode of dissemination of fungus and bacterial diseases of plants. *Ann. Rept. Mich. Acad. Sci.* 20: 357-423.

Gentry, P., Lopez, G., and Mariau, D. (1975). Dégâts de *Pestalotiopsis* induits par des attaques de *Gargaphia* en Colombie. *Oléagineaux* 30: 199-204.

Gillespie, W. H. (1971). The oak wilt situation. The oak symposium, Morgantown, W. Va. p. 124-128.

Gloyer, W. O., and Fulton, B. B. (1916). Tree crickets as carriers of *Leptosphaeria coniothyrium* (Fckl.) Sacc. and other fungi. *N. Y. State Agr. Expt. Sta.* (*Geneva, N. Y.*) *Bull.* No. 50.

Goncalves, C. R. (1940). Observacoes sobre *Pseudococcus comstocki* (Kuw.) attacando citrus na baixada fluminense. *Rodriguesia* 13: 121-129.

Graham, C. C., and McNeill, M. J. (1972). Soybean crown and root damage by *Bradysia coprophila. J. Econ. Entom.* 65: 597-599.

Gray, E. G., Shaw, M. W., and Shiel, R. S. (1975). The role of mites in the transmission of smoulder in narcissus. *Plant Pathol.* 24: 105-107.

Griswold, C. L. (1953). Transmission of the oak wilt fungus by the pomace fly. *J. Econ. Entoml.* 46: 1099-1100.

Griswold, C. L. (1956). Transmission of the oak wilt fungus by *Pseudopityophthorus minutissimus* (Zimm.). *J. Econ. Entomol.* 49: 560-561.

Gruenhagen, R. H., Riker, A., and Richards, C. A. (1947). Burn blight of jack and red pine following spittle insect attack. *Phytopathology* 37: 757-772.

Haddow, W. R. (1941). Needle blight and late fall browning of red pine (*Pinus resinosa* Ait.) caused by a gall midge (Cecidomyiidae) and the fungus (*Pullularia pullulans* (de Bary) Berkhout). *Trans. Roy. Can. Inst.* 23: 161-189.

Haddow, W. R., and Newman, F. S. (1942). A disease of the Scots Pine (*Pinus sylvestris* L.) caused by the fungus *Diplodia pinea* Kick associated with the pine spittle-bug (*Aphrophora parallela* Say). I. Symptoms and etiology. *Trans. Roy. Can. Inst.* 24: 1-17.

Hansen, H. N. (1929). Thrips as carriers of fig-decaying organisms. *Science* 69: 356-357.

Hansen, H. N., and Davey, A. E. (1932). Transmission of smut and molds in figs. *Phytopathology* 22: 247-252.

Hanson, E. W., and Milliron, H. E. (1942). The relation of certain weevils to root rot and basal rot of cereals and grasses. *Phytopathology* 32: 22.

Hanson, E. W., Milliron, H. E., and Christensen, J. J. (1950). The relation of the blue grass billbug, *Calendra parvula* (Gyllenhal), to the development of basal stem rot and root rot of cereals and grasses in north-central United States. *Phytopathology* 40: 527-543.

Hawboldt, L. S. (1945). History of spread of the beech scale, *Cryptococcus fagi* (Baerensprung), an insect introduced into the Maritime Provinces. *Acadian Naturalist,* I, 137-146 (1944); *Rev. Appl. Entomol.* 33: 353-354.

Heald, F. D. (1913). The dissemination of fungi causing disease. *Trans. Am. Microscop. Soc.* 32: 1-29.

Henry, B. W. (1944). *Chalara quercina* n. sp., the cause of oak wilt. *Phytopathology* 34: 631-635.

Henry, B. W., and Moses, C. S. (1943). An undescribed disease causing rapid dying of oak trees. *Phytopathology* 33: 18.

Hepting, G. H., Toole, E. R. and Boyce, J. S., Jr. (1952). Sexuality in the oak wilt fungus. *Phytopathology* 42: 438-442.

Hertert, H. D., Miller, D. L. and Partridge, A. D. (1975). Interaction of bark beetles (Coleoptera:

Scolytidae) and root rot pathogens in grand fir in northern Idaho. *Canad. Entomologist* 107: 899-904.

Herzog, D. C., Thomas, J. W., Jensen, R. L., and Newsom, L. D. (1975). Association of sclerotial blight with *Spissistilus festinus* girdling injury on soybean. *Environm. Entomology* 4: 986-988.

Himelick, E. B., and Curl, E. A. (1955). Experimental transmission of the oak wilt fungus by caged squirrels. *Phytopathology* 45: 581-584.

Hinds, T. E. (1972). Insect transmission of *Ceratocystis* species associated with aspen cankers. *Phytopathology* 62: 221-225.

Ingold, C. T. (1971). Fungal Spores. Their Liberation and Dispersal. Clarendon Press, Oxford, 302 p.

James, J. R., Lucas, L. T., and Campbell, W. W. (1977). Interactions between insect injury and *Fusarium* spp. on root rot of ladino clover. *Proceed. Amer. Phytopath. Soc.* 4, No. S-15.

Jares, T. W., and Van Arsdel, E. P. (1977). Possible insect carriers of the oak decline fungus. *Proceed. Amer. Phytopath. Soc.* 4, No. 179.

Jewell, F. F. (1954). Viability of the conidia of *Endoconidiophora fagacearum* Bretz in the fecal material of certain Nitidulidae. *Plant Disease Reptr.* 38: 53-54.

Kable, P. F. (1969). Brown rot of stone fruits on the Murumbidge Irrigation Areas. 1. Aetiology of the disease in canning peaches. *Aust. J. Agric. Res.* 20: 301-316.

Kilpatrick, R. A., and Dunn, G. M. (1958). Observations of insects and fungi associated with taproot survival of white clover in New Hampshire. *Plant Disease Reptr.* 42: 819-820.

Koehler, B. (1959). Corn ear rots in Illinois. *Ill. Agr. Exp. Sta. Bull.* No. 639.

Koehler, B. (1960). Cornstalk rots in Illinois. *Ill. Agr. Exp. Sta. Bull.* No. 658.

Kryukova, E. A. (1976). Insects and vascular mycosis of oak. *Zashchita Rastenii* 5: 42-43 (In Russian). (*Rev. Pl. Path.* 55: 5364).

Laemmlen, F. F. (1969). The association of the mite *Siteroptes reniformis* and *Nigrospora oryzae* in Nigrospora lint rot of cotton. *Phytopathology* 59: 1036.

Lambert, D. H., and McIlveen, W. D. (1976). *Acylomus* sp. infesting ergot sclerotia. *Ann. Entom. Soc. Amer.* 69: 34.

Langdon, R. F. N., and Champ, B. R. (1954). The insect vectors of *Claviceps paspali* in Queensland. *J. Austrailan Inst. Agr. Sci.* 20: 115-118.

LaPrade, J. C., and Manwiller, A. (1977). Relation of insect damage, vector and hybrid reaction to aflatoxin B_1 recovery from field corn. *Phytopathology* 67: 544-547.

Leach, J. G. (1935). Insects in relation to plant diseases. *Bot. Rev.* 1: 448-466.

Leach, J. G. (1940). Insect Transmission of Plant Diseases. McGraw-Hill, New York.

Leach, J. G. Orr, L. W. and Christensen, C. (1934). The interrelationships of bark beetles and blue-staining fungi in felled Norway pine timber. *J. Agr. Research* 49: 315-342.

Leath, K. Y., and Newton, R. C. (1969). Interaction of a fungus gnat, *Bradysia* sp. (Sciaridae) with *Fusarium* spp. on alfalfa and red clover. *Phytopathology* 59: 257-258.

LePelley, R. H. (1942). The food and feeding habits of *Antestia* in Kenya. *Bull. Entomol. Res.* 33: 71-89.

Libby, R. R., and Ellis, D. E. (1954). Transmission of *Colletotrichum lagenarium* by the spotted cucumber beetle. *Plant Disease Reptr.* 38: 200.

Linn, J. B. (1942). *Cephalosporium* leaf spot of *Dieffenbachia. Phytopathology* 32: 172-175.

Martin, W. J. (1947). Alternaria leaf blight of Hevea rubber trees. *Phytopathology* 37: 609-612.

Maublanc, A., and Roger, L. (1935). La phthiriose du Cafeier au Cameroun. *Rev. Appl. Botan.* 15: 25-32.

Mercier, L. (1911). Sur le rôle des insects comme agents de propagation de l'ergot des graminées. *Compt. Rend. Soc. Biol.* 70: 300-302.

Meyer, R. (1978). A Cladosporium sooty mold and its insect associates in central California. *Plant Dis. Reptr.* 62: 382-385.

Milne, D. L. (1975). Coffee rust consumed by a helpful insect *Mycodiplosis hemileiae. Citrus and Sub-Trop. Fruit J.* 498: 11-12.

Nasr, E. S. A., and Azab, A. K. (1969). *Rhizopus nigricans* Ehr. infection in cotton bolls in relation to bollworm infestation, age of bolls and the insecticides used on cotton plants. *Bull. Entomol. Soc. Egypt. Economic Series* 3: 97-102.

Nath, D. K. (1973). Insect transmission of sooty mould (*Capnodium* sp.) to orange orchards at Darjeeling district, West Bengal. *Science and Culture* 39: 262-263 (*Rev. Pl. Path.* 53: 1817).

Nelson, R. M. (1934). Effect of bluestain fungi on southern pines attacked by bark beetles. *Phytopathol. Z.* 7: 327-353.

Nelson, R. M., and Beal, J. A. (1929). Experiments with bluestain fungi in southern pines. *Phytopathology* 19: 1101-1106.

Norris, D. M., Jr. (1953). Insect transmission of oak wilt in Iowa. *Plant Disease Reptr.* 37: 417-418.

Norris, D. M., Jr. (1955). Natural spread of *Endoconidiophora fagacearum* Bretz to wounded red oaks in Iowa. *Plant Disease Reptr.* 39: 249-253.

Nuorteva, M., and Laine, L. (1968). Über die Möglichkeiten der Insekten als Überträger des Wurzelschwamms (*Fomes annosus* (Fr) Cooke). *Suom. hyont. Aikak* 34: 113-135 (*Rev. Appl. Entom., A,* 59, 1628).

Ogawa, J. M. (1957). The dried fruit beetle disseminates spores of the peach brown rot fungus. *Phytopathology* 47: 530.

Ondrej, M. (1973). The parasitic fungus *Botrytis fabae* Sard. in connection with sucking insects. *Biologia Czechoslavakia, A,* 28: 57-63.

Owen, H. (1956). Further observations on the pathogenicity of *Calonectria rigidiuscula* (Berk. and Br.) Sacc. to *Theobroma cacao* L. *Ann. Appl. Biol.* 44: 307-321.

Parker, E. J. (1974). Beech bark disease. *Forest Record* 96, 15 pp.

Parkin, E. A. (1942). Symbiosis and siricid wood wasps. *Ann. Appl. Biol.* 29: 268-274.

Parmeter, J. R., Jr., Kuntz, J. E., and Riker, A. J. (1956). Oak wilt development in bur oaks. *Phytopathology* 46: 423-436.

Perret, J., and Mussillon, P. (1975). Contribution à la connaissance de *Synanthedon tipuliformis* Clerk, la sésie du groseillier et du cassissier (Sesiidae (=Aegeriidae). *Alexanor* 9: 137-144, 163-171.

Peschken, D. P., and Beecher, R. W. (1973). *Centorhynchus litura* (Coleoptera:curculionidae): biology, and first releases for biological control of the weed Canada thistle (*Circium arvense*) in Ontario, Canada. *Canad. Entomol.* 105: 1489-1494.

Peyronel, B. (1950). A mutualistic association between mites of the genus *Pediculopsis* and some fungus parasites of plants. *Atti Accad. Sci. Torino, Classe Fis. Mat. e Nat.,* 84, Reprint, 9 pp.

Phillips, E. H., Smith, E. H., and Smith, R. E. (1925). Fig smut. *Calif. Univ. Agr. Expt. Sta. Bull.* No. 387.

Porter, D. M., and Smith, J. C. (1974). Fungal colonization of peanut fruit as related to southern corn rootworm injury. *Phytopathology* 64: 249-251.

Poulos, P. L., and Heuberger, J. W. (1952). Relation of wounds to the fruit rot phase of the brown rot disease of peaches. *Plant Dis. Reptr.* 36: 198-200.

Powell, J. M. (1975). Arthropod taxa common to *Cronartium* cankers and the forest litter layer. *North. For. Rec. Center, Edmonton, Alta, Canada. Biomonthly Res. Notes* 31: 23.

Rexrode, C. O. (1968). Tree wounding insects as vectors of the oak wilt fungus. *Forest Science* 14: 181-189.

Rexrode, C. O., and Jones, T. W. (1971). Oak bark beetles carry oak wilt fungus in early spring. *Plant Disease Reptr.* 55: 108-111.

Russo, L. F. (1968). Sul *Rhizoglyphus echinopus* (Fum & Rob) e sui danni da essor prodotti alla coltura del gladiolo. (Engl. summary). *Boll. Lab. Ent. Agr. Filippo Silvestri* 26: 314-328. (*Rev. Appl. Entom., A,* 59, 1128).

Satterthwait, A. F. (1932). How to control billbugs destructive to cereal and forage crops. *U. S. Dept. Agr. Farmers' Bull.* No. 1003.

Schewket (Sevket) N. (1934). Die Feigeninsekten und die wesentlichsten Ursachen der Feigenfruchtfäule. *Anz. Schädlingskunde* 10: 118-119.

Scott, W. M., and Ayres, T. W. (1910). The control of peach brown rot and scab. *U. S. Dept. Agr., Bur. Plant Ind. Bull.*, No. 174.

Shurtleff, M. C. (1973). A Compendium of Corn Diseases. *Amer. Phytopathol. Soc.*, St Paul, Minn. 64 p.

Silow, R. A., (1933). A systemic disease of red clover caused by *Botrytis anthophila* Bond. *Trans. Brit. Mycol. Soc.* 18: 239-248.

Sinadskii, Yu. V. (1967). Bioecological details of the development of the apical birch woodwasp (*Pseudoxiphydria betulae* Ensl.). *Zool. Zh.* 46: 1069-1073 (In Russian). (*Rev. Appl. Entom., A,* 58, 1668).

Skuhravy, V. (1973). Needle blight and needle drop on *Pinus silvestris* L. in Europe and *P. resinosa* in North America (Diptera, Cecidomyiidae). *Zeitschrift für Angewandts Entomologie* 72: 421-428.

Smith, F. F., and F. Weiss. (1942). Relationship of insects to the spread of azalea flower spot. *U. S. Dept. Agr. Tech. Bull.* No. 798, 1-44.

Smith, R. E., and Hansen, H. N. (1931). Fruit spoilage disease of figs. *Calif. Univ. Agr. Expt. Sta. Bull.* No. 506.

Snapp, O. I., Alder, C. H., Roberts, J. W., Dunegan, J. C., and Pressley, J. H. (1927). Experiments on the control of the plum curculio, brown rot, and scab, attacking the peach in Georgia. *U. S. Dept. Agric. Bull.* 1482. 32 p.

Snetsinger, R., and Himelick, E. B. (1957). Observations on witches' broom of hackberry. *Plant Disease Reptr.* 41: 541-544.

Stambaugh, W. J., and Fregus, C. L. (1956). Longevity of spores of the oak wilt fungus on overwintered Nitidulid beetles. *Plant Disease Reptr.* 40: 919-922.

Stephenson, L. W., and Russell, T. E. (1974). The association of *Aspergillus flavus* with hemipterous and other insects infecting cotton bracts and foliage. *Phytopathology* 64: 1502-1506.

Steyaert, R. L. (1946). Plant protection in the Belgian Congo. *Sci. Monthly* 63: 268-280.

Stoyanov, D. (1963). Studies on the raspberry midge *Thomasiana theobaldi* Barnes in Bulgaria. (In Bulgarian, English summary). *Izv. Inst. Zasht. Rast. Kostinbrod* 4: 41-66 (*Rev. Appl. Entom., A,* 53, 170-171).

Summanwar, A. S., and Raychaudhuri, S. P. (1969). Role of Eriophyid mite (*Aceria mangifera*) in the causation of mango malformation. *Indian Phytopathol.* 21: 463-464.

Summers, T. E. (1952). Destruction of maize roots by *Pythium graminicola Subr.* and *Diabrotica undecimpunctata howardi* Barber. *Iowa State Coll. J. Sci.* 26: 294-295.

Talbot, P. H. P. (1977). The Sirex-Amylostereum-Pinus association. *Ann. Rev. Phytopathol.* 15: 41-54.

Tan, G. S. (1974). *Helopltis theivora theorbromae* on cocoa in Malaysia. II. Damage and control. *Malays. Agric. Res.* 3: 204-212.

Tate, K. G., and Ogawa, J. M. (1975). Nitidulid beetles as vectors of *Monilinia fructicola* in California stone fruits. *Phytopathology* 65: 977-983.

Taubenhaus, J. J. (1920). A study of the black and yellow moulds of ear corn. *Texas Agr. Sta. Bull.* No. 270.

Taylor, G. S. (1952). Stalk rot development in corn following the European corn borer. *Phytopathology* 42: 20-21.

True, R. P., Barnett, H. L., Dorsey, C. K., and Leach, J. G. (1960). Oak wilt in West Virginia. *West Va. Univ. Agr. Expt. Sta. Bull.* No. 448T.

Velimirovic, V. (1975). Contribution to the study of the ashen vine moth *Lobesia botrana* (Lepidoptera, Tortricidae) in the region of Crna Gora. *Zastita Bilja* 26: 85-91 (Yugoslavian) (*Rev. Appl. Entom., A,* 64, 5548).

Voight, E. (1972). Damage cuased by *Argyrotaenia pulchellana* Haw. to grape vines in Hungary. *Pflanzenschutzberichte* 43: 13-23.

Wellman, F. L. (1972). Tropical American Plant Disease. Scarecrow Press, Metuchen, N. J. 989 p.

Widstrom, N. W., Sparks, A. N., Lillehoj, E. B., and Kwolek, W. F. (1975). Aflatoxin production and Lepidopteran insect injury on corn in Georgia. *J. Econ. Entomol.* 68: 855-856.

Widstrom, N. W., Lillehoj, E. B., Sparks, A. N., and Kwolek, W. F. (1976). Corn earworm damage and aflatoxin B_1 on corn ears protected with insecticide. *J. Econ. Entomol.* 69: 677-679.

Wienk, J. F. 91964-1965). The transmission of *Aspergillus niger* by adult sisal weevils. *Ann. Rep. Sisal Res. Sta. Mlingana.*

Willison, R. S., and Dustan, G. G. (1956). Fruit flies and fungal wastage in peaches. *Can. J. Agr. Sci.* 36: 233-240.

Windels, C. E., Windels, M. B. and Kommedahl, T. (1976). Association of *Fusarium* species with picnic beetles on corn ears. *Phytopathology* 66: 328-331.

Wood, F. A., and Skelly, J. M. (1968). Nitidulid transmission of *Ceratocystis fagacearum* from wounds on infected red oaks. *Phytopathology* 58: 1247-1249.

Yamaguchi, Y., Takaoka, and Uchida, H. (1970). *Lepidocyrtinus* sp. (Entomobryidae: Collembola) eating the colony of the tobacco powdery mildew (*Erysiphe cichoracearum*) on tobacco leaves. *Kontyu* 38: 168-170 (*Rev. Appl. Entom., A,* 63, 274).

Yamamoto, W. (1951). Studies on the dissemination of sooty moulds by insects. *Mem. Hyogo Univ. Agr.* (*Phytopathol. Ser.* No. 1), 1: 1-50 (Japanese; English summary).

Yarwood, C. E. (1943). Association of thrips with powdery mildews. *Mycologia* 35: 189-191.

Yount, W. L. (1958). Results of root inoculations with the oak wilt fungus in Pennsylvania. *Plant Disease Reptr.* 42: 548-551.

Yount, W. L., Jeffrey, A. R., and Thompson, H. E. (1955). Spores of *Endoconidiophora fagacearum* on the external surfaces of the body of Nitidulids. *Plant Disease Reptr.* 39: 54.

Zuckerman, B. M. (1954). Relation of type and age of wound to infection by *Endoconidiophora fagacearum* Bretz. *Plant Disease Reptr.* 38: 290-292.

Chapter 14

MITES

J. T. Slykhuis

Research Station, Agriculture Canada
Summerland, British Columbia, Canada

14.1	INTRODUCTION	326
14.2	CHARACTERISTICS OF THE ERIOPHYIDS	326
14.3	VECTORS OF AGENTS THAT CAUSE DISEASES OF MONOCOTS	328
	14.3.1 *Eriophyes tulipae* and Wheat Streak Mosaic	329
	14.3.2 *Eriophyes tulipae* and Wheat Spot Mosaic	333
	14.3.3 *Eriophyes tulipae* and Onion Mosaic	334
	14.3.4 *Abacarus hystrix* and Ryegrass Mosaic	334
	14.3.5 *Abacarus hystrix* and Agropyron Mosaic	335
14.4	VECTORS OF AGENTS THAT CAUSE DISEASES OF DICOTS	336
	14.4.1 *Cecidophyopsis ribis* and Black Currant Reversion	336
	14.4.2 *Eriophyes ficus* and Fig Mosaic	340
	14.4.3 *Phytoptus insidiosus* and Peach Mosaic	342
	14.4.4 *Phytoptus inaequalis* and Cherry Mottle Leaf	344
14.5	DISEASES PROBABLY CAUSED BY MITES WITHOUT AN INFECTIOUS AGENT	345
	14.5.1 *Eriophyes sacchari* and Sugarcane Chlorotic Streak	345
	14.5.2 *Eriophyes cajani* and Pigeon Pea Sterility	345
	14.5.3 *Phyllocoptes fructiphilus* and Rose Rosette	346
	14.5.4 *Calacarus citrifolii* and Citrus Concentric Ring Blotch	347
14.6	DOUBTFUL OR ERRONEOUS REPORTS OF MITE TRANSMISSION	347
	14.6.1 *Aculus fockeui* and Prunus Latent Virus	347
	14.6.2 *Colomerus vitis* and Vine Panachure	348
	14.6.3 *Eriophyes mangiferae* and Mango Malformation	348
14.7	DISEASE AGENTS SUSPECTED TO BE TRANSMITTED BY ERIOPHYID MITES	349
	14.7.1 Cadang-Cadang of Coconut	349
	14.7.2 Little Cherry Disease	349
	14.7.3 Hordeum Mosaic Virus	350
14.8	CONCLUSIONS	350
14.9	ACKNOWLEDGMENTS	351
14.10	REFERENCES	351

ISBN 0-12-326450-2

X

14.1 INTRODUCTION

The first two plant diseases reported to be caused by viruses transmitted by eriophyid mites are models of contrast. The first, black currant reversion, has been one of the most difficult of the mite-associated diseases to investigate. It was reported to be caused by a mite-transmitted virus in 1925, but there is still doubt that the cause is a virus. The second, wheat streak mosaic, was known to be caused by a sap-transmissible virus before mites were shown to be the vectors. Many investigators have confirmed that the cause is a virus, and that mites transmit it.

At least 14 other diseases have been reported to be caused by viruses transmitted by eriophyid mites, but a critical review of the evidence reveals that only a few have been confirmed to be caused by viruses. Some of the diseases have been shown to be caused by infectious agents transmitted by eriophyid mites, but the agents have not been proved to be viruses. For others, mites were associated with disease symptoms but there was no proof that an infectious agent was present. For still others, there is no firm evidence that the mites had any specific causal relationship to the disease with which they were associated.

The purpose of this review is to evaluate evidence reported on mites in relation to infectious disease agents. It is hoped that questions raised will stimulate investigations that will clarify many of the uncertainties that linger, and that will reveal other mite-disease relationships that remain undiscovered.

14.2 CHARACTERISTICS OF THE ERIOPHYIDS

All the known acarine vectors of plant disease agents are mites of the family Eriophyidae. Some members of the Nalepellidae and Rhyncaphytoptidae, the other two families of the Eriophyoidea, have been suspected but none have been proved to be vectors.

There are five subfamilies in the Eriophyidae but genera with species reported to be vectors of plant disease agents occur in only three of them. These include *Cecidophyopsis* and *Colomerus* in the Cecidophyinae, *Eriophyes* and *Phytoptus* in the Eriophyinae, and *Abacarus, Aculus, Calacarus* and *Phyllocoptes* in the Phyllocoptinae (Jeppson *et al.,* 1975).

14.2.1 General Features and Life Histories

Eriophyid mites are worm-like and colorless or mildly colored to blend with the background of plant tissue. They are very small, usually less than 250 μm in length, and are difficult to see even with a hand lens. Most species cause no noticeable injury to their hosts and so their presence is usually overlooked.

The eriophyids have only two pairs of legs, very few setae, modified piercing and sucking mouth parts, a greatly reduced cephalothorax, and proximal genitalia that lie just behind the coxae. There are no eyes, spiracles or tracheae. Respiration is believed to occur through the epidermis. An outstanding anatomical feature of the abdomen is the presence of tranverse rings of microtubercles which are thought to control water exchange through the epidermis. The microtubercles vary in number, size, and shape between the dorsal and ventral surfaces, between developmental stages, and between species.

Eriophyid mites have four developmental stages: egg, first nymph, second nymph and adult. Usually eggs are laid on the host plant but there is evidence that a few species are ovoviviparous. Nymphs resemble adults but lack external genitalia and vary in microtuberculation. While first nymphs are much smaller than adults, second nymphs approach adult size before molting.

Parthenogenesis has been suspected but has not been proved in eriophyids. Males hatch from unfertilized eggs and females from fertilized eggs. Eriophyids do not copulate. The males deposit spermatophores on the host surface. When a female crawls over a spermatophore, the sperm are squeezed into her spermatheca.

Most species of eriophyids have a simple life cycle in which males and females occur throughout the year. Feeding and reproduction occur whenever temperatures are favorable. However, many eriophyids on deciduous hosts in temperate and arctic zones have a slightly more complex life cycle. There is a single form of male which does not overwinter, and two forms of females. The protogyne, which is the primary or summer female, lays eggs through the summer. The deutogyne, which is the secondary or winter female, picks up sperm from the spermatophores in late summer or early fall, but cannot lay eggs until after being subjected to a period of cold winter weather. The eggs are laid in spring and give rise to males and protogynes.

Some species of eriophyids can complete a generation from egg to egg in 6 days under ideal conditions in summer, but generally two to three weeks are required. Extremely high populations can develop quickly. Several thousand mites have been counted on a leaf or in a single bud.

14.2.2 Host Range Specificity

All eriophyids are phytophagous. Many are highly host specific. A few are so specialized that they colonize only certain varieties of one plant species. A number of eriophyids live on several plant species within one genus. Others, such as the grass eriophyids, have hosts in several genera. It is rare for these mites to have hosts in more than one plant family. One exception is the wheat curl mite, *Eriophyes tulipae* (Keifer), which breeds on graminaceous and liliaceous plants.

The eriophyids do not have a resistant form that can sustain the species through long periods in the absence of a living host. Most species live on perennial plants. Some can be sustained indefinitely on annuals that grow in an overlapping sequence throughout the year.

14.2.3 Feeding Habits, Mechanisms and Effects

Eriophyid mites feed by a piercing and sucking action provided by a pharyngeal pump and a rostrum. The rostrum, or gnathosoma, is enclosed on each side by pedipalps which hold slender mouth stylets in an anterior groove. The basal segments of the pedipalps contain the pharyngeal pump. At the apex of each pedipalp is a suction cup that holds the mite to the plant so that pressure can be applied to the stylets in penetrating the cells. The segmented pedipalps are also capable of telescopic movement which provides forward impulses to the stylets.

There are five mouth stylets. The two anterior ones are the chelicerae that slowly penetrate the plant tissue by movements of the pedipalps. Most eriophyid chelicerae are between 15 and 40 μm long. Just behind the chelicerae, and just as long, are two auxiliary stylets which supposedly inject digestive enzymes into the plant tissue. The fifth or oral stylet is very thin and small. It moves up and down behind the other four and aids in sucking up plant nutrients.

Eriophyids depend on living tissue and feed primarily on the young, soft, and succulent parts of plants. Frequently, they are found living in and around buds, in leaf sheaths, in axils of petioles and twigs, on young fruit, or on leaves. Because of the small stylets they do little mechanical damage to their hosts, but some species produce salivary secretions that affect plants in various ways. Some cause leaf discoloration, russetting, curling, rolling, blistering or the formation of erinea on leaves. Some cause russetting and deformation of fruit, deformation and enlargement of seed or twisting of terminal growth. Some live in buds and cause bud blasting or galls of various colors, shapes, and sizes. Some of these injuries may be mistakenly attributed to insects or the symptoms of fungus or virus diseases.

14.2.4 Movement

Independent movement of eriophyids is limited by their small size and inability to fly. Some species have been observed to crawl a meter or so, but they must rely on other forces such as wind, insects, and birds to carry them long distances. Under warm, dry conditions, some adult eriophyids will stand erect on their terminal lobes and long caudal setae and wave back and forth as though searching for something to grasp. They will cling to passing insects to be carried to other plants. Adults of some species have been observed to spring from the leaf surface and be carried away by the wind. Wind is probably the principal means of dispersal. Large numbers of eriophyids have been caught like pollen on sticky glass slides placed on tall poles or atop high buildings. Like wind-borne pollen, many do not land on susceptible plants, but some do and initiate colonies on the new plants.

14.3 VECTORS OF AGENTS THAT CAUSE DISEASES OF MONOCOTS

Eriophyids in two genera transmit viruses or virus-like agents that cause diseases of monocotyledonous plants.

Eriophyes tulipae (Keifer), the dry bulb mite or wheat curl mite, was collected first from bulbs of tulips, onions and garlic. It lives between the fleshy bulb leaves and induces drying of the bulbs. It was not recognized as a colonizer of grasses until it was found to transmit wheat streak mosaic virus (WSMV). Now it is known to colonize a number of cereals and other annual and perennial grasses. It was dubbed 'wheat curl mite' because of the inrolling and looping it causes when the tips of growing leaves of infested wheat are trapped in the tightly rolled preceding leaves. In maize, the mites cause kernel red streak, a reddish discoloration that results from the effects of salivary toxins on kernels. *E. tulipae* has been reported to transmit the agents that cause wheat streak mosic, wheat spot mosaic and onion mosaic.

Eriophyes tosichella (Keifer) is the name designated by Keifer (1969) for the eriophyid mite that transmits WSMV in Yugoslavia (Tošič, 1971). It resembles *E. tulipae* in most features but was differentiated principally by having 8-rayed instead of 7-rayed featherclaws on the females, and 7 instead of 6 on males. Perhaps comparative biological studies are required to assess the significance of differences between this mite and *E. tulipae*.

Abacarus hystrix (Nalepa), the cereal rust mite, was collected first on *Agropyron repens* (L) Beauv. but occurs on many other grasses including wheat. It lives in the grooves on the upper leaf surface and feeds on the bulliform epidermal cells at the base of the groove. Prior to feeding, the cuticular leaf wax is removed by the rostrum, then the styliform chelicerae puncture the epidermal cells. On *Lolium multiflorum* Lam., the mites preferred the youngest leaves and the youngest part of the more mature leaves (Gibson, 1974). *A. hystrix* transmits ryegrass mosaic virus (RMV) and Agropyron mosaic virus (AMV).

14.3.1 *Eriophyes tulipae* and Wheat Streak Mosaic

Wheat streak mosaic is caused by a readily sap-transmissible virus (WSMV) with filamentous particles about 700 nm long and 15 nm diameter. It causes a severe mosaic of winter wheat and certain millets, and mild mosaic on some varieties of maize, barley and some other cereals and grasses. Serious losses occur in winter wheat infected in the autumn. The plants may be severely stunted with varying degrees of necrosis resulting in death of tillers, reduced seed set and shrivelled kernels (Brakke, 1971).

14.3.1.1 The Vector Search. Mosaic symptoms, probably caused by WSMV, were noted first on wheat in Nebraska in 1922 and in Kansas in 1929. Collections from Kansas in 1932 were studied by McKinney (1937). Serious losses in wheat were attributed to this virus in Colorado in 1947 (Atkinson, 1949) and western Kansas in 1949 (Fellows, 1949).

A severe outbreak of wheat streak mosaic throughout western Kansas in 1949 resulted in the loss of 50% of the total yield of wheat in one county and about 16% for each of several others (Fellows, 1949). Another outbreak in Kansas in 1959 caused the loss of 46,670,000 bushels of wheat, or about 10% of the wheat crop for that year (King and Sill, 1959).

The rapid field spread and the widespread devastation caused by wheat streak mosaic spurred a sense of urgency to discover the natural means of transmission of the virus. Unlike wheat mosaic in Illinois and some other more easterly States, there was no evidence that the cause of wheat streak mosaic virus was carried in soil. Many tests were done with insects. Aphids were prime suspects, and positive results were reported in Colorado (Atkinson, 1949), but similar results were not obtained in Kansas where at least 23 species of insects were tested (Connin, 1953). The leafhopper *Endria inimica* (Say) was associated with mosaic symptoms on wheat in some tests, but similar results were not obtained in other tests. Even chewing insects such as grasshoppers, and the abrasive effects of wind were tested by some experimenters, occasionally with positive but unrepeatable results.

Vector tests with various insects caught in plots in South Dakota, where the disease was spreading throughout the summers of 1950 and 1951, showed that the leafhopper *E. inimica* transmitted the hitherto unrecognized wheat striate mosaic virus, but none of the insects transmitted WSMV. It was found that the mysterious vector could pass through the screen of cages that excluded aphids and leafhoppers (Slykhuis, 1952). Subsequently, thrips were tested in Alberta where the disease was also found, but the results were negative. Then a creature smaller than thrips was noticed on diseased wheat. It transmitted the virus, and was identified as an eriophyid mite, *Aceria tulipae* (Slykhuis, 1955), now designated *(Eriophyes tulipae* (Jeppson *et al.*, 1975). The presence of WSMV in plants infected by the mites was readily confirmed by sap transmission to healthy wheat. If WSMV was present, the inoculated plants developed symptoms of WSMV in 6 to 10 days.

14.3.1.2 Vector-Virus Relations. At least 1/3 of the mites tested singly transmitted the virus. The mites could transmit when transferred from diseased plants at any stage including first nymph, second nymph, adult or during either the first or second quiescent stages. Eggs transferred from diseased plants into the whorl of healthy wheat seedlings hatched and initiated colonies of mites that did not induce mosaic symptoms. Nymphs from such non-viruliferous colonies acquired the virus by feeding on WSMV-infected wheat plants. Adults were able to transmit the virus only if they acquired it during one of the nymphal stages.

Transmission of WSMV by *E. tulipae* was soon confirmed in Kansas (Connin and Staples, 1957) and Nebraska (Staples and Allington, 1956). Since then the same species has been found to be the vector of WSMV wherever wheat streak mosaic has been investigated in North America and Europe, with the exception of Yugoslavia where the vector was named *E. tosichella* (Keifer, 1969, Tošič, 1971). *E. tulipae* and WSMV have been reported together in most countries surrounding the Black Sea (Pop, 1962; Razvyaskina, 1963; Bremer, 1973; Markov *et al.*, 1975).

Generally, arthropod-borne plant viruses can be grouped as stylet-borne and nonpersistent, or circulative and persistent. Wheat streak mosaic virus has some features of both groups. Like the stylet-borne viruses transmitted by aphids, it has filamentous particles and is readily transmitted mechanically by rubbing juice on the leaves of susceptible plants. Like the circulative viruses, it is acquired by the vector only during long feeds (15 min or longer), and the mites continue to transmit after a molt and remain infective for long periods after removal from a virus source, i.e. up to 9 days at 20-25°C (Slykhuis, 1955; Orlob, 1966).

Densely packed accumulations of WSMV particles were found in ultrathin sections of mites reared on diseased wheat (Paliwal and Slykhuis, 1967). It was not interpreted that the particles multiplied in the mites, but that they accumulated during prolonged feeding. Since particles were not found outside the alimentary canal, it was postulated that transmission might occur either by regurgitation, which seemed unlikely, or by excretion. Takahashi and Orlob (1969) found virus particles in parenchymatous cells around the intestine. They speculated that transmission might occur by passage of the virus through the intestinal wall into parenchymatous tissue, then to the salivary glands and into the plant cells through salivary secretions while feeding, analogous to transmission of some viruses by leafhoppers.

A fluorescence antibody technique was used to detect wheat streak mosaic virus antigens in crude "homogenates" prepared from *E. tulipae* reared on virus-infected wheat plants. The virus antigens also were detected in the body fluid released through the anus of mites when pressed gently with a fine glass rod (Sinha and Paliwal, 1971). This suggests that the virus could be excreted but there were no tests to determine if particles were excreted by living mites, and if so that they were infectious.

Stein-Margolina *et al.* (1969) found large conglomerations of spherical virus-like particles and filamentous particles in the intestine of *E. tulipae* reared on plants infected with WSMV. The origin and relationship of the spherical particles to WSMV was not known. The spherical particles of brome mosaic virus have been found accumulated in *E. tulipae* reared on wheat infected with the virus, but the mites did not transmit it (Paliwal, 1972; Stein-Margolina, 1973). Similarly, a few filamentous particles resembling WSMV were found in *Aculodes mckenziei* (Keifer) and *Abacarus hystrix* reared on wheat infected with WSMV, but neither of these mites has been shown to transmit WSMV (Orlob, 1966).

14.3.1.3 Hosts of E. tulipae *and the Epidemiology of WSMV.* The virus and its mite vector require a continuity of living hosts to survive. About 15 perennials have been infected with WSMV by sap inoculation or by mite transmission in North America, and a few have been found infected in the field (Sill and Connin, 1953; Slykhuis, 1955; Connin, 1956), but none have been found susceptible to the virus and also able to sustain *E. tulipae*. Some plants of *Elymus canadensis* L. were infected with WSMV by mites from diseased wheat but the mites did not survive and reproduce on *E. canadensis*. *E. tulipae* found naturally on *E. canadensis* did not survive on wheat. Del Rosario and Sill (1965) reported that few of the *E. tulipae* found on *Agropyron smithii* Rydb. or on onions (*Allium cepa* L.) survived after transfer to wheat. Some of the survivors reproduced on wheat. After several generations, the mites were adapted to wheat and their ability to transmit the virus increased. Neither *A. smithii* nor onions have been reported as hosts of WSMV. It still seems probable that plants of some perennial species of *Agropyron, Bouteloua,*

Elymus, Eragrostis, Festuca, or other of the grasses must be reservoirs of WSMV and *E. tulipae* capable of infesting wheat in North America. In Europe, several perennials have been reported to be sources of WSMV. These include *Bouteloua curtipendula* (Michx.) Torr., *Festuca ovina* L., *Hordeum nodosum* L. and *Lolium perenne* L. (Razvyaskina *et al.*, 1963; Markov *et al.*, 1975).

Both the virus and the mites can be sustained abundantly without reliance on perennials. Where wheat streak mosaic is important its prevalence and severity are predominantly determined by the characteristics of wheat culture. Wheat is highly susceptible to both virus and mites. Winter wheat infected in the autumn carries the virus and mites over winter. During the warm weather, the next spring and summer, the mites multiply rapidly, sometimes to tremendous numbers on each diseased plant. They are dispersed by wind and infect any susceptible hosts on which they land. The mites are such efficient vectors and wheat is so susceptible that in districts where WSMV occurs all wheat plants found infested with mites also are infected with the virus and become sources of infection for other plants. This includes any immature wheat, whether it is a spring-sown crop, volunteer plants from seed that sprouted in summer-fallow fields or at the edge of fields and along roadways, or the regrowth from hail-damaged crops. If such plants are left, they perpetuate the virus and mites through the period from maturity of one winter wheat crop until the next crop can be completed entirely on wheat. Other annuals like barley, maize and many wild annual grasses including species of *Bromus, Digitaria, Echinochloa, Eragrostis, Panicum* and *Setaria*, also are susceptible to WSMV. They can sustain the virus and some also can support populations of *E. tulipae* between wheat crops (Slykhuis, 1955). The growing of maize near wheat, as often occurs in parts of Ontario and Ohio, favors the perpetuation of wheat streak mosaic. Wheat can be a source of infection for maize in early summer. The infected maize may harbor viruliferous mites in the ears until late autumn when the new crops of wheat are well established (Gates, 1970).

14.3.1.4 Control of WSMV. Generally, serious infections in spring wheat crops can be avoided by avoiding seeding near diseased winter wheat, and by seeding early. Infection of winter wheat crops usually can be avoided by delaying seeding until after susceptible spring crops have been harvested and all immature wheat and other susceptible annuals, including barley and grass weeds, have been eliminated in or near fields where wheat is to be sown (Slykhuis, 1955; Staples and Allington, 1956). No specific seeding date can be expected to be safe without co-ordinated elimination of oversummered wheat and other susceptible annuals (Atkinson and Slykhuis, 1963). There can be problems if wheat is grown near susceptible cultivars of maize which may harbor viruliferous mites in the ears until late fall. Cultural control has been reported difficult to achieve in Kansas where the fall season is frequently long and warm, and plants that harbor WSMV and *E. tulipae* are difficult to eliminate before seeding winter wheat.

Mosaic tolerant cultivars have been credited with reducing losses in Kansas (King and Sill, 1959). No immune or highly resistant wheats have been developed but promising sources of resistance to WSMV and the vector appear to be available for breeding programs in *Triticum-Agropyron* chromosome substitution lines (Larson and Atkinson, 1973).

Although miticides were tested to eliminate or reduce mite populations, none proved sufficiently effective for control of WSMV (Kantack and Knutson, 1958). There has been little evidence of continued interest in chemicals for the control of this disease.

14.3.2. *Eriophyes tulipae* and Wheat Spot Mosaic

Wheat spot mosaic (WSpM) was detected in Alberta in 1952 when *E. tulipae* was first recognized as a vector of WSMV (Slykhuis, 1955, 1956). When mites from naturally diseased wheat were tested singly on wheat to determine the percentage that transmitted WSMV, some of the test plants developed chlorotic spots, severe chlorosis, stunting and necrosis not characteristic of WSMV. No virus could be transmitted from these plants by manual inoculation procedures used to confirm the presence of WSMV. All stages of the mites transferred from diseased plants induced similar symptoms. Eggs transferred from diseased plants and hatched on healthy plants produced mites that did not induce the symptoms unless they were fed on diseased plants before being transferred to test plants. It was therefore concluded that the symptoms were caused by an infectious agent transmitted by the mites.

Different isolates of the infectious agent originating from single mites from naturally diseased wheat caused symptoms that differed in pattern and severity. The initial symptoms of most isolates included a chlorotic and necrotic spotting followed by general yellowing. Other isolates caused a mild mottling and chlorosis. In the field, the diseased plants were severely chlorotic and were also usually infected with WSMV. When mites from a plant infected with both agents were tested singly, 65% transmitted WSpM agent and 34% transmitted WSMV. Some mites transmitted both agents.

The host range of WSpM agent is similar to but not identical with that of WSMV. It induced symptoms on most of the common cereals tested including maize, and several wild annual grasses. Several *Triticum* X *Agropyron* hybrids immune to WSMV developed mild symptoms of WSpM.

A similar but milder agent was transmitted by *E. tulipae* to wheat in Ontario. In Ohio, Nault and Styer (1970) found an agent also transmitted by *E. tulipae*, which caused symptoms they designated 'wheat spot chlorosis'. This name is probably more accurately descriptive than 'wheat spot mosaic' for the disease, but the causal agents appear to be very similar if not identical. All stages of *E. tulipae* except the eggs carried the pathogen which persisted up to 8 days in mites kept on an artificial medium. This pathogen, like WSMV, was acquired by nymphs but was not transmitted by adults unless they acquired it as nymphs.

Electron microscope studies of the ultrastructure of leaves of infected wheat, maize and barley revealed the presence of ovoid, double membrane bodies of 0.1-

0.2 μm diameter in the cytoplasm of parenchyma phloem and epidermal cells (Bradfute *et al.*, 1970). The internal components of these bodies consisted of dispersed fibrils suggesting the presence of nucleic acid. Neither ribosomes characteristic of bacteria, mycoplasma and the psittacosis group of organisms, nor an electron dense nucleoid characteristic of many viruses was commonly evident. The double membrane bodies were not found in healthy leaf tissue, but they have not been positively identified as the infectious agent. The same investigators found similar bodies in fig leaves with fig mosaic. Recently, similar double membrane bodies were found in wheat plants infected with the WSpM agent from Alberta (R.I. Hamilton, Research Station, Agriculture Canada, Vancouver). It appears that these structures represent or are induced by a plant pathogen transmitted by eriophyid mites.

14.3.3 *Eriophyes tulipae* and Onion Mosaic

A mosaic disease has infected up to 32% of the onion seed plants in some fields in the forest district in the Ukraine. Transmission tests showed that *E. tulipae*, which is abundant in the diseased onions, induced the symptoms on healthy onion plants. Virus-like particles 675 x 15 nm were found in sap from affected onions and in suspensions and ultrathin sections of mites from diseased plants. The virus-like particles and *E. tulipae* were also found on garlic in the northern Caucasus (Razvyaskina, 1971). There was no report of the effects of mites reared from eggs transferred to healthy plants. Although *E. tulipae* from onions can colonize wheat with difficulty (Del Rosario and Sill, 1965), there has been no evidence that the cause of onion mosaic could infect wheat or that WSMV or the WSpM agent from wheat could infect onions. We will anticipate further reports confirming the cause of onion mosaic, its characteristics and relationship to *E. tulipae*.

14.3.4 *Abacarus hystrix* and Ryegrass Mosaic

Ryegrass mosaic virus (RMV) has filamentous particles about 700 nm long and 15 nm diameter. It occurs naturally on Italian ryegrass (*Lolium multiflorum*), perennial ryegrass (*L. perenne*) and Cocksfoot (*Dactylis glomerata* L.), and has been sap-transmitted to a number of other grasses. It has been reported from Britain and North America (Slykhuis and Paliwal, 1972).

Mosaic symptoms observed on ryegrasses in England, Scotland and Wales in 1956 were found to be caused by a sap transmissible virus with filamentous particles initially designated ryegrass streak mosaic virus. Similar symptoms were observed on ryegrasses in several countries in Europe, but at least some of these were caused by other viruses. In England, the virus was transmitted by eriophyid mites found on diseased plants (Slykhuis, 1958). Mulligan (1960) who continued the investigations in England segregated and tested three species of mites. He found that *Abacarus hystrix* transmitted the virus but *Vasates dubius* (Nalepa) and a species of *Eriophyes*, also found on diseased plants, did not.

All stages of *A. hystrix* transmitted RMV. Eggs hatched on healthy plants produced non-viruliferous mites. When placed on diseased plants, these mites acquired the virus in 2 hr but more mites could transmit after 12 hr. Mites reared on diseased plants lost the ability to transmit within 24 hr on wheat which is immune to RMV.

Ryegrass mosaic is considered to be the most important of the diseases and pest problems of ryegrass swards in south-eastern England. It is widespread and damaging, particularly to *L. multiflorum*, and it can reduce the herbage yield of perennial ryegrass by up to 30% (Heard and Roberts, 1975).

Cultural control by eliminating nearby sources of infection before new crops are planted, as is effective for the control of WSMV in wheat, is virtually impossible for the control of RMV in most farm situations in Britain. The ryegrasses are extremely widespread. The mites and virus disperse principally during June, July and August in Wales (A'Brook, 1975). In eastern England, the numbers of mites and rate of spread of RMV were found to be greatest in autumn and least in winter. The numbers increased again in spring. Since plantings of the ryegrasses are grown for two to several years, there is ample time for new plantings to become infested with virus-carrying mites from outside sources, and for spread to occur within the plantings. One measure that can reduce rate of spread within plantings is close cutting in autumn (Gibson, 1976). It also appears that applications of Endosulfan to ryegrass can reduce resident mite populations and the rate of spread of RMV within the crops.

In a search for resistance, Gibson and Heard (1976) found 2/108 selections of *L. perenne* and 1/150 of *L. multiflorum* remained free from RMV after 7 to 9 inoculations. Wilkins and Hides (1976) found that symptom severity of 10 populations of *L. multiflorum* infected artificially in the glasshouse and naturally in the field with RMV was strongly correlated. Under sward conditions, yields were reduced 27% in a sensitive genotype and 13% in a more tolerant one. There has been little information on resistance to the mites in ryegrasses.

14.3.5 *Abacarus hystrix* and a Gropyron Mosaic

Agropyron mosaic virus (AMV) has filamentous particles about 717 nm long and 15 nm diameter. It infects only species of Gramineae, and has been found principally in *Agropyron repens* (L) Beauv. and wheat. It is readily transmitted by sap inoculation. It causes a light green to yellow mosaic (Slykhuis, 1973).

The virus was first isolated in Virginia (McKinney, 1937) and has been found in several other States of the U.S.A., in Canada, Europe and Britain (Catherall and Chamberlain, 1975).

A. repens is a perennial reservoir host from which the virus can spread to wheat (Slykhuis, 1962). It has been transmitted artifically to several other grasses. Although most isolates from *A. repens* cause only mild mosaic symptoms, some isolates obtained in Ontario caused a severe yellow mosaic, moderate stunting and yield losses of 25 to 75% in some cultivars of wheat inoculated in the field.

Natural spread of AMV into wheat was demonstrated in plots of winter wheat sown near diseased *A. repens* at different dates from early spring to late fall at Ottawa, Canada. Some of the wheat plants in the early-sown plots became infected with AMV evidently from the *A. repens*. Up to 100% of the plants became infected in plots sown between June and early September indicating that AMV and its vector had multiplied on the earlier sown wheat which then became a potent source of infection for plants in the later seedings. This resembled the spread of WSMV from wheat to wheat in Alberta (Slykhuis *et al.*, 1957), but AMV spread much more slowly than WSMV would have spread under similar circumstances.

The mites *A. hystrix* and *Aculodes mckenziei* (Keifer) were almost always found together on *A. repens* and wheat naturally infected with AMV. *E. tulipae*, although never found on *A. repens*, was sometimes found on wheat. Numerous tests were done to determine if any of these mites or other arthropods found on naturally diseased wheat and *A. repens* transmitted AMV to seedlings of *A. repens* or wheat. Mites were tested from naturally infected plants and from colonies of each species reared on sap inoculated plants. They were transferred by coiling or clipping sections of leaves from mite-infested diseased plants on the test seedlings or by transferring the mites with a single hair mounted on a handle. Sometimes, a few of the test plants to which *A. hystrix* had been transferred developed symptoms of agropyron mosaic. None of the non-infested controls, or plants receiving only *A. mckenziei* or *E. tulipae* from AMV-infected wheat became infected. There remained the suspicion that the rare transmission of AMV associated with *A. hystrix* could have resulted from accidental mechanical transmission from contact of leaves or from touching the plants when transferring the mites.

Satisfactory evidence that *A. hystrix* transmitted the virus was obtained in experiments with pure colonies of each of the mite species reared on young *A. repens* and/or wheat infected with AMV by sap inoculation (Slykhuis, 1969). Pots of these infected source plants infested with the different species of mites or free from mites were placed in separate growth cabinets. Pots of test plants were also placed in the cabinets but separated by space and a screen to assure that there could be no contact between the source and test plants. Small fans were used to blow the mites from the sources toward the test plants. Varying numbers ranging from 1% to 40% of the plants in 11 of 14 different tests developed agropyron mosaic after 9 to 19 days in the growth cabinets with plants infested with *A. hystrix*. No infection resulted in any of 14 separate tests run concurrently with *A. mckenziei* or *E. tulipae*.

Catherall and Chamberlain (1975) reported transmission of AMV by *A. hystrix* in Wales also by blowing the mites from source to test plants with a fan.

In Ontario, all patches of *A. repens* examined were infested with *A. hystrix* and *A. mckenziei*, but AMV was not always present. Although there may be variations of *A. repens* resistant to AMV, the lack of universal infection of *A. repens* with AMV is probably principally related to the low incidence of transmission by the mites.

14.4 VECTORS OF AGENTS THAT CAUSE DISEASES OF DICOTS

14.4.1 *Cecidophyopsis ribis* and Black Currant Reversion

Black Currant Reversion was the first plant disease reported to be caused by a virus transmitted by an eriophyid mite. There has been a long history of investigations of this disease, with many problems in determining whether or not the disease was caused by the mites or by an agent carried by the mites, and in determining whether or not the infectious agent is a virus.

14.4.1.1 The mite. The black currant gall mite (*Cecidophyopsis ribis* (Westw.), previously known under the generic names *Phytoptus* and *Eriophyes*, was recognized as a serious pest of black currants (*Ribes nigrum* L.) in Scotland as far back as 1850 (Massee, 1928). Since then, it has been recognized throughout Europe as well as Britain. The mites spend most of the year in buds. They generally emerge between mid-March and early June and spread to other parts of the same bush by crawling, and are dispersed further by rain, wind, or insects. They show a positive response to light so they generally move toward the branch tips.

14.4.1.2 Recognition of Reversion as an Infectious Disease. Mites that invade new buds of black currants early in the growing season cause them to enlarge and fail to differentiate flowers or true leaves. In Britain, it was noticed that a condition called 'reversion,' because the affected bushes appeared to have reverted to the wild, ancestral form, was frequently associated with the mites. The leaves of the reverted bushes were narrower and smaller than normal. The leaf bases were flattened by a reduction in depth of the basal sinus. There were fewer subsidiary veins than normal. The leaf margins were more coarsely toothed and the leaf surface was more coarsely rugose. The blossoms often appeared almost transparent and many fell before producing fruit. The fruit that formed was usually abnormal, shrivelled and fell before maturity. The cause of reversion was transmitted by grafting and was therefore assumed to be a virus (Lees, 1925; Amos *et al.*, 1927). However, there remained the possibility that mites were present on the scion and could have induced the symptoms.

Massee (1952) demonstrated that reversion symptoms developed on bushes to which large numbers were transferred during the winter each year for three years. Smith (1962a) found that a single mite transferred during the normal period of mite migration in the spring or early summer could start an infestation leading to bud galling and reversion. These experiments confirmed an association between the mites and reversion symptoms but did not differentiate between symptoms caused by the reversion 'virus' and the mites.

Controlled transmission of the cause of currant reversion by the mites and by grafting was demonstrated by Thresh (1964a). Small black currant seedlings were infested in a glasshouse with gall mites from reverted bushes. After 4 days, the seedlings were dipped in 0.05% endrin solution to kill the mites. The seedlings showed no mite injury or galls but within one month some of them developed vein pattern symptoms frequently associated with reversion. The presence of reversion was confirmed by grafting patches of bark on healthy Wellington XXX bushes in an isolated plot free from mites. These plants developed reversion but check plants did not.

Transmission by grafting patches of bark without mites from diseased bushes has made it possible to differentiate the symptoms of currant reversion disease from the complex effects of the mites on black currant bushes and to detect symptoms formerly overlooked. Thresh (1964c) noted that a chlorotic vein pattern was a preliminary, transient phase of the reversion disease and was generally followed by the leaf, blossom and fruit abnormalities recognized in the reversion syndrome. Conversely, a conspicuous asymmetrical abnormality of leaves was included in some of the early descriptions of reversion symptoms. Later, it was attributed to mechanical injury from which the plant soon recovered and was called 'false' reversion. Thresh (1963) associated such symptoms with the presence of the mites in the apex of affected shoots. They were not transmissible by grafting. Their presence during the summer provided a good indication of mite infestation.

The tendency for reverted bushes to bear heavy populations of mites seems to have been an early observation, but the association was not absolute. Sometimes, reversion symptoms occurred on bushes on which mites were not found but was absent on some other bushes with heavy populations of mites (Lees, 1925).

14.4.1.3 Mite-Pathogen Relationships. Although Smith (1963) found that the gall mites were capable of transmitting reversion during all except the first 2 weeks of their migration period from April 1, to June 28, 1960, he was not able to show a close correlation between the development of mite infestations and currant reversion on bushes in a young plantation.

These observations are not entirely consistent with later observations by Thresh (1964c) who noted that reverted bushes are more susceptible to the mites than healthy bushes. The young buds of bushes infected with reversion by graft transmission were 110 times more susceptible to dispersing mites than healthy bushes. It was suggested that virus infection increased the vulnerability and accessibility of apical and axillary buds by decreasing the density of hairs on the young stems and leaves. The heaviest infestations of mites occurred on reverted bushes. Mites rarely caused serious damage when reversion was absent and usually remained at a low population level. Proeseler (1973) did not agree that the increased susceptibility of reverted bushes was due to decreased density of hairs. He suggested that it resulted from physiological changes following infection.

14.4.1.4 The Causal Agent. There have been several reports describing mycoplasma-like structures in thin sections of mites from reverted bushes (Silvere and Tiits, 1969; Protsenko *et al.*, 1975) but there was no conclusive evidence that these structures represented the reversion disease. Silvere and Romeikis (1971) isolated gram positive bacteria from diseased bushes. The bacteria produced slimy white colonies within which were rod-shaped cells 0.5 to 1 X 1.3 to 3 μm multiplying by transverse fusion.

Jacob (1976a, 1976b) stated that under field conditions black currants do not develop typical leaf symptoms, but in growth chambers, under continuous light with an increase of the red part of the spectrum, spotting, banding and ring chlorosis typical for the virus developed on leaves of reversion-diseased black currants. He did not explain why the latter symptoms, which do not occur on reverted bushes in the field, should be considered typical and diagnostic for the disease in growth chambers. Nevertheless, he reported transmission experiments based on the development of these symptoms on Baldwin black currant seedlings grown under the designated light conditions. The virus was transmitted by both nymphal states and adults, but not through eggs. The mites acquired the virus in 3 hr and retained it for 25 days. It was transmitted optimally during 48 hr. There was no detailed description of how inoculation feeds were terminated with certainty. He also reported that the cause of reversion was a virus identical with potato virus Y (PVY). The virus was mechanically transmitted from black currant to *Nictiana glutinosa, N. rustica, Solanum demissum* "A6" and to other hosts of PVY, and was identified serologically and with the electron microscope. Black currants were re-infected with the virus by mechanical inoculation from *N. glutinosa.*

The above results are surprising. There is no previous confirmed record of a virus being transmitted both by aphids and by eriophyid mites. Potato virus Y infects at least 60 species of plants, most of them in the Solonaceae but some in the Chenepodiaceae and Leguminosae (Delgado-Sanchez and Grogan, 1970). It will require careful confirmation to establish *R. nigrum* as another host of PVY.

In a personal communication, J. M. Thresh reported attempts to detect PVY in East Malling sources of reversion. Sap transmission was tested using extracts of leaves and developing buds prepared with materials known to be highly effective for transmitting viruses from black currants. An antiserum prepared against an isolate of PVY from potato was used in the ELISA test to detect the virus in preparations from diseased black and red currants. There was no evidence by either method for the occurrence of PVY in any of the sources of reversion tested. We conclude that the nature of black currant reversion agent remains in doubt.

14.4.1.5 Control. Early recommendations for the control of reversion included removal of reverted bushes to eliminate sources of the disease, and application of lime sulfur sprays in the spring to kill mites emerging from buds (Amos and Hatton, 1928). Similar measures are still the basic recommendations in England. The importance of removal of reverted bushes is emphasized by evidence that diseased bushes are more susceptible and harbour much heavier populations of mites than healthy bushes. Applications of lime sulfur are most effective when applied during the period of greatest emergence of mites from infested buds and penetration of new axillary buds. This occurs after the first flowers open in April until early June. More than one spray is necessary for adequate commercial control. Smith (1962b) achieved 94.1% reduction in mite spread after three applications of 1% lime sulfur plus 0.025% dinonyl sodium sulphosuccinate. Thresh (1968), using four applications at 14-day intervals during the dispersal period, found that endrin at 0.04% gave almost complete control of mite spread and decreased the incidence of virus infection. Similar sprays with 0.05% endosulfan was less, and 1% lime sulfur least effective. Part of the effectiveness of endrin, fluoroacetemide and endosulfan is attributable to killing mites within the young buds (Collingwood and Dicker, 1960). Some plant-growth retardants, by reducing growth of shoots during the mite migration period, also reduced the numbers of buds available for invasion, thus giving appreciable protection against mites. Foliar sprays of benomyl and related systemic fungicides partially protected black currant plants from infection with gall mites. Soil applications of benomyl gave complete protection (Smith, 1975). Entry of mites into new buds was not affected, nor was transmission of the reversion agent. The main effect was probably on the survival of mites after entry into buds before gall formation.

Thresh (1964d) eliminated mite infestations from dormant black currant cuttings by treating in warm water at 40 degrees for 40 min, 42.5 degrees for 30 min, or at 45 degrees for 20 min. Fruiting plantations planted with bushes free from mites and reversion and isolated at least 100 m from old plantings remained healthy for many years.

Some cultivars of black currants were less susceptible than others. The least susceptible of 8 cultivars tested by Thresh (1964a) had the most densely hairy shoots. Tiits (1964) suggested that since reversion was limited to *R. nigrum* resistance might be achieved by crossing with other *Ribes* species, but no promising results have been cited.

14.4.2 *Eriophyes ficus* and Fig Mosaic

14.4.2.1 The Mite. The fig bud mite, *Eriophyes ficus* Cotte, occurs anywhere figs (*Ficus* spp.) are grown (Jeppson *et al.*, 1975). All stages of both sexes can be found on fig trees throughout the year. They spend the dormant season in and around buds. As the buds burst in spring, they move to the leaves. The females lay eggs on petioles and on both leaf surfaces. Most eggs are deposited among the hairs on the lower surface of leaves. The mites can develop from egg to adult in 5 to 7 days in warm weather. In July in California, many mites enter the fruit. The feeding of the mites causes chlorosis, russetting or surface browning of leaves, blasting and distortion of buds and reduction of new growth. When populations are very heavy, they may cause defoliation of branches or trees. Populations of *E. ficus* increase more rapidly on plants supplied with optimum nutrients than on plants deficient in nitrogen or phosphorus.

14.4.2.2 The Disease. Fig mosaic, studied first by Condit and Horne (1933), is widespread in figs in most areas where figs are grown. It occurs on all continents except South America. In some areas, all fig trees and nursery stock are infected. The only sources of mosaic-free fig plants in such areas are young plants grown from seed in locations removed from diseased fig trees (Flock and Wallace, 1955; Blodgett and Gomec, 1967).

The symptoms include a mosaic ranging from varied yellow spots to light green blotches on leaves and chlorotic spots on the fruit. Sometimes, the leaves are malformed and the fruit drops prematurely.

Fig mosaic affected most plants tested of more than 16 species of *Ficus* (Burnett, 1962; Nagaich and Vashisth, 1962). *Cudrania tricuspidata* and mulberry (*Morus indica*), both members of the fig family (Moraceae), were also infected (Vashisth and Nagaich, 1965).

14.4.2.3 Differentiating an Infectious Agent. An early awareness of an association between eriophyid mites, principally *E. ficus* and fig mosaic is revealed by quotations from Condit and Horne (1933): ". . . it is possible that the spots on fig leaves may result from localized feeding of mites when the leaves were young and tender." Then, in the same paper, noting that three fig trees of the Mission variety had heavy infestations of mites but no mosaic symptoms, they commented: "We have suspected fig mites of transmitting the virus of leaf mosaic, but thus far our suspicions are not confirmed or supported by experimental data." They demonstrated that mosaic developed on new growth from buds from disease-free seedlings grafted onto diseased orchard trees, but mites were present which could have induced the symptoms.

Flock and Wallace (1955) reported that mosaic symptoms persisted on fig plants from cuttings that were treated with dusting sulfur to kill the fig bud mites, then rooted and grown in a mite-free environment. Mosaic symptoms also appeared on new growth. In other tests, healthy fig seedlings developed mosaic symptoms after scions from a diseased mite-free source were grafted into them.

Transmission of the fig mosaic agent by the fig mite was demonstrated by transferring various numbers of mites individually with a fine brush to small seedlings. After 3 to 5 days, the seedlings were dusted with sulfur to kill the mites. Most of the seedlings on which mites fed developed mosaic symptoms in 10 to 90 days including 7/10 that received one mite each and 18/20 that received 100 mites each. When mites from colonies developed from eggs transferred to disease-free seedlings were fed on test plants for 5 days to 2 weeks, some of the seedlings developed leaf distortion, slight chlorosis and russetting but no mosaic.

14.4.2.4 Characteristics of the Infectious Agent. Double membrane bodies, similar to bodies found in wheat plants infected with wheat spot mosaic agent, were found in leaves of *Ficus carica* systemically infected with fig mosaic "virus" (Bradfute *et al.*, 1970). The double membrane bodies were accompanied by an open tubulous mass of electron dense granular material which appeared to be structurally related. These structures were not found in healthy leaf tissue but have not been positively identified as infectious agents.

14.4.2.5 Mite-Pathogen Relationships. A more recent report (Proeseler, 1972) confirms that the "virus" could be transmitted from infected to healthy seedlings of *F. carcia* by a single mite. The minimum feeding periods required for acquisition and infection were 5 min. The causal agent was acquired more readily from terminal buds than from leaf blades or petioles and was not acquired from basal parts of stems or from roots. The mites could transmit a few hours after acquisition feeds. Mites from diseased figs remained infectious for 6 to 10 days at room temperature.

14.4.2.6 Control. In many fig-producing areas, losses caused by fig mosaic must be significant but cannot be measured accurately because there are no disease-free trees for comparison. The virus is spread so efficiently by vegetative propagation including grafting, and by the fig bud mite, that even the nursery stock is almost entirely infected. Blodgett and Gomec (1967) suggested that if healthy trees could not be found for propagation, attempts should be made to use heat therapy to obtain virus-free stock. This could be grown as mother trees to supply cuttings to root- and budwood to graft on healthy rootstock which can be grown from seed. Control of the vector in new, disease-free plantings might be possible and profitable.

Ozalp and Kepler (1972) failed to inactivate the virus by heat treatment at 52 to 55°C and 60 to 68% relative humidity for 3 to 10 days. However, 9 of 171 Sarilop fig trees were selected as resistant in the Aegean region, and were to be used for breeding in new nurseries.

14.4.3 *Phytoptus insidiosus* and Peach Mosaic

14.4.3.1 The Mite. *Phytoptus* (*Eriophyes*) *insidiosus* (Wilson and Keifer) was not recognized until it was found to be a vector of the peach mosaic agent. It has been found on peaches and plums in western U.S.A. and Mexico. It occurs beneath the scales of dormant and retarded buds in leaf axils and among rudmentary leaves of expanding buds on vigorous shoots.

14.4.3.2 The Disease. Peach mosaic was found in Texas and Colorado in 1931. It was recognized subsequently in Arizona, Arkansas, California, New Mexico, Oklahoma, and Utah in the U.S.A. and northern Mexico (Pine, 1976). It is economically important on peach in western Colorado and southern California. Many plums, especially wild species, are susceptible and are reservoir hosts. There are no substantiated reports of its occurrence outside North America (Pine and Sutic, 1968).

Symptoms of peach mosaic are varied. They consist of retarded foliation, mottling and deformity of leaves, blossom break, fruit deformity and abnormal tree growth. In some varieties, symptoms may be masked. In others, there may be severe shock during the first year of infection. Some show the most severe symptoms the second year followed by a milder chronic stage.

14.4.3.3 The Causal Agent. The peach mosaic agent is transmitted by tissue grafts using diseased fruit, leaf pieces, shootwood and bark, and rootwood and bark. There has been no evidence of transmission with expressed juice or by dodder. One of the difficulties in assessing transmissibility of the causal agent has been the variability and unpredictability of the incubation period in peaches (Wilson *et al.*, 1955). In graft transmission tests in the greenhouse, symptoms developed on some seedlings in 14 days, but on others not until the next season. In the field, some of the trees that were graft inoculated when breaking dormancy in the spring developed mosaic in 3 to 6 weeks. If the trees were in foliage when inoculated, they often did not show symptoms until after a period of dormancy.

The physical and chemical characteristics of the causal agent are still unknown, hence the agent cannot be designated as a virus with certainty.

14.4.3.4 Host Range of Peach Mosaic. Almond, apricot and other *Prunus* species closely related to peach are susceptible. Several native or naturalized wild plums including *P. americana* Marsh., *P. angustifolia* Marsh, and *P. munsoniana* Wight and Hedr. are hosts and reservoirs of infection. Sweet and sour cherries (*P. avium* L. and P. *cerasus* L.) and most wild cherries, but not *P. besseyi* Bailey, *P. tomentosa* Thunb. and *P. japonica* Thunb. are immune (Pine, 1976).

14.4.3.5 The Vector Search. Spread by means in addition to grafting appeared to be rapid in Texas, Colorado and later in California. Attempts to prevent spread by eradication of all known diseased trees were started early and appeared to be effective in orchards of Elberta and J. H. Hale peaches in Colorado (Bodine, 1934; Bodine and Durrel, 1941) even though the means of natural spread was not known. By exposing peach test trees in sites of rapid spread, either by placing pots of test trees in the orchard for 2-month periods or by planting trees in the orchards on different dates, near Riverside, California, it was shown that spread occurred slowly in March and April, most rapidly in May and June, and continued on through the growing season into October. It appeared that natural spread could occur any time peach foliage was present and green.

After more than 8,000 tests with more than 150 species of insects and mites, transmission by a previously undescribed eriophyid mite was demonstrated (Wilson *et al.,* 1955). The initial transmissions were accomplished by transferring mites from naturally infested mosaic-diseased field peach and plum trees to the buds of 65 potted Rio Oso Gem peach seedlings in the greenhouse. Symptoms of peach mosaic developed in 14 to 100 days on 16/56 of the seedlings infested with mites from diseased peach and 1/9 seedlings infested with mites from hortulan plum. The presence of peach mosaic in these test seedlings was confirmed by graft transmission to other healthy peach seedlings. Test seedlings which received equivalent numbers of mites of the same species from healthy peach and plum trees did not develop symptoms. Mites from eggs laid in buds of diseased trees but hatched on healthy seedlings did not induce mosaic symptoms. The effects of the virus were therefore differentiated from any effects of the feeding of non-infectious mites.

14.4.3.6 Mite-Pathogen Relationships. When the mites were transferred from diseased trees to test trees they transmitted to causal agent only when placed on buds. The mites retained the ability to infect test seedlings after 2 days, but not after 3 days, on dry glass coverslips. By the third day, they were torpid and may not have been able to feed (Jones and Wilson, 1957-58).

When single adults were transferred from diseased peach to healthy peach seedlings, 2 of 80 test seedlings developed mosaic symptoms. When the numbers of mites per plant were increased, the numbers of transmissions increased. Eighteen out of 25 test plants were infected by 50 mites per plant.

Colonies of *P. insidiosus* have been established by rooting cuttings of flowering peach bearing mite-infested buds (Oldfield and Wilson, 1970). This procedure may facilitate rearing large number of *P. insidiosus* in the greenhouse for further investigations of mite-pathogen relationships.

14.4.7 Phytoptus insidiosus in Relation to Control. Peach mosaic has been contained within relatively limited areas, since 1935, through a program of diseased tree removal (Pine, 1976). Infected areas in the U.S.A. are under Federal-State quarantines, and tree removal is coordinated with nursery inspection to prevent the spread of the disease to non-infected areas. Since *P. insidiosus* is more widely distributed than peach mosaic, any introduction of the causal agent to new areas where the vector is established is hazardous.

The use of tolerant peach cultivars is desirable in areas where roguing is not feasible and peach growers must operate in the presence of the disease. Tolerant cultivars are not desirable in areas where roguing is practiced because the disease is more difficult to diagnose in tolerant trees.

Jones *et al.* (1970) reported effective control of the spread of peach mosaic by chemical control of the mites in peach orchards in which the disease was spreading rapidly. Populations of *P. insidiosus* were greatly reduced by a single spray of diazinon at petal fall each year from 1962 through 1965. The rate of spread of peach mosaic in the orchard dropped from 118 newly infected trees found in 1962 to 46 in 1963, 2 in 1966, 1 in 1967 and 0 in 1968. These results indicate that, without removing the diseased trees, control of the vector halted significant spread of the virus.

The peach mosaic agent was not inactivated by heat treatment. Chemotherapy has been attempted without conclusive results.

14.4.4 *Phytoptus inaequalis* and Cherry Mottle Leaf

14.4.4.1 The Mite. *Phytoptus inaequalis* (Wilson and Oldfield, 1966) is similar to *P. insidiosus*. It lives in lateral buds of *Prunus emarginata* (Douglas). The mites leave the old enlarged buds as they dry out in early summer and disperse to new buds. They may move to but seldom have been found in sweet cherry.

14.4.4.2 The Disease. Cherry Mottle Leaf (CML), first observed in Washington in 1917 and Oregon in 1920 (Reeves, 1941), also has been found in British Columbia, California, Idaho, Montana, and Utah. It is most prevalent in orchards near wild bitter cherry (*Prunus emarginata*) (Cheney and Parish, 1976). It causes irregular chlorotic mottling and distortion, puckering, tattering, shot hole and reduced size of leaves, especially of terminal leaves. It also shortens the internodal growth of some cultivars of cherries. The fruit may be reduced in size and lack flavor. The disease has been found in Europe, but only in trees imported from North America (Basca, 1969; Jones *et al.*, 1969). The infectious agent occurs naturally in sweet cherry and peach, but peach is usually symptomless. Some cultivars of sweet cherry, including Bing, are severely affected. Napoleon (Royal Ann) is moderately, while Republican and Van are mildly affected. Thirty-seven of the cherry cultivars tested were symptomless or nearly symptomless. Infection has also been reported in some ornamental flowering cherries and plums.

14.4.4.3 Differentiating an Infectious Agent. The causal agent is transmitted readily by budding and grafting any CML-infected shoot or root tissue to a susceptible host like Bing cherry. Leaf symptoms have been noted in 14 days on test plants grown in a greenhouse, and in 37 days in the field (Reeves, 1941). The causal agent has not been transmitted by seed or mechanically and its physical and chemical characteristics have not been determined.

14.4.4.4 Mite-Pathogen Relations. The causal agent has been transmitted by *Phytoptus (Eriophyes) inaequalis* found in enlarged buds of naturally diseased *P. emarginata* by L. S. Jones and described by Wilson and Oldfield (1966). To test for transmission of the cherry mottle leaf agent, 50 mites from naturally infected *P. emarginata* were transferred to each of 20 newly germinated peach seedlings in a greenhouse. The peach seedlings remained symptomless, but buds were grafted from each peach seedling to a healthy Bing test tree the next spring. Typical cherry mottle leaf symptoms developed on 12 of the Bing cherry trees. In later tests, the virus was transmitted by the same mite from *P. emarginata* directly to Bing cherry (Oldfield, 1970).

There have been no further reports on the relations of the mites to cherry mottle leaf, or of attempts to control the disease by controlling the mites.

14.5 DISEASES PROBABLY CAUSED BY MITES WITHOUT AN INFECTIOUS AGENT

14.5.1 *Eriophyes sacchari* and Sugarcane Chlorotic Streak

E. sacchari was reported to transmit a virus that caused a chlorotic streak of sugarcane in India (Sithananthan *et al.*, 1972). Canes to which mites from diseased sugarcane were transferred developed streak symptoms in 15 to 20 days. The authors implied that the disease was identical with sugarcane streak. The latter disease is caused by the maize streak virus which has isometric particles and is transmitted by the leafhopper *Cicadulina mbila* (Naude) (Boch, 1974). The authors did not present evidence that the streak symptoms associated with *E. sacchari* on sugarcane were caused by an infectious agent, or that the presumed infectious agent was similar to maize streak virus.

14.5.2 *Eriophyes cajani* and Pigeon Pea Sterility

Pigeon pea sterility, which causes mosaic and partial sterility of pigeon pea (*Cajanus cajan* (Linn.) Millsp.) was reported to be caused by a virus transmitted by an eriophyid mite (*Eriophyes cajani* Channa Basavanna) in India (Seth, 1962, 1965). The cause was assumed to be a virus because it was transmitted from diseased plants to 35 of 99 healthy plants of pigeon pea cultivar N.P. 69 by wedge grafting. Attempts to transmit it by sap inoculation, dodder (*Cuscuta reflexa* Roxb.) with leafhoppers, aphids, whiteflies, thrips and two species of mites found on pigeon pea failed. Seven of 30 test plants became diseased after leaves from diseased plants infested with mites (*E. cajani*) were placed on them. In other tests, 5 to 20 of the mites were transferred from leaves of diseased plants to each test plant. After 4 days, the plants were sprayed regularly with 0.1% Ekatox to kill the mites. In 3 to 5 weeks, 31 of 136 test plants developed symptoms of the disease. Similar transfers of mites from healthy plants did not cause the symptoms.

From the evidence presented, there appears to be little doubt that *E. cajani* is associated with the pigeon pea sterility disease, but the evidence is not convincing that the disease symptoms are caused by an infectious agent carried by the mites. The procedures used for graft transmission could have transferred *E. cajani* from diseased to test plants. The use of 0.1% Ekatox sprays may or may not have eliminated all mites. Survivors and their progeny could have induced the symptoms attributed to a virus. There is no certainty that the mites on "healthy" source plants were identical with those on "diseased" plants. Also, the numbers of mites on each could have been significant. Seth (1965) stated: "The population of the eriophyid mite was more on diseased plants as compared to healthy ones." There was no report that non-infectious mites from any source could acquire the infectious agent by feeding on diseased plants that were not already infested with mites. Efforts to colonize the mites on pigeon pea or other plants under controlled conditions were unsuccessful hence there were no colonies of mites originating from eggs transferred from diseased plants to hatch on healthy plants. If such mites did not induce the disease until after they had fed on diseased plants, there would be a strong indication that the disease was caused by an infectious agent.

Both the disease and *E. cajani* depend on pigeon pea for perpetuation. Although pigeon pea is grown as an annual it sometimes grows as a perennial along hedges and may carry mites that can infest new crops nearby. These and other self-sown plants should be eliminated before sowing a new crop of pigeon pea in the vicinity.

Although most cultivars of pigeon pea are affected by the disease (Subramanian *et al.*, 1973), two resistant and 16 tolerant lines were found among 234 that were tested under field conditions (Singh *et al.*, 1975).

14.5.3 *Phyllocoptes fructiphilus* and Rose Rosette

A disease which causes rosetting, misshapen leaves, bud breaking and sometimes dwarfing of young plants, virtually eliminates flowering. It affects many cultivars of different species of roses. It occurs principally where roses are grown near wild roses in some rural and mountainous areas in western North America. It was reported to be caused by a virus transmissible by grafting and by the wild rose bud and fruit mite (*Phyllocoptes fructiphilus* K.) in Nebraska (Allington *et al.*, 1968).

The mites live and breed in protected areas between leaf petiole bases and lateral buds, particularly toward the end of the shoot. They overwinter in any protected place where green surfaces are appressed to each other. They may also move into open ends of fruit and live among the seeds. It is believed by some that the feeding of the mites causes many if not all of the symptoms of rose rosette that have been attributed to a virus (Jeppson *et al.*, 1975).

Evidence that rose rosette is caused by a virus was based on successful graft transmission from diseased to healthy plants (Allington *et al.*, 1968).

Tests for transmission of the cause of rose rosette by the mite *P. fructiphilus* were done by transferring the mites individually by a hair on a handle from cuttings collected from diseased wild or cultivated roses in the field to healthy *Rosa eglanteria* or *R. multiflora* in a greenhouse. In a series of eight trials at different dates, 10 to 150 mites were transferred to each healthy test plant. Some of the test plants developed symptoms in 32 to 146 days in 6 of the eight trials. A total of 11 of 38 plants developed symptoms. The highest infection rate occurred in a trial in which 10 mites were transferred from diseased *R. eglanteria* to each of 11 healthy *R. multiflora*. In 32 to 45 days, 5 of them developed symptoms. None of a number of non-infested *R. eglanteria* and *R. multiflora* plants kept in the same greenhouse developed symptoms.

The above results show that mites induced rose rosette symptoms but they did not prove that the symptoms were caused by an infectious agent. There was no indication that graft transmission had been achieved without mites being present. No experiments were reported to determine if disease symptoms continued to develop on plants after all mites were eliminated with a miticide. There also were no comparisons of the effects of "non-viruliferous" mites. No colonies of mites were developed from eggs hatched on healthy plants to determine if the ability of the mites to induce rose rosette was carried through the eggs.

14.5.4 *Calacarus citrifolii* and Citrus Concentric Ring Blotch

Citrus concentric ring blotch in South Africa was reported to be caused by the feeding of the grey citrus rust mite, *Calacarus citrifoli* K. (Dipenaar, 1958). Later Rossouw and Smith (1963) found that mites from diseased citrus caused the disease on test trees, but mites from healthy ones did not. They suggested that the disease was caused by a virus carried by the mites from diseased trees, but they did not present evidence that the same species of mites was transferred from both sources. It appears likely that the symptoms were caused by the rust mite and not by an infectious agent.

14.6 DOUBTFUL OR ERRONEOUS REPORTS OF MITE TRANSMISSION

14.6.1 *Aculus fockeui* and Prunus Latent Virus

A virus latent in plum was reported to be transmitted by *Aculus (Vasates) fockeui* Nal. from varieties of plum with and without symptoms of virus disease to *Chenopodium faetidum* Schrad. (Proeseler and Kegler, 1966). After 7 to 10 days, the test plants developed ochre-colored local lesions surrounded by necrotic rings. Similar symptoms developed by mechanical inoculation of *C. faetidum* with sap from affected leaves. *C. quinoa* developed gray necrotic local lesions. Neither host developed systemic symptoms. Rigid particles 750 nm long were found in the lesions. It was assumed that the symptoms on the *Chenopodium* plants were caused by a virus latent in plum and transmitted by *A. fockeui* but not by *Diptacus gigantorhynchus* Nal.

Subsequently, it was reported that the virus, which was transmitted from 13 plum varieties and 8 *Prunus* species, could be transmitted by a single mite, but increased numbers of mites increased the numbers of local lesions. The time required for infection was determined by transferring mites from plum to test plants, then killing the mites with metasystox at a specific time. The minimum infection period was a few minutes. For most local lesions, a feeding period of 24 hr was necessary. There were no differences in transmission by larval stages and adults, and the virus persisted in the mites at least 4 days (Proeseler, 1968).

There was no indication that controlled experiments were done to show the effects of non-viruliferous *A. fockeui* on *Chenopodium* test plants, or to demonstrate infection of plum seedlings with virus from local lesions on the test plants.

In the 12 years since the above association was reported, there has been no confirmation or clarification by other workers in Europe where similar host-virus-vector situations would be expected to occur. The most similar situation cited earlier in this review is the report that a virus similar to PVY was the mite-transmitted agent that causes currant reversion (Jacob, 1976b). Both situations involve transferring mites adapted to specific members of the Rosaceae (*Prunus* and *Ribes*) to herbaceous plants in the Chenopodiaceae or Solanaceae, and thereby transmitting viruses between these greatly differing hosts. This seems highly unlikely. Both of these reports need careful confirmation or clarification. Could there be other explanations for the presence of filamentous particles in the herbaceous test plants?

14.6.2 *Colomerus vitis* and Vine Panachure

Vine panachure was reported by Ochs (1958) to be caused by a sap transmissible virus also transmitted by several species of insects, nematodes and by the eriophyid mite *Colomerus vitis*. It appears that there was no experimental evidence for any of these claims (Niemeyer and Bode, 1959). Although the claim was repeated (Ochs, 1960), no details of experimental evidence were presented.

14.6.3 *Eriophyes mangiferae* and Mango Malformation

Mango malformation causes proliferation of vegetative shoots and transformation of the inflorescence of mango leaves into compact masses of sterile flowers in India. Nariani and Seth (1963) reported that after mite-infested bud scales from diseased or apparently healthy trees were pinned on mango seedlings, the latter developed symptoms of malformation. Ginai (1965) suggested that malformation was caused by a virus spread by the mite *Eriophyes mangiferae* (Sayed). Summanwar and Raychaudhuri (1968) reported that a fungus, *Fusarium moniliforme*, caused the malformation disease. The mites carry the fungus on their bodies and the injury caused by the mites provided a means for the entry of the fungus into the host. Spores of the fungus were observed on the mites and the fungus was isolated from mites, but there was no evidence presented to show that they actually caused infection. More recently, Varma *et al.* (1973) reported that *F. moniliforme* is invariably isolated from malformed parts of all cultivars of mango and that the means of natural spread of the fungus remains obscure.

14.7 DISEASE AGENTS SUSPECTED TO BE TRANSMITTED BY ERIOPHYID MITES

14.7.1 Cadang-Cadang of Coconut

Cadang-cadang, a destructive degenerative disease of coconut palms in the Philippines, has been spreading slowly but relentlessly for many years (Price, 1958). Because of the pattern of spread, the cause was suspected to be a mite-transmitted virus (Briones and Sill, 1963), but vector tests with four species of mites from coconut were negative. Price and Begornia (1972) reported that weed control for 5 years had no effect on rate of spread, hence they concluded that spread must occur from tree to tree rather than from weeds.

Although virus-like particles, rickettsia-like organisms and viroids have been suggested as causes of the disease, the cause and its means of transmission are still unknown.

14.7.2 Little Cherry Disease in British Columbia

The little cherry disease (LCD) was discovered in sweet cherry (*Prunus avium* L.) in an orchard surrounded by natural forest in the Kootenay Valley, B.C., Canada, in 1933 (Foster and Lott, 1947; Welsh and Cheney, 1976). The fruit of infected black sweet cherries, such as Lambert, was only 1/2 to 2/3 normal size at picking time. It remained dull red in color after normal fruit had become dark in color. The small fruit was triangular and pointed in shape and flat in flavor. Some cultivars such as Sam, Van, Deacon and Star displayed varying degrees of reddening of the foliage in late summer and early fall.

The disease soon appeared in other orchards near the first one. Once present, it spread throughout an orchard in 2 or 3 years. By 1950, most cherry trees in nearly all fruit growing districts of the Kootenay Valley were infected. It was not found in the Okanagan Valley, 160 kilometers to the west, until 1969 when it was recognized in 49 trees in one orchard. Intensive annual surveys were instituted, and removal of diseased and suspect trees was enforced in an attempt to eradicate the disease. A few additional diseased trees have been found each year since, except in 1972. There were 314 diseased trees found and removed in 39 sites in 1977.

The cause of LCD is readily transmitted by grafting scionwood, buds or small pieces of bark from diseased onto healthy trees. Sam, a sweet cherry cultivar, which develops red leaves in late summer when infected, is used as an indicator host. Sour cherry (*P. cerasus* L.), flowering cherry (*P. serrulata* Lindl.) and several other cherry species are symptomless carriers. A similar agent has been graft-transmitted from ornamental flowering cherry (*P. serrulata*) and the native bitter cherry (*P. emarginata*) (Wilkes and Welsh, 1961).

The rapid spread of LCD suggests transmission by an agent that can be abundant under suitable conditions, and efficient as a vector. Many insects have been tested.

Three species of leafhoppers were associated with transmissions of LCD by Wilde (1960) but subsequent exhaustive tests have failed to confirm this report. Tests with other insects including aphids, thrips, and scale insects have failed to indicate a vector. Mealy bugs are still suspects.

Red leaf symptoms indicative of LCD developed on Sam cherry trees after exposure near diseased sweet cherry and suspected diseased bitter cherry in certain locations in the Kootenay Valley in 1976 and 1977, and after infestation with eriophyid mites on leaves from diseased trees. Eriophyid mites present included the plum rust mite (*Aculus fockeui*), the big-beaked plum mite (*Diptacus gigantorhynchus* (Nal.) and *Phyllocoptes abaenus* K. but sometimes other unidentified mites were observed. All species of eriophyids present on naturally diseased cherries must be tested separately to determine if any of them are vectors of the little cherry agent.

14.7.3 Hordeum Mosaic Virus

Hordeum mosaic virus (HMV) was isolated from barley (*Hordeum vulgare* L.), and the wild perennial grasses *H. jubatun* L., *Agropyron trachycaulum* (Link) and *Agrohordeum (Elymus) macounii* (Vasey) Lepage from Alberta, Canada. This virus resembles wheat streak mosaic virus (WSMV), Agropyron mosaic virus (AMV) and ryegrass mosaic virus (RVM) in particle morphology, sap transmissibility and some host relationships (Slykhuis and Bell, 1966; Slykhuis and Paliwal, 1972). The general similarities between HMV and the other above viruses suggest that it too could be mite-transmitted. Attempts to transmit it with the vector of WSMV (*Eriophyes tulipae*), and the vector of AMV and RMV (*Abacarus hystrix*) did not succeed. No transmission tests have been done with mites found on naturally diseased grasses in Alberta where the virus appears to be indigenous.

14.8 CONCLUSIONS

Wheat streak mosaic virus (WSMV), ryegrass mosaic virus (RMV) and agropyron mosaic virus (AMV) are the only plant disease agents known to be viruses transmitted by eriophyid mites. WSMV is efficiently transmitted by *Eriophyes tulipae*. It has some characteristics of stylet-borne viruses and some characteristics of circulative viruses in relation to its mite vector. RMV is transmitted less efficiently, and AMV with very low efficiency by *Abacarus hystrix*. Large numbers appear to be required to infect a plant. Onion mosaic is suspected to be caused by a virus with flexuous particles transmitted by *E. tulipae*.

Although infectious agents transmitted by eriophyid mites have been shown to be the cause of black currant reversion, wheat spot mosaic, fig mosaic, peach mosaic and cherry mottle leaf, the natures of the infectious agents are still in doubt. The cause of black currant reversion has been described as a mycoplasma, a bacterium and as potato virus Y, but none of these has been confirmed. Vesicular bodies have been found in wheat and maize plants with wheat spot mosaic, but these bodies have not been proved to represent the causal agent.

Pigeon pea sterility, rose rosette, citrus concentric ring blotch, and sugarcane chlorotic streak developed on healthy plants after specific eriophyid mites from diseased plants were transferred to them, but there is no firm evidence that the mites carried an infectious agent. Some of these diseases developed on test plants after scions from diseased plants were grafted on them, but there was no assurance that mites were not present.

There is no confirmation that eriophyid mites were involved as vectors or causes of "vine panachure" or "prune latent virus" reported in Germany.

Mango malformation, once reported to be caused by a mite-transmitted virus, is now recognized to be caused by a fungus. There has been no confirmation that mites spread the fungus.

There is a possibility that eriophyid mites may carry an agent causing cadang-cadang disease of coconuts in the Philippines, but after 20 years there is no evidence to confirm the suspicion.

Unspecified eriophyid mites are suspected to be carriers of an infectious agent that causes the little cherry disease which threatens the sweet cherry industry in British Columbia, Canada.

Because hordeum mosaic virus resembles the three filamentous viruses of grasses known to be transmitted by eriophyid mites, it is suspected to have an eriophyid mite vector.

14.9 ACKNOWLEDGEMENT

The author is indebted to C. V. G. Morgan for assistance in preparing the section on "Characteristics of the Eriophyids," and for advice on taxonomy.

14.10 REFERENCES

A'Brook, J. (1975). Epidemiology and effects of grass viruses in relation to their insect vectors. *Ann. Appl. Biol.* 81: 261-264.

Allington, W. B., Staples, R., and Viehmeyer, G. (1968). Transmission of rose rosette virus by the eriophyid mite, *Phyllocoptes fructiphilus*. *J. Econ. Entomol.* 61: 1137-1140.

Amos, J., and Hatton, R. G. (1928). Reversion in black currants: its incidence and spread in the field in relation to possible control measures. *J. Pomol. Hort. Sci.* 6: 167-183.

Amos, J., Hatton, R. G., Knight, R. C., and Massee, A. M. (1927). Experiments in the transmission of reversion in black currants. *Ann. Rept. East Malling Res. Sta. Kent, Suppl. II,* 126-150.

Atkinson, R. E. (1949). Western wheat mosaic in Colorado and its transmission by the grain aphid, *Toxoptera gramineum*. *Phytopathology* 39: 2.

Atkinson, T. G., and Slykhuis, J. T. (1963). Relation of spring drought, summer rains and high fall temperatures to the wheat streak mosaic epiphytotic in Southern Alberta, 1963. *Can. Plant Disease Survey* 33: 154-159.

Baska, W. (1969). Occurrence of cherry mottle leaf virus in Poland. *Bull. Acad. Pol. Sci. Sér. Sci. Biol.* 17: 613-615.

Blodgett, E. C., and Gömec, B. (1967). Fig mosaic. *Plant Disease Reptr.* 51: 893-896.

Bock, K. R. (1974). Maize streak virus. *Commonw. Mycol. Inst./A.A.B. Descr. Plant Viruses No. 133,* 4 pp.

Bodine, E. W. (1934). Occurrence of peach mosaic in Colorado. *Plant Disease Reptr.* 18: 123.

Bodine, E. W., and Durrell, L. W. (1941). Host range of peach mosaic in western Colorado. *Phytopathology* 31: 322-333.

Bradfute, O. E. Whitmoyer, R. E., and Nault, L. R. (1970). Ultrastructure of plant leaf tissue infected with mite-borne viral-like pathogens. *Proc. Electron Microsc. Soc. Am.* 28: 178-179.

Brakke, M. K. (1971). Wheat streak mosaic virus. *Commonw. Mycol. Inst./A.A.B. Descr. Plant Viruses No. 48.*

Bremer, K. (1973). Comparison of four virus isolates of wheat streak mosaic from Turkey. *Phytopathologia Mediterranea* 12: 67-71.

Briones, M. L., and Sill, W. H., Jr. (1963). Habitat, gross morphology and geographical distribution of four new species of eriophyid mites from Coconuts in the Philippines. *F.A.O. Plant Protect. Bull.* 11: 25-30.

Burnett, H. C. (1962). Additional hosts of the fig mosaic virus. *Plant Disease Reptr.* 46: 693.

Catherall, P. L., and Chamberlain, Judith A. (1975). Occurrence of agropyron mosaic virus in Britain. *Plant Pathology* 24: 155-157.

Cheney, P. W., and Parish, C. L. (1976). Cherry mottle leaf. *In* "Virus Diseases and Noninfectious Disorders of Stone Fruits in North America." *U. S. Dept. Agr., Agr. Handb. No. 437,* 216-218.

Collingwood, C. A., and Dicker, G. H. (1960). A comparison of various chemicals for the control of the black currant gall mite. *Plant Pathology* 9: 39-48.

Condit, I. J., and Horne, W. T. (1933). A mosaic of the fig in California. *Phytopathology* 23: 887-889.

Connin, R. V. (1953). Studies to determine the insect vectors of the wheat streak mosaic virus. *U. S. Dept. Agr. Spec. Rept. W-13,* 10 pp.

Connin, R. V. (1956). The host range of the wheat curl mite, vector of wheat streak mosaic. *J. Econ. Entomol.* 49: 1-4.

Connin, R. V., and Staples, R. (1957). Role of various insects and mites in the transmission of wheat streak mosaic virus. *J. Econ. Entomol.* 50: 168-170.

Delgado-Sanchez, S., and Grogan, R. G. (1970). Potato virus Y. *Commonw. Mycol. Inst./ A.A.B. Descr. Plant Viruses No. 37,* 4 pp.

Del Rosario, Maria S., and Sill, W. H., Jr. (1965). Physiological strains of *Aceria tulipae* and their relationships to the transmission of wheat streak mosaic virus. *Phytopathology* 55: 1165-1175.

Dipenaar, B. J. (1958). Concentric ring blotch of citrus–its cause and control. *S. Afr. J. Agric. Sci.* I: 83-106.

Fellows, H. (1949). A survey of the wheat mosaic disease in western Kansas. *Plant Disease Reptr.* 33: 356-358.

Flock. R. A., and Wallace, J. M. (1955). Transmission of fig mosaic by the eriophyid mite *Aceria ficus. Phytopathology* 45: 52-54.

Foster, W. R., and Lott, T. B. (1947). "Little Cherry," a virus disease. *Scientific Agric.* 27: 1-6.

Gates, L. F. (1970). The potential of corn and wheat to perpetuate wheat streak mosaic in southern Ontario. *Can. Plant Disease Survey* 50: 59-62.

Gibson, R. W. (1974). Studies on the feeding behavior of the eriophyid mite *Abacarus hystrix,* a vector of grass viruses. *Ann. Appl. Biol.* 78: 213-217.

Gibson, R. W. (1976). Effects of cutting height on the abundance of the eriophyid mite *Abacarus hystrix* (Nalepa) and the incidence of ryegrass mosaic virus in ryegrass. *Plant Pathology* 25: 152-156.

Gibson, R. W., and Heard, A. J. (1976). Selection of perennial and Italian ryegrass plants resistant to ryegrass mosaic virus. *Ann. Appl. Biol.* 84: 429-432.

Ginai, M. A. (1965). Malformation of mango inflorescence. *West Pakistan J. Agr. Res.* 3: 248-251.

Heard, A. J., and Roberts, E. T. (1975). Disorders of temporary ryegrass swards in south-east England. *Ann. Appl. Biol.* 81: 240-243.

Jacob, H. (1976a). Untersuchungen zur Übertragung des virösen Atavismus der Schwarzen Johannisbeere (*Ribes nigrum* L.) durch die Gallmilbe *Cecidophyopsis ribis* Westw. *Z. Pfl. Krankh. Pfl. Schutz*. 83: 448-458.

Jacob, H. (1976b). Investigations on symptomatology, transmission, etiology, and host specificity of black currant reversion virus. *Acta Horticulturae* 66: 99-104.

Jeppson, L. R., Keifer, H. H., and Baker, E. W. (1975). Mites Injurious to Economic Plants. Univ. of Calif. Press, Berkeley, Los Angeles, London. 614 pp.

Jones, L. S., and Wilson, N. S. (1958). Transmission experiments with *Eriophyes insidiosus*. Unpublished Project Rept., Fruit Insects Section, Entomol. Res. Div., U. S. Dept. Agr. 1 p.

Jones, L. S., Wilson, N. S., and Anthon, E. W. (1969). Transmission of cherry mottle leaf virus by an eriophyid mite. Entomol. Res. Div., U. S. Dept. Agr., Riverside, Calif. (Unpublished report).

Jones, L. S., Wilson, N. S., Burr, W., and Barnes, M. M. (1970). Restriction of peach mosaic virus spread through control of the vector mite *Eriophyes insidiosus*. *J. Econ. Entomol.* 63: 1551-1552.

Kantack, E. J., and Knutson, H. (1958). Chemical control studies on the wheat curl mite. *J. Econ. Entomol.* 51: 68-72.

Keifer, H. H. (1969). Eriophyid studies. C-3. *U.S. Dept. Agr. Spec. Publ.* 20 pp.

Keifer, H. H., and Wilson, N. S. (1956). A new species of eriophyid mite responsible for the vection of peach mosaic virus. *Bull. Calif. Dept. Agr.* 44: 145-146.

King, C. L., and Sill, W. H., Jr. (1959). 1959 wheat streak mosaic epiphytotic in Kansas. *Plant Disease Reptr.* 43: 1256-1257.

Larson, R. I., and Atkinson, T. G. (1973). Wheat agropyron chromosome substitution lines as sources of resistance to wheat streak mosaic virus and its vector *Aceria tulipae*. *Proc. 4th. Internat. Wheat Genetics Symposium, Colombia, Mo.* pp. 173-178.

Lees, A. H. (1925). Reversion disease of black currants: means of infection. *Ann. Appl. Biol.* 12: 199-210.

Markov, M. P., Kaitazova, P., and Stefanov, I. (1975). Identification of wheat streak mosaic virus in Bulgaria. *"Rasteniev" dni Nauki* 12: 130-137.

Massee, A. M. (1928). The life-history of the black currant gall mite, *Eriophyes ribis* (Westw.) Nal. *Bull. Entomol. Res.* 18: 297-309.

Massee, A. M. (1952). Transmission of reversion of black currants. *Ann. Rept. East Malling Res. Sta. Kent.*, 1951, 162-165.

McKinney, H. H. (1937). Mosaic diseases of wheat and related cereals. *U.S. Dept. Agr. Circ. 442*, 23 pp.

Mulligan, T. E. (1960). The transmission by mites, host range and properties of ryegrass mosaic virus. *Ann. Appl. Biol.* 48: 575-579.

Nagaich, B. B., and Vashisth, K. S. (1962). Mosaic of *Ficus* spp. in India. *Curr. Sci.* 31: 166-167.

Nariani, T. K., and Seth, M. C. (1963). Role of eriophyid mites in causing malformation disease of mango. *Indian Phytopathol.* 15: 231-234.

Nault, L. R., and Styr, W. E. (1970). Transmission of an eriophyid-borne wheat pathogen by *Aceria tulipae*. *Phytopathology* 60: 1616-1618.

Niemeyer, L., and Bode, O. (1959). Über den Virusnachweis bei Reben. *Z. Pflanzenkrankh. Pflanzenschutz* 66: 640-644.

Ochs, G. (1958). Untersuchungen über die Verbreitung der Rebenviren durch Vektoren. *Naturwissenschaften* 45: 193.

Ochs, G. (1960). Übertragungsversuche von drei Rebenviren durch Milben und Insekten. *Z. Angew. Zool.* 47: 485-491.

Oldfield, G. (1970). Mite transmission of plant viruses. *Ann. Rev. Entomol.* 15: 343-380.

Oldfield, G., and Wilson, N. S. (1970). Establishing colonies of *Eriophyes insidiosus*, the vector of the peach mosaic virus. *J. Econ. Entomol.* 63: 1006-1007.

Orlob, G. (1966). Feeding and transmission characteristics of *Aceria tulipae* Keifer as vector of wheat streak mosaic virus. *Phytopathol. Z.* 55: 218-238.

Özalp, M. O., and Heper, E. (1972). Studies of fig mosaic virus and trees resistant to it in the Aegean region. *Bitki Koruma Bülteni* 12: 228-262.

Paliwal, Y. C. (1972). Brome mosaic virus infection in the wheat curl mite *Aceria tulipae*, a nonvector of the virus. *J. Invertbr. Pathol.* 20: 288-302.

Paliwal, Y. C., and Slykhuis, J. T. (1967). Localization of wheat streak mosaic virus in the alimentary canal of its vector *Aceria tulipae* Keifer. *Virology* 32: 344-353.

Pine, T. S. (1976). Peach mosaic. *In* "Virus Diseases and Noninfectious Disorders of Stone Fruits in North America." *U.S. Dept. Agr., Agr. Handb. No. 437*, 61-70.

Pine, T. S. and Šutić, D. (1968). Šarka (Plum Pox) disease in reference to diseases of *Prunus* in the United States. *Plant Disease Reptr.* 52: 250-252.

Pop, I. (1962). Die Strichelvirose des Weizens in der Rumänischen Volksrepublik. *Phytopath. Z.* 43: 325-336.

Price, W. C. (1958). Report to the government of the Philippines on the yellow mottle disease (Cadang-Cadang) of Coconut. *F.A.O. Rept. 850*, 50 pp.

Price, W. C., and Bigornia, A. E. (1972). Evidence of spread of Cadang-Cadang disease of Coconut from tree to tree. *F.A.O. Plant Protect. Bull.* 20: 133-135.

Proeseler, G. (1968). Transmission of a latent *Prunus* virus by the gall mite *Vasates fockeui* Nal. *Tagungsberichte* 97: 297-300.

Proeseler, G. (1972). Beziehungen zwischen Virus, Vektor und Wirtspflanze am Beispiel des Feigen-Mosaik-Virus und *Aceria ficus* Cotte (Eriophyoidea). *Acta Phytopathologica Acad. Scientatum Hungaricae* 7:179-186.

Proeseler, G. (1973). Die Gallmilbe *Cecidophyopsis ribis* (Westw.) als Schädling der Johannisbeeren. *Arch. Phytopathol. u. Pflanzenschutz* 9: 383-394.

Proeseler, G., and Kegler, H. (1966). Übertragung eines latenten Virus von Pflaume durch Gallmilben (Eriophyidae). *Monatsberichte der Deutschen Akademie der Wissenschaften zu Berlin* 8: 472-476.

Protsenko, A. E., Fedotina, V. L., and Sergucheva, N. A. (1975). Mycoplasma-like organisms in the tissues of yellows infected plants. *Abstr. In Rev. Plant Pathol.* 56: 1057.

Razvyazkina, G. M. (1971). Das Zwiebelmosaikvirus und seine Verbreitung im Freiland. *Tagungsbericht Deutsche Akademie der Landwirtschaftswissenschaften zu Berlin No. 115*, 69-76.

Razvyazkina, G. M., Kapkova, E.A., and Belyanchikova, Y. V. (1963). Wheat striate mosaic virus. *Zashch. Rast. Moskva.* 8: 54-55.

Reeves, E. L. (1941). Mottle leaf, a virus disease of cherries. *J. Agr. Res.* 9: 555-572.

Rossouw, D. J., and Smith, A. J. (1963). The relation of *Calacarus citrifolii* Keifer to concentric ring blotch of citrus. *S. Afr. Citrus J.* 354: 7-9.

Seth, M. L. (1962). Transmission of pigeon pea sterility by an eriophyid mite. *Indian Phytopathol.* 15: 225-227.

Seth, M. L. (1965). Further observations and studies on pigeon pea sterility. *Indian Phytopathol.* 18: 317-319.

Sill, W. H., Jr., and Connin, R. V. (1953). Summary of the known host range of the wheat streak mosaic virus. *Trans. Kansas Acad. Sci.* 56: 411-417.

Silvere, A. P., and Romeikis, Mare-Anne. (1971). Isolation of endophytic bacilli from black currant tissues with reversion disease *Eesti NSV Tead. Akad. Toim. biol. Seer.* 20: 365-368.

Silvere, A. P., and Tiits, A. (1969). Mycoplasm-like microorganisms–a new problem in phytopathology. *Biologia* 1969, 228-230.

Singh, K., Dahiya, B. S., and Chohan, J. S. (1975). Evaluation of arhor (*Cajanus cajan*) germplasm lines against the sterility disease in the Punjab. *J. Res. Punjab Agr. Univ.* 12: 327-328.

Sinha, R. C., and Paliwal, Y. C. (1977). Detection of wheat streak mosaic virus antigens in vector mites with fluorescent antibodies. *Phytopathology* 67: 570-572.

Sithanantham, S., Muthuswamy, S., and Bhaskaran, T. L. (1972). A note on the role of the eriophyid mite *Aceria sacchari* Channabasavanna (Eriophyoidea: acarina) in the transmission of streak virus disease of sugarcane. *Sci. and Culture* 38: 248-250.

Slykhuis, J. T. (1952). Virus diseases of cereal crops in South Dakota. *S. Dak. Agr. Expt. Sta. Tech. Bull.* 11, 29 pp.

Slykhuis, J. T. (1955). *Aceria tulipae* Keifer (Acarina:Eriophyidae) in relation to the spread of wheat streak mosaic. *Phytopathology* 45: 116-128.

Slykhuis, J. T. (1956). Wheat spot mosaic caused by a mite-transmitted virus associated with wheat streak mosaic. *Phytopathology* 46: 682-687.

Slykhuis, J. T. (1958). A survey of virus diseases of grasses in northern Europe. *F.A.O. Plant Protect. Bull.* 6: 129-134.

Slykhuis, J. T. (1962). Agropyron mosaic as a disease of wheat in Ontario. *Can. J. Botany* 40: 1439-1447.

Slykhuis, J. T. (1969). Transmission of agropyron mosaic virus by the eriophyid mite *Abacarus hystrix*. *Phytopathology* 59: 29-32.

Slykhuis, J. T. (1973). Agropyron mosaic virus. *Commonw. Mycol. Inst./A.A.B. Descr. Plant Viruses No. 118,* 4 pp.

Slykhuis, J. T., Andrews, J. E., and Pittman, V. J. (1957). Relation of date of seeding winter wheat in southern Alberta to losses from wheat streak mosaic root rot, and rust. *Can. J. Plant Sci.* 37: 113-127.

Slykhuis, J. T., and Bell, W. (1966). Differentiation of agropyron mosaic, wheat streak mosaic, and a hitherto unrecognized hordeum mosaic virus in Canada. *Can. J. Botany* 44: 1191-1208.

Slykhuis, J. T., and Paliwal, Y. C. (1972). Ryegrass mosaic virus. *Commonw. Mycol. Inst./ A.A.B. Descr. Plant Viruses No. 86,* 4 pp.

Smith, B. D. (1962a). Experiments in the transfer of the black currant gall mite (*Phytoptus ribis* Nal.) and of reversion. *Ann. Rept. Agr. Hort. Res. Sta.* Long Ashton Bristol, 1961, 170-172.

Smith, B. D. (1962b). The behavior and control of the black currant gall mite *Phytoptus ribis* (Nal.). *Ann. Appl. Biol.* 50: 327-334.

Smith, B. D. (1963). A field study of the spread of the black currant gall mite (*Phytoptus ribis* Nal.) and of the virus disease reversion. *Ann. Rept. Agr. Hort. Res. Sta.* Long Ashton, Bristol, 1962, 124-129.

Smith, B. D. (1975). Effects of soil applied benzimidazole fungicide on the behavior of *Cecidophyopsis ribis* and on the transmission of reversion virus. *Ann. Appl. Biol.* 80: 205-209.

Staples, R., and Allington, W. B. (1956). Streak mosaic of wheat in Nebraska and its control. *Univ. Neb. Agr. Expt. Sta. Res. Bull. 178,* 41 pp.

Stein-Margolina, V. A. (1973). Bromegrass mosaic virus in the mite *Aceria tritici* (an electron microscope investigation). *IZV. Akad. Nauk. SSR, Ser. Biol.* 1973, 189-195.

Stein-Margolina, V. A., Cherni, N. E., and Razvyazkina, G. M. (1969). Phytopathogenic viruses in cells of the plant and mite vector. *IZV. Akad. Nauk. SSR, Ser. Biol.* 1969, 62-68.

Subramanian, K. S., Samuel, G. S., and Janarthanan, R. (1973). Studies on the varietal resistance of pigeon pea (*Cajanus cajan* L.) to sterility mosaic disease. *Madras Agr. J.* 60: 38-40.

Summanwar, A. S., and Raychaudhuri, S. P. (1958). The role of eriophyid mite (*Aceria mangiferae*) in the causation of mango malformation. *Indian Phytopathol.* 21: 463-464.

Takahashi, Y., and Orlob, G. (1969). Distribution of wheat streak mosaic virus-like particles in *Aceria tulipae*. *Virology* 38: 230-240.

Thresh, J. M. (1963). Abnormal black currant foliage caused by the gall mite *Phytoptus ribis* Nal. *Ann. Rept. East Malling Res. Sta. Kent,* 1962, 99-100.

Thresh, J. M. (1964a). Association between black currant reversion virus and its gall mite vector. *Nature* 202: 1085-1087.

Thresh, J. M. (1964b). Warm water treatments to eliminate the gall mite *Phytoptus ribis* Nal. from black currant cuttings. *Jubilee Ann. Rept. East Malling Res. Sta. Kent,* 1963, 131-132.

Thresh, J. M. (1964c). Black currant reversion disease. *Jubilee Ann. Rept. East Malling Res.*

Sta. Kent, 1963, 184-189.

Thresh, J. M. (1968). Field experiments on the chemical control of black currant reversion virus and its gall-mite vector (*Phytoptus ribis* Nal.). *Ann. Appl. Biol.* 62: 255-264.

Tiits, A. (1964). Some observations on black currant reversion disease transmitted by grafting. *Eesti NSV Tead. Akad. Toim. Ser. Biol.* 13: 267-271.

Tosic, M. (1971). Virus diseases of wheat in Serbia. I. Isolation and determination of the wheat streak mosaic virus and brome mosaic virus. *Phytopathol. Z.* 70: 145-162.

Varma, A., Raychaudhuri, S. P., Lele, V. C., and Ram, A. (1973). Preliminary investigations on epidemiology and control of mango malformation. *Proc. Indian Nat. Sci. Acad. B.* 37: 291-300.

Vashisth, K. S., and Nagaich, B. B. (1965). *Morus indica*–an additional host of fig mosaic. *Indian Phytopathol.* 18: 315.

Welsh, M. F., and Cheney, P. W. (1976). Little cherry, *In* "Virus Diseases and Noninfectious Disorders of Stone Fruits in North America." *U. S. Dept. Agr., Agr. Handb. No. 437,* 231-237.

Wilde, W. H. A. (1960). Insect transmission of the virus causing little cherry disease. *Can. J. Plant Sci.* 40: 707-712.

Wilks, J. M., and Welsh, M. F. (1961). Host range studies of the little cherry disease virus. *Can. J. Plant Sci.* 41: 544-548.

Wilkins, P. W., and Kides, D. H. (1976). Tolerance to ryegrass mosaic virus, its assessment and effect on yield. *Ann. Appl. Biol.* 83: 399-405.

Williams, L. E., Gordon, D. T., Nault, L. R., Alexander, L. J., Bradfute, O. E., and Findley, W. R. (1967). A virus of corn and small grains in Ohio and its relation to wheat streak mosaic virus. *Plant Disease Reptr.* 51: 207-211.

Wilson, N. S., Jones, L. S., and Cochrane, L. C. (1955). An eriophyid mite vector of the peach mosaic virus. *Plant Disease Reptr.* 39: 889-892.

Wilson, N. S., and Oldfield, G. N. (1966). New species of eriophyid mites from western North America, with a discussion of eriophyid mites on *Populus. Ann. Entomol. Soc. Am.* 59: 585-599.

Chapter 15

FATE OF PLANT VIRUSES IN MITE VECTORS AND NONVECTORS*

Y. C. Paliwal
Chemistry and Biology Research Institute,
Research Branch, Agriculture Canada,
Ottawa, Ontario, Canada

15.1 INTRODUCTION 357
15.2 GENERAL CHARACTERISTICS AND ANATOMICAL FEATURES OF ERIOPHYID MITES 358
15.3 BEHAVIOR OF VIRUSES IN THEIR ERIOPHYID VECTORS 359
15.3.1 Wheat streak mosaic virus–*Eriophyes tulipae* system 359
15.3.2 Less understood virus-vector systems 360
15.4 BEHAVIOR OF VIRUSES IN NONVECTOR ERIOPHYIDS 362
15.4.1 Wheat streak mosaic virus in *Abacarus hystrix* 362
15.4.2 Barley stripe mosaic virus in *E. tulipae* 363
15.4.3 Bromegrass mosaic virus in *E. tulipae* 365
15.5 VIRUSES IN TETRANYCHID MITES 365
15.6 MECHANISMS OF VIRUS TRANSMISSION 370
15.7 REFERENCES 372

15.1 INTRODUCTION

Plant parasitic mites have been shown to be vectors of causal agents of at least 9 virus-like plant diseases but adequate evidence for the viral nature of the agents is available only in 4 cases. All confirmed reports about the transmission of these diseases involve the 4 legged, eriophyid mites. About 1000 species of eriophyid mites parasitize cultivated crop plants and trees but only a few serve as vectors of viruses or other plant pathogens. Mite-plant virus interactions with reference to mechanisms of transmission and vector specificity have not been extensively studied. Progress of such studies is slow because the microscopic size of eriophyids and their poorly understood anatomy and physiology, which appear to be quite different from insects, pose specific problems in working with eriophyids. The objective of this chapter is to review the available knowledge on behavior of plant

*Contribution No. 1100, Chemistry and Biology Research Institute.

ISBN 0-12-326450-2

viruses in the body of their mite vectors and nonvectors. Also, the mechanisms of virus transmission by the vectors will be discussed and the possible reasons for certain mites being nonvectors analyzed.

15.2 GENERAL CHARACTERISTICS AND ANATOMICAL FEATURES OF ERIOPHYIDS

Knowledge of morphology, functioning of mouthparts and of the internal anatomy of mites is essential for understanding the fate of viruses in the vectors and nonvectors.

Eriophyid mites that act as virus vectors are elongate, worm-like, minute arthropods, only 200-250 μm long. They have relatively short 7-10 day life cycles (Slykhuis, 1955; Staples and Allington, 1956) and adult life span appears to be only 6-9 days at 25° - 30°C. There is no second nymphal instar which means that functional adulthood is manifest in the second post larval instar (Krantz and Lindquist, 1979). Eriophyids have piercing-sucking mouthparts and feed on epidermal cells by means of 5 stylets contained in their gnathosoma. The two front stylets (chelicerae) and the unpaired median oral stylet penetrate cells, the two auxillary stylets inject salivary enzymes into the cell and the oral stylet then sucks up the sap. Average eriophyid chelicerae can only penetrate plant tissue 15-36 μm deep. The exceedingly thin and small oral stylet can only suck sap and other cytoplasmic contents liquified by digestive salivary enzymes (Jeppson *et al.*, 1975). Anatomical and physiological features of eriophyids are rather primitive, and there are no specialized organs of respiration, blood circulation and excretion as found in insects. The body cavity (hemocoele) contains parenchyma tissue which is composed of moveable loose cells believed to be performing the circulatory, respiratory and excretory functions. Metabolic waste remains in the hemolymph and excretory (guanine) granules and storage glycogen are found in parenchyma (Silvere, 1971, 1976).

Major anatomical features of a typical eriophyid (*Eriophyes tulipae* K.,* = *Aceria tulipae* K.) are depicted in Fig. 1. The alimentary canal is a simple tube with a long, narrow foregut and large sac-like midgut. The hindgut, also a sac, opens as a "uropore" at the anal end. The midgut ends blindly with no connection to the hindgut (Paliwal and Slykhuis, 1967; Paliwal, 1979), a condition common in mites of the suborder Trombidiformes to which eriophyids belong (Hughes, 1959). A pair of salivary glands, each with at least 3 lobes are positioned laterally around the circular neuro-synganglion in the anterior part of the body. Males are rare in *E. tulipae*. Females have a prominent longitudinal oviduct which is filled with developing oocytes and eggs. The much enlarged anterior portion of the oviduct almost entirely fills the body cavity.

*The generic name *Eriophyes* is used here for consistency with another chapter dealing with eriophyids in this volume. However, the International Commission on Zoological Nomenclature has recently placed this mite back under genus *Aceria*. An official communication to this effect is anticipated from the Commission in the near future.

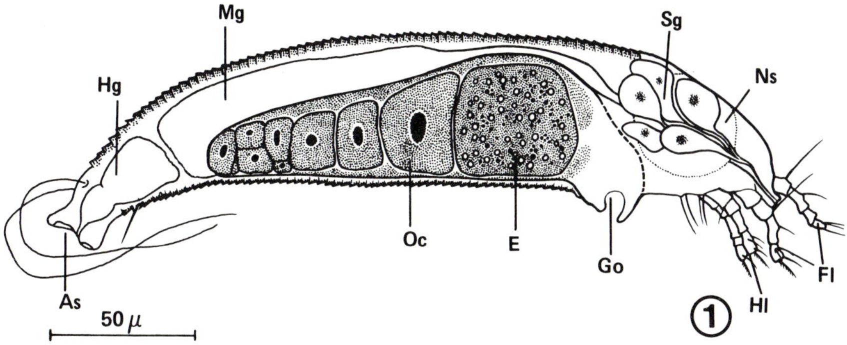

FIG. 1. Diagram of a young adult *Eriophyes tulipae* showing main anatomical features. As, anal sucker; E, egg; Fl, forelegs; Go, genital opening; Hg, hindgut; Hl, hindlegs, Mg, midgut; Ns neurosynganglion; Oc, oocyte; Sg, salivary glands.

15.3 BEHAVIOR OF VIRUSES IN THEIR ERIOPHYID VECTORS

Of the 4 mite transmitted plant viruses, wheat streak mosaic and onion mosaic are transmitted by *Eriophyes tulipae* (Slykhuis, 1955; Razvjazkina, 1971), while another grass-inhabiting eriophyid *Abacarus hystrix* (Nal.) transmits the ryegrass mosaic and Agropyron mosaic viruses (Mulligan, 1960; Slykhuis, 1969). Wheat streak mosaic virus (WSMV) is transmitted in a persistent manner since mites retain infectivity up to 9 days (Slykhuis, 1965). Periods for which the vectors of the other three viruses retain infectivity are not known. Little is known about ingestion by mites and behavior in the vector body of these viruses except WSMV for which information has slowly accumulated over the last two decades.

15.3.1 Wheat Streak Mosaic Virus–*Eriophyes tulipae* System

Orlob's (1966) demonstration that mites retained their ability to transmit WSMV after a molt and after immersion in 1% formaldehyde, a treatment that inactivates the virus in plant extracts, showed that WSMV was ingested by mites and held internally in their bodies. Also, homogenates of mites reared on infected plants showed typical filamentous (700 nm) WSMV particles, caused WSMV infection of wheat plants and reacted positively with WSMV antiserum (Orlob, 1966; Paliwal and Slykhuis, 1967). Direct evidence for occurrence of the virus in the body of *E. tulipae* came from examination of ultrathin sections of mites which revealed a large accumulation of WSMV particles in the alimentary canal (mainly the "hindgut") of mites reared on infected plants (Paliwal and Slykhuis, 1967; Takahashi and Orlob, 1969; Stein-Margolina, 1975). Takahashi and Orlob (1969) also reported retention of the virus in the "hindgut" of mites for at least 4 days after removal from a virus source and observed some virus-like particles in the parenchyma tissue in a few mites. The appearance of these extraintestinal particles was erratic and virus-like particles were not seen in the salivary gland tissue. However, a recent reinvestigation (Paliwal, 1979) showed that the main site of WSMV accumulation in *E. tulipae* was the sac-like posterior midgut (Fig. 2) which was erroneously identified as hindgut earlier (Paliwal and Slykhuis, 1967). The virus was densely packed in the lumen of the posterior midgut forming stream-like patterns (Fig. 2a). In the anterior part of midgut, virus particles were aggregated side to side forming distinct bands (Fig. 2b). No virus particles were found in the hindgut. Consistent with these observations is the fact that mites retain infectivity after molts during which the internal linings of the fore- and hindgut are believed to be shed, which would eliminate any virus present in these areas. It was also noted that large amounts of intact virus particles persisted in the posterior midgut for at least up to 5 days after removal of the mites from the virus source. WSMV particles were also found consistently in the body cavity interspersed in the parenchyma tissue and in the salivary gland cells (Fig. 3) of the mites suggesting that the virus is circulative in the mite.

There is no evidence for multiplication of the virus in the mite body. Transmission ability of older adult mites that have fed on infected plants as nymphs gradually decreases (Orlob, 1966). Also, the bulk of WSMV particles in the mite body occur extracellularly in the gut lumen and body cavity and do not cause infection of mite cells. Intracellular presence of the virus particles in salivary glands may be due to some accumulation of the virus while passing through these tissues although limited multiplication cannot be ruled out. Persistence of the virus in the mite does not necessarily imply virus multiplication. It appears that enough virus is ingested and accumulated by mite nymphs in their gut and preserved in an infective state that transmission is maintained for almost all of the life span of the mites.

15.3.2 Less Understood Virus-Vector Systems

Flexuous rod-shaped particles (675 nm) of onion mosaic virus have been detected in suspensions of its vector mites *E. tulipae* (Razvjazkina, 1971). Virus particles were also found in ultrathin sections of mites reared on infected onion plants but the sites of virus occurrence in the mite body were not identified. Since the extent of persistence of infectivity of the vector is not known and information on behavior of the virus in the mite body is limited, an understanding of the mechanism of transmission of the virus must await further work.

FIG. 2. Part of a longitudinal section of adult *E. tulipae* showing the posterior midgut (Mg) almost filling the body cavity (H) and containing a large amount of wheat steak mosaic virus (WSMV) in its lumen; Hg, hindgut: In, integument; Oc, oocyte; P, parenchyma cells. Bar = 5 μm (Paliwal, 1979).

FIGS. 2a and 2b. WSMV particles of midgut in higher magnification; 2a. densely packed virus particles of the posterior midgut; 2b. side to side aggregated particles forming distinct bands in anterior part of midgut. In both figures bar = 1 μm and arrows indicate particles in cross section or in oblique cuts (Paliwal and Slykhuis, 1967).

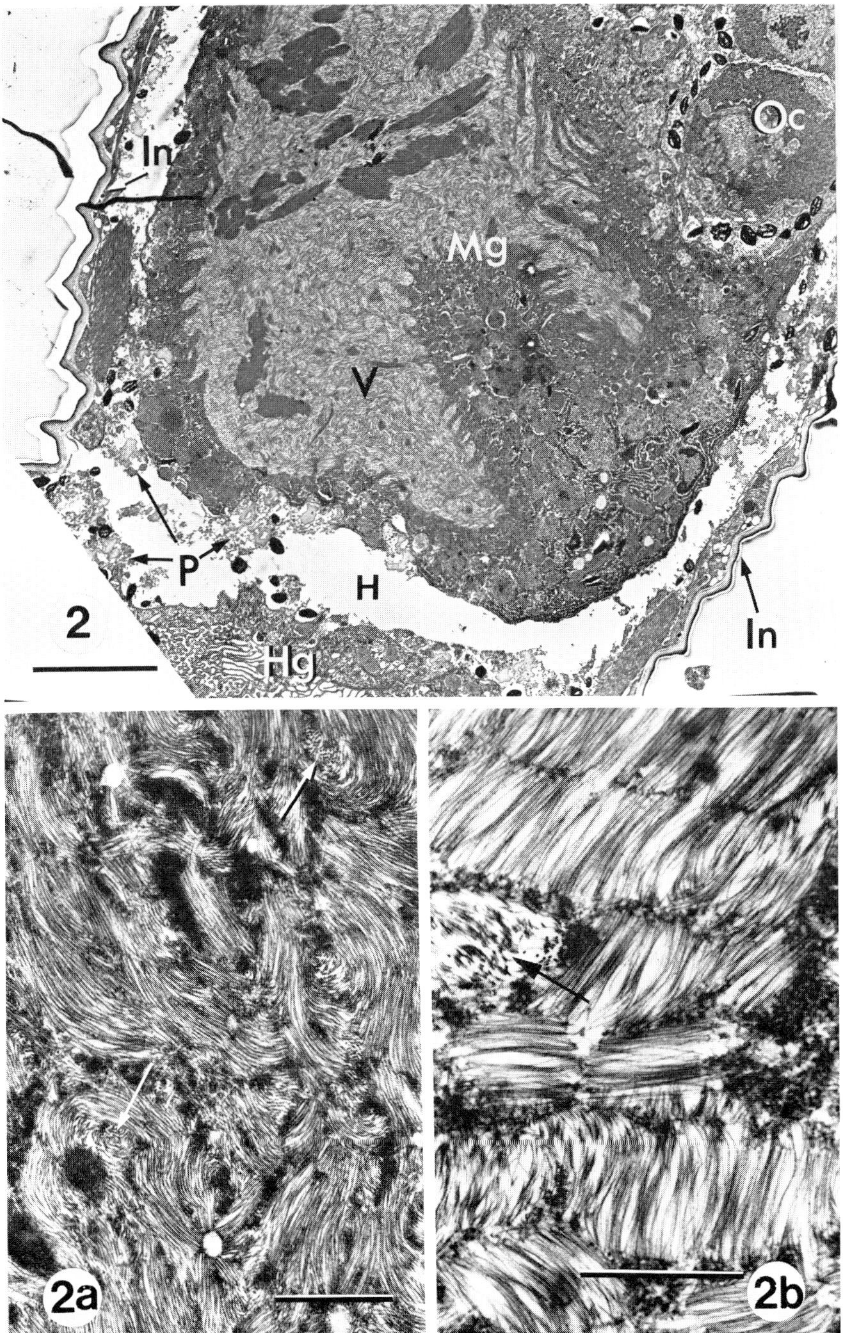
In
Oc
Mg
V
P
H
In
2
Hg
2a
2b

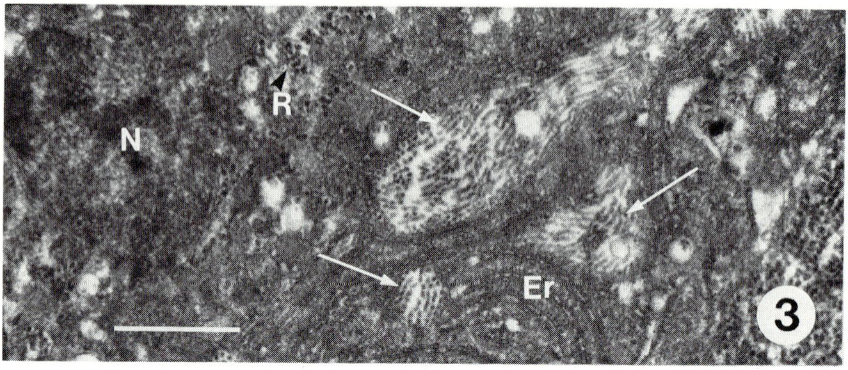

FIG. 3. WSMV particles (arrows) in a salivary gland cell of *E. tulipae*; Er, endoplasmic reticulum; N, nucleus; R, ribosomes, bar = 0.5 μm (Paliwal, 1979).

In preliminary work on the relationship of Agropyron mosaic virus to its mite vector in author's laboratory, several *A. hystrix* reared on the virus-infected wheat plants were sectioned and examined. No virus particles were found in the alimentary canal or elsewhere in the body of the mites. Since large numbers of mites are usually required to obtain transmission of the virus, it would appear that only a small percentage of mites in a population may serve as vectors (Slykhuis, 1969). Even though all mites do not transmit, larvae feeding continually on virus-infected plants would be expected to ingest the virus which should be observed in the gut. Possibilities, such as stylet-borne transmission and little or no ingestion of the virus by the mite should be further examined in order to understand the mechanism of transmission of the virus.

There are no reports in the literature regarding the fate of ryegrass mosaic virus in its vector *A. hystrix*.

15.4 BEHAVIOR OF VIRUSES IN NONVECTOR ERIOPHYIDS

Understanding the behavior of ingested virus in the body of nonvectors can help to elucidate the mechanism of virus transmission and specificity in vectors. The fate of WSMV, barley stripe mosaic (BSMV) and Bromegrass mosaic (BMV) viruses in nonvector eriophyids has been investigated (Paliwal, 1972, 1979). Like WSMV, both BMV and BSMV affect wheat, barley, oats and grasses in nature. Particles of BSMV are short rods (128 nm) and BMV is a polyhedral sphere (27 nm). No arthropod vectors are known for either BSMV or BMV.

15.4.1 WSMV in *Abacarus hystrix*

A. hystrix freely colonizes wheat plants infected with WSMV but does not transmit the virus. Homogenates of mites reared on infected plants were not infective when inoculated to wheat plants although they occasionally showed one or two virus particles in the electron miscroscope. Ultrathin sections of mites cultured on infected plants were examined to determine the behavior of virus that may have been ingested. Virus particles, some degraded, were found in small numbers in the lumen of midgut (Figs. 4, 5) of about 28% of the mites examined but no particles were found in any other part of the mite body (Paliwal, 1979). Obviously, the relationship between WSMV and *A. hystrix* is not conducive to efficient virus ingestion and preservation in the gut, let alone distribution in the mite body for circulative transmission.

15.4.2 Barley Stripe Mosaic Virus in *Eriophyes tulipae*

BSMV was efficiently ingested and accumulated by the mites in their bodies by feeding on infected plants. The virus harbored by mites was infectious and reacted specifically with BSMV antiserum (Paliwal, 1979). The major site of accumulation of BSMV in *E. tulipae* was the midgut (Fig. 6) where large amounts of virus particles

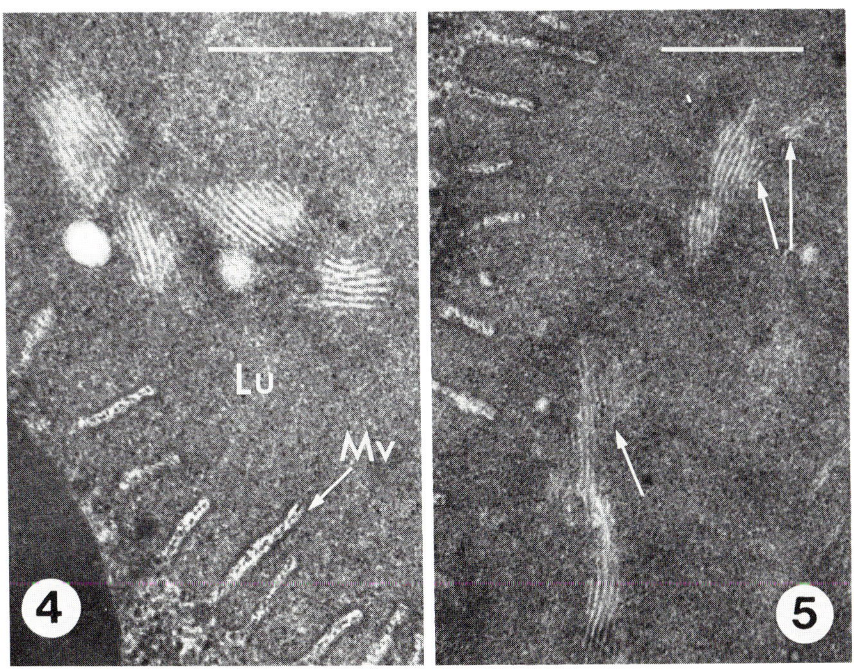

FIGS. 4 and 5. WSMV particles in the midgut lumen of nonvector mite *Abacarus hystrix*. Fig. 4 shows the largest number of particles that were ever observed in a mite. Particles show evidence of some degradation especially in Fig. 5 (arrows), Lu, lumen of the midgut; Mv, microvilli; bars in both figures = 0.5 μm (Paliwal, 1979).

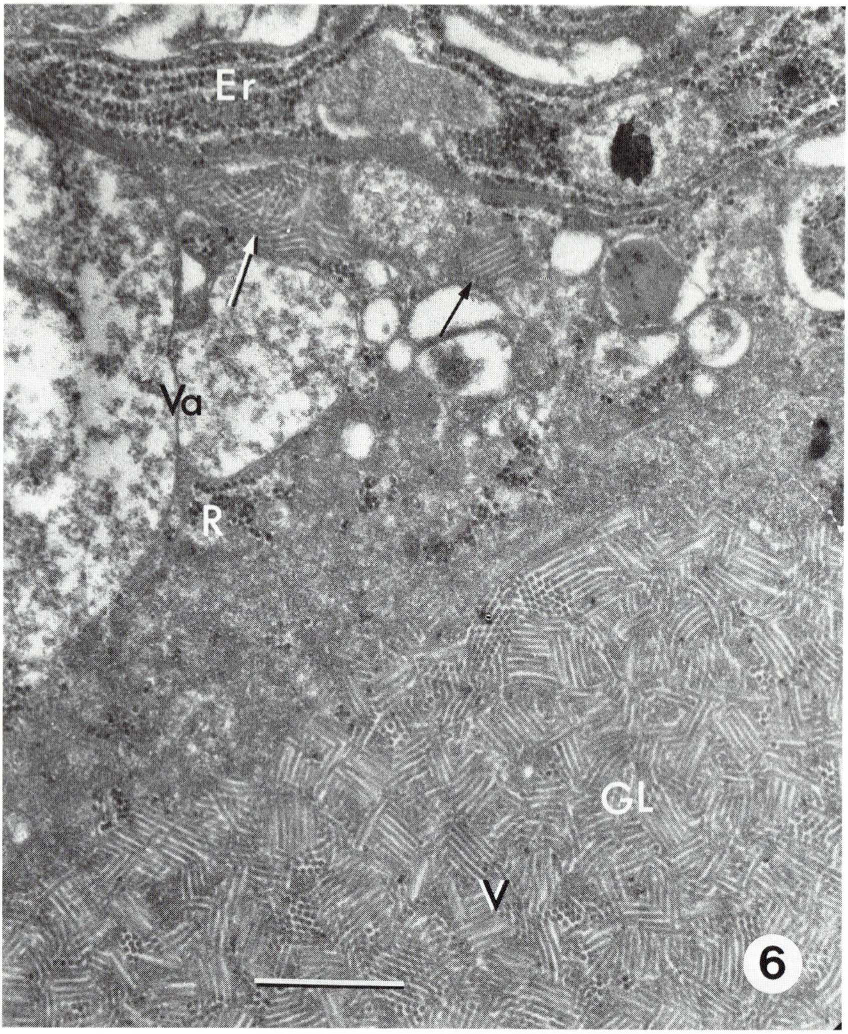

FIG. 6. Barley stripe mosaic virus particles (V) in the lumen (Gl) and in the cytoplasm (arrows) of an epithelial cell of the midgut of a nonvector *E. tulipae*; Er, endoplasmic reticulum; R, ribosomes; Va, vacuole, bar = 0.5 μm (Paliwal, 1979).

were found packed in the lumen. Mites retained intact virus in their gut for at least 4 days after removal from a virus source. Passage of virus particles out of the digestive system was evident by their occurrence in different areas of the mite body cavity. However, the virus was not found in the salivary glands. Thus, it appears that lack of passage of the virus through salivary glands prevents the circulative transmission of BSMV by the mite.

15.4.3 Bromegrass Mosaic Virus in *Eriophyes tulipae*

Behavior of bromegrass mosaic virus (BMV) in *E. tulipae* has several interesting features not observed in any other nonvector mite-virus combination. Paliwal (1972) and Stein-Margolina (1973) showed that the mites ingested the virus during feeding on infected plants and accumulated large amounts of it in their body. The virus persisted in the mite body, in an infective and serologically reactive state, for at least 6 days. BMV particles were densely packed in the midgut lumen and in spherical membrane-bound, lysosome-like structures in the epithelial cells of the gut tissue. The entire cytoplasm of the gut wall cells (single cell thick epithelium) was replaced by such "virus balls" (Fig. 7, 8) and by loose particles filling the vacuoles and often the cisternae of endoplasmic reticulum. Cytopathic effects such as degeneration of gut microvilli, reduction in the number and size of mitochondria and degradation of endoplasmic reticulum were also noted. Clumps of virus particles were localized in the hemocoele in different parts of the mite body, in nurse cells around oocytes (Fig. 11) and in muscle cells associated with the integument. The virus was not found in the salivary glands. Evidence suggesting multiplication of BMV in the gut tissue has also been presented (Paliwal, 1972). Virus-free second larval instars of *E. tulipae* were allowed an acquistion access feeding on infected plants for 15-16 hours, transferred to virus-free plants and their ultrathin sections examined in the electron microscope for BMV particles at different times after the acquisiton access period. A clearly discernible increase with time in virus concentration and crystallization of the particles in the gut cell cytoplasm were observed (Fig. 10, 11). The cytoplasmic infection of mite cells by the virus and occurrence of viroplasms containing the virus particles (Fig. 9), similar to those formed by animal viruses, in gut cells also suggest virus multiplication in the mite.

The relationship between BMV and *E. tulipae* is perhaps the most specific and "intimate" among the other virus-nonvector arthropod combinations studied (Storey, 1933; Bennett and Wallace, 1938; Hutchinson and Matthews, 1963; Orlob, 1968). In these cases, ingested virus is found only in the gut lumen of the nonvector and is excreted in the feces. In some cases, virus is retained in an infectious state in the gut for a few days.

15.5 VIRUSES IN TETRANYCHID MITES

Tetranychid mites are eight-legged larger mites with well-developed, highly efficient piercing-sucking mouth parts. Unlike eriophyids, tetranychids feed deeper (up to mesophyll), dislodge cells and cause visible injury to the leaf tissue. They suck up most of the cell contents since fragments of plant cells, chloroplasts and other organelles are seen in the tetranychid gut (Orlob and Takahashi, 1971; Jeppson *et al.*, 1975).

The two-spotted spider mite, *Tetranychus urticae* Koch [= *T. telarius* (L.)] is the only tetranychid which has been reported to transmit a plant virus, potato virus Y

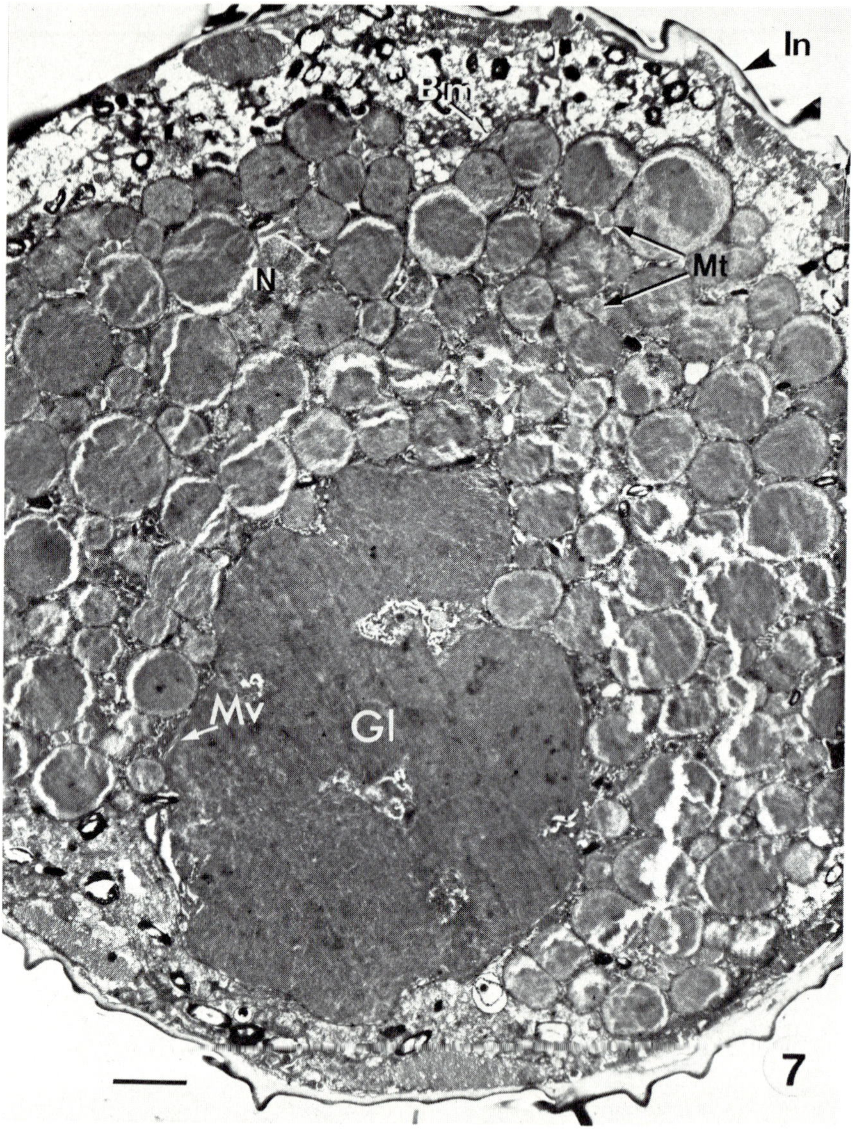

FIG. 7. Cross Section of a mite (*E. tulipae*) reared on bromegrass mosaic virus (BMV) infected plants. The section is through the large, posterior midgut showing that almost all of the gut tissue has been replaced by spherical bodies (virus "balls") packed with virus particles. Very few organelles are present; occasional mitochondria (Mt) and nucleus (N) are seen but endoplasmic reticulum is practically nonexistent. Gut lumen (Gl) is packed with virus; Mv, microvilli; Bm, basement membrane of the gut; In, integument; bar = 5 μm (Paliwal, 1972).

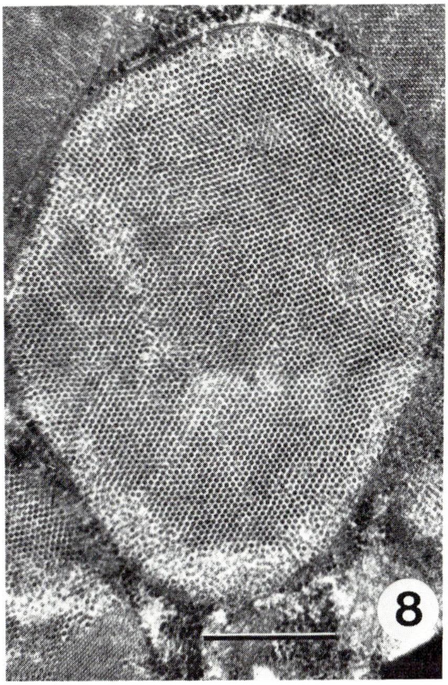

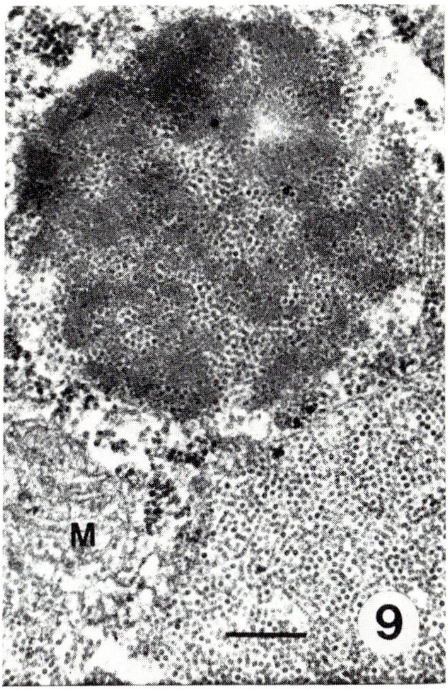

FIG. 8. A virus "ball" packed with BMV particles in a gut cell of nonvector *E. tulipae*, bar = 0.5 μm (Paliwal, 1972).
FIG. 9. BMV particles appearing in a viroplasm (electron dense matrix) and a pocket of loose particles in the cytoplasm of a gut cell of *E. tulipae*, M, mitochondrion; bar = 0.5 μm (Paliwal, 1972).

(Schultz, 1963). However, Fritzsche *et al.*, (1967) and Orlob (1968) failed to confirm the results presented in this report. The mite apparently can ingest and pass into its gut a number of plant viruses, both rod-shaped and spherical but some are degraded in the alimentary canal of the mite. Tobacco mosaic, potato virus X, bromegrass mosaic, tomato bushy stunt, tobacco ring spot, and southern bean mosaic viruses can be detected, in an infective state, in the alimentary canal of *T. urticae* after feeding on infected plants. Also, mites regularly excrete a significant amount of virus in their feces. Beet mosaic and onion yellow dwarf viruses and potato virus Y apparently are degraded in the gut of the mite (Fritzsche, *et al.*, 1967; Orlob, 1968; Orlob and Takahasi, 1971). None of the 9 viruses mentioned above are transmitted by *T. urticae* (Fritzsche *et al.*, 1967; Orlob, 1968). Obviously, there is little or no biological interaction between plant viruses and *T. urticae*; virus ingestion, propulsion through the digestive system and excretion of the virus, along with ingested food material, are passive processes of no consequence towards virus transmission.

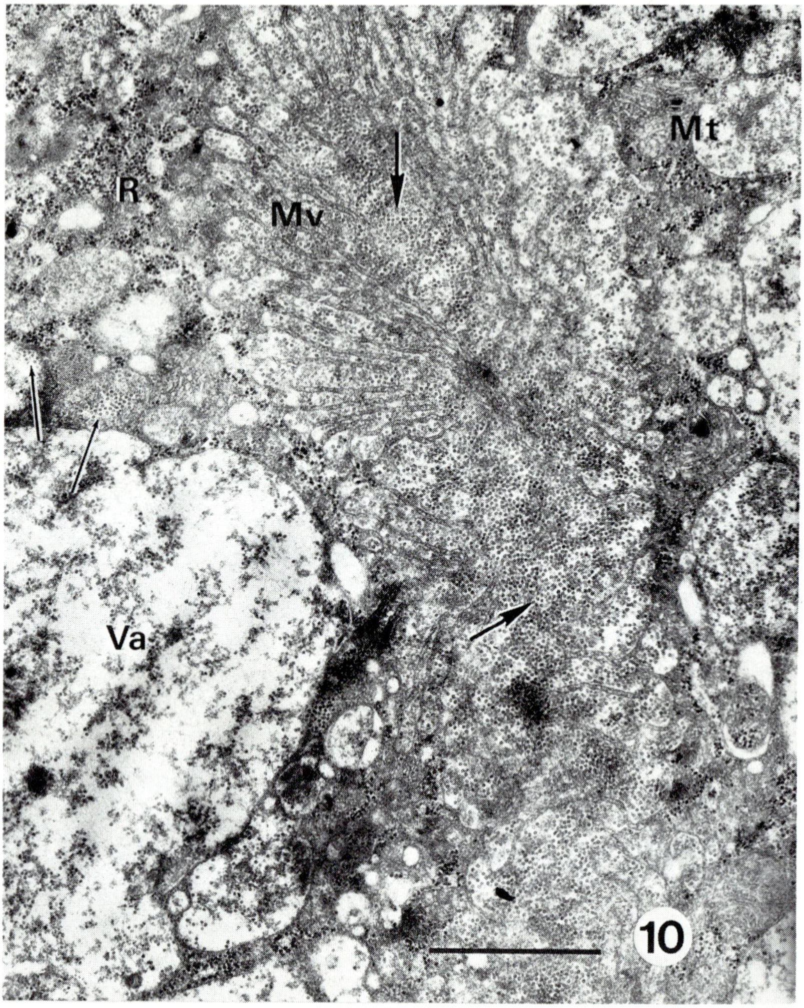

FIG. 10. Part of a section through midgut of an *E. tulipae* fixed immediately after a 15-16 hr acquisition access to a BMV infected plant showing the ingested virus mostly in the gut lumen (large arrows) and occasional clusters of virus particles (small arrows) in vesicles in the cell cytoplasm. Few scattered particles can also be spotted at a number of places in the cytoplasm; Mt, mitochondria; Mv, microvilli; R, ribosomes; Va, vacuole, bar = 1 μm (Paliwal, 1972).

FIG. 11. Midgut tissue and part of genital canal (tangential section, hence no gut lumen seen) of an *E. tulipae* at day 7 after the 15-16 hr acquisition access on BMV infected plants. Great increase in the amount of virus in the gut as compared with that seen immediately after the acquisition access period (Fig. 10) is evidenced by an abundance of virus crystals (Vc) and pockets of unarranged virus particles (large arrows) throughout the gut cell cytoplasm. Note the presence of small virus crystals (small arrows) in the nurse cells (Nc) around the oocyte (Oc); Bm, basement membrane of the gut; In, integunent of the mite; Nu, nucleus; bar = 2 μm The insert shows a virus crystal in higher magnification, bar = 0.5 μm (Paliwal, 1972).

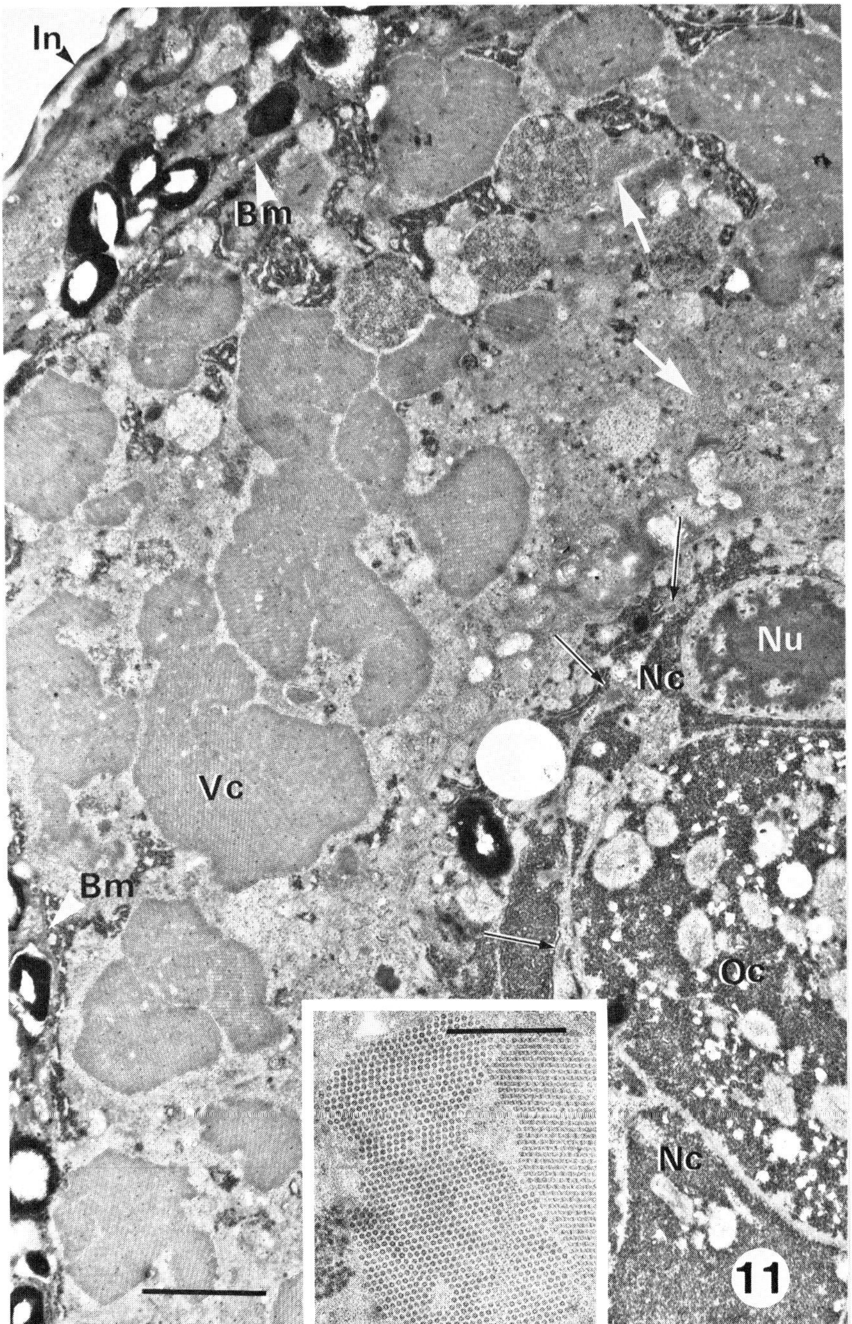
In
Bm
Nu
Nc
Vc
Bm
Oc
Nc
11

15.6 MECHANISMS OF VIRUS TRANSMISSION

WSMV is the only mite-vectored virus transmitted in a persistent manner. Recent work (Paliwal, 1979) suggested that the virus is circulative in the mite body and inoculated to the plants via salivary secretions during mite feeding. While this appears to be the main mode of transmission of the virus, other concurrent modes of transmission are possible. Before the evidence for circulativeness of WSMV in the mite body, it was speculated that the large pool of virus found in the gut of *E. tulipae* may be excreted, smeared on the leaf surface and introduced into the leaf cells through feeding punctures or other injury possibly through setae on the mite body. Alternatively, a backflow of virus through the straight, tubular mid- and foregut to the mouthparts and introduction into the leaf cells during feeding were visualized (Paliwal and Slykhuis, 1967). Orlob (1968) observed that large amounts of infectious tobacco mosaic virus (TMV) excreted by *Tetranychus urticae* in feces on the leaves was not transmitted by feeding and locomotory activity of these mites. However, purified TMV or potato virus X sprayed on leaves were readily transmitted by the mites through feeding punctures. Tetranychids usually excrete a solid waste which is unlikely to be easily smeared on the leaf surface because of its consistency and lack of natural means of smearing it. Eriophyids on the other hand eliminate an almost clear liquid which would easily smear because mites drag the rear of their body and their setae on the leaf surface while moving. Eriophyids feed by making small, neat punctures with little or no damage to the cell (Gibson, 1974). This type of gentle feeding would be more favorable to inoculation of the virus present on the leaf surface than the tetranychid feeding which causes considerable cell damage. However, a lack of communication between the hindgut and the midgut at least in adults of *E. tulipae* has been noted and no virus particles were found in the hindgut of such mites reared on WSMV infected plants (Paliwal, 1979). Therefore, if any virus is excreted in liquid feces by *E. tulipae*, perhaps by larvae, it could be transmitted through feeding punctures but such transmission would have to be rather erratic.

Beetle vectors are known to transmit plant viruses by regularly egesting (regurgitating) virus from their gut during feeding (Walters, 1969). Egestion of ingested food from the gut is known in several other arthropods such as leafhoppers (Harris, 1977, 1978, 1979; Harris *et al.*, 1980), termites and hymenopterous insects (Kloft, 1976). Aphids regularly egest fluids into artificial feeding media and during plant feeding, and the sucking pump of aphids apparently can function normally in either direction (Garrett, 1973; Harris and Bath, 1973; Kloft, 1977). The egestion of fluids serves to clear clogged food canals (Harris, 1977). Certain nonpersistent and semipersistent virus transmissions by aphids can be logically explained on the basis of a virus ingestion-egestion hypothesis of noncirculative transmission. The observed transmission characteristics of such viruses fit this hypothesis (Harris, 1977, 1978). Eriophyids have an exceedingly narrow food canal and clogging of the food canal and egestion of ingested fluids to clear the canal are highly probable. A functional pharyngeal pump to effect suction and possibly ejection, also exists in eriophyids (Shevtchenko and Silvere, 1968). If eriophyids do egest virus-containing fluids in this manner from their anterior midgut which has plenty of virus particles, virus transmission could be easily effected during feeding. WSMV particles have not been observed in the foregut, pharynx and food canal areas of viruliferous *E. tulipae*. Little or no virus is likely to stay back in these areas if they are regularly flushed, like in aphids (Harris, 1977), by pressurized ingestion-egestion of fluids effected by the pharyngeal pump. A similar reason for a "clean" pharynx in the mite *T. urticae* was visualized by Orlob and Takahashi (1971). However, no efforts have been made to serially section the mite gnathosoma and cephalothorax region to locate the few particles that may be left behind during the ingestion-egestion flushes in the food canal-pharynx-oesophagus areas and may have so far been missed in electron microscopic examinations.

Bimodal and possibly trimodal transmission of WSMV by *E. tulipae*, therefore, is conceivable although "noncirculative" transmission through an ingestion-egestion mechanism or transmission of excreted virus, if any, through feeding punctures are likely to be inconsistent and infrequent.

Transmission of a virus by an eriophyid mite results from a highly specific biological interaction between the mite and the virus. For circulative transmission, ingestion, retention and survival of the virus in an infective state in the alimentary canal of mites are basic requirements. The virus must then release from the gut to circulate in the body cavity of the mite and finally pass into the salivary glands to be introduced into the plant cells via salivary secretions—a circulative cycle first demonstrated for leafhopper-transmitted viruses (Sinha, 1968). Nonfulfillment of any of these requirements and of this circulative cycle would prevent a mite species from serving as a vector. Interestingly enough, examples of virus-mite combinations fulfilling none, one, two or all of the conditions of circulative transmission are available. Agropyron mosaic virus particles (700 nm, flexuous rods) were not found in the alimentary canal or elsewhere in the body of its vector *A. hystrix* reared on infected plants (Paliwal, unpublished) indicating that the virus is "noncirculative" in the mite and other modes of transmission may be operational. In case of WSMV, massive amounts of virus particles accumulate in the vector *E. tulipae* but only a small number of virus particles are found in the gut of some individuals of the nonvector *A. hystrix* (Paliwal, 1979). The oral opening of eriophyids is extremely small and is estimated to be 500-1000 nm only (Oldfield, 1970). While the feeding mechanism of *E. tulipae* allows 700 nm long WSMV particles to pass along with sap, it is possible that the oral opening and food canal of *A. hystrix* are narrower than that of *E. tulipae* permitting only the particles with their long axis aligned with the oral opening or broken ones to pass.

The behavior of BSMV in *E. tulipae* represents one step further (than WSMV in *A. hystrix*) towards fulfillment of requirements for circulative transmission. In addition to the efficient ingestion, accumulation and retention of the virus in an infective state in the gut, the virus is distributed in the body cavity of the mite. BMV behaves similarly in *E. tulipae* but its interaction with the mite is much greater in that it infects the cells of alimentary canal, ovaries and the integument and multiplies at least in the gut tissue. However, in both cases, the last link of the circulative cycle, i.e. passage of the virus into salivary glands, appears to be missing. The WSMV-*E. tulipae* system apparently meets all the requirements of the circulative cycle leading to an efficient virus transmission.

The flexuous rod-shaped particles of WSMV are found in the salivary glands of *E. tulipae*. Although, particles of both BMV and BSMV are considerably smaller than those of WSMV and are abundant in the body cavity, they are not found in the salivary glands. Obviously, much more than the size-related ease of passage through salivary gland membranes is involved. A specific interaction governed by (1) reciprocity between recognition sites on virus coat protein and on salivary gland membranes as suggested in case of aphids (Rochow, 1969; Harris, 1977, 1979), (2) ability of a virus to survive in an infective state in the gland cell "environment" and (3) buildup of a concentration of virus particles in the salivary gland cells conducive to an active transport of particles in the salivary secretions would appear to determine the salivary transit of virus from the body cavity to the mouth parts.

15.7 REFERENCES

Bennett, C. W., and Wallace, H. E. (1938). Relation of the curly top virus to the vector, *Eutettix tenellus*. *J. Agri. Res.* 56: 31-51.

Fritzsche, R., Schmelzer, K., and Schmidt, H. B. (1967). Prüfung der Eignung von *Tetranychus unticae* Koch als Vektor planzenpathogener Viren. *Arch. Pflanzenschutz* 3: 89-100.

Garrett, R. G. (1973). Non-persistent aphid-borne viruses. *In* "Viruses and Invertebrates" (A. J. Gibbs, Ed.), pp. 476-492. North Holland Publishing Company, Amsterdam and London.

Gibson, R. W. (1974). Studies on the feeding behavior of the eriophyid mite *Abacarus hystrix*, a vector of grass viruses. *Ann. app. Biol.* 78: 213-217.

Harris, K. F. (1977). An ingestion-egestion hypothesis of noncirculative virus transmission. *In* "Aphids as Virus Vectors" (K. F. Harris and K. Maramorosch, Eds.), pp. 165-220. Academic Press, New York.

Harris, K. F. (1978). Aphid-borne viruses: ecological and environmental aspects. *In* "Viruses and Environment" (E. Kurstak and K. Maramorosch, Eds.), pp. 311-337. Academic Press, New York.

Harris, K. F. (1979). Leafhoppers and aphids as biological vectors: vector-virus relationships. *In* "Leafhopper Vectors and Plant Disease Agents" (K. Maramorosch and K. F. Harris, Eds.), pp. 217-308. Academic Press, New York.

Harris, K. F., and Bath, J. E. (1973). Regurgitation by *Myzus persicae* during membrane feeding: its likely function in transmission of nonpersistent plant viruses. *Ann. Entomol. Soc. Am.* 66: 793-796.

Harris, K. F., Treur, B., Tsai, J., and Toler, R. (1980). Observations on ingestion-egestion behavior of a leafhopper vector during membrane feeding. In manuscript.

Hughes, T. E. (1959). "Mites or the Acari." The Athlon Press, Univ. of London.

Hutchinson, P. B., and Matthews, R. E. F. (1963). The fate of turnip yellow mosaic virus in the nonvector aphid *Hyadaphis brassicae* (L.). *Virology* 20: 169-175.

Jeppson, L. R., Keifer, H. H., and Baker, E. W. (1975). "Mites Injurious to Economic Plants." University of California Press, Berkeley.

Krantz, G. W., and Lindquist, E. E. (1979). Evolution of Phytophagous mites. *Annu. Rev. Entomol.* 24: 121-158.

Kloft, W. J. (1976). Revised "Applied Part" in "Laboratory Training Manual on the Use of Isotopes and Radiation in Entomology," 2nd edition. Int. Atomic Energy Agency, Vienna.

Kloft, W. J. (1977). Radioisotopes in aphid research. *In* "Apids as Virus Vectors" (K. F. Harris and K. Maramorosch, Eds.), pp. 291-310. Academic Press, New York.

Mulligan, T. (1960). The transmission by mites, host range and properties of ryegrass mosaic virus. *Ann. appl. Biol.* 48: 575-579.

Oldfield, G. N. (1970). Mite transmission of plant viruses. *Annu. Rev. Entomol.* 15: 343-380.

Orlob, G. B. (1966). Feeding and transmission characteristics of *Aceria tulipae* Keifer as vector of wheat streak mosaic virus. *Phytopathol. Z.* 55: 218-238.

Orlob, G. B. (1968). Relationship between *Tetranychus urticae* Koch and some plant viruses. *Virology* 35: 121-133.

Orlob, G. B., and Takahashi, Y. (1971). Location of plant viruses in the two-spotted spider mite *Tetranychus urticae* Koch. *Phytopathol. Z.* 72:21-28.

Paliwal, Y. C. (1972). Brome mosaic virus infection in the wheat curl mite *Aceria tulipae*, a non-vector of the virus. *J. Invertebr. Pathol.* 20: 288-302.

Paliwal, Y. C. (1980). Relationship of wheat streak mosaic and barley stripe mosaic viruses to vector and nonvector eriophyid mites. *Arch. Virol.* 63: 123-132.

Paliwal, Y. C., and Slykhuis, J. T. (1967). Localization of wheat streak mosaic virus in the alimentary canal of its vector *Aceria tulipae* Keifer. *Virology* 32: 344-353.

Razvjazkina, G. M. (1971). Das Zwiebelmosaikvirus und seine Verbreitung im Freiland. *Tag.-Ber. Dt. Akad. Landwirtsch. -wiss. Berlin* 115: 69-76.

Rochow, W. F. (1969). Specificity in aphid transmission of a circulative plant virus. *In* "Viruses, Vectors, and Vegetation" (K. Maramorosch, Ed.), pp. 175-198. Wiley-Interscience, New York.

Schulz, J. T. (1963). *Tetranychus telarius* L., new vector of virus Y. *Plant Disease Reptr.* 47: 594-596.

Shevtchenko, V., and Silvere, A. P. (1968). The feeding organs of the four-legged mites, *Acarina, Eriophyoidea. Izvestiya Akad. Nauk Estonsk. SSR–Biologia* 17: 248-264.

Silvere, A. P. (1971). Origin and evolution of gall mites (Eriophyoidae) in the light of their internal anatomy. *Proc. 3rd Int. Congress Acarol., Prague* 1971: 247-249.

Silvere, A. P., and Stein-Margolina, V. (1976). "Tetrapodili." Akademii Nauk Estonskoi SSR, Tallin USSR.

Sinha, R. C. (1968). Recent work on leafhopper transmitted viruses. *Adv. Virus Res.* 13: 181-223.

Slykhuis, J. T. (1955). *Aceria tulipae* Keifer (Acarina:Eriophyidae) in relation to the spread of wheat streak mosaic. *Phytopathology* 45: 116-128.

Slykhuis, J. T. (1965). Mite transmission of plant viruses. *Adv. Virus Res.* 11: 97-137.

Slykhuis, J. T. (1969). Transmission of Agropyron mosaic virus by the eriophyid mite, *Abacarus hystrix. Phytopathology* 59: 29-32.

Slykhuis, J. T. (1980). Mites. *In* "Vectors of Plant Pathogens" (K. F. Harris and K. Maramorosch, Eds.). pp. 325-356. Academic Press, New York.

Staples, R., and Allington, W. B. (1956). Streak Mosaic of Wheat in Nebraska and its control. Univ. Nebraska Agri. Exp. St. Res. Bull. 178: 40p.

Stein-Margolina, V. A. (1973). Bromegrass mosaic virus in the mite *Aceria tritici* (Electron microscopical investigation). *Izvestiya Akad. Nauk. USSR* 2: 189-195.

Stein-Margolina, V. A. (1975). Phytopathogenic viruses of cereals in the gall mite *Aceria tritici* Shev. *Tag.-Ber. Akad. Landwirtsch.-Wiss. DDR Berlin* 134: 181-198.

Storey, H. H. (1933). Investigations on the mechanism of the transmission of plant viruses by insect vectors–I. *Proc. Roy. Soc. London, Ser. B.* 113: 463-485.

Takahashi, Y., and Orlob, G. B. (1969). Distribution of wheat streak mosaic virus-like particles in *Aceria tulipae. Virology* 38: 230-240.

Walters, H. J. (1969). Beetle transmission of plant viruses. *Adv. Virus Res.* 15: 339-363.

Chapter 16

NEMATODES

C. E. Taylor

Scottish Horticultural Research Institute
Invergowrie, Dundee
Scotland

16.1	INTRODUCTION	375
16.2	VIRUSES AND DISEASES	376
	16.2.1 Nepoviruses	377
	16.2.2 Tobraviruses	382
16.3	NEMATODE VECTORS	383
	16.3.1 Geographical Distribution and Hosts	384
	16.3.2 Distribution in Soil	387
	16.3.3 Feeding and Pathogenicity	389
	16.3.4 Life Cycles	391
16.4	VIRUS TRANSMISSION CHARACTERISTICS AND MECHANISMS	392
	16.4.1 Transmission	393
	16.4.2 Persistence	393
	16.4.3 Sites of Retention	394
	16.4.4 Specificity of Transmission	397
	16.4.5 Mechanism of Transmission	399
	16.4.6 Non-persistent Associations	401
16.5	CONTROL	401
16.6	REFERENCES	404

16.1 INTRODUCTION

It has long been recognized that virus diseases can be spread by agents in the soil (Beijerinck, 1898; Behrens, 1899) and the possibility that nematodes may be vectors was not overlooked. Percival (1895) claimed that nettle-head disease of hops (*Humulus lupulus*) was spread by the hop cyst nematode (*Heterodera humuli*) but Duffield (1925) later disproved this; it has since been shown that the disease is caused by a strain of arabis mosaic virus (Bock, 1966) and is transmitted by *Xiphinema diversicaudatum* (Valdez *et al.*, 1974). Johnson (1945) showed that chemical fumi-

ISBN 0-12-326450-2

gation of wheat mosaic-infected soil prevented transmission of the virus and surmised from this that nematodes might be vectors, but later investigations implicated a fungus, *Polymyxa graminis* (Rao and Brakke, 1969). Attempts to transmit lettuce big vein disease with several nematode species also yielded negative results (Allen, 1948) and again this later proved to be transmitted by a fungus, *Olpidium brassicae* (Campbell and Fry, 1966).

The first experimental evidence of an association between a nematode and a plant virus was provided by Hewitt *et al.* (1958) when they demonstrated the transmission of grapevine fanleaf virus by *X. index*. Soon afterwards it was shown that the virus could be transmitted to a range of herbaceous host plants by mechanical inoculation of plant sap (Cadman *et al.*, 1960) thus allowing comparisons to be made with other similar viruses which were also found to have nematode vectors. These viruses have many physical, chemical and biological properties in common and form the nepovirus group (Cadman, 1963; Harrison and Murant, 1977a); they are vectored by species of *Xiphinema* and *Longidorus* (Table I). *Trichodorus* and *Paratrichodorus* species are vectors of another distinct group, the tobraviruses (Table II). The subject of nematode transmitted viruses has been reviewed extensively: Cadman, 1963; Harrison, 1960, 1964a, 1977; Hewitt and Raski, 1962; Lamberti *et al.*, 1975; Taylor, 1971, 1972a; Taylor and Cadman, 1969.

16.2 VIRUSES AND DISEASES

Diseases caused by nematode-transmitted viruses typically occur as patches of infected plants within a crop, with infection spreading slowly from the periphery of the patch, reflecting the horizontal distribution and relative immobility of the nematode vectors in the soil (Fig. 1a). Infection is also characterized by persistence in the soil for long periods, even during a fallow or when virus-resistant crops are grown. Although some viruses can persist in their nematode vectors for several months, or even years, survival in infected seeds is an important means of virus perennation, especially with some nepoviruses transmitted by *Longidorus* spp. (Murant and Taylor, 1965; Murant and Lister, 1967).

In general, both the viruses and their nematode vectors have wide host ranges among wild and cultivated plants; grapevine fanleaf virus is a notable exception in having no natural host other than grapevine (*Vitis* spp.). Most of the records of the occurrence of nematode-transmitted viruses are from Western Europe and North America, mainly because of the early research interests in those regions, but there are recent records from Australia (Stubbs, 1971; Meagher *et al.*, 1976), New Zealand (Thomas and Procter, 1972), Nigeria (Caveness *et al.*, 1975), Japan (Komuro *et al.*, 1970; Iwaki and Komuro, 1971, 1974; Yagita and Komuro, 1972), Russia (Romanenko, 1971; Ivanova, 1972; Teploukhova, 1974) and it is probable that they occur in cultivated soils in most developed countries.

TABLE I Nepo and Other Polyhedral Viruses and Their Nematode Vectors

Virus	Vector	Reference
Nepo Viruses		
Arabis mosaic type strain	*Xiphinema diversicaudatum*	Harrison and Cadman, 1959; Jha and Posnette, 1959
	X. coxi	Fritzsche, 1964
	X. bakeri	Iwaki and Komuro, 1974
hop strain	*X. diversicaudatum*	Valdez *et al.*, 1974
Artichoke Italian latent	*Longidorus apulus*	Rana and Roca, 1975
Cacao necrosis	*Longidorus sp.* suspected	Kenton, 1977
Cherry leaf roll	*X. diversicaudatum*	Fritzsche and Kegler, 1964
	X. coxi	Fritzsche, 1964
	X. vuittenezi	Flegg, 1969
Cherry rasp leaf	*X. americanum*	Nyland *et al.*, 1969
Grapevine chrome mosaic	*X. index*	Mali, *et al.*, 1975
	X. vuittenezi suspected	Martelli and Sarospataki, 1969
Grapevine fanleaf	*X. index*	Hewitt *et al.*, 1958
	X. italae	Cohn *et al.*, 1970
Mulberry ringspot	*L. martini*	Yagita and Komuro, 1972
Myrobalan latent ringspot	*Longidorus* sp. suspected	Dunez and Dupont, 1976
Peach rosette mosaic	*X. americanum*	Dias and Cation, 1974
	Criconemoides xenoplax	Klos *et al.*, 1967
Raspberry ringspot type strain	*L. elongatus*	Taylor, 1962
English strain	*L. macrosoma*	Harrison, 1964b
	L. elongatus	Taylor and Murant, 1969
cherry strain	*L. macrosoma*	Fritzsche and Kegler, 1968
	X. diversicaudatum	Fritzsche and Kegler, 1968
Strawberry latent ringspot	*X. diversicaudatum*	Lister, 1964
	X. coxi	Putz and Stocky, 1970
Tobacco ringspot	*X. americanum*	Fulton, 1962
	X. coxi	van Hoof, 1971
Tobacco ringspot satellite	*X, americanum*	McGuire and Schneider, 1973
Tomato black ring English strain	*L. attenuatus*	Harrison *et al.*, 1961
Scottish strain	*L. elongatus*	Harrison *et al.*, 1961
Tomato ringspot	*X. americanum*	Breece and Hart, 1969
	X. brevicolle	Fritzsche and Kegler, 1968
Other Polyhedral Viruses		
Brome mosaic	*X. diversicaudatum*	Schmidt *et al*, 1963
	L. macrosoma	Fritzsche, 1970
Carnation ringspot	*X. diversicaudatum*	Fritzsche and Schmelzer, 1967
Cowpea mosaic	*X. basiri*	Caveness *et al.*, 1975
Prunus necrotic ringspot	*L. macrosoma*	Fritzsche and Kegler, 1968

16.2.1 Nepoviruses

Nepoviruses have isometric particles about 28 nm in diameter, with angular outlines. They have bipartite single-stranded RNA genomes, with the two parts occurring in different particles that have identical protein coats. Both parts are needed to

TABLE II Tobraviruses and Their Nematode Vectors

Virus	Isolate	Vector	Reference
Pea early-browning	Dutch	*Paratrichodorus pachydermus*	van Hoof, 1962
		P. teres	van Hoof, 1962
	English	*P. anemones*	Harrison, 1967
		P. pachydermus	Gibbs and Harrison, 1964
		Trichodorus viruliferus	Gibbs and Harrison, 1964
	Italian	*T. viruliferus*	van Hoof *et al.*, 1966
Tobacco rattle	Dutch	*P. anemones*	van Hoof, 1968
		P. nanus	van Hoof, 1968
		P. pachydermus	Sol and Seinhorst, 1961
		P. teres	van Hoof, 1968
		T. cylindricus	van Hoof, 1968
		T. sparsus	van Hoof, 1968
		T. viruliferus	van Hoof, 1968
Gladiolus notch-leaf	Dutch	*T. similis*	Cremer and Kooistra,1964
	English	*P. pachydermus*	Gibbs and Harrison,1964
		T. primitivus	Mowat and Taylor, 1962
	German	*T. primitivus*	Sänger, 1961
	Japanese	*P. minor*	Komuro *et al.*, 1970
	North American		
	(California)	*P. allius*	Ayala and Allen, 1966
		P. minor (= christiei)	Ayala and Allen, 1966
		P. porosus	Ayala and Allen, 1966
	(Oregon)	*P. allius*	Jensen and Allen, 1964
	(Wisconsin)	*P. minor*	Walkinshaw *et al.*, 1961

produce an infection (Harrison *et al.*, 1972; Quacquarelli *et al.*, 1976; Randles *et al.*, 1977) but each has different biological functions. In raspberry ringspot and tomato black ring viruses, the larger RNA-1 carries determinants for host range, seed transmissibility and symptom expression, whereas RNA-2 carries determinants for serological specificity and nematode transmissibility (Harrison *et al.*, 1974a; Hanada and Harrison, 1977; Harrison and Murant, 1977b).

As a group nepoviruses are found throughout the world, but individual viruses have a restricted geographical distribution determined by their nematode vector. In most instances the viruses are not disseminated to the limits of their vector distribution, the exception being grapevine fanleaf virus which has been spread together with its vector, *Xiphinema index*, to most areas of the world where grapevines are grown. The 19 nepoviruses that have so far been characterized (Harrison and Murant, 1977a) can be divided into those that commonly occur and originate in Europe and those from North America; together with mulberry ringspot virus found only in Japan (Tsuchizaki, 1975) and cacao necrosis virus found only in West Africa and for which a nematode vector has only tentatively been identified (Kenten, 1977).

Tobacco and tomato ringspot viruses are largely confined to North America, where they are associated with *X. americanum sensu lato.** Despite its name tomato

*Since the writing of this chapter, Lamberti (1979) (*Nematol. medit*. 7: 51-106) has described 15 new species of *X. americanum sensu lato*. He considers *X. americanum sensu stricto* to be geographically restricted to the eastern part of the North American continent; he regards *X. americanum* as described from California as *X. californicum* n.sp. One can speculate that *X. californicum* is a vector of tomato ringspot virus but not of tobacco ringspot virus, whereas *X. americanum sensu stricto* is a vector of both viruses.

FIG. 1. (a) An outbreak of strawberry latent ringspot virus, transmitted by *Xiphinema diversicaudatum*, in a plantation of raspberry cv. Malling Jewel. (b) *Longidorus elongatus* feeding at the root tip of *Loluim perenne*.

ringspot virus is less serious in tomatoes than in woody hosts such as raspberry and blackberry in which it causes mosaic or ringspot symptoms, peach in which it causes yellow bud mosaic and stem pitting, or grapevine in which it causes yellow vein disease. Tomato ringspot virus has also been implicated as the cause of apple union necrosis and decline, which is prevalent in Pennsylvania, Maryland, New York and other fruit-growing areas of the eastern United States (Stouffer, 1977; Stouffer *et al.*, 1977). It is discontinuous in its distribution, occurring along the Pacific coast from California to southern British Columbia and then in the region of the Great Lakes and in the Niagara peninsula of Canada. Tomato ringspot virus is endemic in wild hosts along the California coast (Frazier *et al.*, 1961) and because these are symptomlessly infected whereas cultivated hosts show varying symptoms of disease, it is suggested that this is indicative of its region of origin (Stace-Smith, 1978). Tobacco ringspot virus is virtually absent from the Pacific coast but is generally prevalent elsewhere in the United States and southern Canada. It is the cause of diseases in cherry and blueberry but is primarily a problem in annual crops such as tobacco, soybean, eggplant and cucurbits.

Peach rosette mosaic virus appears to be largely restricted to Michigan State, USA, with a few reports of its occurrence in peach from New York State (Hildebrand, 1941). In addition to infecting peach, the virus also causes a disorder of grapevine (*Vitis labrusca* cv. Concord) with symptoms of delayed spring growth, poor yield and sometimes death (Ramsdell and Myers, 1974). The virus has been transmitted erratically by *X. americanum* (Dias and Cation, 1976) and *Criconemoides xenoplax* has also been reported as a vector of the virus in peach (Klos *et al.*, 1967). Cherry rasp leaf is another North American virus transmitted by *X. americanum* but with a distribution apparently restricted to the west of the Rocky Mountains from California, Colorado and Utah to southern British Columbia. It causes two diseases, cherry rasp leaf and flat apple, and has been found in several weeds growing under peach and cherry trees (Stace-Smith and Hansen, 1976).

Arabis mosaic was the first of the European viruses to be associated with a nematode vector, *X. diversicaudatum* (Harrison and Cadman, 1959; Jha and Posnette, 1959). The virus was first isolated from a single plant of *Arabis* growing in a glasshouse at Cambridge, England (Smith and Markham, 1944) but subsequently it was found to be the cause of yellow dwarf disease of raspberry, mosaic and yellow crinkle of strawberry, stunt of cucumber and diseases of varying severity in sugar beet, white clover, rose, narcissus and in association with viruses of the Prunus necrotic ringspot type causes one form of rasp-leaf of cherry. Most of the isolates from different hosts seem to be closely related to the type strain of the virus (Cadman, 1960; Hollings, 1963) but the strain from hop is serologically distinguishable (Bock, 1966) and also has a more restricted host range. Arabis mosaic virus infects grapevine and is also distantly related to grapevine fanleaf virus (Cadman *et al.*, 1960).

Strawberry latent ringspot virus is also transmitted by *X diversicaudatum*. It infects strawberry and raspberry causing mottle and decline (Fig. 1a), causes strap-leaf of celery, and in rose is associated with chlorotic ringspotting and stunting (Murant, 1974). Many other woody and herbaceous hosts are infected and as with other viruses in this group strawberry latent ringspot virus is transmitted through the seed of some species. Isolates from olive, peach and raspberry in Italy differ serologically from the type strain of the virus (Martelli, personal communication) which was described from strawberry and other crop plants in Great Britain (Lister, 1964).

Cherry leaf roll virus was described from cherry (Cropley, 1961) but different strains of the virus have also been identified from elm, golden elderberry and rhubarb (Jones and Murant, 1971). The virus is also present in North America and although it may have been imported in golden elderberry from Europe and perpetuated by propagation, its presence in native dogwood (*Cornus florida*) and elm (*Ulmus americana*) suggests that an isolate of the virus may have originated there (Stace-Smith, 1978). In Europe, the cherry strain of the virus is reported to be transmitted by *X. diversicaudatum*, *X. coxi* and *X. vuittenezi* (Fritzsche and Kegler, 1964; Flegg, 1969a). The first two species are known in North America, but it would seem that there, as perhaps also in Europe, the main mechanism for the spread of cherry leaf roll virus is through infected seed and pollen (Fulton and Fulton, 1970; Jones and Murant, 1971; Stace-Smith, 1978).

Grapevine fanleaf virus should perhaps be designated as an Old World virus, as there is some evidence that with its major vector, *X. index*, it originated in ancient Persia (Hewitt, 1968; Weischer, 1975). From there infected vines, carrying *X. index* on soil attached to the roots, were distributed throughout Europe during the time of the Roman occupation and later to the New World and to countries such as Australia in the southern hemisphere. Specific strains of the virus cause fanleaf, yellow mosaic and veinbanding diseases of grapevine, and some isolates are associated with leaf enations, bark and wood pitting, and flat trunk disease. Loss of yield due to infection with grapevine fanleaf virus can be of the order of 50% (Vuittenez, 1970; Dias, 1977). Grapevine fanleaf is uniquely linked with grapevine, which is its own natural host, but many other nematode transmitted viruses infect grapevine and are responsible for economic losses. These include arabis mosaic, strawberry latent, raspberry ringspot, tomato black ring and grapevine chrome mosaic among the European viruses, and tomato ringspot, tobacco ringspot and peach rosette mosaic of the North American viruses.

Grapevine chrome mosaic virus was described from Hungary where it was associated with chrome-yellow mosaic, stunting and decline of grapevine (Martelli and Quacquarelli, 1972) and believed to be spread by *X. vuittenezi* (Martelli and Sarospataki, 1969). In England, an isolate of the virus causes celery yellow-vein disease (Hollings *et al*., 1969).

Four other viruses of economic importance in Europe are transmitted by *Longidorus* spp. One of the strains of raspberry ringspot virus, transmitted by *L. elongatus*, causes ringspot and leaf-curl disease of raspberry and in the Netherlands a similar isolate is the cause of spoonleaf disease of red currant. A serologically distinct strain of the virus causes diseases in raspberry and blackberry in England and is transmitted by *L. macrosoma*. Tomato black ring virus also exists as several different strains. The so-called beet ringspot strain is transmitted by *L. elongatus* and the lettuce ringspot, potato bouquet, celery yellow vein and the type strain are transmitted by *L. attenuatus*. Both raspberry ringspot and tomato black ring viruses cause diseases in a wide range of crop plants and also infect many weed hosts, usually symptomlessly, in many of which they are seed-borne. Artichoke Italian latent virus, transmitted by *L. apulus* (in early references to transmission identified as *L. attenuatus*), has so far been recorded only from southern Italy where it causes stunting and yellowing in artichoke and chicory (Vovlas *et al.*, 1971; Martelli and Rana, 1975), and from Bulgaria where it was isolated from grapevine showing fan-like symptoms and poor growth (Jankulova *et al.*, 1976). Myrobalan latent ringspot virus is latent in myrobalan, but causes short internodes and rosetting in peach and enations on the leaves of bird cherry (Dunez and Dupont, 1976). It has been found only in southwest France and nematode transmission by *Longidorus* sp. is suspected, although not established, because of its distant serological relationship with tomato black ring virus.

Mulberry ringspot and cacao necrosis are nepoviruses with apparently restricted distribution and host range (Tsuchizaki, 1975; Kenton, 1977). The former has been found only on mulberry in Japan and is transmitted by *L. martini* (Yagita and Komuro, 1972). The latter causes leaf necrosis, defoliation and dieback of cocoa and occurs in West Africa; its distant serological relationship with tomato black ring virus suggests *Longidorus* sp. as a vector and *Longidorus* spp. have been found in soils with outbreaks of cocoa necrosis disease in Ghana (Kenton, 1977). Potato black ringspot virus obtained from cultivated potato in Peru is distantly related serologically to tobacco ringspot virus (Salazar and Harrison, 1977), but no nematode vector has yet been identified.

Brome mosaic and carnation ringspot viruses are not classified as nepoviruses but both are reported to be transmitted by nematodes. Brome mosaic infects bromegrass and other wild grasses in central North America, eastern Europe and South Africa, and it has been transmitted by *Xiphinema* spp. in the laboratory (Schmidt, Fritzsche and Lehmann, 1963; Fritzsche, 1975). Carnation ringspot virus causes leaf mottle, ringspots, stunting and distortion and occurs wherever carnations are grown in temperate regions; it is reported to be transmitted by *L. macrosoma* and *X. diversicaudatum*. (Fritzsche and Schmelzer, 1967).

16.2.2 Tobraviruses

Tobraviruses have straight tubular particles of two predominant lengths, long 180-210 nm and short 45-115 nm, with an overall diameter of approximately 23 nm. Their single stranded RNA genomes are in two parts and, as with nepoviruses, occur in the different particles. Both particles are necessary to establish infection. There are only two viruses in the group, tobacco rattle and pea early-browning, and their natural vectors are species of *Paratrichodorus* and *Trichodorus*. Both viruses infect a wide range of wild and cultivated plants, tobacco rattle virus having been isolated from 100 species representing 50 different families of dicotyledons and monocotyledons.

Tobacco rattle was first described by Quanjer (1943) who recognized it as the causative agent of "Streifen-und Kräuselkrankheit" (Boning, 1931) and "Mauche" disease (Behrens, 1899). Nowadays these diseases are no longer a major problem and tobacco rattle is better known as the cause of economically important diseases in bulbous flower crops in the Netherlands and to some extent in Great Britain, and of one type of potato spraing disease (= corky ringspot of tubers) in Europe and the United States. The symptoms of spraing disease consist of arcs of corky tissue in the tuber flesh and when infected tubers are planted a few of the progeny plants develop stem-mottle disease. In turn, tubers from the diseased plants may be misshapen and contain brown spots or may develop spraing-like lesions and be of fairly normal shape (Harrison and Robinson, 1978). In several bulbous ornamentals, tobacco rattle virus induces mottling, distortion and stunting; in gladiolus it causes notched leaf symptoms (Cremer and Schenk, 1967). Several different strains of the virus have been described including those from potato in Scotland (type strain) and Oregon; and from pepper in Brazil and California (Harrison, 1970).

Early-browning disease of pea was first recognized in the Netherlands in 1949 and was subsequently shown to be caused by a soil-borne virus that was similar to, but distinct from, tobacco rattle virus (Bos and Van der Want, 1962). The virus has also been isolated from pea in eastern England(Gibbs and Harrison, 1964) and from populations of infected *T. viruliferus* in Italy (van Hoof *et al.*, 1966; Roca *et al.*, 1977), but it has not been reported from the United States. Symptoms of infection in pea range from scattered necrotic areas on the leaves and stipules of a few plants, to severe stunting and premature death of plants in large areas of a crop. In the Netherlands, the virus has been recorded in 25% of the seed from infected mother plants (Bos and Van der Want, 1962) but in Britain the virus was only 1-2% seed-borne (Harrison, 1973).

16.3 NEMATODE VECTORS

Nematodes that transmit plant viruses are included in two sub-orders of the Dorylaimida, class Adenophorea, whereas most other plant parasitic nematodes are in the order Tylenchida, class Secernentea (Hooper, 1975). Within the order Dorylaimina there are two divergent lines (Maggenti, 1971), one being the sub-order Dorylaimina containing *Longidorus* and *Xiphinema* vectors of nepoviruses in the family Longidoridae, and the other the sub-order Diphtherophorina containing *Trichodorus* and *Paratrichodorus* vectors of tobraviruses in the family Trichodoridae.

Longidorus and *Xiphinema* are large nematodes, 2-12 mm long in the adult stage, and are characterized by the elongate, hollow odontostyle or spear with which they pierce the plant roots and feed on the cell contents (Fig. 2). *Longidorus* species have a short cheilostome and the guiding sheath is not folded when the odontostyle is retracted; whereas in *Xiphinema* the cheilostome is much longer, resulting in a posteriorly situated guide ring with the guiding sheath remaining partly folded beyond the guide ring when the odontostyle is retracted. The genera also differ in the morphology of the odontophore which in *Xiphinema* is more sclerotized than in *Longidorus*, and has a flanged or knobbed base.

Trichodorus and *Paratrichodorus* are short, stout nematodes, 0.5-1.5 mm long (Fig. 2). The stoma is armed with a non-axial, ventrally curved mural tooth (onchio-

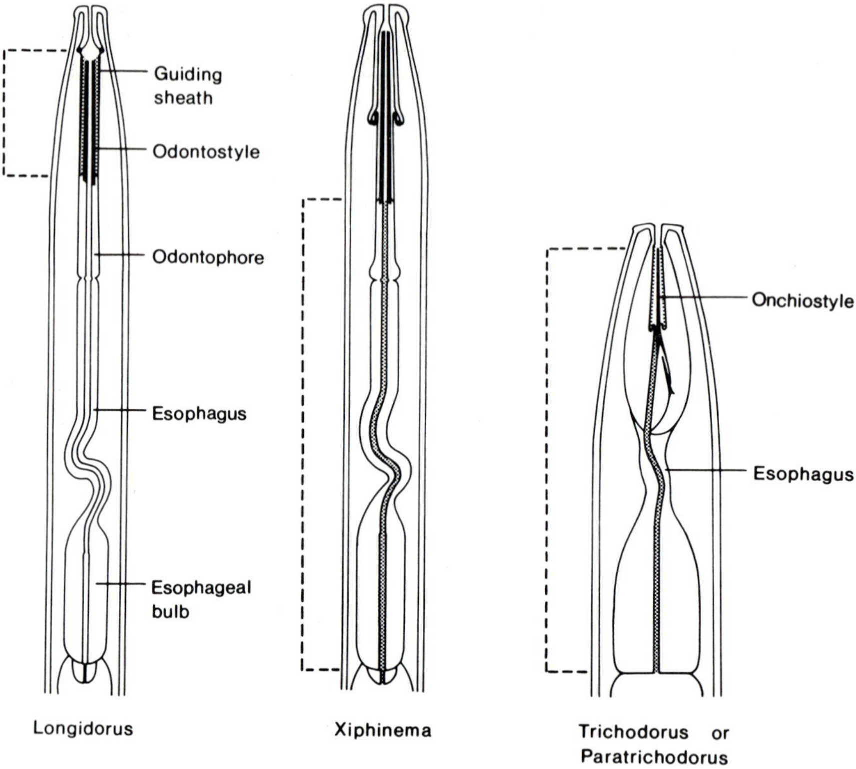

FIG. 2. Sites of specific retention of viruses in vector nematodes. The virus is associated with the cuticle of the alimentary tract in the areas indicated (hatching and dotted lines).

style), the middle region of which fuses with the pharyngeal wall. The solid onchiostyle is used to penetrate the epidermal cell walls of the root tip and to disturb the cell contents during ingestion, and is not used as a feeding tube as in longidorids or tylenchids (see later).

16.3.1 Geographical Distribution and Hosts

Interest in the taxonomic relationship of dorylaimid nematodes was stimulated by the discovery of their role as virus vectors. So much so that 80-90% of the present *Xiphinema* and *Longidorus* species and over 50% of the *Trichodorus* and *Paratrichodorus* species have been described since 1958. More than 60 species of *Xiphinema* are recognized, predominantly from the tropics and sub-tropics, but few species have been implicated as virus vectors; 35 species of *Longidorus* were recognized by Lamberti (1975) and Loof (1975) lists 45 described species of *Trichodorus* or *Paratrichodorus*.

X. americanum is common and widely distributed in agricultural and forest soils in the United States (Adams and Eichenmuller, 1964) and is also recorded from northern India (Siddiqi, 1959), Sri Lanka (Loos, 1949), South America (Thorne and Schieber, 1962), Korea (Lee and Han, 1976), Japan (Iwaki and Komuro, 1974) and from Russia (Ivanova, 1972; Milkus *et al.*, 1975). Early records of its occurrence in Europe and the mediterranean regions probably all relate to *X. pachtaicum*. In the United States, *X. americanum* infestations have been associated with the stunting of ornamental spruce (Griffin and Epstein, 1964), necrosis and destruction of feeder roots of laurel oak (Christie, 1952), unthriftiness of red clover (Norton, 1967) among the wide range of host plants that are attacked.

The geographical distribution of *X. index* is worldwide but closely related to that of its natural host, grapevine. It is present throughout the mediterranean area and has been exported with its host to California, Argentina, Chile, South America and Australia. Although in nature *X. index* is closely associated with grapevine (Weischer, 1975), it also multiplies successfully on other *Vitis* spp., fig and rose (Raski and Lider, 1959; Radewald, 1962) and cultures have been bred in England on tomato, potato and *Solanum nigrum* (Southey, 1973).

X. italiae has also been shown to be a vector of grapevine fanleaf virus, in addition to *X. index* (Table I) but whilst it is widely distributed throughout most of the mediterranean basin and also occurs in some southeastern European countries and in South Africa (Smith, 1975), it does not occur in North or South America. In France *X. italiae* populations mostly are limited to vineyards of the mediterranean region where it co-exists with *X. index* (Dalmasso, 1970) but in Italy and Sicily it has been found in association with citrus, olive, mulberry and conifers, as well as grapevine (Martelli and Raski, 1963) and in Bulgaria on apricot, plum and peach (Choleva, 1975).

X. diversicaudatum is widespread in Europe and of frequent occurrence in parts of the United States, Asia, Australia and New Zealand. It has a wide host range among woody and herbaceous crops and weeds, and can be damaging to strawberry (Schindler and Braun, 1957), rose (Coolen and D.Herde, 1970), raspberry and other rosaceous crops. It mainly occurs in temperate regions and appears to be restricted to cool, humid areas, e.g. in France it is abundant in the northern part along river banks or along the Channel or Atlantic coasts but not near the drier mediterranean coast (Dalmasso, 1970); in Germany, Fritzsche (1966) found *X. diversicaudatum* was most prevalent in moist, shady sites in the vicinity of water or in soils with a high water table.

X. coxi is recorded as a vector of several nepoviruses (Fritzsche, 1964; van Hoof, 1971; Putz and Stocky, 1970). It was first recorded in Florida, USA (Tarjan, 1964) and has since been found in France, Germany, the Netherlands and in a few widely scattered localities in England (Taylor and Brown, 1976). In Europe, *X. coxi* is considered to be a denizen of wet uncultivated soils and because it adapts badly to cultivation (Dalmasso, 1970) it is thus unlikely to be important as a vector of viruses in crop situations. European populations, however, differ morphologically from those described from Florida (Dalmasso, 1970; Taylor and Brown, 1976). *X. vuittenezi* is a possible vector of cherry leaf roll and grapevine chrome mosaic viruses. It is widespread in Germany and eastern France, most frequently in association with grapevines, and is said to be the commonest *Xiphinema* species in Poland (Dalmasso, 1970); it has been recorded at a few localities in Great Britain probably having been introduced with rooted plants from the Continent.

X. pachtaicum is suspected as a vector of stem pitting disease of peach in Italy (Giunchidi and Tacconi, 1974); and grapevine fanleaf virus in Spain (Alfaro Garcia, 1971). It was formerly known as *X. mediterranean* (Martelli and Lamberti, 1967) and as the name implies is chiefly a mediterranean species being common and widely distributed in a variety of habitats such as cultivated land, shrubland, forest and permanent pasture. It has been recorded only on six occasions in southeast England (Flegg, 1969) but these are regarded as fugitive populations from the Continent, particularly as the species is notably absent in Belgium and the Netherlands (Taylor and Brown, 1976).

X. vuittenezi, reported as a vector of cherry leaf roll virus (Flegg, 1969) is widespread in Germany and eastern France, extending throughout central Europe where in some areas it is the commonest *Xiphinema* species. Except in Great Britain, where there are only a few scattered populations, it is often found in association with grapevines. Other *Xiphinema* species suspected as virus vectors include *X. brevicolle*, again often found in association with grapevines in Europe and South Africa, *X. bakiri* found in western United States, British Columbia and Japan, and *X. basiri* which has been associated with cowpea mosaic in Nigeria.

Longidorus species are found throughout the world, though the virus vector species are mostly from northern Europe. An exception is *L. martini* which is the vector of Mulberry ringspot virus in Japan (Yagita and Komuro, 1972). *L. elongatus* is the most widely distributed of the species occurring in temperate climates and has been recovered throughout northern and central Europe, in Canada and in northern United States. *L. macrosoma* is fairly common in Continental Europe west of the Rhine but in England is limited to the southern half of the country. *L. attenuatus* is common in light, sandy well-drained soils in eastern England and has been recorded locally from France and Germany. *L. attenuatus* has also been recorded from Italy but those populations previously associated with artichoke Italian latent virus are now referred to *L. apulus* and those from Sardinia are referred to *L. protae* (Lamberti and Bleve Zachev, 1977).

Trichodorid nematodes are widespread in North America and Europe and have been recorded from all the main continental land masses of the world and from many islands (Winfield and Cooke, 1975). Eleven species in the genera *Trichodorus* and *Paratrichodorus* (Siddiqi, 1973) have been shown to transmit tobraviruses. A few species such as *P. minor* and *P. porosus* are widely distributed among many countries, and *P. pachydermus* and *T. primitivus* are prevalent in Europe. Some species, however, are fairly locally distributed although there is evidence that they may from time to time successfully be distributed to new localities.

16.3.2 Distribution in Soil

The geographical distribution of many vector species can be correlated with soil type. Pore space and moisture holding capacity are probably the most important characteristics of a soil in determining its suitability as a habitat for nematodes. These characteristics can be deduced from soil type, which is an expression of the relative amounts of sand, silt and clay particles constituting the particular soil.

In Europe, *X. diversicaudatum* has often been reported from clay soils (Harrison and Winslow, 1961; Fritzsche, 1966; Jones *et al.*, 1969) but a recent survey in Great Britain (Taylor and Brown, 1976) showed that it can occur in a wide range of soils, and most commonly in those ranging from loam to silty clay loam. The survey was based on several thousand soil samples from five types of vegetation, and led to the general conclusion that in Great Britain, *Longidorus* and *Xiphinema* species are found most frequently in lighter soils with the greatest proportion of all the species in the sandy loam to loam categories. Analysis of soil samples of single populations of *X. diversicaudatum* , *i.e.* when no other *Longidorus* or *Xiphinema* species were present, indicated a modal distribution in soils with an average of 53% sand; a similar analysis for *L. elongatus* showed a modal distribution in soils with an average of 68% sand. In southern Germany, *X. vuittenezi* is more frequent in loess than in sandy soils whereas *X. index* occurs mainly in lighter soils (Ruedel, 1971) although soil type does not appear to be a limiting factor for its geographical distribution (Weischer, 1966; 1974). In Israel, *X. pachtaicum* are predominant in clay soils whereas the larger longidorids prevail in lighter soils (Cohn, 1969); in other countries it occurs in soils ranging from light to heavy textured (Arias, 1975; Choleva, 1975).

Most of the observations on the distribution of *Longidorus* and *Xiphinema* in relation to soil types are based on samples from cultivated soils and fail to indicate whether the species present are a residue of the original fauna or have recently been introduced. A study of natural habitats may help to identify the dominant factors influencing distribution and the characteristics of cultivation which are favorable, or otherwise, to nematodes. *X. index* is an obvious example of a nematode which is favored by cultivations and has only been reported once from a natural habitat (Sturhan quoted in Weischer, 1975). On the contrary, *X. coxi* is restricted to uncultivated or marginal soils and tends to disappear when the soil is cultivated (Weischer, 1975; Taylor and Brown, 1976). Rau (1975) investigating virus vector nematodes in undisturbed natural vegetation biotypes in northwest Germany found that some species were strictly confined to certain habitats. For example, on a ridge of carboniferous limestone, *X. diversicaudatum* and *L. macrosoma* were common in the shallow soil covering the slopes and top, where the plant community was characterized by beech with grass (*Melica uniflora*); *L. elongatus* was absent there but found in the deeper soils at the base of the ridge in an oak/hornbeam habitat (Fig. in Weischer, 1975). Taylor *et al.* (1978) have speculated that underlying rocks may influence the distribution of *L. macrosoma* in Great Britain noting that the species is recorded most frequently in clay soils overlying chalk outcrops in southern England.

Trichodorid nematodes are almost entirely restricted to free-draining sandy soils and are rarely found in clay or silt soils. In the Netherlands, *P. teres* is found in marine, sandy soils, whereas *P. pachydermus* is generally widespread, but usually in sandy soils rather than those with much clay (van Hoof, 1962) and Rossner (1969) recorded trichodorids in light, sandy forest-tree nurseries in West Germany but not in clay soils. In a survey of sugar beet fields prone to Docking disorder in eastern England, trichodorid nematodes were most abundant in soils with an 80% or more sand fraction but were not found in fields with an average of 56% sand and in which the disorder rarely occurred (Cooke, 1973). From a survey undertaken in Great Britain (Alphey and Boag, 1976) 50% of the trichodorid populations were found in soils with a sand fraction greater than 80% and a silt fraction of less than 10%. In Scotland, Cooper (1971) found virus-carrying trichodorids in 68% of the freely drained podsols sampled, but did not recover them from heavy, badly drained soils. It has been postulated by Jones *et al*. (1969) that trichodorids are confined to soils with 80-90% sand (Alphey and Boag, 1976) because they are unable to penetrate the densely packed soils containing much clay, silt or fine sand with particle diameters of less than 50 μm. However, trichodorids probably cannot colonize very coarse sands, such as dune sand (Yeates, 1967) although populations may become established when the dunes are stabilized (Cooper and Harrison, 1973) and are capable of retaining sufficient moisture for survival and movement of the nematodes.

Nematodes in general are distributed irregularly in the soil both horizontally and vertically. Because the nematode vector species are ectoparasites feeding mainly at the root tips, their distribution vertically in the soil generally will correspond with that of suitable roots. For example, the bulk of *L. elongatus* populations were found in the top 10 cm of the soil in association with the roots of its preferred herbaceous hosts (Taylor, 1967) and in strawberry plantations in Poland *L. elongatus* was most numerous at 10-12 cm depth (Szczygiel and Hasior, 1972). On the other hand, *X. diversicaudatum* feeds on a wide variety of plants ranging from surface-rooting herbaceous crops and weeds to deeper-rooting woody hosts such as raspberry, and populations may be distributed to a depth of 60 cm or more (Harrison and Winslow, 1961; Taylor and Thomas, 1968; Cotten, 1976). The largest populations of *X. index* were recorded at 30-60 cm depth in vineyards established on the stony, shallow soils of the Cote d'Or, France, but in the more fertile and deeper soils of the California Napa Valley populations were distributed to a depth of 3.6 m (Raski *et al*., 1965). In Israel, it was found that the maximum numbers of *Xiphinema* or *Longidorus* occurred in the 0-30 cm zone, irrespective of species, with numbers decreasing with increasing depth (Cohn, 1969). Flegg (1968a) found that *L. macrosoma* increased with depth up to 70 cm and Cotten (1975) recovered more than 800 *L. macrosoma*/litre at a depth of 60 cm in raspberry plantations.

Both the horizontal and vertical distribution of trichodorid species seems to be more variable than that of longidorids. Fluctuations in population densities are related to the susceptibility of trichodorids to dry conditions and to their ability to reproduce rapidly under suitable conditions and if suitable host plants are available. When the soil remains at or above field capacity for long periods of time large populations can be maintained, especially where host roots are abundant. Under drought conditions numbers decrease rapidly, but below the cultivated layer of soil (20-30 cm depth) the nematodes are not subject to moisture or mechanical stress and more stable populations are maintained (Winfield and Cooke, 1975). Hijink and Kuiper (1966) found that in the majority of sites sampled, the greatest numbers of trichodorids were below the cultivated layer (20 cm). However, Kuiper and Loof (1962) found most *P. teres* at 5-10 cm depth in a sugar beet crop; there were fewer nematodes at 15-30 cm where there was little root growth, but numbers increased again below 30 cm. Richter (1969) concluded that generally trichodorids inhabit deeper soil layers than other plant parasitic nematodes, but noted differences between species in vertical distribution.

16.3.3. Feeding and Pathogenicity

The feeding apparatus and feeding behavior of longidorid and trichodorid nematodes are quite distinct, but both groups feed at or near the root tips and to a greater or lesser extent cause galling and necrosis. Christie and Perry (1951) provided the first evidence of parasitism by trichodorids when they observed *P. minor* (= *christiei*) injuring the roots of beet, celery and sweet corn in Florida, USA. Horner and Jensen (1954) were the first to associate *Longidorus* with root damage when they demonstrated that *L. elongatus* was the cause of severe necrosis or "die-back" of the roots of peppermint in Oregon, USA; and a few years later Schindler (1957) gave a detailed account of the pathogenicity of *X. diversicaudatum*.

Xiphinema species differ in their selection of feeding sites. Feeding by *X. americanum* is rarely at the root tips and is associated wtih necrosis and discoloration of the cortical tissues, sometimes leading to complete destruction of the roots. These symptoms have been reported on several woody hosts (Christie, 1952; Griffin and Epstein, 1964; Di Sanzo and Rohde, 1969), on strawberry (Perry, 1958) and periwinkle (Epstein and Barker, 1966). *X. diversicaudatum* feeds almost exclusively at root tips causing prominent terminal or subterminal swellings as reported on rose (Schindler, 1957), strawberry (Schindler and Braun, 1957), celery (Harrison and Winslow, 1961) and several other herbaceous plants (Thomas, 1970). Similar symptoms have been reported for *X. bakeri* on forest tree seedlings (Sutherland, 1970) and raspberry (McElroy, 1972); and for *X. index* on grapevine (Raski and Radewald, 1958).

Longidorus species feed exclusively at the root tips, as far as is known, and they appear to cause relatively less severe galling than *Xiphinema* (Thomas; 1969; 1970, Fig. 1b). Nevertheless, feeding by *Longidorus* results in galling of the root tips and stunting of the lateral and sometimes tap roots, but little or no necrosis is usually evident on the roots except at the tips (Cohn, 1975). Damage of this sort has been reported for *L. elongatus* on strawberry (Sharma, 1965; Seinhorst, 1966; Wyss, 1970a; McElroy, 1972), sugar beet, carrot (Whitehead *et al.*, 1971) and clover (Thomas, 1969); *L. attenuatus* on strawberry (Wyss, 1970a) and sugar beet (Whitehead and Hooper, 1970); *L. africanus* on lettuce (Radewald *et al.*, 1969) and

L. vineacola on onion and celery (Cohn and Ausher, 1971).

The galls induced by *Longidorus* and *Xiphinema* are regarded as evidence of the discharge by the nematodes of some substance into the plant during feeding that triggers off a sequence of events leading to gall formation (Cohn, 1975; Trudgill, 1976; Towle and Doncaster, 1977). Salivation, *i.e.* depletion of the ducts of the dorsal gland cells within the oesophageal bulb, has been observed in *X. index* (Wyss, 1977a, 1977b) and *X. diversicaudatum* (Trudgill, 1976). Histological changes in root-tip galls induced by the nematodes include hyperplasia, swelling of cell nuclei, clusters of small cells with enlarged nuclei (Cohn and Orion, 1970), formation of giant cells (Davis and Jenkins, 1960), enlargement of cells, enlargement of nucleoli, accumulation of protein, cell division and cell wall thickening (McElroy and van Gundy, 1968). Weischer and Wyss (1976) speculated that the modified cells forming the gall tissue might function as highlý modified nurse cells and are selected by the nematodes in preference to other tissues in the roots (Towle and Doncaster, 1978; Wyss, 1978).

Feeding behavior in *Xiphinema* and *Longidorus* may be described in several consecutive phases (Wyss, 1975; 1978; Trudgill, 1976; Towle and Doncaster, 1978). Before penetration of the roots with the odontostyle, the nematodes explore the root surface with their lips. Penetration is achieved by rapid thrusting of the odontostyle and ingestion may start as soon as the first cell has been penetrated (Sutherland, 1969; Trudgill, 1976) although in some cases penetration may continue until the stele has been reached (Trudgill, 1976; Towle and Doncaster, 1977). Ingestion is achieved by pulsation of the oesophageal bulb and is typically intermittent, *i.e.* periods of bulb pulsations alternating with short quiescent periods (Cotten, 1973; Cohn, 1970; Weischer and Wyss, 1976). Salivation occurs immediately after ondontostyle thrusting has ceased during the penetration phase and also during the periods of pumping quiescence (Wyss, 1978).Most nematodes cease feeding and retract their odontostyles only after full protraction has been reached (Trudgill, 1976). Pulsation of the oesophageal bulb continues after the withdrawal of the odontostyle from the root.

Feeding of trichodorid nematodes is restricted to the epidermal cells of the root tip producing a condition descriptively termed "stubby-root" (Christie and Perry, 1951); necrosis of the root cells may also occur to a greater, or lesser degree (Pitcher and Flegg, 1965; Wyss, 1975). Winfield and Cooke (1975) refer to many examples of damage to crops caused by trichodorids. Trichodorids aggregate at the tips of young, growing roots and by their feeding cause decline and finally cessation of growth; in turn, new lateral root tips are attacked and they become stubby and sometimes necrotic. Russell and Perry (1966) suggest that the stubby-root condition may result from the injection of oesophageal secretions by the nematodes into the meristem. Hoff (1964) showed that *P. minor* feeding on onion caused an increase in the size of the cortex both radially and longitudinally and that the early phases of this stubby-root damage are a result of cell maturation nearer the apical meristem than would normally be the case.

Wyss (1975; 1978) has made a detailed cinematographic analysis of feeding by

T. similis. The nematode explores the root surface by rubbing the lips over the surface of the epidermal cells and then follows by rapid thrusts of the onchiostyle to penetrate the cell wall, which usually takes only a few seconds. Thrusting of the onchiostyle continues less rapidly during the next salivation phase when secretions from the oesophageal glands are injected into the cell by pulsations of the bulb and perhaps aided by the movement of the onchiostyle. Within 5-10 seconds of initial penetration, cytoplasmic streaming within the cell stops and cytoplasm is drawn towards the point of the onchiostyle where it accumulates in a large mass. The ingestion phase is initiated by a few slow thrusts of the onchiostyle into the mass of cytoplasm and large part of it is then withdrawn by the nematode. When most of the accumulated cytoplasm has been ingested, the thrusting of the onchiostyle is discontinued and the stoma is detached from the feeding tube formed by the nematode during salivation and anchored in the cell wall (see Wyss 1975 for illustrations); the nematode then moves to another cell.

16.3.4 Life Cycles

Xiphinema and *Longidorus* species have long life cycles, from several months to two years depending on species, and individual nematodes may survive for several years. In temperate climates, egg laying occurs only in spring and early summer when host plants are actively producing new roots. Egg hatching usually occurs shortly after they are laid but larval development then fails to show a clear relation with season and all stages may be found throughout the year (Flegg, 1966; Dalmasso, 1970). However, in climates with severe winters only eggs and a few larvae may survive in frozen soils (Norton, 1963; Griffin and Barker, 1966).

Laboratory rearing provides some indication of the length of life cycles under near ideal conditions of temperature, soil type, moisture and host status. In California, *X. index* reared on grapevine at 30°C completed its development from oviposition to the adult stage in 22-27 days (Radewald and Raski, 1962) but in Italy the life cycle took 2-4 months at 20-22°C (Prota *et al.*, 1977) and in Israel 7-9 months at 20-23°C (Cohn and Mordechai, 1969). *L. elongatus* developed from egg to adult in 16-23 weeks at 20°C (Yassin, 1969) and in 9 weeks at 30°C (Wyss, 1970b). Optimum temperature for development (usually recorded as rate of population increase) differs between species, with *X. americanum* at 20°C (Griffin and Barker, 1966), *X. index* at 24°C (Radewald and Raski, 1962), *L. elongatus* at 25°C (Wyss, 1970b) and *L. africanus* at 30°C (Lamberti, 1969).

Trichodorids are capable of multiplying rapidly and in general have much shorter life cycles than longidorids. Rohde and Jenkins (1957a,b) found that *P. minor* completed its life cycle in 21-27 days at 22°C and 16-17 days at 30°C with a tenfold or more increase in population in 60 days; Morton and Perry (1968) recorded 17-18 days at 27°C. The life cycle of *T. viruliferus* on apple roots took about 45 days from egg to adult at 15-20°C (Pitcher, 1967; Pitcher and McNamara, 1970). Optimum temperatures for growth and reproduction have been investigated for very few species, with 25°C for *P. minor* (Malek *et al.*, 1965) and 24°C for *P. allius* and *P. porosus* (Ayala *et al.*, 1970).

16.4 VIRUS TRANSMISSION CHARACTERISTICS AND MECHANISMS

The appearance of a patch of diseased plants in a crop is often the first indication of the existence of nematode-transmitted virus infection. The slow spread of infection from the edge of the patch and the persistence of the disease during crop rotations are further clues to soil-borne infection as distinct from viruses transmitted by aerial vectors such as aphids or leafhoppers. The problem of associating nematode vectors with the causative viruses has been approached in different ways but several procedures and techniques are now routinely and widely used in investigations of soil-borne virus outbreaks.

The soil-borne nature of many viruses can be demonstrated by growing healthy plants in soil taken from around the roots of infected plants. Herbaceous seedlings are used as "bait plants" and exposed for 3-6 weeks before being assayed for virus by inoculating sap to suitable test plants. This test does not diagnose the type of vector involved, other than it is present in the soil. If nematodes are suspected as vectors they may be extracted from soil samples from infected areas and hand-picked specimens added to bait plants growing in sterilized potting compost.

Transmission may be affected by the way in which nematodes are extracted from the soil (Taylor and Thomas, 1968), by the potting mixture used in test pots (Ayala and Allen, 1968), by the size of the test pots (Fritzsche, 1967; Valdez, 1972a), by the moisture in the potting mixture (Fritzsche, 1967; van Hoof, 1976), the age and kind of bait plants used (Cadman and Harrison, 1960), and the time of year (Fritzsche and Kegler, 1968) or temperature (Douthit and McGuire, 1974, 1975; van Hoof, 1975) at which the experiment is performed. Such sources of variation and error must be considered in evaluating the performance and ability of nematodes as virus vectors. Transmission is a sequence of interactions between virus, nematode and plant, the probability and efficiency of transmission depending on each event in the sequence being successfully completed (Table III).

TABLE III Sequence of Interactions Between Plant, Virus, and Nematode in the Transmission Process

Sequence	Process	Interaction[a]	Identification[b]
Plant	Multiplication	P/V	mechanical inoculation
Plant/nematode	Feeding	P/N	root galls; nematode inoculation
Nematode	Retention	V/N	electron microscopy
Nematode /plant	Inoculation	P/N/V	root galls
Plant	Infection	V/P	symptoms; bait tests

[a]P = plant; N = nematode; V = virus.

[b]Techniques used to identify that the process and interaction have occurred.

16.4.1 Transmission

Many different nematodes acquire viruses when they feed on infected plants which suggests that viruses are readily accessible in the roots. However, the efficiency and frequency with which nematode vectors transmit nepo- and tobra- viruses increases with increasing times of access to both infector donor plants and to the healthy acceptor or bait plants. The nematodes often spend some time exploring the roots before commencing feeding and also they feed intermittently, so that the increase in transmission efficiency with increasing access time undoubtedly reflects the increasing proportion of nematodes that have fed. This is readily seen in experiments in which different numbers of nematodes are used for transmission (Trudgill and Brown, 1978).

In some early experiments *X. index* was shown to acquire grapevine fanleaf virus (Hewitt and Raski, 1962) and *X. diversicaudatum* to acquire arabis mosaic virus (Jha and Posnette, 1961) when given access of 24 hours to infected plants. It has since been shown that nematode vectors can acquire virus from infected plants in much shorter access periods, *e.g.* tomato ringspot virus by *X. americanum* within an hour (Teliz *et al.*, 1966); tobacco rattle virus by *P. allius* (Ayala and Allen, 1968) within an hour; grapevine fanleaf virus by *X. index* within 15 minutes (Das and Raski, 1968) or within less than 5 minutes (Alfaro and Goheen, 1974). Teliz *et al.* (1966) found that the transmission rate of tomato ringspot virus by *X. americanum* was low and variable with access periods up to 48 hours but high rates of transmission were obtained with 4 day access periods. Using single *X. americanum*, McGuire (1964) obtained few transmissions of tobacco ringspot virus after 24 hour access periods but the rate of transmission increased proportionately with time up to 10 days. It seems probable that nematode vectors acquire virus as soon as sap is ingested from infected roots and that there is no latent period within the nematode before transmission can occur. Inoculation of a healthy plant by a virus-carrying vector can probably occur in a single, brief feed and there are grounds for believing that this may be the most efficient way of virus transmission, particularly in the case of trichodorid nematodes which tend to destroy the plant cells on which they feed (Wyss, 1971; Taylor and Robertson, 1975).

16.4.2 Persistence

Nematode vectors kept in fallow soil without access to plants lose their infectivity slowly, the period of retention much exceeding the *in vitro* longevity of the virus in plant sap. The usual way of investigating the virus retention by vectors is to take nematodes from a highly infective population and add these in batches to pots of sterilized compost in which bait plants are planted at intervals to assay the rate of transmission. Alternatively, samples of field soil can be planted at intervals, but this method is subject to the possible error of germinating weed seeds acting as sources of virus infection (Lister and Murant, 1967). Laboratory experiments done in these ways show that viruses transmitted by *Xiphinema* and trichodorid nematodes may be retained for several months in their vectors and generally for longer periods than those transmitted by *Longidorus*. In any group of vectors, the rate of transmission of an infective population decreases with the period of starvation (Taylor, 1972; McGuire, 1973).

Bergeson *et al.* (1964) obtained transmission of tobacco ringspot virus after 49 weeks of starvation at 10°C and similarly McGuire (1973) showed transmission after 9 months' starvation at 8°C; Taylor and Thomas (1968) showed that *X. diversicaudatum* could transmit strawberry latent ringspot and arabis mosaic viruses after 84 and 112 days starvation, respectively. Viruses are also retained for long periods in *Xiphinema* vectors allowed to feed on virus-immune hosts: *X. index* retained grapevine fanleaf virus for 12 weeks when given access to virus-immune fig (Das and Raski, 1968), and *X. diversicaudatum* given access to Malling Jewel raspberry, which is immune to arabis mosaic virus, retained the virus for 8 months (Harrison and Winslow, 1961). Van Hoof (1970) reported transmission of tobacco rattle virus by *P. pachydermus* in soil stored for about 2 years, and Ayala and Allen (1968) found that the Californian strain of the virus persisted in *P. allius* for 20 weeks in soil without host plants, but for 27 weeks when virus-immune sweetpea was grown in the soil. Pea early-browning virus is also likely to persist for long periods in trichodorid vectors (Harrison, 1966). *Longidorus* vectors retain their associated viruses for shorter periods, usually for no longer than 12 weeks. *L. elongatus* kept at 20°C without plants transmitted raspberry ringspot and tomato black ring viruses for about 8 weeks (Lister and Murant, 1967; Taylor, 1970) and Debrot (1964) found that *L. macrosoma* retained the English strain of raspberry ringspot virus for 34 days but did not experiment further.

Although nematode vectors retain viruses for long periods, there is no experimental evidence of a biological association similar to that between persistent or circulatory viruses and aphids. Indeed, it is now accepted that nematode-transmitted viruses do not persist through the molt, nor pass through nematode eggs (Harrison and Winslow, 1961; Taylor and Raski, 1964; Ayala and Allen, 1968; Das and Raski, 1968). By analogy with aphids, this suggests that the viruses are retained in a non-persistent manner, probably extra-cellularly attached to some part of the feeding apparatus which is shed during the molt and with the virus slowly and intermittently released from sites of retention during feeding. This hypothesis is supported by the identification of specific sites of retention within each of the vector genera (see later).

16.4.3 Sites of Retention

The presence of virus in nematodes can readily be demonstrated by mechanically inoculating comminuted specimens to test plants. Tobacco rattle virus was detected in *Trichodorus* in this way using *Chenopodium amaranticolor* test plants (Sanger *et al.*, 1962). Similarly, grapevine fanleaf virus has been detected in *X. index* (Raski and Hewitt, 1963) and raspberry ringspot and tomato black ring viruses in *L. elongatus* (Taylor and Murant, 1969). Unfortunately, such demonstrations of infectivity give few clues to the location of transmissible virus in nematodes. Indeed, as viruses can often be demonstrated in nematodes that are not vectors (van Hoof, 1967) it suggests that the virus detected in this way is probably in the intestine and unlikely to play any part in the transmission process. Injecting *X. index* with grapevine fanleaf virus failed to make them infective (Betto and Raski, 1966) and although Roggen (1966) reported some morphological and physiological differences in the hypodermal chords between *X. index* cultured on virus-infected and virus-free plants, these can be explained as dietary effects. Earlier work on electron microscopy of *Xiphinema* species failed to provide any clues to the location of virus retained for transmission (Wright, 1965; Lopez-Abella *et al.*, 1967; Roggen *et al.*, 1967) but later studies would appear to have established the feeding apparatus as the locus for virus retention in each of the vector genera.

Electron microscopy of thin sections of *L. elongatus* cultured on plants infected with raspberry ringspot or tomato black ring viruses revealed the presence of virus-like particles between the odontostyle and the guiding sheath (Taylor and Robertson, 1969; Figs. 2 and 3a). The particles were not found in association with other tissues elsewhere in the nematode and hence it was concluded that in *L. elongatus* the two viruses which it transmits are specifically retained by adsorption to the cuticular guiding sheath. In *L. macrosoma*, vector of the English strain of raspberry ringspot virus, particles were found in association with the inner surface of the odontostyle (Taylor and Robertson, 1975) and this also is the site of retention for artichoke Italian latent virus in *L. apulus* (Taylor *et al.*, 1976). Thus in the *Longidorus* vectors so far examined, viruses appear to be retained by association with the odontostyle or guiding sheath.

In *X. diversicaudatum* carrying arabis mosaic or strawberry latent ringspot viruses (Taylor and Robertson, 1970a, 1975), *X. index* carrying grapevine fanleaf virus (Raski *et al.*, 1973) or *X. americanum* carrying tobacco ringspot virus (McGuire *et al.*, 1970, the site of retention is the same irrespective of the specific association between virus and vector (Figs. 2a and 3d). In each vector, virus particles associate in a monolayer with the cuticle lining of the lumen of the odontophore, slender oesophagus and the oesophageal bulb (Figs. 2 and 3). In *Trichodorus* and *Paratrichodorus* carrying tobacco rattle virus, the particles associate with the pharyngeal walls (but not with the onchiostyle) and with the lining of the oesophageal lumen (Taylor and Robertson, 1970b; Figs. 2, 3b, and c). Thus, the sites of retention in *Xiphinema* and trichodorid nematodes are similar in their association with the oesophagus but differ markedly from the sites of retention in *Longidorus*, and these differences are paralleled by the differences in periods of retention of viruses in the vector groups.

It seems reasonable to assume that the virus particles are selectively and specifically adsorbed at the sites of retention as the virus-containing plant sap passes from the plant to the nematode's intestine during the ingestion phase of feeding. Other viruses present in the sap but not transmitted by the nematode are not retained at these sites, although exceptionally strains of a virus may be retained and not transmitted (as discussed later). Thus, no virus particles were found associated with any

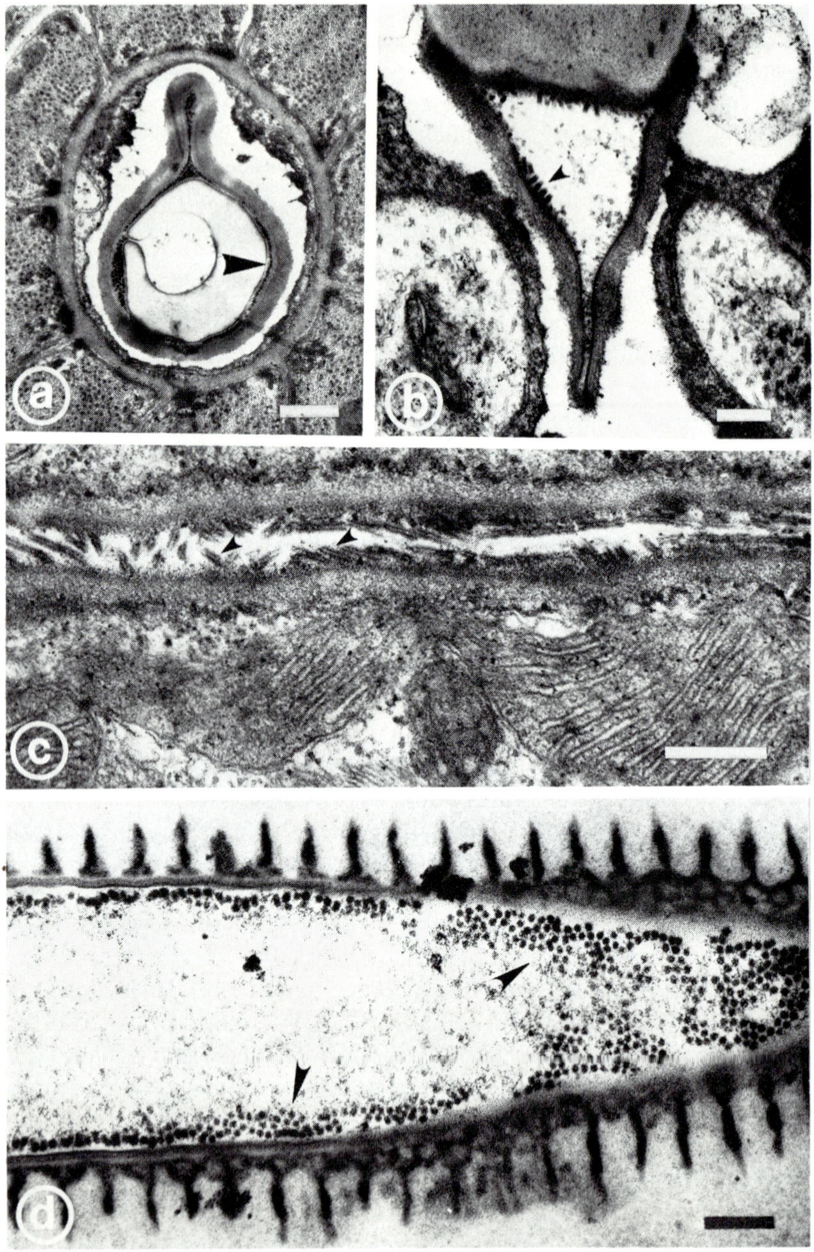

FIG. 3. (a-d) Virus particles at sites of retention in vector nematodes. a) Transverse section of the odontostyle of *Longidorus elongatus* with tomato black ring virus particles (arrowed) between odontostyle and guide ring. b) Transverse section of the pharynx of *Paratrichodorus*

part of the feeding apparatus in *L. elongatus* fed on plants infected with arabis mosaic virus, which it does not transmit, although the presence of the virus could be demonstrated in the intestine (Taylor and Robertson, 1969). This is an example of specificity of association between vector and virus which is particularly evident between nepoviruses and *Xiphinema* and *Longidorus* and seems to reflect the specific association between the properties of the protein coat of the virus and those of the cuticle at the site of retention within the nematode vector (Harrison, 1964; Harrison *et al.*, 1974b).

16.4.4 Specificity of Transmission

Two major groups of viruses, nepo- and tobra-viruses, are transmitted by two corresponding groups of nematode vectors. This may be regarded as the first level of specificity. Within each nematode genus there is a second level of specificity whereby a given virus can be transmitted by one vector but not by another. For example, *L. elongatus* transmits raspberry ringspot and tomato black ring viruses but not arabis mosaic virus (Table I).

A third level of specificity which is most evident among the nepoviruses, indicates that serologically distinct strains of a particular virus require different, although closely related, species of the same nematode genus for their transmission. Thus, *X. diversicaudatum* transmits arabis mosaic virus but not grapevine fanleaf virus, which serologically is distantly related to arabis mosaic virus. As a further example, Scottish strains of raspberry ringspot and tomato black ring viruses are transmitted by *L. elongatus*., but the English strains, which share about half of their antigenic groups with the Scottish strains, have *L. macrosoma* and *L. attenuatus*, respectively, as vectors (Table I). However, when virus variants show only slight serological differences they may be transmitted by the same vector species. *X. americanum* can transmit both the type strain and the serologically distinguishable grapevine yellow vein strain of tomato ringspot virus (Teliz *et al.*, 1966) and single specimens of *X. americanum* have been shown to transmit two serologically distinct strains of tobacco ringspot virus (Sauer, 1966).

American and European isolates of tobacco rattle virus are serologically distinguishable (Cadman, 1963; Oswald and Bowman, 1958) and are transmitted by different species of trichodorid nematodes. However, each isolate of the virus can be transmitted by more than one vector species, and conversely, a single *Trichodorus* or *Paratrichodorus* species can transmit more than one isolate (Table II, Ayala and Allen, 1968; van Hoof *et al.*, 1966). Isolates of pea early browning virus are also serologically distinguishable and are transmitted by several trichodorid species (Gibbs and Harrison, 1964; van Hoof, 1962). Although this suggests that specificity of transmission is not well developed between the tobraviruses and their vectors,

pachydermus with tobacco rattle virus. c) Longitudal section of the oesophagus of *P. pachydermus* with tobacco rattle virus. d) Longitudinal section of the odontophore of *Xiphinema diversicaudatum* with particles of strawberry latent ringspot virus. Bar represents 200 nm (in b, c, d) 500 nm (in a).

van Hoof (1968) points to a fourth level of specificity which is becoming apparent with increasing and sophisticated experimentation. In comparing transmission of tobacco rattle virus by nine *Trichodorus* or *Paratrichodorus* species from the Netherlands he found that transmission occurred only when the nematode was given access to the virus isolate with which it was associated in the field. Populations of *X. diversicaudatum* have also been found to differ in their ability to transmit arabis mosaic (van Hoof, 1966; Dalmasso *et al.*, 1971; Discussion in Martelli, 1975) and *X. coxi* to transmit tobacco ringspot virus (van Hoof, 1971), which again suggests a high degree of specificity with local nematode populations becoming infective only with local virus isolates. It is salutary to note that specificity of transmission is apparent in nature but that anomalies arise from the results of laboratory experiments. Nevertheless, although the question of specificity remains somewhat of an open one, it does seem that where it can be identified it is most strongly developed where the viruses are the most distantly related serologically, and decreasing with increasing antigenic similarity. This is shown in Table IV by comparing the relative transmissibility by the same nematode species of serologically related viruses. Evidence of a different kind derives from experiments with natural and pseudo-recombinant isolates of raspberry ringspot virus (Harrison *et al.*, 1974b) and tomato black ring virus (Harrison and Murant, 1977b). Pseudo-recombinants of two strains of raspberry ringspot virus were readily transmitted by *L. elongatus* when the RNA-2 was derived from the Scottish strain, of which *L. elongatus* is the natural vector; but pseudo-recombinant isolates with RNA-2 derived from the English strain of the virus, for which *L. macrosoma* is the natural vector, were rarely transmitted (Fig. 4). Similar experiments with tomato black ring virus (Harrison and Murant, 1977b) clearly indicate that the RNA-2, which determines the protein coat of the virus (as discussed earlier), also critically affects transmissibility. This in turn suggests that transmission by the vector is dependent on the nature of the protein surface of the virus particles.

It may seem paradoxical that although serologically distinctive strains of a virus

TABLE IV. Nematode Transmission of Serologically Related (A and B) Viruses[a]

	Viruses Compared[b] A	B	Transmission[c] A	B	Serological[d] Relationship
Trichodorus anemones	PEBV-B	PEBV-D	+++	–	1
Xiphinema index	GFV	AMV	+++	–	8
X. diversicaudatum	AMV	GFV	+++	–	8
Longidorus elongatus	TBRV-S	TBRV-E	+++	? +	12
L. macrosoma	RRV-E	RRV-S	++	+	25
L. elongatus	RRV-S	RRV-E	+++	++	25
L. elongatus	RRV-S	RRV-LG	+++	+++	90

[a] Information from Harrison *et al.*, 1974b.
[b] PEBV = pea early browning; GFV = grapevine fanleaf; AMV = arabis mosaic; TBRV = tomato black ring (E = English strain; S = Scottish strain); RRV (E, S stratus; LG = Lloyd George strain).
[c] Frequency of transmission +++ = frequent; + = a few transmissions; – = no transmission.
[d] Per cent cross reacting antibody.

may have different vectors serologically unrelated viruses may have the same vector (*e.g.* arabis mosaic and strawberry latent ringspot viruses transmitted by *X. diversicaudatum*). This is possibly because the features of the particle surface that are involved in serological specificity may not be identical with those that determine specific retention, so that a major change occurring in one type of feature often, but not always, results in a change in the other type (Harrison *et al.*, 1974b). If surface charge density determines the adsorption of virus particles at the sites of retention, as has been speculated (Taylor and Robertson, 1978), this may be similar in two serologically unrelated viruses; but also, a major change in an antigenic (serological) determinant could lead to an alteration in the surface charge density of a virus. It is interesting to note that surface interactions are prominent in the specific association between tobacco necrosis virus and the zoospores of the fungal vector, *Olpidium brassicae* (Temmink *et al.*, 1970).

16.4.5 Mechanism of Transmission

The process of transmission by nematodes may be stated simply as adsorption of virus particles at the site of retention and their later dissociation. Dissociation seems

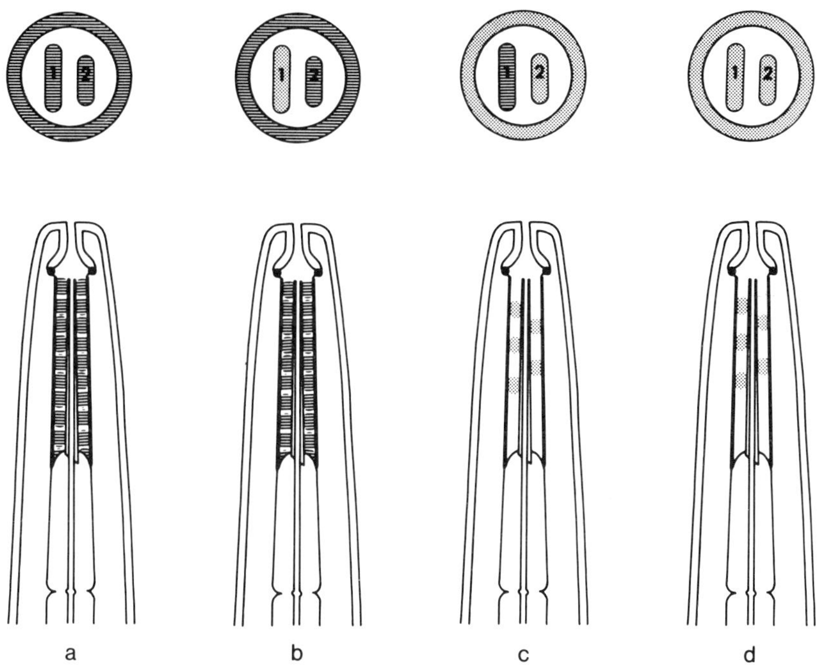

FIG. 4. Diagram illustrating the likelihood of transmission of natural and pseudo-recombinant isolates of raspberry ringspot virus by *Longidorus elongatus*. "Particles" at site of retention indicates greater likelihood of retention and transmission when the protein coat is derived from RNA-2 of the Scottish type.

most likely to occur when saliva, secreted by the oesophageal glands, passes through the lumen of the oesophagus into the plant cell during the initial phases of feeding. This may bring about changes in pH or ionic conditions within the lumen of the oesophagus which alters the surface charge of the virus particles and cause their dissociation from the site of retention (Fig. 5). Other factors may be involved and dissociation mediated in other ways, but whatever the mechanism involved it is likely to result in slow release of the particles from the site of retention because viruses can persist for long periods in their vectors and in many instances it has been shown that a single nematode can transmit serially to several plants (McGuire, 1973).

Whilst specificity of transmission has been viewed mainly in terms of association by specific adsorption there is some evidence that specificity may also be determined by the mechanism of dissociation. When *L. macrosoma* were exposed to plants infected with the English or the Scottish strain of raspberry ringspot virus, the former strain was transmitted but not the latter, as expected (Table I). However, electron microscopic examination of sections of the odontostyle showed virus particles adsorbed to it in both sets of *L. macrosoma.* A possible interpretation of these observations is that the Scottish strain of raspberry ringspot virus becomes adsorbed to the odontostyle in *L. macrosoma* but does not become specifically dissociated, as does the English strain, and hence is not transmitted. Considered in terms of surface charge density this would mean that the change in pH brought about by salivation sufficiently altered the charge on the particles of the English strain to cause its dissociation, but not of the Scottish strain. Changes in mobility of viruses at different pHs, as observed in gel electrophoresis, (M.A. Mayo, personal communication) give some support for this hypothesis.

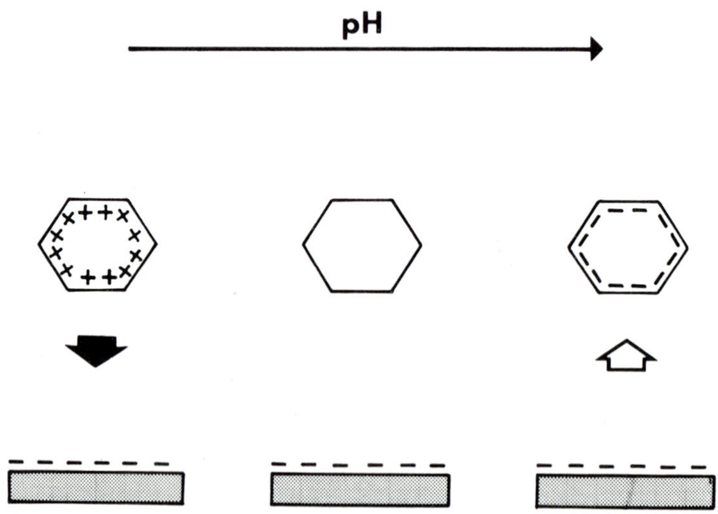

FIG. 5. Influence of pH on the surface charge density of virus particles and on adsorption and release of particles at the sites of retention in the nematode vectors.

16.4.6 Non-persistent Associations

The anomalous results of several laboratory experiments suggest that in some circumstances viruses may be transmitted in a non-specific manner. For example, Valdez (1972b) transmitted the English strain of raspberry ringspot virus with *X. diversicaudatum* and also with *L. caespiticola* and *L. leptocephalus*, the last two species not having been associated with any virus infection in field situations. Prunus necrotic ringspot virus has been transmitted by *L. macrosoma* (Fritzsche and Kegler, 1968); carnation ringspot virus by *X. diversicaudatum* (Fritzsche and Schmelzer, 1967) and brome mosaic virus by *X. diversicaudatum* and *L. macrosoma* (Fritzsche, 1975; Schmidt *et al.*, 1963). McElroy *et al.* (1977) tested *Paralongidorus maximus* as a vector of 20 viruses, recovering nine of them directly from the nematode, and possibly transmitting the English and Scottish strains of raspberry ringspot virus, although they realized that they had not completely avoided possible contamination in their experiments. Taylor and Robertson (1969; 1975) have published electron micrographs showing virus particles in the cheilostome of *L. elongatus* exposed to plants infected with arabis mosaic virus and of clumps of virus particles lodged in the posterior cheilostomes of *L. macrosoma* and *X. diversicaudatum*. They suggest that there may be occasions when these viruses are transmitted, particularly in experimental situations when nematodes are transferred quickly from infector to test plants. McNamara (1978) has shown that the apparent transmission of raspberry ringspot virus by *X. diversicaudatum*, as reported by Valdez (1972b), may be accounted for by nematodes entangled in the roots of the bait plants and hence comminuted with the roots for inoculation to virus indicator plants. Thus some doubt must attach to reports of virus transmission, particularly in the absence of corroboratory field associations, when they are based only on experiments in which the roots of bait plants were tested and not the above ground parts in which the virus might be systemically translocated.

16.5 CONTROL

The distribution and use of infected planting material is probably the most important way in which nematode-transmitted viruses are distributed to areas that may or may not already be infested with vector species. Viruses associated with grapevine, strawberry, potato and bulbous ornamentals such as tulip and narcissus have been distributed in commerce in this way. Nepoviruses are typically transmitted through a large proportion of the seed of many species, including wild plants, and to a lesser extent tobraviruses are also seed-borne (Lister and Murant, 1967). Particularly weed seeds may provide an effective means of disseminating viruses but there is no direct evidence of this. However, there is little doubt that infected weed seeds are the most important overwintering sources of ringspot and tomato black ring viruses which rarely survive during winter fallow periods in their vector, *L. elongatus* (Murant and Taylor, 1965; Taylor and Murant, 1968). Infected weed seeds appear to play no comparable role in the survival and dissemination of viruses transmitted by *Xiphinema* or trichodorid nematodes. The viruses may persist for several months in the long-lived adults, but over long periods reservoirs of virus may remain in root pieces of plants remaining after a crop has been removed (Raski *et al.*, 1965).

Although viruses can be readily distributed, the nematode vectors are relatively immobile, moving at most only a few meters per year under their own volition. However, they may be carried in the soil adhering to the roots of transplants, or on farm machinery, or occasionally on the feet of birds or other animals. Establishment at a new site is then dependent on the suitability for the nematode species of environmental conditions such as soil texture, moisture, temperature and host plant. Because of the long life cycle of *Xiphinema* and *Longidorus* species, undisturbed conditions are preferable and hence grassland, orchards and woodlands, or hedgerows, become important as ecological niches in which populations can become established.

Apart from measures that may be taken to prevent viruses and vectors reaching new sites (*e.g.* the use of certified virus-free planting material), control measures are largely dependent on the eradication of virus from land already infected. One of the most important ways of controlling many soil-borne pathogens is by adopting suitable crop rotations, but this is precluded for most nematode-transmitted viruses because of the wide host range of virus and vectors. Periods of fallow and cultivation of the soil result in some decline of nematode populations, but are relatively ineffective because of the perennation of the viruses in root pieces and weed seeds. Weed control delayed infection of strawberry with raspberry ringspot and tomato black ring viruses (Taylor and Murant, 1968) but surprisingly increased the incidence of tobacco rattle virus in potato, presumably because the trichodorid vectors had no alternative to potato as a host (Cooper and Harrison, 1973). Weed control may effectively prevent the spread of infection in a susceptible crop when the crop itself is a poor host for the vector. For example, *L. elongatus* not only fails to reproduce on raspberry (Taylor, 1967), but the nematode populations rapidly decline if the crop residues of canes and roots are plowed into the soil, apparently because these release a water-soluble nematicidal chemical (Taylor and Murant, 1966).

Particularly with long-term perennial crops, the use of resistant cultivars is an attractive solution to the problem of nematode-transmitted virus infection. But although crops may be planted that are immune to virus infection they may nevertheless be a suitable host for the vector and allow it to multiply. Hence not only does this increase the chances of infection in subsequent susceptible crops but also increases the damage due to the nematodes feeding. More remotely, using virus-resistant cultivars may lead to the evolution of resistant-breaking strains of the virus. Resistant-breaking strains of raspberry ringspot virus have been found locally in Scotland (Murant *et al.*, 1968) but on the other hand no resistance-breaking strains of tomato black ring virus have been reported although it is transmitted by the same vector and frequently occurs with raspberry ringspot virus. Resistance to pea early-browning virus is known in pea, but cultivars were susceptible at some sites indicating the presence of resistance-breaking strains of the virus (Harrison, 1966). Similarly, potato cultivars differed greatly in their reaction to tobacco rattle virus at different test sites in England (Richardson, 1973). In neither of these situations with tobraviruses was there evidence that the resistance-breaking strains had arisen as a response to the growing of resistant cultivars, and the strains presumably were already established in other hosts at the sites (Harrison, 1977).

The use of chemical nematicides generally has proved the most satisfactory method of dealing with nematode vectors and preventing the infection of newly-planted crops with the viruses they transmit. Because of various problems associated with the application of chemicals to the soil complete elimination of nematode populations is rarely achieved in practice; chemical control is in the first instance a problem of bringing chemical and nematode together. However, because of the low rate of multiplication of nematode vector populations, chemical treatment can provide adequate protection for annual or short-term perennial crops. For long-term perennial crops, chemical treatment may have to be combined with a period of fallowing or growing immune hosts to eliminate virus from the nematode populations and to decrease their numbers. Fallowing combined with soil fumigation with D-D proved effective for the control of nettlehead of hop caused by arabis mosaic virus and transmitted by *X. diversicaudatum* (Thresh and Pitcher, 1971; Pitcher and Thresh, 1978).

The chemical nematicides used fall into two groups, fumigants and non-fumigants. Of the fumigants, D-D (dichloropropane-dichloropropene) has been extensively used in many crop situations. Cooper and Thomas (1971) obtained 99% control of trichodorid nematodes and decreased the incidence of tobacco rattle in potato crops by over 90%; moreover, the treatment remained effective for two successive years. D-D also proved effective against *L. attenuatus* and trichodorid nematodes causing Docking disorder of sugar beet in eastern England (Whitehead and Tite, 1967; Whitehead, 1971) and against *X. diversicaudatum* and *L. elongatus* infesting strawberry crops in England and Scotland (Harrison, *et al.*, 1963; Murant and Taylor, 1965; Taylor and Murant, 1968). In France, good control of *X. index* transmitting grapevine fanleaf virus was obtained with D-D or methyl bromide (Vuittenez, 1961; Boubals, 1968) but in California the treatments were less effective because of the survival of nematodes at depths below 1.2 m (Hewitt *et al.*, 1962). The effectiveness of treatment with D-D may be improved by the incorporation of dazomet granules into the surface layer (Seinhorst, 1973).

Specialist equipment is needed for the application of fumigants so that the recent introduction of granular systemic pesticides which can be applied with ordinary farm machinery is attractive. The two most widely used systemic nematicides are the oxime carbamates, aldicarb and oxamyl. These chemicals disrupt the nervous system and behavior of nematodes and at commercial rates of use their action is reversible and does not necessarily cause death. Hence, systemic nematicides primarily act by decreasing the efficiency of the vector and only to a lesser extent by decreasing the numbers of vectors present. Alphey *et al.* (1975) obtained good control of tobacco rattle virus in potato with oxamyl although the numbers of trichodorid vectors were not significantly decreased by the treatment. However, the systemic nematicides have half-lives of less than 14 days (Bromilow, 1973) and hence effective control is of short duration. Oxamyl gave little control of raspberry ringspot virus in raspberry

(Trudgill and Alphey, 1976) and granular systemic nematicides may be regarded as generally unsuitable for control of nematode vectors associated with perennial crops. The fungicide quintozene (pentachloronitrobenzene) appears to act on nematodes in a similar manner to the oxime carbamates but has a much longer persistence in the soil. It has been used for the control of raspberry ringspot transmission to raspberry and strawberry by *L. elongatus* (Murant and Taylor, 1965; Taylor and Murant, 1968).

Sampling of the soil prior to planting can be used to detect the presence of vector nematodes and bait testing of the soil can be used to assay virus infection. The results from these tests, together with knowledge of the ecology of the vector and its association with the viruses it transmits, can help to decide the need for chemical treatment or other control measures. The cheapest decision, of course, will be to avoid planting susceptible crops on land with virus-carrying nematodes.

16.6 REFERENCES

Adams, R. E., and Eichenmuller, J. J. (1964). Studies with *Xiphinema americanum* (abstr.). *Nematologica* 10: 70.

Alfaro Garcia, A. (1971). Presencia en España del virus del 'fanleaf' de la vid. *An. Inst. Nac. Invest. Agr. Ser. Prot. Veg.* 1: 71-80.

Allen, M. W. (1948). Relation of soil fumigation, nematodes and inoculation techniques to big vein disease of lettuce. *Phytopathology* 38: 612-627.

Alfaro, A., and Goheen, A. C. (1974). Transmission of strains of grapevine fanleaf virus by *Xiphinema index Pl. Dis. Reptr.* 58: 549-552.

Alphey, T. J. W., and Boag, B. (1976). Distribution of trichodorid nematodes in Great Britain. *Ann. appl. Biol.* 84: 371-381.

Alphey, T. J. W., Cooper, J. I., and Harrison, B. D. (1975). Systemic nematicides for the control of Trichodorid nematodes and of potato spraing disease caused by tobacco rattle virus. *Pl. Path.* 24: 117-121.

Arias, M. (1975). Geographical distribution of *Xiphinema* in Spain. *In* "Nematode vectors of plant viruses" (F. Lamberti, C. E. Taylor, and J. W. Seinhorst, Eds.). Plenum Press, New York and London.

Ayala, A., and Allen, M. W. (1966). Transmission of the California tobacco rattle virus by three species of the nematode genus *Trichodorus* (abstr.). *Nematologica* 12: 87.

Ayala, A., and Allen, M. W. (1968). Transmission of the California tobacco rattle virus (CTRV) by three species of the nematode genus *Trichodorus. J. Agric. Univ. P. Rico.* 52: 101-125.

Ayala, A., Allen, M. W., and Noffsinger, E. M. (1970). Host range, biology and factors affecting survival and reproduction of the stubby root nematode. *J. Agric. Univ. P. Rico* 54: 311-369.

Behrens, J. (1899). Weitere Beiträge zur Kenntnis der Tabakpflanze. XIV. Die Mauche (Manke) des Tabaks. *Landwirtsch. Vers. Sta.* 52: 442-447.

Beijerinck, M. W. (1898). Über ein Contagium vivum fluidum als Ursache der Flecken-Krankheit der Tabaksblätter. *Verhandl. Kon. Akad. Wetenschap. Amsterdam II.* 6: 1-22.

Bergeson, G. B., Athow, K. L., Laviolette, F. A., and Thomasine, M. (1964). Transmission, movement and vector relationships of tobacco ringspot virus in soybean. *Phytopathology* 54: 723-728.

Betto, E., and Raski, D. J. (1966). Attempts to inoculate *Xiphinema index* with grape fanleaf virus by micro-inoculation. *Nematologica* 12: 453-461.

Bock, K. R. (1966). Arabis mosaic and *Prunus* necrotic ringspot viruses in hop (*Humulus lupulus* L.). *Ann. appl. Biol.* 57: 131-140.

Boning, K. (1931). Zur Ätiologie der Streifen-und Kräuselkrankheit des Tabaks. *Z. Parasitenk* 3: 103-114.

Bos, L., and Van der Want, J. P. H. (1962). Early browning of pea, a disease caused by a soil- and seed-borne virus. *Tijdschr. Plantenziekten* 68: 368-390.

Boublas, D. (1968). Résultats de traitements du sol contre les nematodes vecteurs de la dégénérescence inféctieuse, virose de la vigne. *C. R. 8th Symp. Int. Nematol., Antibes, 1965,* 112-113.

Breece, J. R., and Hart, W. H. (1959). A possible association of nematodes with the spread of peach yellow bud mosaic virus. *Pl. Dis. Reptr.* 43: 489-490.

Bromilow, R. (1973). Breakdown and fate of oxime carbamate nematicides in crops and soils. *Ann. appl. Biol.* 75: 473-479.

Brezki, M. W., and Szczygiel, A. (1974). Nicienie z rodzaju *Trichodorus* Cobb (Nematoda: Trichodoridae) w faunie Polski. *Roczniki Nauk Rolniczych* 4: 59-66.

Cadman, C. H. (1960). Studies on the relationships between soil-borne viruses of the ring-spot type occurring in Britain and continental Europe. *Virology* 11: 653-654.

Cadman, C. H. (1963). Biology of soil-borne viruses. *Ann. Rev. Phytopathol.* 1: 143-172.

Cadman, C. H., and Harrison, B. D. (1960). Studies on the behavior in soils of tomato black ring, raspberry ringspot and arabis mosaic viruses. *Virology* 11: 1-20.

Cadman, C. H., Dias, H. F., and Harrison, B. D. (1960). Sap-transmissible viruses associated with diseases of grapevine in Europe and North America. *Nature* 187: 577-579.

Campbell, R. N., and Fry, P. R. (1966). The nature of the association between *Olpidium brassicae* and lettuce big-vein and tobacco necrosis viruses. *Virology* 29: 222-233.

Caveness, F. E., Gilmer, R. M., and Williams, R. J. (1975). Transmission of cowpea mosaic by *Xiphinema basiri* in Western Nigeria. *In* "Nematode Vectors of Plant Viruses" (F. Lamberti, C. E. Taylor, and J. W. Seinhorst, Eds.). Plenum Press, London and New York.

Choleva, B. (1975). Nematodes of the family Longidoridae in Bulgaria. *In* "Nematode Vectors of Plant Viruses" (F. Lamberti, C. E. Taylor, and J. W. Seinhorst, Eds.). Plenum Press, London and New York.

Christie, J. R. (1952). Some new nematode species of critical importance to Florida growers. *Proc. Soil Sci. Soc. Florida* 12: 30-39.

Christie, J. R., and Perry, V. G. (1951). A root disease of plants caused by a nematode of the genus *Trichodorus. Science* 113: 491-493.

Cohn, E. (1969). The occurrence and distribution of species of *Xiphinema* and *Longidorus* in Israel. *Nematologica* 15: 179-192.

Cohn, E. (1970). Observations on the feeding and symptomatology of *Xiphinema* and *Longidorus* on selected host roots. *J. Nematol.* 2: 167-173.

Cohn, E. (1975). Relations between *Xiphinema* and *Longidorus* and their host plants. *In* "Nematode Vectors of Plant Viruses" (F. Lamberti, C. E. Taylor, and J. W. Seinhorst, Eds.). Plenum Press, New York and London.

Cohn, E., and Ausher, R. (1971). Seasonal occurrence of *Longidorus vineacola* on celery in Israel and its control. *Israel J. agr. Res.* 21: 23-25.

Cohn, E., and Mordechai, M. (1969). Investigations on the life cycles and host preference of some species of *Xiphinema* and *Longidorus* under controlled conditions. *Nematologica* 15: 295-302.

Cohn, E., and Orion, D. (1970). The pathological effect of representative *Xiphinema* and *Longidorus* species on selected host plants. *Nematologica* 16: 423-428.

Cohn, E., Tanne, E., and Nitzany, F. E. (1970). *Xiphinema italiae*, a new vector of grapevine fanleaf virus. *Phytopathology* 60: 181-182.

Cooke, D. A. (1973). The effect of plant parasitic nematodes, rainfall and other factors on Docking disorder of sugar beet. *Pl. Path.* 22: 161-170.

Cooke, D. A., Dunning, R. A., and Winder, G. H. (1974). The effect of nematicides, applied to the seed rows in spring, on growth and yield of sugar beet in Docking-disorder-affected fields. *Ann. appl. Biol.* 76: 289-298.

Cooke, D. A., Winfield, A. L., Atkinson, K. M., John, M. E., and Wilson, W. R. (1976). Fumigation with D-D or Telone on soils prone to docking disorder. II. Effects on numbers of *Trichodorus* spp. and *Longidorus* spp. Nematoda). *Ann. appl. Biol.* 82: 109-115.

Coolen, W. A., and D'Herde, C. J. (1972). Nematodes associated with glasshouse roses. *Proc. 9th int. Nem. Symp. Warsaw, 1967,* 259-265.

Cooper, J. I. (1971). The distribution in Scotland of tobacco rattle virus and its nematode vectors in relation to soil type. *Pl. Path.* 20: 51-58.

Cooper, J. I., and Harrison, B. D. (1973). The role of weed hosts and the distribution and activity of vector nematodes in the ecology of tobacco rattle virus. *Ann. appl. Biol.* 73: 53-66.

Cooper, J. I., and Thomas, P. R. (1971). Chemical treatment of soil to prevent transmission of tobacco rattle virus to potatoes by *Trichodorus* spp. *Ann. appl. Biol.* 69: 23-34.

Cotten, J. (1973). Feeding behavior and reproduction of *Xiphinema index* on some herbaceous test plants. *Nematologica* 19: 516-520.

Cotten, J. (1975). Virus vector species of *Xiphinema* and *Longidorus* in relation to certification schemes for fruits and hops in England. *In* "Nematode Vectors of Plant Viruses." (F. Lamberti, C. E. Taylor, and J. W. Seinhorst, Eds.). Plenum Press, New York and London.

Cotten, J. (1976). Observations of life-cycle, population development and vertical distribution of *Longidorus macrosoma* on raspberry and other crops. *Ann. appl. Biol.* 83: 407-412.

Cremer, M. C., and Schenk, P. K. (1967). Notched leaf in *Gladiolus* spp. caused by viruses of the rattle virus group. *Neth. J. Pl. Pathol.* 73: 33-48.

Cropley, R. (1961). Cherry leaf-roll virus. *Ann. appl. Biol.* 49: 524-529.

Dalmasso, A. (1970). Influence directe de quelques facteurs écologiques sur l'activité biologique et la distribution des espèces françaises de la famille des Longiforidae (Nematoda-Dorylaimida). *Ann. Zool. Ecol. anim.* 2: 162-300.

Dalmasso, A., Munck-Cardin, M. C., and Legin, R. (1972). Preliminary results on role of *Xiphinema diversicaudatum* on transmission of arabis mosaic virus to vine. *Ann. Phytopath.* 4: 410.

Das, S., and Raski, D. J. (1968). Vector efficiency of *Xiphinema index* in the transmission of grapevine fanleaf virus. *Nematologica* 14: 55-62.

Davies, R. A., and Jenkins, W. R. (1960). Nematodes associated with roses and the root injury caused by *Meloidogyne hapla* Chitwood, 1949, *Xiphinema diversicaudatum* (Micoletzky, 1927) Thorne, 1939, and *Helicotylenchus nannus* Steiner, 1945. *Bull. Md. Agr. Exptl. Sta.* A. 106, 16 pp.

Debrot, E. A. (1964). Studies on a strain of raspberry ringspot virus occurring in England. *Ann. appl. Biol.* 54: 183-191.

Dias, H. F. (1977). Does Ontario grape industry need virus certification program. *The Great Lake Fruit Growers News* 16: 26.

Dias, H. F., and Cation, D. (1976). The characterization of a virus responsible for peach rosette mosaic and grape decline in Michigan. *Canad. J. Bot.* 54: 1228-1239.

Di Sanzo, C. P., and Rohde, R. A. (1969). *Xiphinema americanum* associated with maple decline in Massachusetts. *Phytopathology* 59: 279-284.

Douthit, L. B., and McGuire, J. M. (1974). Effects of temperature on *Xiphinema americanum* and tobacco ringspot virus infection on cucumber. *Proc. Amer. Phytopath. Soc.* 1: 158 (abstr.).

Douthit, L. B., and McGuire, J. M. (1975). Some effects of temperature on *Xiphinema americanum* and inflation of cucumber by tobacco ringspot virus. *Phytopathology* 65: 134-138.

Duffield, C. A. W. (1925). Nettlehead in hops. *Ann. appl. Biol.* 12: 536-543.

Dunez, J., and Dupont, G. (1976). Myrobalan latent ringspot virus. *C. M. I./A. A. B. Descriptions of Plant Viruses* No. 160.

Epstein, A. H., and Barker, K. R. (1966). Pathogenicity of *Xiphinema americanum* on *Vinca minor. Pl. Dis. Reptr.* 50: 420-422.

Fisher, J. M., and Raski, D. J. (1967). Feeding of *Xiphinema index* and *X. diversicaudatum. Proc. Helm. Soc. Wash.* 34: 68-72.

Flegg, J. J. M. (1966). Once-yearly reproduction in *Xiphinema vuittenezi. Nature* 212: 741.

Flegg, J. J. M. (1968a). The occurrence and depth distribution of *Xiphinema* and *Longidorus* species in southeastern England. *Nematologica* 14: 189-196.

Flegg, J. J. M. (1968b). Life-cycle studies of some *Xiphinema* and *Longidorus* species in south-eastern England. *Nematologica* 14: 197-210.

Flegg, J. J. M. (1969). Tests wtih potential nematode vectors of cherry leaf roll virus. *Rep. E. Malling Res. Stn. for 1968*: 155-157.

Forer, L. B., Trudgill, D. L., and Alphey, T. J. W. (1975). Some effects of oxamyl on the virus-vector nematodes *Longidorus elongatus* and *Xiphinema diversicaudatum*. *Ann. appl. Biol.* 81: 207-214.

Frazier, N. W., Yarwood, C. E., and Gold, A. H. (1961). Yellow bud virus endemic along Cali-coast. *Pl. Dis. Reptr.* 45: 649-651.

Fritzsche, R. (1964). Untersuchungen über die Virusübertragung durch Nematoden. *Wiss. Z. Univ. Rostock* 13: 333-347.

Fritzche, R. (1966). Beiträg zur Ökologie von *Xiphinema diversicaudatum* (Micoletzky) Thorne. *Nachr. Deutsch. Pflanzensch.* 70: 8-11.

Fritzsche, R. (1967). Methoden der Übertragung pflanzenpathogen Viren durch Nematoden. *Biol. Zbl.* 86: 753-759.

Fritzsche, R. (1970). Beiträge zum Übertragungsmechanismus pflanzen-pathogen Viren durch Nematoden. *Proc. 9th Int. Nem. Symp. Warsaw, 1967*: 293-300.

Fritzsche, R. (1975). Übertragung des Trespenmosaik- und Arabis-Mosaik-Virus durch Nematoden in Abhängigkeit von der Infektiosität der Wurzeln der Wirtspflanzen. *Archiv für Phytopathologie und Pflanzenschutz* 11: 197-201.

Fritzsche, R., and Kegler, H. (1964). Die Übertragung des Blattrollvirus der Kirsche (cherry leafroll virus) durch Nematoden. *Inst. Phytopathol. Akad. Landwirt. Berlin Buchbesprechungen* 12: 299.

Fritzsche, R., and Kegler, H. (1968). Nematoden als Vektoren von Viruskrankheiten der Obstgewächse. *TagBer. dt. Akad. Landw-Wiss. Berl.* 97: 289-295.

Fritzsche, R., and Schmelzer, K. (1967). Übertragbarkeit des Nelkenringflecken-Virus durch Nematoden. *Naturwissenschaften* 54: 498-499.

Fulton, J. P. (1962). Transmission of tobacco ringspot virus by *Xiphinema americanum*. *Phytopathology* 57: 535-537.

Fulton, J. P., and Fulton, R. W. (1970). A comparison of some properties of elm mosaic and tomato rinspot viruses. *Phytopathology* 60: 114-115.

Gibbs, A. J., and Harrison, B. D. (1964). A form of pea early browning virus found in Britain. *Ann. appl. Biol.* 54: 1-11.

Giunchidi, L., and Tacconi, R. (1974). Deperimento di peschi in rapporto alla virosi "butteratura del legno." *Inform. Fitopat.* 24: 5-8.

Griffin, G. D., and Barker, K. R. (1966). Effects of soil temperature and moisture on the survival and activity of *Xiphinema americanum*. *Proc. Helminthol. Soc. Wash.* 33: 126-130.

Griffin, G. D., and Epstein, A. H. (1964). Associaton of dagger nematode *Xiphinema americanum*, with stunting and winter-kill of ornamental spruce. *Phytopathology* 54: 177-180.

Hanada, K., and Harrison, B. D. (1977). Effects of virus genotype and temperature on seed transmission of nepoviruses. *Ann. appl. Biol.* 85: 79-92.

Harrison, B. D. (1960). The biology of soil-borne plant viruses. *Advances Virus Res.* 7: 131-161.

Harrison, B. D. (1964a). The transmission of plant viruses in soil. *In* "Plant Virology" (M. K. Corbett and H. D. Sisler, Eds.). Univ. Fla. Press, Gainesville.

Harrison, B. D. (1964b). Specific nematode vectors for serologically distinctive forms of raspberry ringspot and tomato black ring viruses. *Virology* 22: 544-550.

Harrison, B. D. (1966). Further studies on a British form of pea early-browning virus. *Ann. appl. Biol.* 57: 121-129.

Harrison, B. D. (1970). Tobacco rattle virus. *C. M. I./A. A. B. Descriptions of Plant Viruses*, No. 12.

Harrison, B. D. (1973). Pea early-browning virus. *C. M. I./A. A. B. Descriptions of Plant Viruses*, No. 120.

Harrison, B. D. (1977). Ecology and control of viruses with soil-inhabiting vectors. *Ann. Rev. Phytopathol.* 15: 331-360.

Harrison, B. D., and Cadman, C. H. (1959). Role of a dagger nematode (*Xiphinema* sp.) in outbreaks of plant disease caused by arabis mosaic virus. *Nature* 184: 1624-1626.

Harrison, B. D., and Cooper, J. I. (1974). Studies on the nature of the resistance of potato cultivars to tobacco rattle virus. *Potato Res.* 17: 348-349.

Harrison, B. D., and Murant, A. F. (1977a). Nepovirus Group. G. M. I./A. A. B. *Descriptions of Plant Viruses,* No. 185.

Harrison, B. D., and Murant, A. F. (1977b). Nematode transmissibility of pseudo-recombinant isolates of tomato black ring virus. *Ann. appl. Biol.* 86: 209-212.

Harrison, B. D., and Robinson, D. J. (1978). The Tobraviruses. *Adv. Virus Res.* 23: 25-27.

Harrison, B. D., and Winslow, R. D. (1961). Laboratory and field studies on the relation of arabis mosaic virus to its nematode vector *Xiphinema diversicaudatum* (Micoletzky). *Ann. appl. Biol.* 49: 621-633.

Harrison, B. D., Mowat, W. P., and Taylor, C. W. (1961). Transmission of a strain of tomato black ring virus by *Longidorus elongatus* (Nematoda). *Virology* 14: 480-485.

Harrison, B. D., Murant, A. F., and Mayo, M. A. (1972). Evidence for two functional RNA species in raspberry ringspot virus. *J. Gen. Virol.* 16: 339-348.

Harrison, B. D., Murant, A. F., Mayo, M. A., and Roberts, I. M. (1974a). Distribution of determinants for symptom production, host range and nematode transmissibility between the two RNA components of raspberry ringspot virus. *J. Gen. Virol.* 22: 233-247.

Harrison, B. D., Peachey, J. E., and Winslow, R. D. (1963). The use of nematicides to control the spread of arabis mosaic virus by *Xiphinema diversicaudatum* (Micol.). *Ann. appl. Biol.* 52: 243-255.

Harrison, B. D., Robertson, W. M., and Taylor, C. E. (1974b). Specificity of retention and transmission of viruses by nematodes. *J. Nematology* 6: 155-164.

Harrison, R. E., and Smart, G. C. Jnr. (1975). Vertical distribution of *Trichodorus christiei* and *T. proximus* relative to soil moisture. (abstr.). *J. Nematology* 7: 324.

Hewitt, W. B. (1968). Viruses and virus diseases of the grapevine. *Rev. Appl. Mycol.* 47: 433-435.

Hewitt, W. B., and Raski, D. J. (1962). Nematode vectors of plant viruses. *In* "Biological transmission of disease agents" (K. Maramorosch, Ed.). Academic Press, New York.

Hewitt, W. B., Goheen, A. C., Raski, D. J., and Gooding, G. V. (1962). Studies on virus diseases of the grapevine in California. *Vitis* 3: 57-83.

Hewitt, W. B., Raski, D. J., and Goheen, A. C. (1958). Nematode vector of soil-borne fanleaf virus of grapevine. *Phytopathology* 48: 586-595.

Hijink, M. J., and Kuiper, K. (1966). Waarnemingen over de verdeling van aaltjes in de grond. *Meded. Rijks. Landbouwwetensch., Gent* 31: 558-571.

Hildebrand, E. M. (1941). A new case of rosette mosaic on peach. *Phytopathology* 31: 353-355.

Hoff, J. K. (1964). Studies on the pathogenicity of *Trichodorus christiei* with special references to the distribution of mitotic divisions and cellular differentiation in the *Allium cepa* root tip. *Diss. Abstr.* 25: 17.

Hollings, M. (1963). Cucumber stunt mottle, a disease caused by a strain of arabis mosaic virus. *J. hort. Sci.* 38: 138-149.

Hollings, M., Stone, O. M., and Martelli, G. P. (1969). Celery yellow vein and Hungarian chrome mosaic viruses. *Rep. Glasshouse Crops Res. Inst. 1968*: 102.

Hooper, D. J. (1975). Virus vector nematodes–taxonomy and general introduction. *In* "Nematode Vectors of Plant Viruses" (F. Lamberti, C. E. Taylor, J. W. Seinhorst, Eds.). Plenum Press, New York and London.

Horner, C. E., and Jensen, H. J. (1954). Nematodes associated with mint in Oregon. *Pl. Dis. Reptr.* 38: 39-41.

Ivanova, T. S. (1972). Ectoparasitic nematodes in Tadzhikistan. *Zashchita Rastenii* 7: 21 (in Russian).

Iwaki, M., and Komuro, Y. (1971). Viruses isolated from narcissus (*Narcissus* spp.). *Ann. Phytopath. Soc. Japan* 37: 108-116 (in Japanese).

Iwaki, M. and Komuro, Y. (1974). Viruses isolated from narcissus (*Narcissus* spp.) in Japan. V.

Arabis mosaic virus. *Ann. Phytopath. Soc. Japan.* 40: 344-353 (in Japanese).

Jankulova, M., Savino, A., Gallitelli, D., Quacquarelli, A., and Martelli, G. P. (1976). Isolation of artichoke Italian latent virus from the grapevine in Bulgaria. *Proc. 6th Conf. Intern. Counc. Grapevine Viruses Cordova* 1976: 24.

Jensen, H. J., and Allen, T. C. (1964). Transmission of tobacco rattle virus by a stubby root nematode, *Trichodorus allius. Pl. Dis. Reptr.* 48: 333-334.

Jha, A., and Posnette, A. F. (1959). Transmission of a virus to strawberry plants by a nematode. *Nature* 184: 962-963.

Jha, A., and Posnette, A. F. (1961). Transmission of arabis mosaic virus by the nematode *Xiphinema diversicaudatum* (Micol.) *Virology* 13: 119-123.

Johnson, F. (1945). The effect of chemical soil treatments on the development of wheat mosaic. *Ohio J. Sci.* 45: 125-128.

Jones, A. T. (1973). A comparison of some properties of four strains of cherry leaf roll virus. *Ann. appl. Biol.* 74: 211-217.

Jones, F. G. W., Larbey, D. W., and Parrott, D. M. (1969). The influence of soil structure and moisture on nematodes, especially *Xiphinema, Longidorus, Trichodorus* and *Heterodera* spp. *Soil. Biol. Biochem.* 1: 153-165.

Kenten, R. H. (1977). Cacao Necrosis Virus. *C. M. I./A. A. B. Description of Plant Viruses* No. 173.

Klos, E. J., Fronek, F., Knierim, J. A., and Cation, D. (1967). Peach rosette mosaic transmission and control studies. *Mich. agric. Exp. Stat. Quart. Bull.* 49: 287-293.

Komuro, Y., Yoshing, M., and Ichinohe, M. (1970). Tobacco rattle virus isolated from aster showing ringspot syndrome and its transmission by *Trichodorus minor* Colbran. *Ann. phytopath. Soc. Japan*, 36: 17-26.

Kuiper, K., and Loof, P. A. A. (1962). *Trichodorus flevensis* n. sp. (Nematoda: Enoplida). A plant nematode from new polder soil. *Versl. PlZiekt. Dienst. Wageningen* 136: 193-200.

Lamberti, F. (1969). Effect of temperature on the reproduction rate of *Longidorus africanus. Pl. Dis. Reptr.* 53: 531.

Lamberti, F. (1975). Taxonomy of *Longidorus* (Micoletzky) Filipjev and *Paralongidorus* Siddiqi, Hooper and Khan. *In* "Nematode Vectors of Plant Viruses" (F. Lamberti, C. E. Taylor, and J. W. Seinhorst, Eds.). Plenum Press, New York and London.

Lamberti, F., and Bleve Zacheo, T. (1977). Two new species of *Longidorus* (Nematodi: Longidoridae) from Italy. *Nematol. medit.* 5: 73-83.

Lamberti, F., Taylor, C. W., and Seinhorst, J. W., Eds. (1975), "Nematode Vectors of Plant Viruses." NATO ASI., Ser. A: Life Sci. Vol. 2. Plenum Press, New York and London.

Lee, Y. B., and Han, S. C. (1976). The nematode genus *Xiphinema* (Dorylaimida: Longidoridae) from Korea. *Korean J. Pl. Protection* 15: 17-21.

Linford, M. B., and McKinney, H. H. (1954). Occurrence of *Polymyxa graminis* in roots of small grains in the United States. *Pl. Dis. Reptr.* 38: 711-713.

Lister, R. M. (1964). Strawberry latent ringspot: a new nematode-borne virus. *Ann. appl. Biol.* 54: 167-176.

Lister, R. M., and Murant, A. F. (1967). Seed transmission of nematode-borne viruses. *Ann. appl. Biol.* 54: 167-176.

Loof, P. A. A. (1975). Taxonomy of Trichodoridae. *In* "Nematode Vectors of Plant Viruses" (F. Lamberti, C. E. Taylor, J. W. Seinhorst, Eds.). Plenum Press, New York and London.

Loos, C. A. (1949). Notes on free-living and plant parasitic nematodes from Ceylon. Nos. 5 and 6. *J. zool. Soc. India* 1: 23-29, 30-36.

Lopez-Abella, D., Jimenez-Millan, F., and Garcia-Hidalgo, F. (1967). Electron microscope studies of some cephalic structures of *Xiphinema americanum. Nematologica* 13: 283-286.

Maggenti, A. R. (1971). Nemic relationships and origins of plant parasitic nematodes. *In* "Plant Parasitic Nematodes" (B. M. Zuckerman, W. F. Mai and R. A. Rohde, Eds.). Academic Press, New York and London.

Malek, W.R., Jenkins, W.R., and Powers, E.M. (1965). Effect of temperature on growth and reproduction of *Criconemoides curvatum* and *Trichodorus christiei* (abstr.). *Nematologica* 11:41.

Mali, V. R., Vanek, G., and Bojňanský, V. (1975). Transmission by nematodes of some grapevine viruses occurring in Czechoslovakia and Hungary. *Pol 'nohospodárstvo*, A. No. 3: 1-130.

Martelli, G. P. (1978). Nematode-borne viruses of grapevine. *In* "Plant Disease Epidemiology" (P. R. Scott and Bainbridge, Eds.). Blackwells Sc. Publ., London.

Martelli, G. P., and Lamberti, F. (1967). Le specie di *Xiphinema* Cobb, 1913 travate in Italia e commenti sulla presenza di *Xiphinema americanum* Cobb (Nematoda, Dorylaimoidea). *Phytopath. medit.* 6: 68-85.

Martelli, G. P., and Quacquarelli, A. (1972). Hungarian chrome mosaic and tomato black ring: two similar but unrelated plant viruses. *Ann. Phytopath., h. ser.*, 123-141.

Martelli, G. P., and Rana, G. L. (1975). Virus e virosi del Carciofo e del cardo (abstr.). *2nd Congr. Intern. Carciofo, Bari, 1973*: 135-136.

Martelli, G. P., and Raski, D. J. (1963). Osservazioni su *Xiphinema index* Thorne e Allen, fico e "degenerazione infettiva" della vite. *Inftorm. fitopatol.*, 13: 416-420.

Martelli, G. P., and Sarospataki, G. (1969). Nematodes of the family Longidoridae (Thorne 1935, Meyl 1960) found in Hungarian vineyards and virus transmission trials with *Xiphinema index*, Thorne and Allen. *Phytopath. medit.* 8: 1-7.

Mayo, M. A., Murant, A. F., Harrison, B. D., and Goold, R. A. (1974). Two protein and two RNA species in particles of strawberry latent ringspot virus. *J. gen. Virol.* 24: 29-37.

McElroy, F. D. (1972). Studies on the host range of *Xiphinema bakeri* and its pathogenicity to raspberry. *J. Nematol.* 4: 16-22.

McElroy, F. D., and van Gundy, S. D. (1968). Observations on the feeding processes of *Hemicycliophora arenaria. Phytopathology* 58: 1558-1565.

McElroy, F. D., Brown, D. J. F., and Boag, B. (1977). The virus-vector and damage potential, morphometrics and distribution of *Paralongidorus maximum. J. Nematol.* 9: 122-130.

McGuire, J. M. (1964). Efficiency of *Xiphinema americanum* as a vector of tobacco ringspot virus. *Phytopathology* 54: 799-801.

McGuire, J. M. (1973). Retention of tobacco ringspot virus by *Xiphinema americanum. Phytopathology* 63: 324-326.

McGuire, J. M., and Douthit, L. B. (1975). Acquisition and transmission of tobacco ringspot virus to soybean by *Xiphinema americanum* (abstr.). *Proc. Amer. Phytopath. Soc.* 2: 43.

McGuire, J. M., Kim, K. S., and Douthit, L. B. (1970). Tobacco ringspot virus in the nematode *Xiphinema americanum. Virology* 42: 212-216.

McGuire, J. M., and Schnedier, I. R. (1973). Transmission of satellite of tobacco ringspot virus by *Xiphinema americanum. Phytopathology* 63: 1429-1430.

McNamara, D. G. (1978). Studies on the ability of the nematode *Xiphinema diversicaudatum* (Micol.) to transmit raspberry ringspot virus and to survive in plant-free soil. *Ph. D. thesis, Univ. Reading.*

Meagher, J. W., Brown, R. H., Taylor, R. H., and Harris, A. R. (1976). The distribution of *Xiphinema index* and other parasitic nematodes associated with grapevines in northeastern Victoria. *Aust. J. Exp. Agric. and Anim. Husb.* 16: 932-936.

Milkus, B. N., Shterenberg, P. M., and Cherevkova, D. S. (1975). Nematodes of the genus *Xiphinema* (Longidoridae) in vineyards of South Ukraine. *Zool. Zhurnal* 54: 1248-1250 (in Russian).

Mircetich, S. M., and Civerolo, E. L. (1972). Relationship between stem-pitting in peach and in other *Prunus* species. *Phytopathology* 62: 1294-1302.

Morton, H. V., and Perry, V. G. (1968). Life cycle and reproductive potential of *Trichodorus christiei* (abstr.). *Nematologica* 14: 11.

Mowat, W. P., and Taylor, C. E. (1962). Nematode-borne viruses. *Ann. Rpt. Scott. Hort. Res. Inst. 1961-62*, 68.

Murant A. F. (1974). Strawberry latent ringspot virus. *CMI/AAB Descriptions of Plant Viruses.* No. 126.

Murant, A. F., and Lister, R. M. (1967). Seed-transmission in the ecology of nematode-borne viruses. *Ann. Appl. Biol.* 59: 63-70.

Murant, A. F., and Taylor, C. E. (1965). Treatment of soil with chemicals to prevent transmission of tomato black ring and raspberry ringspot viruses by *Longidorus elongatus* (de Man). *Ann. appl. Biol.* 55: 227-237.

Murant, A. F., Taylor, C. E., and Chambers, J. (1968). Properties, relationships and transmission of a strain of raspberry ringspot virus infecting raspberry cultivars immune to the common Scottish strain. *Ann. appl. Biol.* 61: 175-186.

Nemec, S., and Malek, R. B. (1975). Effects of nematicides and strawberry growth on nematodes, especially *Xiphinema americanum*, in root-rot sites in Illinois (abstr.). *J. Nematol.* 7: 328.

Norton, D. C. (1963). Population fluctuations of *Xiphinema americanum* in Iowa. *Phytopathology* 53: 66-68.

Norton, D. C. (1967). *Xiphinema americanum* as a factor in unthriftiness or red clover. *Phytopathology* 57: 1390-1391.

Nyland, G., Lownsbery, B. F., Lowe, B. K., and Mitchell, J. F. (1969). The transmission of cherry rasp leaf virus by *Xiphinema americanum. Phytopathology* 69: 1111-1112.

Oswald, J. W., and Bowman, T. (1958). Studies on a soil-borne potato virus disease in California (abstr.). *Phytopathology* 48: 396.

Percival, J. (1895). The eelworm disease of hops. 'Nettleheaded' or 'skinkly' plants. *J. S. E. agric. Coll. Wye*, No. 1, pp. 5-9.

Perry, V. G. (1958). Parasitism of two species of dagger nematodes (*Xiphinema americanum* and *X. chambersi*) to strawberry. *Phytopathology* 48: 420-423.

Pitcher, R. S. (1967). The host-parasite relations and ecology of *Trichodorus viruliferus* on apple roots, as observed from an underground laboratory. *Nematologica* 13: 547-557.

Pitcher, R. S., and Flegg, J. J. M. (1965). Observations of root-feeding by the nematode *Trichodorus viruliferus* (Hooper). *Nature* 207: 317.

Pitchers, R. S., and McNamara, D. G. (1970). The effect of nutrition and season of year on the reproduction of *Trichodorus viruliferus. Nematologica* 16: 99-106.

Pitcher, R. S., and Thresh, J. M. (1978). Control of nettlehead and related diseases of hop by fallowing and fumigation. *In* "Plant Disease Epidemiology" (P. R. Scott and A. Bainbridge, Eds.). Blackwells Sc. Publ., London.

Prota, U., Bleve-Zacheo, T., Garau, R., and Lamberti, F. (1977). Studies on the life cycle of *Xiphinema index. Nematol. medit.* 5: 329-333.

Putz, C., and Stocky, G. (1970). Premières observations sur une souche de strawberry latent ringspot virus transmise par *Xiphinema coxi* Tarjan et associée a une maladie du framboisier en Alsace. *Ann. Phytopathol.* 2: 329-347.

Quacquarelli, A., Gallitelli, D., Savino, V., and Martelli, G. P. (1976). Properties of grapevine fanleaf virus. *J. gen. Virol.* 32: 349-360.

Quanjer, H. M. (1943). Bijdrage tot de kennis van de in Nederland voorkomende ziekten van Tabak en van de Tabaksteelt op kleigrond. *Tijdschr. PlZiekt.* 49: 37-51.

Radewald, J. D. (1962). The biology of *Xiphinema index* and the pathological effect of the species on grape. *Ph. D. thesis Univ. California, Davis.*

Radewald, J. D., and Raski, D. J. (1962). A study of the life cycle of *Xiphinema index. Phytopathology* 52: 748.

Radewald, J. D., Osgood, J. W., Mayberry, K. S., Paulus, A. O., and Shibuya, F. (1969). *Longidorus africanus*, a pathogen of head lettuce in the Imperial Valley of Southern California. *Pl. Dis. Reptr.* 53: 381-384.

Ramsdell, D. C., and Myers, R. L. (1974). Peach rosette mosaic virus, symptomatology and nematodes associated with grapevine 'degeneration' in Michigan. *Phytopathology* 64: 1174-1178.

Rana, G. L., and Roca, F. (1975). Trasmissione con nematodi del virus, "latente italiano del carciofo" (AILV). *Atti 2nd Congr. Int. Carciofo, Bari, 1974*: 855-858.

Randles, J. W., Harrison, B. D., Murant, A. F., and Mayo, M. A. (1977). Packaging and biological activity of the two essential RNA species of tomato black ring virus. *J. Gen. Virol.* 36: 187-193.

Rao, A. S., and Brakke, M. K. (1969). Relation of soil-borne wheat mosaic virus and its fungal vector, *Polymyxa graminis*. *Phytopathology* 59: 581-587.

Raski, D. J., and Hewitt, W. W. (1963). Plant parasitic nematodes as vectors of plant viruses. *Phytopathology* 53: 39-47.

Raski, D. J., and Lider, L. (1959). Nematodes in grape production. *California Agr.* 13: 13-15.

Raski, D. J., and Radewald, J. D. (1958). Reproduction and symptomatology of certain ectoparasitic nematodes on roots of Thompson seedless grapes. *Pl. Dis. Reptr.* 52: 941-943.

Raski, D. J., Hewitt, W. B., Goheen, A. C., Taylor, C. E., and Taylor, R. H. (1965). Survival of *Xiphinema index* and reservoirs of fanleaf virus in fallowed vineyard soil. *Nematologica* 11: 349-352.

Raski, D. J., Maggenti, A. R., and Jones, N. O. (1973). Location of grapevine fanleaf and yellow mosaic virus particles in *Xiphinema index*. *J. Nematol.* 5: 208-211.

Rau, J. (1975). Das Vorkommen virusübertragender Nematoden in ungestörten Biotopen Niedersachsens. *Thesis Dipl. Ing. agr. Techn. Univ. Hannover.*

Richardson, D. E. (1973). An external ringspot of potatoes. *Pl. Pathol.* 22: 120-122.

Richter, E. (1969). Zur vertikalen Verteilung von Nematoden in einem Sandboden. *Nematologica* 15: 44-54.

Roca, F., Conti, M., and Lamberti, F. (1977). *Trichodorus viruliferus* (Nematoda, Trichodoridae), vettore del virus "rattle" del Tobacco (TRV) nell'Astigiano. *Nematol. medit.* 5: 185-194.

Roca, F., Martelli, G. P., Lamberti, F., and Rana, G. L. (1975). Distribution of *Longidorus attenuatus* Hooper in Apulian artichoke fields and its relationship with artichoke Italian latent virus. *Nematol. medit.* 3: 91-101.

Roggen, D. R. (1966). On the morphology of *Xiphinema index* reared on grape fanleaf virus-infected grapes. *Nematologica* 12: 287-296.

Roggen, D. R., Raski, D. J., and Jones, N. O. (1967). Further electron microscopic observations of *Xiphinema index*. *Nematologica* 13: 1-16.

Rohde, R. A., and Jenkins, W. R. (1957a). Host range of a species of *Trichodorus* and its host-parasite relationships on tomato. *Phytopathology* 47: 295-298.

Rohde, R. A., and Jenkins, W. R. (1957b). Effect of temperature on the life cycle of stubby-root nematodes (abstr.). *Phytopathology* 47: 29.

Romanenko, N. D. (1971). Distribution of nematode-vectors of viruses on fruit and berry crops. *Plodovodye i Yadodov Nechernoz Polosy* 3: 383-386 (in Russian).

Romascu, E., and Zinca, N. (1974). *Xiphinema index* Thorne et Allen, 1950. A new damaging pest of grapevines in Romania (Nematoda: Dorylaimidae). *Ann. I. C. P. P.* 20: 273-283.

Rossner, J. (1971). Einfluss der Austrocknung des Bodens auf wandernde Wurzelnematoden. *Nematologica* 17: 127-144.

Roy, T. K. (1973). Notes on the pathogenicity of *Xiphinema basiri* Siddiqi, 1959, and its host records. *Indian J. Nematol.* 3: 161-163.

Rüdel, M. (1971). Vorkommen einiger Arten der Gattung *Xiphinema* (Nematoda: Dorylaimidae) in Pfalz und Rheinhessen. *Weinberg und Keller* 18: 505-520.

Rüdel, M., Querfurth, G., and Paul, H. L. (1977). Natürliches Vorkommen des "carnation ringspot virus" in *Stellaira media* (L.) Cyrill. aus Weinbergen erstmals nachgewiesen. *Nachr. Deutsch. Pflanzenschutzd.* 29: 59-60.

Russell, C. C., and Perry, V. G. (1966). Parasitic habit of *Trichodorus christiei* on wheat. *Phytopathology* 56: 357-358.

Salazar, L. F., and Harrison, B. D. (1977). Two previously undescribed potato viruses from South America. *Nature*, London 265: 337-338.

Sanger, H. L. (1961). Untersuchungen über schwer übertragbare Formen das Rattelvirus. *Proc. 4th Conf. Potato Virus Dis. Braunschweig 1960*, 22-28.

Sanger, H. L., Allen, M. W., and Gold, A. H. (1962). Direct recovery of tobacco rattle virus from its nematode vector (abstr.). *Phytopathology* 52: 750.

Sauer, N. I. (1966). Simultaneous association of strains of tobacco ringspot virus within *Xiphinema americanum*. *Phytopathology* 56: 862-863.

Schindler, A. F. (1957). Parasitism and pathogenicity of *Xiphinema diversicaudatum*, an ectoparasitic nematode. *Nematologica* 2: 25-31.

Schindler, A. F., and Braun, A. J. (1957). Pathogenicity of an ectoparasitic nematode, *Xiphinema diversicaudatum*, on strawberries. *Nematologica* 2: 91-93.

Schmidt, H. B., Fritzsche, R., and Lehmann, W. (1963). Die Übertragung des Weidelgrasmosaik-Virus durch Nematoden. *Naturwissenschaften* 50: 386.

Seinhorst, J. W. (1966). *Longidorus elongatus* on *Fragaria vesca. Nematologica* 12: 275-279.

Seinhorst, J. W. (1973). The combined effects of dichloropropane-dichloropropene mixture injected into the soil and of dazomet applied to the soil surface. *Neth. J. Pl. Path*. 79: 194-206.

Sharma, R. D. (1965). Direct damage to strawberry by *Longidorus elongatus* (de Man, 1876) Thorne and Swanger, 1936. *Meded. LandbHoogesch. OpzoekStns Gent* 30: 1437-1443.

Siddiqi, M. R. (1959). Studies on *Xiphinema* spp. (Nematoda: Dorylaimoidea) from Aligarh (North India), with comments on the genus *Longidorus* Micoletzky, 1922. *Proc. helm. Soc. Wash*. 26: 151-163.

Siddiqi, M. R. (1973). Systematics of the genus *Trichodorus* Cobb, 1913 (Nematoda: Dorylaimida), with descriptions of three new species. *Nematologica* 19: 259-278.

Smith, K. M., and Markham, R. (1944). Two new viruses affecting tobacco and other plants. *Phytopathology* 34: 324-329.

Smith, P. C. (1975). Distribution of *Xiphinema* spp. in South African vineyards. *In* "Nematode Vectors of Plant Viruses" (F. Lamberti, C. E. Taylor and J. W. Seinhorst, Eds.). Plenum Press, New York and London.

Smith, S. H., Stouffer, R. F., and Soulen, D. M. (1973). Induction of stem-pitting in peaches by mechanical inoculation with tomato ringspot virus. *Phytopathology* 63: 1404-1406.

Southey, J. F. (1973). Identification of *Xiphinema* species. *In* "The Longidoridae"–Association of Applied Biologists Workshop Manual, 14 June 1973. Rothamsted Exp. Stn., Harpenden.

Stace-Smith, R. (1978). Characteristics of nematode-borne plant viruses. *Proc. Amer. Phytopath. Soc.* 4: (in press).

Stace-Smith, R., and Hansen, A. J. (1976). Some properties of cherry rasp leaf virus. *Acta Hortic.* 67: 193-197.

Stouffer, R. F. (1977). Tomato ringspot virus/graft union necrosis. Is there a link? *Amer. Fruit Grower* 97: 16, 31.

Stouffer, R. F., Smith, S. H., Lewis, F. H., and Soulen, D. M. (1972). Soil transmission of *Prunus* stem-pitting (abstr.). *Phytopathology* 62: 672-673.

Stouffer, R. F., Hickey, K. D., and Welsh, M. F. (1977). Apple union necrosis and decline. *Pl. Dis. Reptr.* 61: 20-24.

Stubbs, L. L. (1971). Plant Pathology in Australia. *Ann. Rev. Pl. Path*. 50: 46-78.

Sutherland, J. R. (1969). Feeding of *Xiphinema bakeri. Phytopathology* 59: 1963-1965.

Sutherland, J. R. (1970). Some forest nursery seedling hosts of the nematode, *Xiphinema bakeri*. *Can. J. Plant Sci*. 50: 588-590.

Szczygiel, A., and Hasior, H. (1972). Vertical distribution of plant parasitic nematodes in the soil of strawberry plantations. *Ekologia Polska* 20: 493-506.

Tarjan, A. C. (1964). Two new American dagger nematodes (*Xiphinema*: Dorylaimidae) associated with citrus, with comments on the variabilty of *X. bakeri* Williams, 1961. *Proc. helminth. Soc. Wash*. 31: 65-76.

Taylor, C. E. (1962). Transmission of raspberry ringspot virus by *Longidorus elongatus* (de Man) (Nematoda: Dorylaimidae). *Virology* 17: 493-494.

Taylor, C. E. (1967). The multiplication of *Longidorus elongatus* (de Man) on different host plants with reference to virus transmission. *Ann. appl. Biol*. 59: 275-281.

Taylor, C. E. (1970). The association of *Longidorus elongatus* with raspberry ringspot and tomato black ring viruses. *Proc. 9th Int. Nematol. Symp. Warsaw 1967*, 283-289.

Taylor, C. E. (1971). Nematodes as vectors of Plant Viruses. *In* "Plant Parasitic Nematodes"

(B. M. Zuckerman, W. F. Mai and R. A. Rohde, Eds.). Vol. 2. Academic Press, New York.

Taylor, C. E. (1972a). Transmission of viruses by nematodes. *In* "Principles and Techniques in Plant Virology" (C. I. Kado and H. O. Agrawal, Eds.). Van Nostrand-Reinhold, New York.

Taylor, C. E. (1972b). Nematode transmission of plant viruses. *PANS* 18: 269-282.

Taylor, C. E., and Brown, D. J. F. (1976). The geographical distribution of *Xiphinema* and *Longidorus* nematodes in the British Isles and Ireland. *Ann. appl. Biol.* 84: 383-402.

Taylor, C. E., and Cadman, C. H. (1969). Nematode Vectors. *In* "Viruses, Vectors and Vegetation" (K. Maramorosch, Ed.). Interscience Publ., New York.

Taylor, C. E., and Murant, A. F. (1966). Nematicidal activity of aqueous extracts from raspberry canes and roots. *Nematologica* 12: 488-494.

Taylor, C. E., and Murant, A. F. (1968). Chemical control of raspberry ringspot and tomato black ring viruses in strawberry. *Pl. Pathol.* 17: 171-178.

Taylor, C. E., and Murant, A. F. (1969). Transmission of strains of raspberry ringspot and tomato black ring viruses by *Longidorus elongatus* (de Man). *Ann. appl. Biol.* 64: 43-48.

Taylor, C. E., and Raski, D. J. (1964). On the transmission of grape fanleaf by *Xiphinema index*. *Nematologica* 10: 489-495.

Taylor, C. E., and Robertson, W. M. (1969). The location of raspberry ringspot and tomato black ring viruses in the nematode vector, *Longidorus elongatus*. *Ann. appl. Biol.* 64: 233-237.

Taylor, C. E., and Robertson, W. M. (1970a). Sites of virus retention in the alimentary tract of the nematode vectors, *Xiphinema diversicaudatum* (Micol.) and *X. index* (Thorne and Allen). *Ann. appl. Biol.* 66: 375-380.

Taylor, C. E., and Robertson, W. M. (1970b). Location of tobacco rattle virus in the nematode vector, *Trichodorus pachydermus* Seinhorst. *J. gen. Virol.* 6: 179-182.

Taylor, C. E., and Robertson, W. M. (1975). Acquisition, retention and transmission of viruses by nematodes. *In* "Nematode Vectors of Plant Viruses" (F. Lamberti, C. E. Taylor and J. W. Seinhorst, Eds.). Plenum Press, New York and London.

Taylor, C. E., and Robertson, W. M. (1978). Virus vector relationships and mechanics of transmission. *Proc. Amer. Phytopath. Soc.* 4: 20-29.

Taylor, C. E., and Thomas, P. R. (1978). The association of *Xiphinema diversicaudatum* (Micoletsky) with strawberry latent ringspot and arabis mosaic viruses in a raspberry plantation. *Ann. appl. Biol.* 62: 147-157.

Taylor, C. E., Robertson, W. M., and Roca, F. (1976). Specific assocaition of artichoke Italian latent virus with the odontostyle of its vector *Longidorus attenuatus. Nematol. medit.* 4: 23-30.

Taylor, C. E., Alphey, T. J. W., and Brown, D. J. F. (1978). The distribution of nematode virus-vectors in Great Britain. *In* "Plant Disease Epidemiology" (P. R. Scott and A. Bainbridge, Eds.). Blackwells Sc. Publ., London.

Teliz, D., Grogan, R. G., and Lownsbery, B. F. (1966). Transmission of tomato ringspot, peach yellow bud mosaic, and grape yellow vein viruses by *Xiphinema americanum. Phytopathology* 56: 658-663.

Temmink, J. H., Campbell, R. N., and Smith, P. R. (1970). Specificity and site of *in vitro* acquisition of tobacco necrosis virus by zoospores of *Olpidium brassicae. J. gen. Virol.* 9: 201-213.

Teploukhova, T. N. (1974). Some data on the nematode vectors of phytopathogenic viruses in the USSR. *Trudy vses. nauchno-issled. Inst. Zashch. Rast.* 41: 40-45 (in Russian).

Thomas, P. R. (1969). Crop and weed plants compared as hosts of viruliferous *Longidorus elongatus* (de Man). *Pl. Path.* 18: 23-28.

Thomas, P. R. (1970). Host status of some plants for *Xiphinema diversicaudatum* (Micol.) and their susceptibility to viruses transmitted by this species. *Ann. appl. Biol.* 65: 169-178.

Thomas, W., and Procter, C. H. (1972). Arabis mosaic virus in *Cyphomandra betacea* Sendt. *N. Z. J. Agric. Res.* 15: 395-404.

Thorne, G., and Schieber, E. (1962). American dagger nematode (*Xiphinema americanum*) on coffee in Guatemala, with suggestions for nematode control in nurseries. *Pl. Dis. Reptr.* 46: 857.

Thresh, J. M., and Pitcher, R. S. (1971). The spread and control of nettlehead and other diseases of hop associated with arabis mosaic virus. *Proc. 6th Br. Insectic. Fungic. Conf.* (1971), 314-318.

Tomlinson, J. A., and Walkey, D. G. A. (1967). The isolation and identification of rhubarb viruses occurring in Britain. *Ann. appl. Biol.* 59: 415-426.

Towle, A., and Doncaster, C. C. (1978). Feeding of *Longidorus caespiticola* on rye grass, *Lolium perenne. Nematologica* 24: 277-285.

Trudgill, D. L. (1976). Observations on the feeding of *Xiphinema diversicaudatum. Nematologica* 22: 417-423.

Trudgill, D. L., and Alphey, T. J. W. (1976). Chemical control of the virus vector nematode *Longidorus elongatus* and of *Pratylenchus crenatus* in raspberry plantations. *Pl. Path.* 25: 15-20.

Trudgill, D. L., and Brown, D. J. F. (1978). Ingestion, retention, and transmission of two strains of raspberry ringspot virus by *Longidorus macrosoma. J. Nematol.* 10: 85-89.

Tsuchizaki, T. (1975). Mulberry Ringspot Virus. *CMI/AAB Descriptions of Plant Viruses* No. 142.

Valdez, R. B. (1972a). A micro-container technique for studying virus transmission by nematodes. *Pl. Path.* 21: 114-117.

Valdez, R. B. (1972b). Transmission of raspberry ringspot virus by *Longidorus caespiticola, L. leptocephalus* and *Xiphinema diversicaudatum* and of arabis mosaic virus by *L. caespiticola* and *X. diversicaudatum. Ann. appl. Biol.* 71: 229-234.

Valdez, R. B., McNamara, D. G., Ormerod, P. J., Pitcher, R. S., and Thresh, J. M. (1974). Transmission of the hop strain of arabis mosaic virus by *Xiphinema diversicaudatum. Ann. appl. Biol.* 76: 113-122.

Van Hoof, H. A. (1962). *Trichodorus pachydermus* and *T. teres*, vectors of the early-browning virus of peas. *Tijdschr. Plantenziekten* 68: 391.

Van Hoof, H. A. (1966). Nematode populations active and inactive with regard to transmission of Nepo viruses. *Nematologica* 12: 615-618.

Van Hoof, H. A. (1967). Het mechanisme van de virusoverbrenging door nematoden. *Neth. J. Plant Path.* 73: 193-194.

Van Hoof, H. A. (1968). Transmission of tobacco rattle virus by *Trichodorus* species. *Nematologica* 14: 20-24.

Van Hoof, H. A. (1970). Some observations on retention of tobacco rattle virus in nematodes. *Neth. J. Pl. Path.* 76: 329-330.

Van Hoof, H. A. (1971). Viruses transmitted by *Xiphinema* species in the Netherlands. *Neth. J. Pl. Path.* 77: 30-31.

Van Hoof, H. A. (1975). The effect of temperature on the transmission of tobacco rattle virus in tulips by *Trichodorus*, using the 'bait-leaf' method. *Nematologica* 21: 104-108.

Van Hoof, H. A. (1976). The effect of soil moisture content on the activity of trichodorid nematodes. *Nematologica* 22: 260-264.

Van Hoof, H. A., Maat, D. Z., and Seinhorst, J. W. (1966). Viruses of the tobacco rattle virus group in Northern Italy: their vectors and serological relationships. *Neth. J. Pl. Path.* 72: 253-258.

Vovlas, C., and Roca, F. (1975). Nematode transmission of the causal agent of chicory chlorotic ringspot, a strain of artichoke Italian latent virus. *Nematol. Medit.* 3: 83-90.

Vovlas, C., Martelli, G. P., and Quacquarelli, A. (1971). Le virosi delle piante ortensi in Puglia. VI. Il complesso delle maculature anulari della Cicoria. *Phytopath. medit.* 10: 244, 254.

Vuittenez, A. (1961). Les nématodes vecteurs de virus et le probleme de la dégénérescence infectieuse de la vigne. *Les Nematodes C. N. R. A. Versailles* 1961, 55-78.

Vuittenez, A. (1970). Fanleaf of grapevine. *In* "Virus Diseases of Small Fruits and Grapevines" (N. W. Frazier, ed.). Univ. Calif. Divn Agric. Sci., Berkeley.

Walkinshaw, C. H., Griffin, G. D., and Larson, R. H. (1961). *Trichodorus christiei* as a vector of potato corky ringspot (tobacco rattle) virus. *Phytopathology* 51: 806-807.

Weischer, B. (1966). Ein Beitrag zur geographischen Verbreitung und Oekologie von Arten der Gattung *Xiphinema* und *Longidorus. Mitt. Biol. Bundeanst. Dahlem.* 118: 100-106.

Weischer, B. (1974). *Xiphinema*–Arten europäischer Weinberge. *Weinberg und Keller* 21: 61-76.

Weischer, B. (1975). Ecology of *Xiphinema* and *Longidorus*. *In* "Nematode vectors of plant viruses" (F. Lamberti, C. E. Taylor and J. W. Seinhorst, Eds.). Plenum Press, New York and London.

Weischer, B., and Wyss, U. (1976). Feeding behavior and pathogenicity of *Xiphinema index* on grapevine roots. *Nematologica* 22: 319-325.

Whitehead, A. G. (1971). Nematicides for field crops. *Br. Insect. Fung. Conf (6th) Brighton, 1971*, 15-18.

Whitehead, A. G., and Hooper, D. J. (1970). Needle nematodes (*Longidorus* spp.) and stubby-root nematodes (*Trichodorus* spp.) harmful to sugar beet and other field crops in England. *Ann. appl. Biol.* 65: 339-350.

Whitehead, A. G., and Tite, D. J. (1967). Small doses of D-D soil fumigant to control free-living nematodes injurious to sugar beet. *Plant Path.* 16: 107-109.

Whitehead, A. G., Dunning, R. A., and Cooke, D. A. (1971). Docking disorder and root ecto-parasitic nematodes of sugar beet. *Rep. Rothamsted exp. Stn. 1970*, 219-236.

Winfield, A. L., and Cooke, D. A. (1975). The ecology of *Trichodorus*. *In* "Nematode Vectors of Plant Viruses" (F. Lamberti, C. E. Taylor and J. W. Seinhorst, Eds.). Plenum Press, New York and London.

Wright, K. A. (1965). The histology of the oesophageal region of *Xiphinema index* Thorne and Allen, 1950, as seen with the electron microscope. *Can. J. Zool.* 43: 689-700.

Wyss, U. (1970a). Parasitierungsvorgang und Pathogenität wandernder Wurzelnematoden an *Fragaria vesca* var. semperflorens. *Nematologica* 16: 55-62.

Wyss, U. (1970b). Untersuchungen zur Populationsdynamic von *Longidorus elongatus*. *Nematologica* 16: 55-62.

Wyss, U. (1971). Der Mechanismus der Nahrungsaufnahme bei *Trichodorus similis*. *Nematologica* 17: 508-518.

Wyss, U. (1975). Feeding of *Trichodorus, Longidorus* and *Xiphinema*. *In* "Nematode Vectors of Plant Viruses" (F. Lamberti, C. E. Taylor and J. W. Seinhorst, Eds.). NATO Advanced Study Institute Series, Series A: Life Sciences Vol. 2, Plenum Press, New York and London, 203-221.

Wyss, U. (1978). Feeding processes of virus-transmitting nematodes. *Proc. Amer. Phytopath. Soc.* 4: 30-41.

Yagita, H., and Komuro, Y. (1972). Transmission of mulberry ringspot virus by *Longidorus martini* Merny. *Ann. Phytopath. Soc. Japan* 38: 275-283.

Yassin, A. M. (1969). Glasshouse and laboratory studies on the biology of the needle nematode. *Longidorus elongatus. Nematologica* 15: 169-178.

Yeates, G. W. (1967). Studies on nematodes from dune sands. 9. Quantitative comparison of the nematode faunas of six localities. *New Zealand J. Sc.* 10: 927-948.

Chapter 17*

FUNGI

David S. Teakle

Department of Microbiology
University of Queensland
St. Lucia, Australia

17.1 INTRODUCTION 417
17.2 VECTORS 418
17.2.1 Chytrids 418
17.2.2 Plasmodiophorids 420
17.3 VIRUSES 421
17.3.1 Isometric Viruses 421
17.3.2 Straight Tubular Viruses 423
17.3.3 Filamentous Viruses 424
17.3.4 Agents of Undetermined Morphology 424
17.4 VIRUS-VECTOR RELATIONSHIPS 425
17.4.1 Evidence of Virus Transmission by Fungi 425
17.4.2 Resolving Virus-Vector Relationships 427
17.4.3 Present Concepts and Terminology 429
17.5 SUSPECTED FUNGUS TRANSMISSION 429
17.6 CONCLUDING REMARKS 433
17.7 ACKNOWLEDGMENTS 434
17.8 REFERENCES 435

17.1 INTRODUCTION

The fungi which have been confirmed as being vectors of plant viruses are certain lower fungi belonging to the chytrids (family Olpidiaceae) or the plasmodiophorids (family Plasmodiophoraceae). These fungi are obligate plant parasites which commonly infect roots under cool, wet soil conditions. As well as transmitting a number of more or less well characterized viruses, these vectors transmit several agents whose

*Dedicated to Cecil E. Yarwood on his 73rd birthday, September 16, 1979.

ISBN 0-12-326450-2

viral nature is assumed but not yet fully established.

Two distinctly different virus-vector relationships have been shown by these fungi. On the one hand there is the *persistent* relationship, with the viruses surviving inside the resting spores of the fungi and being internally-borne by the zoospores; and on the other hand there is the *non-persistent* relationship, with the viruses *not* surviving in the resting sporangia and being externally-borne by the zoospores.

In addition to the more or less well established virus-vector combinations, there are others for which limited experimental evidence of fungal transmission has been presented but for which independent confirmation is usually lacking. These combinations include suspected fungal vectors belonging to families other than the Olpidiaceae and Plasmodiophoraceae. In the hope of stimulating further interest, these will also be discussed.

This review follows four earlier ones dealing specifically with the fungus transmission of plant viruses (Grogan and Campbell 1966; Teakle 1967, 1969, 1972). The earlier reviews, and others such as that of Harrison (1977) which deal partly with fungal transmission, complement the present contribution and can be consulted for additional information.

17.2 VECTORS

Only the well-established chytrid and plasmodiophorid vectors will be discussed in this section. A summary of the dimensions and some other properties of these fungi is given in Table I. Suggested life cycles of these fungi, and also of certain suspected vectors, are given in Karling (1968) or Teakle (1969). Some aspects of the life cycles remain controversial because of difficulties in studying these obligate plant parasites.

17.2.1 Chytrids

Olpidium brassicae and *O. radicale* (syn. *O. cucurbitacearum*) are well-established vectors of at least three isometric viruses, while *O. brassicae* also transmits the virus-like agents causing tobacco stunt and lettuce big-vein diseases (Table I). Both species have two types of zoosporangia, thin-walled sporangia and thick-walled resting sporangia. The thin-walled sporangia are formed in the outer cells of roots from naked, holocarpic thalli; when mature they are readily stimulated by irrigation to release zoospores through exit tubes or papillae. The zoospores are usually smaller in *O. brassicae* than in *O. radicale* (Table I); each is propelled by a single, posterior, whiplash flagellum. Zoospores encyst on plant roots preferentially in the zone of elongation behind the root tip, the flagellum is withdrawn, and penetration of the host cell wall by the cyst protoplast may be complete after 2 - 4 hr. Inside the host the fungus protoplast constitutes a naked thallus which increases in size for 2-3 days before completing the cycle by forming a thin zoosporangial wall.

Whereas thin-walled sporangia are produced by single zoospores, resting sporangia

TABLE I. Some Properties of the Main Fungal Vectors[1]

Property	*Olpidium brassicae*	*O. radicale* (cucurbit strain, syn. *O. cucurbitacearum*)	*Polymyxa graminis*	*P. betae*	*Spongospora subterranea*
Zoospore					
Size (μm)	2.8-5.6	4-5 x 8	2.5-5.6	4.5-5	2.5-4.6
No. of flagella	1	1	2	2	2
Length of flagella (μm)	13-22	20-21	16-20, 4-5	18-25, 6.7-7.5	13.7, 4.4
Host penetration by	zoospore protoplast	zoospore protoplast	whole zoospore or protoplast	zoospore protoplast	whole protoplast
Thin-walled sporangium					
Size (μm)	10 x 10 to 25 x 134	14 x 14 to 34 x 138	variable; not given	variable; not given	variable; not given
Resting structure					
Type	sporangium	sporangium	spore in a cystosorus	spore in a cystosorus	spore in a cystosorus
Size (μm)	10-34	10-34	4-7	4-5	3.5-4.5
Outer wall appearance	ridged to stellate	smooth	smooth	smooth	smooth
Host range	many families	Cucurbitaceae	mainly Graminae	mainly Chenopodiaceae	mainly Solanaceae
Viruses (agents) transmitted	tobacco necrosis, satellite, lettuce big-vein, tobacco stunt	cucumber necrosis	barley yellow mosaic, oat mosaic, rice necrosis mosaic, wheat mosaic, wheat spindle streak mosaic, wheat yellow mosaic	beet necrotic yellow vein	potato mop-top

[1] Data from Garrett and Tomlinson (1967), Barr (1968), and Karling (1968).

are probably formed following the fusion of two zoospores or thalli. The thalli destined to become resting sporangia initially resemble those of thin-walled sporangia. However, after a few days they develop a thick wall which is stellate in appearance in *O. brassicae* (Hiruki and Alderson, 1976) and smooth in *O. radicale* (Barr, 1968; Lange and Insunza, 1977). Resting sporangia survive for many years in air-dried soil or roots; their germination to produce zoospores is erratic and the conditions involved are not well understood.

Brief descriptions of methods which can be used to handle cultures of *O. brassicae* have been published (Teakle, 1967, 1972). Cultures are sensitive to a number of factors, including the type and vigor of the host plant, the soil moisture level, and the temperature. When doing laboratory studies with zoospore suspensions, low temperatures of 10-20°C and a suitable suspending medium favor retention of zoospore motility, which is required for *in vitro* infection (Teakle, 1962). Claims that zoospore suspensions can infect roots in soil after heating 55°C for 10 minutes (Campbell and Grogan, 1964) have been withdrawn, since heat-resistant resting sporangia apparently contaminated their zoospore suspensions (Campbell and Lin, 1976).

The favorable effect of frequent irrigation or rain, mild temperatures and vigorous growth of host plants is reflected in the common natural occurrence of *O. brassicae* in roots of crop plants in many temperate agricultural areas. It is also frequently found in glasshouses. Its activity in subtropical and tropical countries is adversely affected by high summer temperatures and in such areas as northern Queensland it apparently infects favored crop plants such as lettuce only in the winter months (unpublished data). *O. radicale* probably also has a wide geographical distribution, but has been less studied than *O. brassicae*.

The taxonomic position of the vector chytrids has been of continuing interest. Sahtiyanci (1962) transferred root-infecting *Olpidium* spp. to the genus *Pleotrachelus* because their sporangia sometimes have more than one exit tube; however, this move has not been generally accepted. Lange and Insunza (1977) recently transferred *O. cucurbitacearum* (Barr, 1968) to *O. radicale* Schwartz and Cook, which they have redescribed. *O. radicale* now includes root inhabiting, olpidiaceous species with smooth-walled resting sporangia and relatively large, elongated zoospores which have a slow, steady way of swimming in contrast to the rapid, jerky motion of zoospores of most chytrids. Strains of *O. radicale* which they studied had wider host ranges than the cucurbit strain, formerly called *O. cucurbitacearum*, which is restricted to the Cucurbitaceae (Table I).

17.2.2 Plasmodiophorids

The three species involved, *Polymyxa graminis, P. betae* and *Spongospora subterranea,* transmit a number of straight tubular or filamentous viruses (Table I). While all three species have zoospores in the same size range as those of *Olpidium* spp., they differ in that the zoospores have two flagella of unequal length. Reports differ on the mode of penetration of zoospores into the host plant; zoospore encystment and subsequent penetration by the protoplast only is reported for *P. betae* (Keskin, 1964), penetration of the whole zoospore is claimed for *S. subterranea* (Kole, 1954), while both methods of penetration apparently occur with *P. graminis* (Ledingham, 1939; D'Ambra, 1967).

The thalli of plasmodiophorids like those of the vector chytrids are typically naked plasmodia, although a thin membrane has been reported to surround the developing zoosporangium in *P. graminis* (Ledingham, 1939). Zoosporangia are somewhat characteristic in appearance, occurring singly or in loose groups in *S. subterranea,* but as elongate, lobed, irregular, septate or non-septate structures with long exit tubes in *Polymyxa* spp.

The mature resting stages of the plasmodiophorids consist of cystosori, comprising irregular aggregates of small spores in *Polymyxa* spp. and hollow spore balls in *S. subterranea.* After the cystosori are released from roots, the individual spores germinate to produce single zoospores.

Handling cultures of the plasmodiophorids is more difficult than of the chytrids. Zoospore discharge from zoosporangia is usually less abundant when infected roots are irrigated, and the multiplication of contaminating fungi, including *O. brassicae,* may be a problem. Germination of resting spores is erratic; it is favored by air-drying the infested material before irrigation.

The natural host ranges of the plasmodiophorid vectors are somewhat restricted, with each species being commonly found in a different plant family (Table I). However, the fungi are inconspicuous in roots and are probably often overlooked when they infect non-preferred hosts. Therefore one should not be too surprised that *P. graminis* is reported to infect clover in addition to grasses and *S. subterranea* several non-solanaceous hosts (Gerdemann, 1955; Karling, 1968).

The temperature relations of the three vector species are probably similar. *P. graminis* develops most freely at about 18°C and zoospores rapidly become immobile at 28°C (Ledingham, 1939). *S. subterranea* has an optimum temperature for root hair infection of 14-20°C and a maximum of 22-25°C (Kole, 1954). As might be expected, these fungi are found mostly in temperate countries; possibly the major period of activity of *P. graminis* in subtropical areas such as Florida and Louisiana is during the winter, as it is for *S. subterranea* in subtropical Queensland (unpublished data).

17.3 VIRUSES

Three groups of viruses transmitted by chytrid or plasmodiophorid vectors can be recognized on the basis of a general similarity in particle morphology. In addition, a fourth group can be erected to contain those virus-like agents of undetermined morphology which have a persistent relationship with *O. brassicae* (Table II).

17.3.1 Isometric Viruses

Included in this group are satellite, tobacco necrosis and cucumber necrosis viruses (Table II). Although these three viruses differ in particle size and are serologically

TABLE II. Properties of Some Soil-borne Plant Viruses

Group	Virus (or agent)	Properties of virus particles: Shape	Properties of virus particles: Size (nm)	Vector	Vector relations	References[c]
1	Satellite	Isometric	17	*Olipidium brassicae*	Non-persistent	15
	Tobacco necrosis	Isometric	26	*Olipidium brassicae*	Non-persistent	14
	Cucumber necrosis	Isometric	31	*O. radicale* (syn. *O. cucurbitacearum*)	Non-persistent	82
	Pea stem necrosis	Isometric	25	? *Olpidium* sp.	?	Nakano and Iemura, 1977
	Red clover necrotic mosaic	Isometric	27	? *O. radicale*	?	Lange and Insunza, 1977
	Tomato bushy stunt	Isometric	30	–[d]	–	69
	Cymbidium ringspot	Isometric	30	–[d]	–	178
	Galinsoga mosaic	Isometric	27	–[d]	–	Behncken, 1970 Shukla *et al.*, 1979
2	Beet necrotic yellow vein	St. tubular	65–105+270+390x20	*Polymyxa betae*	Persistent	144
	Potato mop-top[a]	St. tubular	150+300x20	*Spongospora subterranea*	Persistent	138
	Wheat mosaic[a]	St. tubular	110–160+300x20	*P. graminis*	Persistent	77
	Peanut clump	St. tubular	190+245 x 20	? *Polymyxa* sp.	?	Thouvenel *et al.*, 1976
	Broad bean necrosis[a]	St. tubular	150+250x25	–	–	Nakasone and Inouye, 1977
	Oat golden stripe	St. tubular	150+300x20	? *P. graminis*	?	Plumb and Macfarlane, 1976
	Tobacco mosaic[a]	St. tubular	300x18	–[d]	–	151
3	Barley yellow mosaic[b]	Filamentous	275+550x13	*P. graminis*	Persistent	143
	Oat mosaic	Filamentous	600–750x13	*P. graminis*	Persistent	145
	Rice necrosis mosaic[b]	Filamentous	275+550x13	*P. graminis*	Persistent	172
	Wheat spindle streak mosaic[b]	Filamentous	190–1975x13	*P. graminis*	Persistent	167
	Wheat yellow mosaic[b]	Filamentous	200–300+600x13	–	–	Usugi and Saito, 1976
	Sugarcane mosaic	Filamentous	750x13	–[d]	–	88; Bond and Pirone, 1970
4	Lettuce big-vein	Unknown	Unknown	*O. brassicae*	Persistent	Campbell and Grogan, 1964
	Tobacco stunt	Unknown	Unknown	*O. brassicae*	Persistent	Hiruki *et al.*, 1975
	Freesia leaf necrosis	Unknown	Unknown	? *O. brassicae*	? Persistent	Van Dorst, 1975

[a-b] Those viruses having the same letter have been shown to be serologically related. [c] Numbers refer to the CMI/AAB Descriptions of Plant Viruses numbers. [d] There is experimental evidence of mechanical transmission in soil.

unrelated, they are similar in certain other respects. All may be acquired *in vitro* by zoospores of their vectors and are the only viruses so far shown to have this property. Following mechanical inoculation both tobacco necrosis and cucumber necrosis viruses generally produce necrotic local lesions in their hosts. Satellite virus differs in that it has no characteristic symptoms of its own, since it depends on tobacco necrosis virus for replication; usually its presence reduces the size of local lesions produced by tobacco necrosis virus (Kassanis, 1970).

As might be expected from its smaller particles, satellite virus has a smaller genome size than the two other isometric, fungus-transmitted viruses, i.e. 0.4×10^6 daltons compared with $1.3\text{-}1.7 \times 10^6$ daltons. However, the genomes of all three viruses are unipartite and are composed of single-stranded RNA which comprises 15-20% of the particle weight. These properties also apply to the other isometric viruses which are soil-borne, and which may or may not be chytrid transmitted (Table II). Tobacco necrosis and satellite viruses have a similar translational strategy (Salvato and Fraenkel-Conrat, 1977), but that of the other isometric viruses has not yet been determined.

17.3.2 Straight Tubular Viruses

Recent work has shown that two of the three fungus-transmitted viruses in this group, potato mop-top and wheat mosaic viruses, have close affinities with each other and with tobacco mosaic and broad bean necrosis viruses (Powell, 1976; Randles *et al.,* 1976; Nakasone and Inouye, 1977). As well as having a similar particle shape, structure, and percentage of RNA, these four viruses are serologically interrelated (Table II). Thus, the inclusion of potato mop-top and wheat mosaic viruses as possible members of the Tobamovirus Group, of which tobacco mosaic virus is type member (Fenner, 1976), is fully justified. The fungus-transmitted tobamoviruses differ, however, from tobacco mosaic virus in their lability, being more subject to breakage and unwinding from the end (Gumpf, 1971; Harrison, 1974). This fragility, together with the tendency of particles to aggregate, makes the determination of particle length and the function of different particle size classes, more difficult than for the more stable tobacco mosaic virus. Present evidence indicates that all tobamoviruses have 250-300 nm long particles, which are essential for infectivity; shorter particles of 110-160 nm length are probably also essential for the infectivity of wheat mosaic virus (Tsuchizaki *et al.,* 1975), but with tobacco mosaic virus shorter particles are not required, although sometimes enhancing infection. Such non-essential particles associated with tobacco mosaic virus may represent encapsidated portions of the genome which are produced as such during replication of the viral RNA (Beachy and Zaitlin, 1977).

Beet necrotic yellow vein and peanut clump viruses are similar to the fungus-transmitted tobamoviruses in general particle morphology and probably structure, but no serological relationship has been demonstrated. Further, three or four predominant particle lengths are characteristic of beet necrotic yellow vein virus, but at present there are no data indicating which particles are infectious or whether the virus has a multipartite genome (Putz, 1977).

17.3.3 Filamentous Viruses

These viruses have somewhat unstable particles and their properties are poorly understood; for instance, the type of nucleic acid is not yet reported. Particle lengths of the different viruses frequently vary, probably in part because of aggregation and fragmentation. A bimodal size frequency is reported for partially purified preparations of particles of several viruses (Table II), but a trimodal size frequency of 275-300, 530-560 and 770-820 nm, may occur with leaf-dip preparations of two of these viruses, wheat and barley yellow mosaic viruses (Usugi and Saito, 1976). Despite these differences in particle length, which at least in part reflect differences in preparative technique, all have a particle diameter of about 13 nm. Further, all induce infected cells to produce cylindrical inclusion bodies, which in cross-section appear as pinwheels. Thus, these viruses appear to resemble both the potyviruses, which are aphid-transmitted, and certain mite-transmitted viruses such as wheat streak mosaic virus.

17.3.4 Agents of Undetermined Morphology

Tobacco stunt and lettuce big-vein diseases are caused by agents which have long been presumed to be viruses. However, especially in the case of lettuce big-vein, there is no conclusive evidence that a virus is involved. Although virus-like particles and even microorganisms have been associated with these diseases, their causal relationship remains to be established.

With lettuce big-vein disease, Haeske (1958) and Chod *et al.* (1976) reported the presence of rigid, rod-shaped particles of 250 and 244 nm, respectively, in the roots of affected lettuce plants. However, the particles were not found in extracts of diseased leaves, and no evidence of their infectivity was presented. Ragozzino and Furia (1974) reported the presence of rickettsia-like microorganisms in the phloem of leaves and roots of big-vein plants, and occasionally also found membrane-bound groups of filamentous, virus-like particles about 24 nm diameter. Again, evidence that these cause big-vein disease is lacking.

Tobacco stunt disease differs from lettuce big-vein disease in that the causal agent is sap transmissible, particularly if 0.001 *M* 1-phenyl-thiosemicarbazide in 0.01 *M* phosphate buffer, pH 6.8, is present during tissue maceration and inoculation (Hiruki, 1975). When suspended in this medium the tobacco stunt agent had a moderately wide host range (41 species in 9 families), remained infective for 60 hours, had a thermal inactivation point of 75-80°C and had a dilution end point of 10^{-2} - 10^{-3}. When suspended in phosphate buffer alone the extract was infective, but only immediately after extraction.

Attempts to visualize the tobacco stunt agent by electron microscopy of preparations from infected plants or of stunt-transmitting *O. brassicae* have given conflicting results. Spherical, virus-like particles with diameters of 18 nm (Hidaka, 1956), 25-30 nm (Hidaka, 1969) or 40-60 nm (Hiruki *et al.*, 1975) have been found, as well as chlamydia-like bodies of 100-500 nm diameter with a double membrane or a membrane and a wall (Hiruki *et al.*, 1975). Possibly some of these could represent pathogens of *O. brassicae* rather than the tobacco stunt agent.

The most convincing evidence that a virus is indeed the agent causing tobacco stunt is provided by its reaction to chemical treatments. It is resistant to treatment with tetracycline, chloramphenicol or penicillin antibiotics, completely susceptible to ribonuclease but not affected by deoxyribonuclease and moderately affected by pronase and bentonite treatments. Thus, it is probably not a microorganism but an RNA-containing virus with a loosely packed capsid (Hiruki *et al.*, 1975).

17.4 VIRUS-VECTOR RELATIONSHIPS

17.4.1 Evidence of Virus Transmission by Fungi

Various types of evidence have been used to show that at least three isometric, three straight tubular, four filamentous viruses, and two virus-like agents of undetermined morphology, are transmitted by either chytrid or plasmodiophorid vectors (Table II). This evidence has also led to a partial understanding of the virus-vector relationships. Some of the main types of evidence are discussed in the following sections.

17.4.1.1 Constant Association of a Fungus with Virus-infected Plants in Nature. The detection of these associations has often been an initial difficulty with both chytrid- and plasmodiophorid-transmitted viruses. For each disease there is an interesting story surrounding the establishment beyond reasonable doubt of a constant association. Wheat rosette and mosaic disease caused by wheat mosaic virus may be taken as an example. The disease was shown to be important in the USA by the 1920s when the possibility of a fungus vector was raised. Two fungi, *Septoria tritici* and *Helminthosporium sativum,* were found associated in late stages of the disease. However, neither was consistently present when the disease first appeared in the spring (McKinney, 1923), and heavy inoculations with *H. sativum* failed to cause rosette or mosaic leaf symptoms (McKinney *et al.*, 1925). The real vector, *Polymyxa graminis,* was at that time overlooked, probably because it was undescribed, confined to roots, rather inconspicuous and an obligate parasite. Eventually it was described in Canada by Ledingham (1935) without its vector role being recognized. Its occurrence in the USA and possible association with wheat mosaic virus was reported many years later (Linford and McKinney, 1954), but this work was discontinued following Linford's untimely death. A revival of interest and the establishment of a constant association of *P. graminis* with wheat mosaic-affected plants (Estes and Brakke, 1966) followed the demonstration that *O. brassicae* was associated with plant infection by other viruses and virus-like agents (Hidaka, 1960; Teakle, 1960; Campbell *et al.*, 1961; Tomlinson *et al.*, 1962).

Because the chytrid and plasmodiophorid vectors are all obligate parasites, they must be identified microscopically in stained or unstained, infected roots. Fortunately the resting spores and the thin-walled sporangia are somewhat distinctive (Table I), and experienced workers can usually distinguish them from the fruiting structures of the numerous other fungi found in roots. Further, the size and motility of zoospores of the vector fungi are somewhat distinctive, and a tentative identification can be made on the basis of observations of zoospores present in root washings.

One difficulty in trying to establish an association of a fungus with virus infection in the field is that vector populations may fluctuate widely depending on environmental conditions. Thus, it may be useful to promote high populations of a fungus by growing infected plants in a glasshouse or growth room where favorable conditions of moisture and temperature can be provided. Another difficulty is that two or more possible vector fungi may be present in the roots of virus-infected plants. For instance, both *Olpidium brassicae* and *Polymyxa graminis* occur in cereal roots infected by wheat mosaic virus (Canova, 1966) or oat mosaic virus (Macfarlane *et al.,* 1968); although *O. brassicae* may often be the more abundant, tests have shown that *P. graminis* is the vector in each case.

17.4.1.2 Constant Association of a Fungus with Virus Transmission. This can be established in a number of ways. Firstly, experiments can be done in soil. "Bait" plants may be exposed to virus-infective soil for limited periods of time and then examined microscopically for fungus infection and indexed for virus infection. This examination and indexing can be done immediately or after replanting into a steamed potting mix (Slykhuis and Barr, 1978).

Secondly, the association can be established without soil, which may interfere somewhat with the microscopic examination. Infected roots can be washed free of most of the adhering soil and placed in water or other suitable suspending medium to promote zoospore discharge of the vector fungus. Bait plants raised on germinating paper or in sterile sand are placed in the suspension either at the same time as the infected roots or shortly after they have been removed; viruses may be added to the suspensions or the zoospores can be taken from virus-infected roots.

Correlation of virus transmission with zoospore infection of roots can be done in several ways. In the case of *O. brassicae* and tobacco necrosis virus, virus transmission to mung bean results in the production of necrotic local lesions (Teakle, 1967); these develop most abundantly in the zone of elongation behind the root tip or, in young seedlings, just below the hypocotyl, where penetration of zoospores can be seen to be most abundant. In the case of viruses, such as wheat mosaic virus, which are symptomless in the roots of their host plants, the roots must be indexed mechanically or by electron microscopy for virus infection or the plants must be incubated under suitable conditions until systemic symptoms appear in the foliage (Estes and Brakke, 1966).

17.4.1.3 Lack of Transmission Following Selective Elimination of a Vector Fungus. Even if the presence of a fungus is correlated with virus infection, this is not proof of vector ability. The vector fungi so far recognized are obligate root parasites, and have always been more or less contaminated with other soil- or root-inhabiting microorganisms. Once it has been shown that a soil suspension of zoospores, or other fungus-containing preparation is virus-transmitting, it is possible to carry out treatments which more or less selectively inhibit the suspected vector, and then to correlate the effect of these treatments on fungus and virus infection. This technique was used to show that very mild treatments, such as filtration, use of fungicidal chemicals or mild heat treatments which eliminated or immobilized zoospores of *O. brassicae* prevented the transmission of the lettuce big-vein agent (Campbell and Grogan, 1963) or tobacco necrosis virus (Teakle, 1962). Thus the size, chemical sensitivity or heat sensitivity was shown to correlate well with that of the postulated vector. Such experiments can be carried out under any convenient set of conditions, and it is not necessary to carry them out under optimum conditions for survival as assumed by some workers (Grogan and Campbell, 1966; Campbell and Lin, 1976).

17.4.1.4 Transmission Despite Selective Elimination of Non-vector Organisms. In contrast to the delicate nature of the fungal zoospore, the resting spore or resting sporangium of the vector fungus may be resistant to certain harsh treatments which eliminate many other organisms, including fungi. For instance, Campbell (1962) treated preparations of resting sporangia of *O. brassicae* from big-vein lettuce plants with hydrochloric acid without destroying their ability to transmit the big-vein agent. This treatment, which involved holding the preparations at pH 0.5 for 2 hr, could be expected to destroy most microorganisms. Similarly, air-drying eliminates some microorganisms and free viruses in infective soil but has little effect on the resting sporangia or spores of the vector chytrids and plasmodiophorids, or on the transmission of certain viruses.

Mechanical means can also be used to selectively eliminate non-vector organisms. Single sporangia of *O. brassicae* can be obtained by blending infected roots, removing host material by sieving, streaking the resultant suspension over water agar and cutting out small agar blocks containing individual, mature sporangia with the aid of a microscope (Lin *et al.*, 1970). These sporangia can then be washed individually into beakers of lettuce seedlings to establish single-sporangium isolates which continue to transmit viruses. No doubt this method results in the elimination of many other microorganisms and strengthens the evidence that *O. brassicae* is the vector, but no claim has been made that all contaminants are removed.

17.4.2 Resolving Virus-vector Relationships

When a specific fungus is constantly associated with virus infection and transmission (Section 17.4.1), the nature of the association should be investigated. At present only two virus-vector relationships are recognized; the non-persistent relationship of the isometric viruses and the persistent relationship of all other fungus-transmitted viruses and agents (Table II). Techniques which may be used to resolve the relationships are discussed below.

17.4.2.1 Chemical Treatment of Resting Spores. As shown in section 17.4.1.4, resting spores may survive harsh treatments which would be expected to eliminate viruses. The continued ability of resting sporangia of *O. brassicae* to transmit the lettuce big-vein agent despite treatment with strong hydrochloric acid is taken as evidence of internal carriage and the protection of the agent by the fungus (Campbell, 1962). Similarly, treatment of resting spores of *S. subterranea* for 16 hours with 5% sodium dodecyl sulphate does not prevent transmission of potato mop-top virus although it inactivates almost all the virus in leaf extracts within 20 minutes. This virus therefore also seems to be carried internally (Jones and Harrison, 1969). In contrast, tobacco necrosis virus is not transmitted by resting sporangia of *O. brassicae* treated with hydrochloric acid and thus is apparently not internally-borne (Campbell and Fry, 1966).

17.4.2.2 In Vitro Virus Acquisition by Zoospores. When any fungus-transmitted isometric virus is added *in vitro* to zoospores of its vector *Olpidium* sp., the virus is acquired and later transmitted. In contrast, addition of any of the elongate viruses or agents of unknown morphology to zoospores of their chytrid or plasmodiophorid vectors has so far failed to result in transmission, presumably because they fail to acquire the viruses. These latter viruses are acquired by their vector fungi only within plant roots (*i.e. in vivo*).

17.4.2.3 Recovery of Virus from Infective Zoospores. When zoospores of the lettuce (vector) strain *Olpidium brassicae* are exposed to tobacco necrosis virus, washed, disrupted and assayed for virus, a moderate amount of virus is recovered. Comparative procedures using the crucifer (non-vector) strain of the fungus yield much less virus (Teakle and Hiruki, 1964). Thus, apparently the lettuce strain, but not the crucifer strain, of *O. brassicae*, is able to adsorb tobacco necrosis virus *in vitro.*

17.4.2.4 Antiserum Treatment of Zoospores. The use of virus antisera has been helpful in establishing virus-vector relationships. The virus antisera can be used at different strengths, at different stages in the life cycle of the fungus and at different times relative to the addition of the virus. With tobacco necrosis and *O. brassicae* the following results have been obtained: 1) addition of antiserum before acquisition of virus by zoospores prevents transmission, 2) virus already acquired by motile zoospores is partially protected against inactivation by dilute antiserum but not concentrated antiserum, 3) once zoospores have acquired virus and encysted on roots, the addition of antiserum has no effect on virus transmission; therefore, after encystment the virus is probably inside the zoospore (Teakle and Gold, 1963; Kassanis and Macfarlane, 1964; Campbell and Fry, 1966).

17.4.2.5 Negative-stain Electron Microscopy of Zoospores. The adsorption of isometric viruses by zoospores of their vector species of *Olpidium* can be observed in negatively-stained preparations. A suitable technique involves washing away free virus by two or three centrifugations, fixing zoospores in droplets on specimen grids with osmic acid vapor and negatively staining with uranyl acetate (Temmink *et al.,* 1970). The adsorption of virus is stronger with efficient vectors than inefficient vectors, and is not observed with non-vectors.

17.4.2.6 Thin-section Electron Microscopy of Vector Fungi. Although there have been many attempts to detect viruses in vector fungi by means of thin-section electron microscopy, few have been successful. Exceptions to this generalization are the detection of isometric particles resembling tobacco necrosis virus inside encysted zoospores of *O. brassicae* (Temmink *et al.*, 1970) and of straight tubular particles of beet necrotic yellow vein virus inside zoospores and possibly resting spores of *Polymyxa graminia* (Tamada *et al.*, 1974; Tamada, 1975; Fujisawa and Sugimoto, 1977).

17.4.3 Present Concepts and Terminology

17.4.3.1 Concepts. As a result of work such as that outlined in sections 17.4.1 and 17.4.2, two distinct relationships of the viruses and their vectors have been described. In one case, exemplified by *Olpidium brassicae* and tobacco necrosis virus, the zoospore acquires the virus as a surface contaminant either from the soil solution or *in vitro*. The virus may enter the protoplast at flagellum withdrawal and later be transmitted when the zoospore cyst discharges its protoplast into an epidermal cell of a root. Consequent virus infection of the roots leads to the release into the soil of free virus which can be acquired by other zoospores of *O. brassicae*. In this case there is no convincing evidence that zoospores carry the virus internally when they are released from sporangia.

In the other case, exemplified by *O. brassicae* and the tobacco stunt virus, the fungus acquires the virus only inside the infected plant root and cannot acquire it from the soil solution or *in vitro*. The virus is apparently harbored by both the thin-walled sporangia and the resting sporangia, and zoospores released by them can transmit the virus. The persistence of the virus-infectivity of air-dried soil for many years is attributed to the survival of the virus inside resting sporangia.

17.4.3.2 Terminology. At present all viruses which are *zoospore-acquired* are also able to be *acquired in vitro*, are *externally-* or *surface-borne* by the zoospores and have a *non-persistent* relationship with their chytrid vectors. Those viruses or agents which are *thallus-acquired* are also *acquired in vivo*, are *internally-borne* by zoospores and other stages, and have a *persistent* relationship with their chytrid or plasmodiophorid vectors.

17.5 SUSPECTED FUNGUS TRANSMISSION

In addition to the well-established virus-vector combinations described above, there are a number of viruses or agents for which at least some evidence of fungus transmission is available.

17.5.1.1 Pea Stem Necrosis Virus and Olpidium sp. This soil-borne virus has caused a serious disease of peas in the Wakayawa Prefecture of Japan since 1973 (Nakano *et al.*, 1976). The virus has isometric particles with a diameter of approximately 25 nm (Inouye *et al.*, 1977) but so far has failed to react with antisera to other soil-borne viruses with isometric particles, including tobacco necrosis and cucumber necrosis viruses. Evidence of transmission by zoospores of *Olpidium* sp. has recently been obtained (Nakano and Iemura, 1977), but no information on the virus-vector relationship is available.

17.5.1.2 Red Clover Necrotic Mosaic Virus and Olpidium radicale. This virus causes severe mottle, distortion, and necrotic areas on leaves and moderate to severe stunting in red clover in certain European countries. The virus has isometric particles of about 27 nm diameter, is highly stable, is exuded into soil from roots of infected *Nicotiana clevelandii* and red clover, and can infect healthy *Nicotiana clevelandii* bait seedlings planted in such soil presumably by mechanical means (Hollings and Stone, 1977). However, an *Olpidium* sp. or *O. radicale* has been associated with infected plants in Sweden and England, and could also be involved in transmission (B. Gerhardson, personal communication, 1978; R. Bowen, personal communication, 1978; Lange and Insunza, 1977).

17.5.1.3 Tomato Bushy Stunt Virus. This virus has been shown to be soil-borne (Lovisolo *et al.*, 1965; Hollings and Stone, 1975) and can infect mechanically-inoculated roots (Roberts, 1950). The idea of a soil-borne fungal vector arose from the observation that addition of 0.2% captan to infective soil reduced recovery of the virus from roots by 98%, whereas this fungicide had no inactivating effect on the virus during mechanical inoculation (Lovisolo, 1966). However, attempts to implicate a specific vector have failed; tests with *Olpidium brassicae* and *Lagenocystis radicicola* which may be present in the roots of bait plants growing in infective soil, have been negative (Teakle and Gold, 1963; Campbell, 1968; Campbell *et al.*, 1975).

17.5.1.4 Galinsoga Mosaic Virus. This virus has been found only in Queensland, Australia, and studies to date indicate that it is a possible member of the Tombusvirus Group, although serologically unrelated to tomato bushy stunt virus (Behncken, 1970; Shukla *et al.*, 1979). As with tomato bushy stunt and cymbidium ringspot viruses (Lovisolo *et al.*, 1965; Hollings *et al.*, 1977), there is strong evidence for soil transmission by mechanical means. However, an intriguing aspect is that galinsoga mosaic virus usually occurs in association with the French bean stipple streak disease, caused by tobacco necrosis virus and transmitted by *Olpidium brassicae*. Since galinsoga mosaic virus is not transmitted by *O. brassicae* (Behncken, 1970; Shukla *et al.*, 1979), it is possible that soil conditions suitable for the stipple streak disease could also favor either mechanical transmission of galinsoga mosaic virus or transmission by an unidentified vector.

17.5.1.5 Peanut Clump Virus and (?) Polymyxa sp. Peanut clump virus causes marked stunting of peanut in Senegal and Upper-Volta in Africa (Thouvenel *et al.*, 1976), and also naturally infects grass hosts such as *Sorghum arundinaceum* (Dollet *et al.,* 1976). The virus has straight tubular particles of two predominant lengths, 190 and 245 nm, and is relatively stable in infected sap. The virus is not mechanically transmitted when infected plant material is incorporated in soil. However, a plasmodiophorid resembling a *Polymyxa* sp. is associated and may be a vector. Since air-dried soil retained its infectivity for four months, the virus may be carried internally by the resting spores of this fungus (Thouvenel, personal communication, 1978).

An interesting aspect of this disease is that it occurs in tropical countries, whereas other plasmodiophorid-transmitted viruses, such as wheat mosaic and potato mop-top viruses, have been found largely in temperate regions with temperatures more suitable for their viruses and vectors. Peanut clump virus has the high optimum temperature for multiplication of 35°C (Thouvenel *et al.,* 1976), but that of the suspected vector is below 27°C (Thouvenel, personal communication, 1978).

17.5.1.6 Broad Bean Necrosis Virus. This virus is soil-borne in Japan, and has straight, tubular particles of 150 and 250 nm x 25 nm. Serological tests show that it is distantly related to wheat mosaic, cucumber green mottle and odontoglossum ringspot viruses, but not to tobacco mosaic virus (Nakasone and Inouye, 1977). Its mode of transmission in soil has not been reported.

17.5.1.7 Oat Golden Stripe Virus and Polymyxa graminis. A straight tubular virus with particles of two lengths, 150 and 300 nm, and a width of 20 nm, has been found associated with bright yellow stripes on the flag and second leaves of oats cv. Peniarth in England (Macfarlane *et al.,* 1968; Plumb and Macfarlane, 1977). The disease occurs in patches, often in poorly drained areas, and is associated with oat mosaic virus and *Polymyxa graminis.* When infected roots containing zoosporangia of *P. graminis* and the two viruses were placed in water with oat seedlings as bait, both viruses were transmitted. Thus, this plasmodiophorid is a probable vector of the oat golden stripe virus as well as oat mosaic virus.

17.5.1.8 Wheat Yellow Mosaic Virus. This virus, which occurs in Japan, has unstable filamentous particles with physical properties similar to those of barley yellow mosaic virus of Japan and wheat spindle streak mosaic virus of North America (Usugi and Saito, 1976; Usugi, personal communication, 1978). Serologically wheat yellow mosaic virus is closely related to barley yellow mosaic virus and rice necrosis mosaic virus, and appears to be identical to wheat spindle streak mosaic virus (Usugi, personal communication, 1978). Although it has been argued that soil transmission may occur mechanically from particles adsorbed on clay (Miyamoto, 1959), a soil-borne vector is also possible.

17.5.1.9 Freesia Leaf Necrosis and Olpidium brassicae. This disease affects freesias in the Netherlands and possibly also in Germany and England (van Dorst,

1973). The leaves of affected plants develop characteristic chlorotic spots and stripes, which later become necrotic, and the production of flowers is slightly reduced in susceptible cultivars. A much more severe necrotic disease of foliage, flowers and corms occurs in plants doubly infected with freesia leaf necrosis and freesia mosaic virus.

Freesia leaf necrosis disease is soil-borne and *Olpidium brassicae* is associated. The causal agent resembles that of lettuce big-vein disease in that it has not been seen in the electron microscope or transmitted by sap-inoculation (van Dorst, 1975). However, soil which induces big-vein of lettuce does not induce leaf necrosis of freesias (van Dorst, 1976). It is possible, therefore, that the freesia leaf necrosis and lettuce big-vein agents are similar but not identical.

17.5.1.10 Potato Virus X and Synchytrium endobioticum. Both the virus and the fungus are presumably of long association with the potato plant, since both are present in the Andes of South America, the probable home of this plant (Torres *et al.,* 1970), as well as in many other countries. However, the evidence implicating *S. endobioticum* as a vector of potato virus X is derived from laboratory studies in Europe only. Nienhaus and Stille (1965) found that zoospore suspensions taken from potato plants infected by both the wart fungus and potato virus X transmitted the virus to shoots of healthy tubers. No transmission of the virus was achieved if the zoospores were inactivated before testing, or if zoospores from virus-free plants were mixed with a potato virus X suspension.

These results indicate that the virus may be thallus-acquired by the potato wart fungus and carried internally in the zoospores. The soil-borne tobamoviruses and potyviruses which have this relationship can persist in the resting spores of their vectors for years. Such a persistent relationship would have a useful function in the soil survival of the rather unstable particles of potato virus X, although the common carry-over of the virus in tubers would reduce the importance of this. However, no attempts to detect the natural survival of potato virus X in wart-infested soil have been reported, and a recent attempt to repeat the glasshouse transmission by the wart-fungus was unsuccessful (Lange, 1978).

17.5.1.11 Pea False Leaf Roll and Pythium ultimum. This suspected virus is unusual in being reported to be transmissible by sap, seed, *Myzus persicae* and *Pythium ultimum* (Thottappilly and Schmutterer, 1968; Thottappilly, 1972). Evidence for soil transmission is as follows: 1) When a mixture of healthy and diseased seed was planted in the same pot, the disease incidence was increased from 10-12% to 50-60%. 2) Replanting infested soil without prior steaming resulted in 40% disease incidence, compared with no infection in steamed soil. 3) Seed treatment with certain fungicides greatly reduced the spread of the disease. 4) The disease spread from diseased to healthy pea seedlings placed in water on opposite sides of a petri dish. 5) *Pythium ultimum* caused the disease when its mycelium infected healthy seedlings.

This work is doubly intriguing because *Pythium* spp. can apparently support the multiplication of two plant viruses, tobacco mosaic and tobacco necrosis viruses

(Brants, 1969, 1971; Nienhaus and Mack, 1974). The work on transmission of pea false leaf roll virus would be strengthened if the presumed virus were adequately characterized, and its relationship with *Pythium ultimum* and mode of transmission to roots established.

17.5.1.12 Tobacco Mosaic Virus and Rusts and Mildews. Virus-infected plants are often also infected by fungi which have air-borne spores. Early workers questioned whether plant viruses might be transmitted by such fungal spores, but produced neither circumstantial field evidence nor experimental evidence supporting such a hypothesis (see review by Teakle, 1969). Recently Yarwood and associates have re-opened the question with some experimental data suggesting that powdery and downy mildews and rusts may act as vectors (Yarwood, 1971; Nienhaus, 1971; Nienhaus and Yarwood, 1972; Yarwood and Hecht-Poinar, 1973). Some of the evidence is as follows: (a) When conidia of the oak powdery mildew were dusted heavily onto leaves of *Chenopodium quinoa,* local lesions and systemic chlorotic areas were sometimes produced. (b) A strain of tobacco mosaic virus was sometimes recovered from the affected *C. quinoa* plants and shown to be slightly different from the common strain of this virus. (c) The tobacco mosaic virus could be recovered directly from oak tissues or from spore suspensions of oak powdery mildew provided that inhibitors of infection were removed.

Yarwood and associates recognize that this work falls short of proof of virus transmission by air-borne spores because of the possibility that some sort of contamination may be involved. For instance, leaf hairs and other cellular debris in conidial preparations of oak mildew might release virus which could infect leaves mechanically through natural wounds. Further, no evidence indicating either the nature of the virus-vector relationship or the mode of transmission has yet been presented.

17.6 CONCLUDING REMARKS

As our knowledge of soil-borne fungus-transmitted viruses and their vectors has increased, a striking pattern of relationships has emerged. All straight tubular viruses of the tobamovirus type and all filamentous viruses of the soil-borne potyvirus type have so far had plasmodiophorid vectors, while all isometric viruses and agents of unknown morphology have had chytrid vectors. This pattern allows predictions to be made concerning the mode of transmission of additional soil-borne viruses. For instance, broad bean necrosis virus and wheat yellow mosaic virus, which have straight tubular and filamentous particles respectively, can be predicted to have plasmodiophorid vectors; tomato bushy stunt virus with isometric particles could have a chytrid vector, if one occurs.

Since none of the fungus-transmitted viruses has yet been found to multiply in its fungal vector, the viruses are regarded as being plant viruses rather than mycoviruses. However, there is evidence that tobacco mosaic and tobacco necrosis viruses can multiply in *Pythium* spp. which are not vectors of these viruses (Brants, 1969, 1971; Nienhaus and Mack, 1974). Further, the Tobamovirus group includes a virus of the green alga, *Chara coralloides* (Gibbs *et al.*, 1975). Thus, an origin of these viruses in microorganisms rather than plants is conceivable although not yet supported by convincing evidence. Whatever the origin of the three groups of fungus-transmitted viruses, it is possible to speculate on evolutionary trends among these viruses and their relatives.

Firstly, wheat mosaic virus, transmitted by *Polymyxa graminis*, and potato mop-top virus, transmitted by *Spongospora subterranea*, are regarded as being soil-borne tobamoviruses because of their morphological similarity to and serological relationship with tobacco mosaic virus. All the presently known vectors of the soil-borne tobamoviruses are plasmodiophorids (Table II), so fungus transmission may be a primitive character. It is conceivable that evolution occurred from an unstable, fungus-transmitted form to stable, mechanically-transmitted forms such as tobacco mosaic virus.

Secondly, tobacco necrosis and cucumber necrosis viruses, transmitted by *Olpidium* spp., have isometric particles with properties somewhat similar to those of members and possible members of the Tombusvirus group. Evolution from a moderately stable, locally-infecting, fungus-transmitted form to highly stable forms able to cause systemic infection following mechanical transmission is possible.

Thirdly, filamentous viruses transmitted by *Polymyxa graminis* have particles and pinwheel inclusions similar to those of the potyviruses, which are aphid-transmitted. Since the plasmodiophorids are probably more primitive than the aphids, soil-transmission could have pre-dated aerial transmission. Yarwood (1959) has previously suggested virus evolution in roots, but on bases other than those given above.

Major deficiencies in our present knowledge include information on the nature of the agents causing lettuce big-vein and tobacco stunt. Although these agents have been used in detailed studies of vector relationships, their morphology, composition and affinities with virus groups have not yet been determined.

Another deficiency involves our understanding of the bases for vector and virus specificity. Why are some viruses adsorbed by *Olpidium* zoospores while others are not? Why does the crucifer strain of *O. brassicae* apparently not act as a vector? What determines *in vivo* acquisition of some viruses by chytrid and plasmodiophorid vectors and precisely when does it occur? Why are certain viruses and agents labile in sap extracts but able to survive for years inside resting spores of their vectors? Providing answers to such questions concerning well-established vectors, and the investigation of possible vectors in fungal groups outside the chytrids and plasmodiophorids are challenges which will ensure continued interest in the field.

17.7 ACKNOWLEDGMENTS

I wish to thank the many people who provided advice and information during the writing of this review, including V. D'Ambra, R. Bowen, H. Dias, H. van Dorst,

B. Gerhardson, C. Hiruki, T. Inouye, L. Lange, A. Ragozzino, G. Shanks, J. -C. Thouvenel, T. Usugi and C. Yarwood.

17.8 REFERENCES

Barr, D. J. S. (1968). A new species of *Olpidium* parasitic on cucumber roots. *Can. J. Bot.* 46: 1087-1091.

Beachy, R. N., and Zaitlin, M. (1977). Characterization and *in vitro* translation of the RNAs from less-than-full-length, virus-related, nucleoprotein rods present in tobacco mosaic virus preparations. *Virology* 81: 160-169.

Behncken, G. M. (1970). Some properties of a virus from *Galinsoga parviflora. Aust. J. Biol. Sci.* 23: 497-501.

Bond, W. P., and Pirone, T. P. (1970). Evidence for soil transmission of sugarcane mosaic virus. *Phytopathology* 60: 437-440.

Brants, D. H. (1969). Tobacco mosaic virus in *Pythium* spec. *Neth. J. Plant Pathol.* 75: 296-299.

Brants, D. H. (1971). Infection of *Pythium sylvaticum in vitro* with tobacco mosaic virus. *Neth. J. Plant Pathol.* 77: 175-177.

Campbell, R. N. (1962). Relationship between the lettuce big-vein virus and its vector, *Olpidium brassicae. Nature* 195: 675-677.

Campbell, R. N. (1968). Transmission of tomato bushy stunt virus unsuccessful with *Olpidium. Plant Dis. Rep.* 52: 379-380.

Campbell, R. N., and Fry, P. R. (1966). The nature of the associations between *Olpidium brassicae* and lettuce big-vein and tobacco necrosis viruses. *Virology* 29: 222-233.

Campbell, R. N., and Grogan, R. G. (1963). Big-vein virus of lettuce and its transmission by *Olpidium brassicae. Phytopathology* 53: 252-259.

Campbell, R. N., and Grogan, R. G. (1964). Acquisition and transmission of lettuce big-vein virus by *Olpidium brassicae. Phytopathology* 54: 681-690.

Campbell, R. N., Grogan, R. G., and Purcifull, D. E. (1961). Graft transmission of big-vein of lettuce. *Virology* 15: 82-85.

Campbell, R. N., and Lin, M. T. (1976). Morphology and thermal death point of *Olpidium brassicae. Amer. J. Bot.* 63: 826-832.

Campbell, R. N., Lovisolo, O., and Lisa, V. (1975). Soil transmission of petunia asteroid mosaic strain of tomato bushy stunt virus. *Phytopathol. Mediter.* 14: 82-86.

Canova, A. (1966). Ricerche sulle malattie da virus delle Graminacee III. *Polymyxa graminis* Led. vettore del virus del mosaico frumento. *Phytopathol. Mediter.* 5: 53-58.

Chod, J., Polak, J., Kindela, V., and Jokes, M. (1976). Finding of lettuce big-vein virus in Czechoslovakia. *Biol. Plant.* 18: 63-66.

D'Ambra, V. (1967). Osservazioni sulla biologia de *Polymyxa graminis* Ledingham. *Atti Ist. Ven. Sc. Lett. ed Arti* 125: 325-336.

Dollet, M., Fauquet, C., and Thouvenel, J. -C. (1976). *Sorghum arundinaceum,* a natural host of peanut clump virus in Upper-Volta. *Plant Dis. Rep.* 60: 1076-1080.

Dorst, H. J. M. van (1973). Two new disorders in freesias. *Neth. J. Plant Pathol.* 79: 130-137.

Dorst, H. J. M. van (1975). Evidence for a soil-borne nature of freesia leaf necrosis. *Neth. J. Plant Pathol.* 81: 45-48.

Dorst, H. J. M. van (1976). Virus diseases in freesias. *In Annual Report, Glasshouse Crops Research and Experiment Station, Naaldwijk, The Netherlands,* p. 82.

Estes, A. P., and Brakke, M. K. (1966). Correlation of *Polymyxa graminis* with transmission of soil-borne wheat mosaic virus. *Virology* 28: 772-774.

Fenner, F. (1976). Classification and nomenclature of viruses. Second Report of the International Committee on Taxonomy of Viruses. S. Karger, Basel.

Fujisawa, I., and Sugimoto, T. (1977). Transmission of beet necrotic yellow vein virus by

Polymyxa betae. Ann. Phytopath. Soc. Japan 43: 583-586.

Garrett, R. G., and Tomlinson, J. A. (1967). Isolate differences in *Olpidium brassicae. Trans. Brit. Mycol. Soc.* 50: 429-435.

Gerdemann, J. W. (1955). Occurrence of *Polymyxa graminis* in red clover roots. *Plant Dis. Rep.* 39: 859.

Gibbs, A. J., Skotnicki, A. H., Gardiner, J. E. Walker, E. S. and Hollings, M. (1975). A tobamovirus of a green alga. *Virology* 64: 571-574.

Grogan, R. G. and Campbell, R. N. (1966). Fungi as vectors and hosts of viruses. *Annu. Rev. Phytopathol.* 4: 29-52.

Gumpf, D. J. (1971). Purification and properties of soil-borne wheat mosaic virus. *Virology* 43: 588-596.

Haeske, E. (1958). Untersuchungen über das Blattnervenmosaik des Salats. Dissertation L 470, University Giessen, 86 pp.

Harrison, B. D. (1974). Potato mop-top virus and its vector, *Spongospora subterranea. Proc. 1st Intersectional Congress of International Association of Microbiological Societies* 3: 303-312.

Harrison, B. D. (1977). Ecology and control of viruses with soil-inhabiting vectors. *Annu. Rev. Phytopathol.* 15: 331-360.

Hidaka, Z. (1956). Studies on the tobacco stunt disease III. On symptoms and causal agent of the tobacco stunt disease. *Hatano Tob. Exp. Sta. Spec. Bull.* 40: 19-23.

Hidaka, Z. (1960). The behaviour of tobacco stunt virus in soils, particularly supposing *Olpidium brassicae* (Wor.) Dang. as the vector. *Proc. Symp. Soil-borne Viruses,* Dundee, Scotland.

Hidaka, Z. (1969). Jacket photo: electron micrograph of arrangements of tobacco stunt virus particles in a zoospore of *Olpidium* vector. *In* K. Maramorosch (ed.) "Viruses, Vectors, and Vegetation." Interscience, New York, 666 pp.

Hiruki, C. (1975). Host range and properties of tobacco stunt virus. *Can. J. Bot.* 53: 2425-2434.

Hiruki, C., and Alderson, P. G. (1976). Morphology and distribution of resting sporangia of *Olpidium brassicae* in tobacco roots. *Can. J. Bot.* 54: 2820-2826.

Hiruki, C., Alderson, P. G., Kobayashi, N., and Furusawa, I. (1975). The nature of the infectious agent of tobacco stunt in relation to its vector, *Olpidium brassicae. Proc. 1st Intersectional Congress of the International Association of Microbiological Societies* 3: 297-302.

Hollings, M., and Stone, O. M. (1975). Serological and immunoelectrophoretic relationships among viruses in the tombusvirus group. *Ann. Appl. Biol.* 80: 37-48.

Hollings, M., and Stone, O. M. (1977). Red clover necrotic mosaic virus. *C.M.I./A.A.B. Descriptions of Plant Viruses* No. 181.

Hollings, M., Stone, O. M., and Barton, R. J. (1977). Pathology, soil transmission and characterization of cymbidium ringspot, a virus from cymbidium orchids and white clover (*Trifolium repens*). *Ann. App. Biol.* 85: 233-248.

Inouye, T., Osaki, T., Nakasone, W., and Shohara, K. (1977). Physicochemical properties of pea stem necrosis virus (PSNV). *Ann. Phytopathol. Soc. Japan* 43: 371.

Jones, R. A. C., and Harrison, B. D. (1969). The behaviour of potato mop-top virus in soil, and evidence for its transmission by *Spongospora subterranea* (Wallr.) Lagerh. *Ann. Appl. Biol.* 63: 1-17.

Karling, J. S. (1968). "The Plasmodiophorales." Hafner Publ. Co., N.Y. and London. 256 pp.

Kassanis, B. (1970). Satellite virus. *C.M.I./A.A.B. Descriptions of Plant Viruses* No. 15.

Kassanis, B., and Macfarlane, I. (1964). Transmission of tobacco necrosis virus by zoospores of *Olpidium brassicae. J. Gen. Microbiol.* 36: 79-93.

Keskin, B. (1964). *Polymyxa betae* n. sp., ein Parasit in den Wurzeln von *Beta vulgaris* Tournefort, besonders während der Jungendentwicklung der Zuckerrübe. *Arch. Mikrobiol.* 49: 348-373.

Kole, A. P. (1954). A contribution to the knowledge of *Spongospora subterranea* (Wallr.) Lagerh., the cause of powdery scab of potatoes. *Tijdschri. Plziekt.* 60: 1-65.

Lange, L. (1978). *Synchytrium endobisticum* and potato virus X. *Phytopathol. Z.* 92: 132-142.

Lange, L., and Insunza, V. (1977). Root-inhabiting *Olpidium* species: the *O. radicale* complex. *Trans. Brit. Mycol. Soc.* 69: 377-384.

Ledingham, G. A. (1939). Studies on *Polymyxa graminis,* n. gen. n. sp., a plasmodiophoraceous root parasite of wheat. *Can. J. Res.* 17: 38-51.

Lin, M. T., Campbell, R. N., Smith, P. R., and Temmink, J. H. M. (1970). Lettuce big-vein transmission by single-sporangium isolates of *Olpidium brassicae. Phytopathology* 60: 1630-1634.

Linford, M. B., and McKinney, H. H. (1954). Occurrence of *Polymyxa graminis* in roots of small grains in the United States. *Plant Dis. Rep.* 38: 711-713.

Lovisolo, O. (1966). Indagini su virosi di piante ornamentali. *Atti I. Congr. Un. Fitopat. Medit.,* Bari-Napoli, 26 September-1 October 1966, 574-584.

Lovisolo, O., Bode, O., and Völk, J. (1965). Preliminary studies on the soil transmission of petunia asteroid mosaic virus (="petunia" strain of tomato bushy stunt virus). *Phytopathol. Z.* 53: 323-342.

Macfarlane, I., Jenkins, J. E. E., and Melville, S. C. (1968). A soil-borne virus of winter oats. *Pl. Path.* 17: 167-170.

McKinney, H. H. (1923). Investigation on the rosette disease of wheat and its control. *J. Agr. Res.* 23: 771-800.

McKinney, H. H., Webb, R. W., and Dungan, G. H. (1925). Wheat rosette and its control. *Illinois Agric. Exp. Sta. Bull.* 264: 275-296.

Miyamoto, Y. (1959). The nature of soil transmission in soil-borne plant viruses. *Virology* 7: 250-251.

Nakano, A., and Iemura, H. (1977). Studies on the stem necrosis disease of peas. I. Soil transmission of the causal virus. *Abstract of paper presented at the Meeting of the Kansai Division of the Phytopathological Society of Japan,* Matsue, November, 1977.

Nakano, A., Iemura, H., and Kishi, K. (1976). Occurrence of pea stem necrosis, a newly named disease, and the causal virus. *Ann. Phytopathol. Soc. Japan* 42: 82-83.

Nakasone, W., and Inouye, T. (1977). Serological relationships between broad bean necrosis virus and tobamoviruses. *Abstract of paper presented at Meeting of the Kansai Division of the Phytopathological Society of Japan*, Matsue, November, 1977.

Nienhaus, F. (1971). Tobacco mosaic virus strains extracted from conidia of powdery mildews. *Virology* 46: 504-505.

Nienhaus, F. and Mack, C. (1974). Infection of *Pythium arrhenomanes in vitro* with tobacco mosaic virus and tobacco necrosis virus. *Zeitschrift für Pflanzenkrankheiten und Pflanzenschutz* 81: 728-731.

Nienhaus, F. and Stille, B. (1965). Übertragung des Kartoffel-X-Virus durch Zoosporen von *Synchytrium endobioticum. Phytopathol. Z.* 54: 335-337.

Nienhaus, F. and Yarwood, C. E. (1972). Transmission of virus from oak leaves fractionated with Sephadex. *Phytopathology* 62: 313-315.

Plumb, R. T., and Macfarlane, I. (1977). A 'new' virus of oats. *In Rothamsted Experimental Station Report for 1976*, Part 1, pp. 256-257.

Powell, C. A. (1976). The relationship between soil-borne wheat mosaic virus and tobacco mosaic virus. *Virology* 71: 453-462.

Putz, C. (1977). Composition and structure of beet necrotic yellow vein virus. *J. Gen. Virol.* 35: 397-401.

Ragozzino, A., and Furia, A. (1974). Presence of bacteria-like organisms and virus-like particles in lettuce tissues infected by "big-vein." *Ann. Fac. Sci. Agr. Univ. Studi. Napoli. Portici* 4: 265-273.

Randles, J. W., Harrison, B. D., and Roberts, I. M. (1976). *Nicotiana velutina* mosaic virus: purification, properties and affinities with other rod-shaped viruses. *Ann. Appl. Biol.* 84: 193-204.

Roberts, F. M. (1950). The infection of plants by viruses through roots. *Ann. Appl. Biol.* 37: 385-396.

Sahtiyanci, S. (1962). Studien über einige wurzelparasitäre Olpidiaceen. *Arch. Mikrobiol.* 41: 187-228.

Salvato, M. S., and Fraenkel-Conrat, H. (1977). Translation of tobacco necrosis virus and its satellite in a cell-free wheat germ system. *Proc. Natl. Acad. Sci. USA* 74: 2288-2292.

Shukla, D. D., Shanks, G. J., Teakle, D. S. and Behncken, G. M. (1979). Mechanical transmission of galinsoga mosaic virus in soil. *Aust. J. Biol. Sci.* 32 (in press).

Slykhuis, J. T., and Barr, D. J. S. (1978). Confirmation of *Polymyxa graminis* as a vector of wheat spindle streak mosaic virus. *Phytopathology* 68: 639-643.

Tamada, T. (1975). Beet necrotic yellow vein virus. *C.M.I./A.A.B. Descriptions of Plant Viruses* No. 144.

Tamada, T., Abe, H., and Baba, T. (1974). Beet necrotic yellow vein virus and its relation to the fungus, *Polymyxa betae. Proc. 1st Intersectional Congress of International Association of Microbiological Societies* 3: 313-320.

Teakle, D. S. (1960). Association of *Olpidium brassicae* and tobacco necrosis virus. *Nature* 188: 431-432.

Teakle, D. S. (1962). Transmission of tobacco necrosis virus by a fungus, *Olpidium brassicae. Virology* 18: 224-231.

Teakle, D. S. (1967). Fungus transmission of plant viruses, pp. 369-391. *In* K. Maramorosch and H. Koprowski (ed.) "Methods in Virology," Vol. 1. Academic Press, New York. 4 vols.

Teakle, D. S. (1969). Fungi as vectors and hosts of viruses, pp. 23-54. *In* K. Maramorosch (ed.) "Viruses, Vectors, and Vegetation." Interscience, New York. 666 pp.

Teakle, D. S. (1972). Transmission of plant viruses by fungi, pp. 248-266. *In* C. I. Kado and H. O. Agrawal (ed.) "Principles and Techniques in Plant Virology." Van Nostrand-Reinhold, New York. 688 pp.

Teakle, D. S., and Gold, A. H. (1963). Further studies of *Olpidium* as a vector of tobacco necrosis virus. *Virology* 19: 310-315.

Teakle, D. S., and Hiruki, C. (1964). Vector specificity in *Olpidium. Virology* 24: 539-544.

Temmink, J. H. M., Campbell, R. N., and Smith, P. R. (1970). Specificity and site of *in vitro* acquisition of tobacco necrosis virus by zoospores of *Olpidium brassicae. J. Gen. Virol.* 9: 201-213.

Thottappilly, G. (1972). Weitere Übertragungsversuche mit dem Falschen Blattrollvirus der Erbse. *Zeitschrift für Pflanzenkrankheiten und Pflanzenschutz* 79: 10-14.

Thottappilly, G., and Schmutterer, H. (1968). Zur Kenntnis eines mechanisch-, samen-, pilz- und insektenübertragbaren neuen Virus der Erbse. *Zeitschrift für Pflanzenkrankheiten und Pflanzenschutz* 75: 1-8.

Thouvenel, J. -C., Dollet, M., and Fauquet, C. (1976). Some properties of peanut clump, a newly discovered virus. *Ann. Appl. Biol.* 84: 311-320.

Tomlinson, J. A., Smith, B. R., and Garrett, R. G. (1962). Graft transmission of lettuce big-vein. *Nature* 193: 599-600.

Torres, H., French, E. R., and Nielsen, L. W. (1970). Potato diseases in Peru. *Plant Dis. Rep.* 54: 315-318.

Tsuchizaki, T., Hibino, H., and Saito, Y. (1975). The biological functions of short and long particles of soil-borne wheat mosaic virus. *Phytopathology* 65: 523-532.

Usugi, T., and Saito, Y. (1976). Purification and serological properties of barley yellow mosaic virus and wheat yellow mosaic virus. *Ann. Phytopathol. Soc. Japan* 42: 12-20.

Yarwood, C. E. (1959). Evolution of viruses in roots. *Plant Dis. Rep.* 43. 638-639.

Yarwood, C. E. (1971). Erysiphaceae transmit virus to *Chenopodium. Plant Dis. Rep.* 55: 342-344.

Yarwood, C. E., and Hecht-Poinar, E. (1973). Viruses from rusts and mildews. *Phytopathology* 63: 1111-1115.

INDEX

A

Abacarus, 326
Abacarus hystrix, 329, 331, 334–336, 350, 359, 360, 362–363, 371
Abelmoschus esculentus, 40, 53, 63
  yellow vein mosaic, 54–55
Abelmoschus manihot, 63
Abelmoschus moschatus, 60
Abelmoschus tuberculatus, 63
Abides grandis, 296
Abutilon, 59
  mosaic virus, 51, 58, 66–68, 70
Abutilon darwini tesselatum, 66
Abutilon indicum, 67
Abutilon mauritianum, 67
Abutilon megapotamicum, 67
Abutilon mulleri, 67
Abutilon rotundifolia, 67
Abutilon sellovianum, 67
Abutilon striatum thompsoni, 66
Abutilon striatum var. *spurium,* 67
Abutilon tesselatum, 67
Abutilon thompsonii, 67
Acacia, 152, 154
Acacia aneura, 155
Acacia harpophylla, 155
Acacia leucophloea, 153, 155
Acacia pendula, 155
Acalymma, 122
Acalymma thiemei, 117
Acalymma trivittata, 117, 118, 119
Acalymma vittata, 117, 121, 208, 209, 210, 257, 260
Acalymma vittatum, 117
Acalypha indica, yellow mosaic of, 45
Acanthospermum hispidum, 61, 64
  yellow mosaic of, 71
Acarus siro, 264
Acceratogallia gilletei, 182
Aceratagallia sangiunolenta, 182
Aceria, 301
Aceria mangifera, 301
Aceria tulipae, 330, 358
Acinopterus, 181
Acgyrotaenia pulchellana, 311
Acquilegia, 66
Acrostalgamus, 157
Acrostermum hilare, 229, 314
Aculodes mckenziei, 331, 336
Aculus fockeui, 347–348, 350
Adansonia digitata, 19, 30
Adelges, 231
Adelges cooleyi, 227
Adelphocoris rapidus, 228, 236
Aedes aegypti, 168
Aegeria vespiformis, 274
Aerobacter, 227
Aerobacter aerogenes, 227, 228
Aeroglyphus robustus, 264
Agallia ablbidula, 182
Agallia constricta, 182
Agallia deleta, 182
Agapostemon splendens, 188
Agaricus, 263, 272
Agave sisalensis, 298
Ageratum conyzoides, 54, 57, 63, 64, 68
  yellow vein mosaic of, 55
Ageratum crinitus, 54
Ageratum esculentus, 54
Ageratum monihot var. *pungens,* 54
Agrillus bilineatus, 304
Agriotes mancus, 212, 254
Agrobacterium, 221
Agrobacterium pseudotsugae, 203, 227
Agrobacterium tumefaciens, 202, 227
Agrohordeum (Elymus) macounii, 350
Agromyza parvicornis, 171
Agropyron, 331, 333
Agropyron mosaic virus, (AMV), 329, 335–336, 350, 359, 362, 371

Agropyron repens, 329, 335, 336
Agropyron smithii, 331
Agropyron trachycaulum, 350
Ahasverus advena, 265
Albipennis albipennis, 235
Aldicarb, 48
Aleurodicus, 180
Aleurodicus jamaicensis, 181
Aleurodicus ornatus, 181
Aleuroplatus, 181
Aleurotrachelus, 181
Aleurotrachelus floccosus, 41
Aleurotrachelus socialis, 41, 63, 65
Aleyrodidae, 41
Alfalfa, 110, 296
  dwarf, 9
  wilt, 231
Allagoptera arenaria, 194
Allium cepa, 259, 331
Allograpta cubana, 189
Allotrichoma, 249
Almond, 342
  leaf scorch, 9
Alocothrips hadocerus, 155, 156
Alphitobius diaperinus, 314
Alternaria, 54, 156, 157, 307, 316
Alternaria tenuis, 312
Althaea, 63
Althaea escutenlus, 60
Althaea officinalis, 67
Althaea rosea, 45, 53, 54, 57, 59, 60, 67, 71
Althaea tuberculatus, 60
Althaea venosum, 67
Amaranthus, 110
Amaranthus caudatum, 110
Ambithion, 51
Ambrosia, 94
Amyelois transitella, 267
Amylostereum areolatum, 302
Anacentrinus deplanatus, 274
Anacentrinus subnudus, 298
Anacentrinus deplanatus, 295
Anada wrightii, 67
Anaphothrips euphorbiae, 152
Anaphothrips involvens, 152
Anaphothrips obscurus, 151, 255
Anaphothrips silvarum, 152
Anasa tristis, 258
Anastrepha ludens, 276
Andrena, 239
Andrena varians, 239
Aneurothrips, 156
Anisopus fenestralis, 273
Anogeissus, 155
Anopheles annulipes, 168
Ant, 215, 217, 309
Antestia faceta, 313
Antestia lineaticolis, 313
Antillixius, 181
Anthocoris nemorum, 299
Anthonomus grandis, 312
Anthriscus yellows virus, 5
Anthurium andraenum, 65
Antiteuchus tripterus, 311
Anuraphis roseus, 237
Aphid, 2–8, 16, 116, 167–169, 172, 173, 180, 215–217, 306, 330, 345, 371, 392
  apple, 237
  cabbage, 265
  cotton, 258
  green, 238
  green peach, 168, 232, 238, 265
  melon, 258
  palm, 180, 189, 195
  potato, 255
  rosy apple, 237
  sugarcane, 270
  vector-pathogen interaction, 3–8
  walnut, 267
  wooly apple, 228, 238, 300, 301
Aphis avenae, 238
Aphis gossypii, 18, 19, 258
Aphis mali, 237
Apis mellifera, 188, *239, 241,* 258
Apis mellifica, 239
Aphis pomi, 216–218, 220, 237
Aphis sacchari, 270
Aphis sorbi, 237
Aphrophora parallela, 299
Aphrophora salicis, 273
Aphrophora saratogensis, 299
Aphrophora spumaria, 273
Apion, 122
Apion aestivum, 117, 124
Apion aethiops, 117, 124
Apion apricans, 117
Apion varipes, 117
Apion vorax, 117, 124
Apium graveolens, 246, 272
Aplanobacter stewarti, 252
Apophylia, 125
Apple, 110, 220
  bacterial rot, 203, 222–224
  canker disease, 299
  fire blight, 214
  gall, 221

perennial canker, 300
red bug, 236
rot, 261
tree borer, 234
tree cricket canker, 300
twig blight, 233
Apricot, 342
Arabis mosaic virus, 375, 377, 380, 393, 394, 395, 397, 401, 403
Arachis hypogaea, 67
Archips podanus, 299
Arenga engleri, 185, 194
Argyroploce leucotreta, 230
Arikuryoba schizophylla, 185, 194
Armillaria, 296
Armyworm, fall, 271
Arrhenodes minuta, 303–304
Arrhenothrips dhumrapaksha, 155
Arrhenothrips ramakrishnae, 155, 156
Artichoke Italian latent virus, 377, 382, 395
Artriplex, 110
Artriplex calotheca, 102
Atriplex nitens, 102
Atriplex halimus, 102
Atriplex hastata, 102
Atriplex hortensis, 102
Atriplex patula, 102
Atriplex tatarica, 102
Aschiza, 166
Ascia monusta, 266
Ascobacterium luteum, 221, 222
Ash, 139
canker, 262
Aspen, *Ceratocystis* canker, 299
Aspergillus, 157, 313, 316
Aspergillus candidus, 54
Aspergillus flavus, 297, 312, 314
Aspergillus niger, 157, 298, 310, 314
Aspergillus tamarii, 54
Aspidiotus nerii, 273
Asplenium nidus, 152
Aster, 57, 187
leafhopper, 233, 251
yellows, 9
Asystasis coromandeliana, 63
Atherigona, 245
Atherigona tibiseta, 269
Atrachya, 122
Atrachya menetris, 119
Augochlora pura, 189
Austrothrips cochinchinensis, 155
Autographa brassicae, 265
Avena, 260
Azalea, flower spot, 158, 309
Azenophosmethyl, 166

B

Bacillus, 206
Bacillus aeroida, 248
Bacillus amylovorus, 233
Bacillus avenae, 260
Bacillus campestris, 265
Bacillus carotovorus, 245, 246, 248
Bacillus gossypina, 227, 228, 229, 231, 259, 268
Bacillus phytophthorus, 245, 246, 248
Bacillus pini, 231
Bacillus pumilus, 228, 231
Bacillus solanacearum, 263
Bacillus solanisaprus, 248
Bacillus thuringiensis, 167
Bacillus tracheiphilus, 257
Bacteria, 149, 156–157, 167
plant pathogen, 201–226, 227–276
apple rot, 222–224
corn wilt, 211–214
cucurbit wilt, 208–211
fire blight, 214–220
olive knot, 220–222
soft rot, 203–208
southern wilt, 224–225
thrip borne, 156–157
Bacterium angulatum, 259
Bacterium aptatum, 259
Bacterium dianthi, 156
Bacterium dianthi chester, 231
Bacterium juglandis, 267
Bacterium maculicolum, 261
Bacterium malvacearum, 229, 267
Bacterium melophthora, 261
Bacterium pseudotsugae, 227
Bacterium savastanoi, 261
Bacterium solanacearum, 263
Bacterium stewarti, 252
Bacterium tumefaciens, 227
Bacterium vascularum, 269
Balclutha flavescens, 182
Balclutha guajanea, 182
Balclutha hebe, 182
Balclutha incisa, 182
Baliothrips biformis, 151
Balsam, leaf curl of, 60
Banana bluggoe, 224
green soft rot, 271
Moko disease, 224, 225, 262

rhizome rot, 245
root borer, 245, 262
tip over disease, 245
Barley, 350, 362
blight, 269
stripe mosaic virus, (BSMV), 362, 363, 364
yellow dwarf, 8
yellow mosaic virus, 419, 422, 424, 431
yellow straite mosaic, 10
Basswood, 139
Bastardia viscosa, 53
Baythion, 51
Bean, 160
bacterial blight, 268
curly dwarf mosaic virus, 117, 122
double yellow virus, 53
dwarf mosaic virus, 58, 70
French pod twist, 203
golden mosaic virus, 52
golden yellow mosaic virus, 51, 72
haloblight, 157, 261
lima, 314
mild mosaic virus, 120, 122
mosaic disease, 16
mosaic virus, 119, 121, 123, 125, 367
pod mottle virus, 114, 122
pod twist, 260
pseudocurly top, 94
rugose mosaic virus, 123
severe mosaic virus, 119
stipple streak disease, 430
thrip, 232
yellow stipple virus, 119
Bee, 224, 225, 239, 241, 309
honey, 214, 215, 216, 219, 220, 258
Beech bark disease, 298
pathogen, 251
Beet, fodder, 102
garden, 110
latent rosette disease, 97
leafcurl virus, 97, 98, 101–110
leafhopper, 168
mosaic virus, 169, 367
necrotic yellow vein virus, 419, 422, 423
pseudo yellow virus of, 65–66
western yellows, 8
Beetle, 115–132, 167, 215, 216
banded cucumber, 257
bark, 296, 302
bean leaf, 123
blister, 225
brassy flea, 253
brentid, 303
chripomelid, 122, 124
Colorado potato, 225, 232, 257, 263
confused flour, 275
confused grain, 316
cucumber, 234, 258, 267
desert corn flea, 253
dried fruit, 310
elm bark, 133–147
flea, 121, 211, 225, 254
flour, 314
foreign grain, 265
fruit-bark, 234
larder, 248
leaf, 121
leaf-feeding, 116
longhorned, 135
Mexican bean, 122, 123
nitidulid, 300, 303, 310, 311
oak bark, 303
picnic, 313
potato flea, 247, 254, 258, 263
powder-post, 304
red flour, 265
rove, 300
scolytid, 303
spotted cucumber, 231, 253, 257, 307
squarenosed fungus, 265
squash, 258
striped blister, 122
striped cucumber, 121, 208–210, 257, 260
sweet potato flea, 252
toothed flea, 211, 252
twelve-spotted cucumber, 210, 212
twig girdler, 304–305
virus association, 126–127
transmission mechanism, 127–128
types transmitted, 116–121
vectoring, 122–126
western spotted cucumber, 258
Belladonna mottle virus, 118
Bemisia bahiana, 41
Bemisia costa-limai, 41
Bemisia fascialis, 41
Bemisia goldingi, 41
Bemisia gossypiperda, 59, 63
Bemisia gossypiperda var. *mosaicivectura,* 41, 70
Bemisia hibisci, 41
Bemisia inconspicua, 41, 60
Bemisia longispina, 41
Bemisia manihotis, 41

Bemisia nigeriensis, 41, 70
Bemisia rhodesiaensis, 41
Bemisia signata, 41
Bemisia tabaci, 268, 269
  leaf curl vector, 57–65
  other diseases, vector of, 66–71
  yellow mosaic vector, 45–57
Bemisia tabaci race *sidae,* 48, 50
Bemisia tuberculata, 41, 63, 65
Bemisia voyssierci, 41
Beta, 101, 111
Beta macrocarpa, 66, 102
Beta procumbens, 102
Beta trigyna, 102
Beta vulgaris, 57, 65, 66, 94, 169, 259, 274
Beta vulgaris var. *alba,* 102
Beta vulgaris var. *altissima,* 102
Beta vulgaris var. *conditiva,* 102
Beta vulgaris spp. *maritima,* 102
Beta vulgaris var. *vulgaris,* 102
Bhendi, yellow vein mosaic, 54–55
Bibio albipennis, 235
Bibio albipennis albipennis, 217
Bidrin, 141
Billbug, 295
  bluegrass, 274
Birch, constriction disease, 301
Black current reversion, 350
Blackberry, 380, 381
Blackgram mottle virus, 120
Blackleg, potato 203–205
Black root, 110
Blastophaga psenes, 310
Blight, apple twig, 233
  barley, 269
  blossom, 233
  canker, 233
  fire, 214–220, 233
  halo bean, 261
  oat blade, 203
  pear, 233
  red clover blossom, 308
  red pine needle, 307
  rubber tree *Aeternaria,* 307
  Stewart's leaf, 211
  walnut, 203
Blumea neilgherrensis, yellow vein mosaic of, 55
Blycine hispida, 52
Bobone disease, 10
Boletus, 297
Bollworm, 230
  cotton, 312
  pink, 230, 312
Bombus hortorum, 240
Bombus terrestris, 240
Borassus flabellefer, 194
Borer, shothole, 234
Bornetina, 297
Bostaera balli, 182
Bothriocera, 182
Botryodiplodia theobromae, 298, 311
Botrytis, 316
Botrytis anthophila, 308
Botrytis cinerea, 311
Botrytis fabae, 315
Bouteloua, 331
Bouteloua curtipendula, 332
Brachiaria ramosa, 50
Brachycera, 165, 166, 168
Brachypeplus pilosellus, 311
Bradysia, 250, 296
Bradysia caprophila, 296
Brassica chinensis, 63
Brassica juncea, 63
Brassica oleracea botrytis, 261
Brassicaceae, black rot, 203
  brown rot, 265
  decay, 274
  leaf blight, 265
  soft rot, 204, 227, 245, 264, 272
  stump rot, 265
Brevicoryne brassicae, 265
Broad bean, mold, 315
  mottle virus, 119
  necrosis virus, 422, 423, 431
  stain virus, 117
Broccoli necrotic yellow, 8
Brome mosaic virus, 119, 377, 382
Bromegrass, 382
  mosaic, (BMV), 362, 365, 367, 371, 372
Bromus, 332
Browallia elata, 55
Bug, apple red, 236
  brown stink, 229
  cacao, 311
  green stink, 229, 268
  lace, 110, 112
  Lygus, 215, 312
  pale damsel, 270
  pale legume, 236
  rapid plant, 236
  spittle, 299

squash, 258
stink, 312, 314
tarnished plant, 208, 231, 232, 237, 254
tingid, 180
Butterfly, 242
Byctothrips ayyari, 155

C

Cabbage, maggot, 245, 274
seedstalk curculio, 245
worm, 266
Cacao necrosis virus, 377, 382
Cactoblastis cactorum, 244, 276
Cactobrosis fernaldialis, 244
Catophagus spinolae, 276
Cactus, bacterial necrosis, 244
Cajanus cajan, 46, 50, 53, 345
Calacarus, 326
Calacarus citrifolii, 347
Calendra parvula, 295
Caliothrips fasciatus, 151, 232, 275
Caliothrips indicus, 151
Caliothrips sudanensis, 151
Callestephus chinensis, 187
Calliphora augur, 269
Callistephus chinensis, 66
Calocoris rapidus, 229
Calonectria (Fusarium) rigidiuscula, 298
Calonyction aculeatum, 50
Calycopteris floribundus, 153, 154, 155, 156
Camegiea giganta, 244
Camponotus acvapimensis, 311
Campylomma verbasci, 236
Canavalia ensiformis, 51, 52, 53
Canker, 203, 216, 217, 218
caratocystis, 299–300
citrus, 266
nectria, 298–299
sclerodis, 302
tree, 298–303
white pine blister rust, 302
Canovalia marithima, 53
Cantharis andersoni, 233
Cantaris livida, 233
Cape-gooseberry, leaf curl of, 64
Capsella bursa-pastoris, 66
Capsicum, 246
Capsicum annum, 61, 63
leaf curl of, 57–58
Capsicum frutescens, 58, 187
Capsicum microcarpum, 58
Capsicum pubescens, 58
Capsicum sinensis, 58
Carica papaya, 63
leaf curl of, 58
Carlavirus, 3
Carloduvica palmata, 185
Carnation, anther smut, 308
bacterial disease, 231
bud rot disease, 308
ringspot virus, 377, 382, 401
Carneocephala flaviceps, 182
Carneocephala sagittifera, 181, 182
Carpocapsa pomonella, 242
Carpophilus dimidiatus, 228
Carpophilus freemani, 299
Carophilus hemipterus, 299, 310
Carpophilus lugubris, 300, 303
Carpophilus mutilatus, 228, 310
Carrot, 187, 389
proliferation disease, 90
soft rot, 246
Cassava, bacteriosis, 268
mosaic disease, 69–70
Casearia tomentosa, 155, 156
Castanea, 67
Castor bean, leaf blight, 269
Casuarina stricta, 154, 155
Catharanthus roseus, 88, 186, 189
Cauliflower, spot disease, 203
pathogen, 251
peppery leaf spot, 261
Caulimovirus, 3
Cedicophyopsis, 326
Cecidophyopsis ribis, 336–340
Cedusa, 180
Cedusa inflata, 184
Ceiba pentandra, 19
Celery, 167, 169, 170, 173, 174, 187, 390
black heart, 272
heart rot, 168, 203, 207–208, 246
mosaic virus, (CeMoV), 169, 174
yellow vein disease, 381
Centorhynchus litura, 307
Centrosema, 46
Cephalosporium, 297
Cephalosporium acremonium, 313
Cephalosporium dieffenbachia, 307
Cephalosporium (Phialophora) diospyri, 304
Cerataphia variabilis, 189
Cerataphis, 180
Cerataphis variabilis, 180
Ceratitis annonae, 271
Ceratitis capitata, 316
Ceratocystis, 302

Ceratocystis alba, 299
Ceratocystis crassivaginata, 299
Ceratocystis fagacearumn, 303, 304
Ceratocystis fimbriata, 299, 300, 305
Ceratocystis minor, 299
Ceratocystis moniliformis, 299
Ceratocystis pilifera, 299
Ceratocystis pini, 302
Ceratocystis populina, 299
Ceratocystis tremulo-aurea, 299
Ceratocystis ulmi, 133–147
Ceratomegilla fuscilabris, 254
Ceratostomella, 302
Cercospora beticola, 110
Cereal mosaci, northern, 10
  tillering disease, 10
Cereus giganteus, 244
Cerotoma, 122
Cerotoma arcuatus, 117
Cerotoma atrofasciata, 117
Cerotoma ruficornis, 117, 119, 123, 124, 268
Cerotoma trifurcata, 117, 119, 124, 125, 126, 127
Cerotoma variegata, 117
Ceutorhynchus quadridens, 245
Chaelanophothrips orchidii, 158, 180
Chaetocnema, 122
Chaetocnema abyssinica, 120
Chaetocnema confinis, 252
Chaetocnema denticulata, 211, 212, 213, 252
Chaetocnema ectypa, 253
Chaetocnema kenyensis, 120
Chaetocnema pulicaria, 120, 211–214, 253
Chaetocnema pulla, 120, 124
Chaetonerius lalifemur, 311
Chalcodermus ebeninus, 268
Chara coralloides, 434
Chelinidea canyona, 244
Chelinidea tabulata, 244
Chelinidea vittiger, 244
Chemes cooleyi, 227
Chenopodium, 110
Chenopodium album, 66, 102
Chenopodium amaranticolor, 66, 102, 168, 394
Chenopodium capitatum, 66, 102
Chenopodium faetidum, 347, 348
Chenopodium ficifolium, 102
Chenopodium foliosum, 102
Chenopodium giganteum, 102
Chenopodium multifictum, 102
Chenopodium murale, 66, 102, 168
Chenopodium opulifolium, 102
Chenopodium quinoa, 102, 347, 433
Chenopodium serotinum, 102
Chenopodium virgatum, 102
Cherry, 343
  flowering, 349
  leaf roll virus, 377, 381, 386
  gummosis, 260
  mottle leaf, (CML), 344–345, 350
  native bitter, 349
  rasp leaf virus, 377, 380
  sour, 349
  wild, 139
Cheyletus, 275
Cheyletus eruditus, 264
Cheyletus trouessarti, 264
Chicory, 160
Chilli, leaf curl, 57–58
Chilo partellus, 252
Choriona slossonae, 182
Chlorochroa ligata, 229
Chlorochroa sayi, 229, 312
Chlorotettix, 180, 181
Chlorotettix minimus, 181, 182
Chlorotettix rugicollis, 182
Chlorotettix viridius, 181, 182, 255
Chlorotic ring spot mosaic, 40
Chortophila brassicae, 274
Chromaphis juglandicola, 267
Chrysalidocarpus cabadae, 185, 194
Chrysobothris femorata, 304
Chymomyza procnemoides, 299
Chytrid, 418–420, 421, 425, 429
Cicadula divisa, 255
Cicadulina mbila, 345
Cichorium endiva, 66
Cionus, 122
Cionus alauda, 118
Cionus hortulanus, 118
Cionus scrophulariae, 118
Cionus tuberculosis, 118
Circulifer tenellus, 168
Citrullus fistula, 55
Citrus, 297
  canker, 203, 266
  concentric ring blotch, 347, 351
  decline, 89
  disease, 272
  greening disease, 89
  little leaf disease, 9
  stubborn, 9
  vein-phloem degeneration, 89
  yellow shoot, 89
Citrus paradisi, 276

Cladosporium cladosporoides, 315
Cladosporium herbaru, 313
Clastoptera undata, 184
Claviceps purpurea, 308
Claytonia perfoliate, 66
Closterovirus, 3
Clover, 110, 389
  club leaf, 9
  red, 276
    blossom blight, 308
    necrotic mosaic virus, 422, 429
  root borer, 276
  root curculio, 296
  rot, 295
  white, 380
Cockroach, 296
Cocksfoot, 334
  mottle virus, 120, 121, 125, 127
Cocoa, black pod, 311
  Ceylon virus, 17, 18
  mottle leaf virus, (CMLV), 17, 18, 19, 23, 25, 26, 30, 31
  necrosis disease, 382
  necrosis virus, (CNV), 17
  swollen shoot virus, 17–32
  Trinidad virus, (CTV), 17, 18, 20, 23, 24, 27, 28, 29, 31
  yellow mosaic virus, (CYMV), 2, 17
Coconut palm, 177–200, 272
  cadang-cadang, 349, 351
  Kaincope disease, 195
Cocos, 259
Cocos nucifera, 177, 194, 272
Coenosia tricolor, 250
Coffee, 297
  bean, 203
    rot, 313
  Kivu, 271
  leaf rust, 158
  rust, 306
Coffee, 271
Cola chlamydantha, 19
Cola gigantea var. *glabrescens*, 19
Colaspis, 122
Colaspis flavida, 119
Colaspis lata, 117
Coleopterus morio, 300
Coleopterus (Colastus) unicolor, 300
Colletotrichum, 297, 316
Colletotrichum falcatum, 298
Colletotrichum lagenarium, 307
Colomerus, 326
Colomerus vitis, 348
Colopterus niger, 303
Colopterus semitectus, 303
Colopterus truncatus, 300
Colpoptera, 180
Colpoptera elevans, 181
Conifer, wood-stain disease, 302–303
Conium maculatum, 66
Conotrachelus nenuphar, 310
Corchorus, 46
Corchorus aestuans, 53
Corchorus capsularis, 46
Corchorus facicularis, 59
Corchorus olitorius, 46
Corchorus triocularis, 46
Cordia obliqua, 156
Cordia suaveolens, 156
Coreus tristis, 258
Coriandrum sativum, 264
Corky ringspot, 383
Corn, 110, 125
  bacterial wilt, 202, 211–214
  earworm, 313, 314, 316
  European borer, 247, 255, 297, 313, 316
  mold, 314
  Nigrospora disease, 315
  rot, 295, 313–314
  stalk rot, 297–298
  Stewart's wilt, 252
  stunt, 9
  sweet, 211
Cornus florida, 381
Corynebacterium isidiosum, 231
Corynebacterium michiganense, 231
Corynebacterium sepedonicum, 203, 232
Corypha elata, 194
Cosmopolites sordidus, 245, 262
Cosmos bipinnatus, 50
Cossus cossus, 273
Cotton, angular leaf spot, 267
  aphid, 258
  blackarm, 267
  boll rot, 227, 228, 231, 268, 312–313
  bollweevil, 312
  bollworm, 312
  internal boll disease, 312–313
  leaf crumple, 60
  leaf curl of, 58–60
  leaf mottle virus, 68
  lint rot, 312
  mosaic of, 68
  sakel, 59
  stigmatomycosis, 312–313
Cowpea, 123, 125, 314
  bacterial blight, 268
  chlorotic mottle virus, 119, 124, 125

mosaic virus, 117, 123, 125, 127, 377
yellow mosaic of, 53, 122
Crataegus, 233
Crematogaster striatula, 311
Crepidodera cucumeris, 257
Cricket, tree, 300
Criconemoides xenoplax, 377, 380
Cronartium commandrae, 302
Cronartium ribicola, 302
Crotalaria juncea, 53
Croton lobatus, 49
Croton sparsiflorus, 54
yellow vein mosaic of, *55*
Crotonothrips coorgensis, 155
Crotonothrips gallarum, 155
Crown gall, 228
Cryptarch ampla, 300, 303
Cryptocephalus, 122
Cryptocephalus chalybeipennis, 120
Cryptocephalus nigrum, 120
Cryptococcus fagi, 298
Cryptohypnus abbreviatus, 248
Cryptophlebia leucotreta, 230
Cryptorhynchus lapathi, 227, 252
Cryptothrips floridensis, 157
Cucumber, angular leaf spot, 260
beetle, 234
green mottle virus, 431
mosaic virus, 5
necrosis virus, 419, 421, 422, 430, 434
pathogen, 251
stunt, 380
wilt, 208, 209, 210
yellow vein mosaic of, 55
Cucumis melo, 66
Cucumis sativus, 55, 66, 260
yellow vein mosaic of, 56
Cucumovirus, 3
Cucurbit anthracnose, 307
Cucurbita moschata, 55, 66
Cucurbita pepo, 173
yellow vein mosaic of, 55
Cucurbitaceae, wilt, 257
Cudrania tricuspidata, 340
Culex, 168
Culex annulirostris, 168
Culex quinquefasciatus, 168
Curly top virus, 110
Currant
black reversion, 326, 336–340
canker, 301
red spoonleaf disease, 381
Cuscuta campestris, 189–190
Cuscuta reflexa, 345
Cuspicona simplex, 270
Cyamopsis tetragonolobus, 52, 67
Cyclarrhapha, 165, 166, 168
Cymbidium ringspot virus, 422, 430
Cynodon dactylon, 188
Cynomyia cadaverina, 235
Cynomyopsis cadaverina, 217, 235
Cyphomandra betacea, 67
Cytospora, 301

D

DDT, 141, 143, 166
Dacryomyces, 302
Dactylis glomerata, 213, 334
Dactylispa, 122
Dactylispa bayoni, 120
Dacus carota, 187
Dacus ferrugineus tryoni, 272
Dacus oleae, 168, 220–224, 261
Dahlia, 160
Danothrips theivorus, 152
Damping off, 52
Dashen mosaic virus (*DMV*), *17*
Dasineura crataegi, 235
Dasyneura crataegi, 235
Datura jerox, *68*
Datura stramonium, 48, 49, 50, 57, 61, 62, 63, 67, 94
Daucus carota, 66
Daucus carota sativa, 246
Delia florilega, 248
Delococcus tafoensis, 18
Delphacodes, 180
Delphacodes propinqua, 182
Deltocephalus, 180
Deltocephalus flavicosta, 182
Deltocephalus sonories, 182
Dendroctonus engelmani, 302
Dendroctonus frontalis, 302
Dendroctonias ponderosae, 302
Dermestes lardarius, 248
Desmodium yellow mottle virus, 118
Desmodium aculeatum, 52
Diabrotica, 122, 258
Diabrotica adelpha, 117, 119, 125
Diabrotica balteata, 117, 119, 123, 125
Diabrotica duodecimpunctata, 210, 231, 253, 257, 260
Diabrotica graminea, 225, 263
Diabrotica laeta, 117
Diabrotica longicornis, 117, 119, 120, 253, 295
Diabrotica saccharalis, 298

Diabrotica undecimpunctata, 117, 119, 120
Diabrotica undecimpunctata howardi, 208–213, 231, 253, 257, 260, 295, 304, 306
Diabrotica undecimpunctata undecimpunctata, 258
Diabrotica virgifera, 117, 120, 253
Diabrotica vittata, 257, 260
Dianthus caryophyllus, 231
Diaphorina citri, 87, 89
Diaprepes abbreviatus, 268
Diarthrothrips coffeeae, 151
Diazinon, 166, 344
Dicladispa, 122
Dicladispa paucispina, 120
Dicladispa veridicyanae, 120
Dicranotropis hamata, 10
Dictyosperma album, 194
Didymella applanata, 300
Dieffenbachia leaf spot, 307
Digitalis ambigua, 152
Digitaria striate mosaic, 10
Digitaria, 332
Diloba caeruleocephala, 242
Dimenthioate, 55
Diparopsis watersi, 230
Diphaulaca meridae, 117
Diplodia, 297
Diplodia maydis, 313
Diplodia pinea, 299
Diplodia recifensis, 305
Diplodia zeae, 297
Diptacus gigantorhynchus, 347, 350
Disonycha glabrata, 253
Distantiella theobroma, 298
Dixothrips onerosus, 155
Dodder, 189
Dogwood, 381
Dolichos biflorus, 46, 50, 51
Dolichos lab-lab, 40, 52
  yellow mosaic of, 47
Dorymyrmex pryamicus, 189
Doryphera decemlineata, 263
Draeculacephala bradleyi, 182
Draeculacephala inscripta, 182
Draeculacephala portola portola, 182
Draeculacephala producta, 182
Drapetis ciliatocosta, 269
Drosophila, 225, 249, 262, 304, 306
Drosophila busckii, 204, 245, 246, 249
Drosophila funebris, 235, 249
Drosophila melanogaster, 205, 235, 249, 299, 310, 312
Dutch Elm disease, 133–147
  control, 141–143
  transmission, 136–137
Dysaphis plantaginea, 237
Dysaphis sorbi, 216, 237
Dysdercus, 230, 312, 315
Dysdercus fasciatus, 229
Dysdercus honestris, 229
Dysdercus intermedius, 229
Dysdercus mendesi, 230
Dysdercus nigrofasciatus, 230
Dysdercus ruficollis, 230
Dysdercus suturellus, 230
Dysmicoccus brevipes, 16, 18, 24, 29
Dysmicoccus neobrevipes, 16

E

Earias biblaga, 230
Earias insulana, 230, 312
Earwig, 311
Earworm, corn, 230
Eccoptogaster rugulosus, 234
Echinochloa, 332
Echinocysts (Micrampeles) lobata, 210
Echtes Ackerbohnemosaik virus, 117
Eclipta alba, 50
Edessa meditabunda, 229
Edwarsiana rosae, 215, 238
Eggplant, 94, 160, 225
  mosaic virus, 118
  wilt, 263
Elachiptera cornuta, 272
Elachiptera costata, 168, 207, 246
Elaeis guineensis, 193
Elderberry, golden, 381
Elm, 381
  bark beetle, 133–147
  rock, 136
Elymus, 332
Elymus canadensis, 331
Emilia sonchifolia, 159, 160
Empoa rosae, 238
Empoasca, 180, 268
Empoasca canariesis, 215, 238
Empoasca fabae, 215, 216, 238, 251, 255
Empoasca kraemeri, 182
Empoasca mali, 238
Empoasca plebeia, 182
Empoasca unicolor, 238
Endoconidiophora coerulescens, 303
Endosulfan, 335
Endria inimica, 330

Eothrips coimbatorensis, 155
Epicauta, 263
Epicauta pennsylvanica, 232
Epilachna, 117, 122
Epilachna borealis, 258
Epilachna chrysomelina, 117, 122
Episema coeruleocephala, 242
Epilachna varivestis, 117, 119, 120, 122
Epicanta vittata, 117, 122
Epitrimerus, 181
Epitrix, 118, 122
Epitrix atropae, 118
Epitrix cucumeris, 232, 247, 254, 258, 263
Epitrix hirtipennis, 117
Epitrix parvula, 263
Epuraea, 300
Epuraea aestiva, 228
Epuraea depressa, 228
Eragrostis, 332
Erodium cicutarium, 66
Ergot, cereal, 308
  grass, 308
Eriophyes, 326, 336
Eriophyes cajani, 345–346
Eriophyes fici, 275
Eriophyes ficus, 340–342
Eriophyes mangiferae, 348
Eriophyes sacchari, 345
Eriophyes tosichella, 329, 330
Eriophyes tristriatus var. *erinae*, 266
Eriophyes tulipae, 327, 328, 329–334, 336, 350, 358–371
Eriophyid, 326–356
  characteristics, 326–328, 358
  virus vector interaction, 357–373
Eriosoma lanigera, 238
Eriosoma lanigerum, 227, 238, 300
Eristalis tenax, 236
Erwinia, 259
Erwinia amylovora, 202, 214–220, 233, 243
Erwinia aroidea, 228, 243, 245, 251, 259
Erwinia atroseptica, 245
Erwinia cacticida, 243
Erwinia carnegieana, 244
Erwinia carotovora, 168, 202–208, 228, 243, 245–247, 251, 259
Erwinia carotovora var. *atroseptica*, 203
Erwinia carotovora var. *carotovora*, 203, 205
Erwinia carotovora var. *zeae*, 252
Erwinia chrysanthemi, 228, 252, 259
Erwinia espinosa, 203, 252
Erwinia herbicola, 228, 252, 259
Erwinia phytophora, 248
Erwinia salicis, 252, 261
Erwinia stewartii, 202, 211–214, 252
Erwinia tracheiphila, 202, 208–211, 257
Erysimum latent virus, 120
Erysiphe cichoracearum, 306
Erythroneura, 182, 238
Escherichia coli, 259
Euchistus conspersus, 229
Euchistus impectiventris, 229
Eugenia axillaris, 155
Euides weedi, 182
Eumeris strigatus, 205, 246, 247, 250
Eumyndus, 195
Euonyraus, 67
Eupatorium, pseudo mosaic of, 68
Eupatorium odoratum, 68
Euphorbia cyparissias, 152
Euphorbia geniculata, 63
Euphorbia heterophylla, 67
Euphorbia hirta, 63, 67
Euphorbia prunifolia, 67
  yellow mosaic of, 47–48
Euphysothrips minozii, 157
Euphysothrips (Megaphysothrips) Okasis, 158
Eupurea, 303
Eurynchothrips flavicornis, 155
Euryophthalmus convivus, 261
Euryophthalmus succinctus, 230
Euscelis bicolar, 255
Euura atra, 252
Euxesta notata, 247
Euzophera semifuneralis, 299
Exitianua exitiosus, 182

F

Fagraea litoralis, 156
Fannia, 250
Ferrisia virgata, 16, 18, 21–31, 34, 180
Festuca, 332
Festuca ovina, 332
Ficus, 152, 275, 340
Ficus carica, 341
Ficus benjamina, 155, 156
Ficus retusa, 155
Fig, 385
  decay, 275
  fungus disease, 309–310
  mosaic, 340–342, 350
  Smyrna, 310
  souring, 275
Fir, Douglas, 203, 227
  grand, 296

Fire blight, 233
  raspberry, 243
Fleahopper, 168
Fly, 165–176, 215–217, 225
  black onion, 205
  carrot rust, 247
  cluster, 235
  Diptera, 165–170
  drone, 236
  false crane, 251
  false stable, 235
  fruit, 245, 247, 249
  Hessian, 274, 295
  hornfly, 168
  house, 235, 250, 316
  leafminer, 166–174
  Liromyza, 170–172
  Mediterranean fruit, 316
  Mexican fruit, 276
  olive fruit, 168, 220, 221, 261
  onion bulb, 246, 247, 250
  stable, 168
  tsetse, 168
  walnut husk, 267
  white, 306, 345
Fomes, 296
Fomes annosus, 296
Forficula auricularia, 311
Formica, 240
Formica fusca, 240
Formica pallidefulva, 240
Formica pallide-fulva nitidiventris, 240
Frankliniella, 260, 275
Frankliniella cephali, 275
Frankliniella fusca, 159, 160, 161
Frankliniella intonsa, 152, 158
Frankliniella moultoni, 157, 158, 242
Frankliniella accidentalis, 158–161, 242
Frankliniella schultzei, 158, 159, 160
Frankliniella tritici, 157, 158, 275
Fraxinus, 67, 262
Freesia leaf necrosis virus, 422, 431, 432
French bean, yellow mosaic, 52
Fringe wings, 149–164
Fruitworm, green, 242
Fungal, plant pathogens, 293–324
  blossom disease, 308–309
  bud disease, 308–309
  foliar disease, 305–307
  fruit disease, 309–316
  seed disease, 309–316
  rust, 305–306
Fungicide, 432
  captan, 430
Fungus, 134–144, 149, 157–158, 179, 180, 376, 417–438, *see also* Mold
  blue-stain, 302
  coffee rust, 306
  oak wilt, 303, 304
  potato wart, 432
  powdery mildew, 306
  sooty mold, 306
  thrips borne, 157–158
  vector, 419
  virus vector, 419
Fusarium, 180, 295–297, 316
Fusarium graminerarum, 297
Fusarium moniliforme, 297, 301, 312–314, 348
Fusarium moniliforme f. s. *fici,* 309
Fusarium oxysporum, 314
Fusarium poae, 308
Fusarium roseum, 314
Fusarium roseum var. *graminearum,* 314
Fusarium solani, 314
Fusarium tricinctum, 314

G

Galinsoga mosaic virus, 422 , 429
Galium verum, 152
Gall, black current, 337
  Douglas fir, 227
  formation, thrip, 150–156
  grape, 227
  pine, 231
  root-tip, 389–390
  warty, 300, 301
Gallmite, 181
Ganoderma, 296
Gardenia, 259
Gargaphia, 307
Garlic, mite, 328
Gaussia attenuata, 194
Geranium, leaf curl of, 58
  leaf spot, 268
Geranium dissectum, 66
Geranium pelargonium, leaf curl of, 58
Germania bulbicola, 304
Gerhardiella delicatus, 229
Gibberella, 295, 297
Gibberella roseum f. sp. *cerealis,* 313
Gibberella zeae, 297
Gibberellic acid, 61, 64
Gilah injection, 68

Gladiolus, 296
notch-leaf virus, 378, 383
Gladiolus, 261
Glischrochilus, 303
Glischronchilus fasciatus, 234
Glischronchilus quadrisignatus, 313
Gloeosporium perennans, 300
Glycine max, 46, 47, 48, 50, 52, 53, 67
crinkle mosaic, 58
Glycyphagus destructor, 264
Gnat, fungus, 296
Gnorimoschema, 247
Gomphrena globosa, 57, 66, 68
Gooseberry, canker, 301
Gossypium, 267
Gossypium arboreum, 59
Gossypium barbadense, 59
Gossypium herbaceum, 59
Gossypium hirsutum, 53, 59, 60, 67, 71
leaf curl, 58–60
Gossypium peruvianum, 59
Grain, *nigrospora* disease, 315
Graminella, 180, 181, 186
Graminella albovenosa, 182
Graminella cognita, 182, 187–188
Graminella floridana, 187–188
Graminella nigrifrons, 182, 187–188
Graminella villica, 182
Grape, 297
gall, 227
gray mold, 311
Grapefruit, rot, 276
Grapevine, 380, 382, 386, 389, 391
chrome mosaic virus, 377, 381
fanleaf virus, 376, 378, 381, 394, 395, 397, 403
Pierce's disease, 9, 10
yellow vein disease, 380
Graphium ulmi, 134
Graphocephala coccinea, 308
Grapholitha molesta, 310, 316
Grass
Bahia, 187
Bermuda, 188
Eastern gama, 213
Kentucky blue, 213
orchard, 212, 213
perennial, 350
St. Augustine, 184, 185, 186, 187
Grasshopper, 168, 255
migratory, 233
twostriped, 233
Grub, white, 212, 227
Gynaikothrips flaviatennatus, 155, 156
Gynaikothrips uzeli, 155
Gynodiabrotica variabilis, 117
Gypsonoma aceriana, 274

H

Hackberry, witches' broom, 301
Haemogamasus pontiger, 264
Haemolaelaps casalis, 264
Halticus citri, 254
Hamamelis, 252
Hamamelistes spinosus, 252
Hamileia vastatrix, 158
Haplaxius crudus, 181, 182, 184–186, 188, 192
Haplothrips aculeatus, 152
Haplothrips gowdeyi, 260
Haplothrips tritici, 151
Haptonchus luteolus, 310
Helianthus annuus, 264
Helichrysum monstrosum, 63
Helicobia australis, 270
Helicomyia saliciperda, 252
Heliconia, 224
Heliothis armigera, 230
Heliothis zea, 230, 312–314, 316
Heliothrips fasciatus, 232, 275
Heliothrips femoralis, 255, 261
Heliothrips haemorrhoidalis, 180
Hellula phidilealis, 245
Helminthosporium, 295, 297
Helminthosporium sativum, 425
Helopeltis theivora theobromae, 311
Hemileia vastatrix, 306
Hercinothrips femoralis, 156, 157, 158, 255, 261
Heterocordylus malinus, 236
Heterodera humuli, 375
Heteropeza pygmaea, 272
Heterothrips azaliae, 158
Hibiscus abelmoschus, 54
Hibiscus brasifensis, 53
Hibiscus cannabinus, 46, 54, 59, 67, 71
Hibiscus esculentus, 54, 55, 59, 61, 63, 67
Hibiscus manihot, 60
Hibiscus moscheutas, 54
Hibisicus panduraeformis, 54
Hibiscus rosasinensis, 63
leaf curl of, 60
Hibiscus sabdoriffa, 46, 54, 59
Hibiscus tetraphyllus, 54

Hibiscus vitifolius, 54
Hollyhock, 58
  yellow mosaic of, 45
Hololepta yucateca, 244
Holticus citri, 168
Homalodisca insolita, 182
Homoneura bispina, 235
Homoneura indesisa, 269
Hop, nettle-head disease, 375, 403
  strain virus, 377
Hoplocampa testudinea, 241
Hordeum mosaic virus, (*HMV*), 350, 351
Hordeum, 269
Hordeum jubatum, 350
Hordeum nodosum, 332
Hordeum vulgare, 350
Hormodendrium, 157
Hornworm, tobacco, 263
Horsegram yellow mosaic, 46–47
Hortensia similis, 181, 182, 186–187
Horvea belomoreana, 194
Humerobates rostrolamellatus, 299
Humulus lupulus, 375
Hybopharsa, 180
Hylastes brunneaus, 296
Hylastinus obscurus, 276
Hylemia, 245, 248
Hylemyia cilicrura, 246, 247, 254
Hylemia lipsea, 234
Hylemyia trichodactyla, 248
Hylemya, 205, 248
Hylemya antiqua, 205, 206, 234, 247
Hylemya brassicae, 204, 227, 245, 264, 272, 274
Hylemya florilega, 204, 248
Hylemya platura, 204, 205, 212, 246–248, 254
Hylobius abieties, 296
Hylurgopinus rufipes, 136, 138
Hyophorbe vershaffeltii, 194
Hypocryphalus mangiferae, 305
Hypselonotus fulvus, 228

I

Idiocerus populi, 273
Idiocerus scurra, 273
Idioderma virescens, 184, 189
Ilnacora malina, 236
Impatients balsamina, 60
Insect, bacterial pathogen transmission, 201–226, 227–276
Insecticide, 166, 173, 190
  systemic, 6
  see also Specific compound
Ipomoea angustifolia, 50
Ipomoea batatas, 68–69
Ipomoea bonanon, 69
Ipomoea crassucaulis, 50
Ipomoea dissecta, 50
Ipomoea mosicana, 69
Ipomoea purpurea, 69
Ipomoea tiliaceae, 50
Ipomoea triloba, 50
Ips acuminatus, 302
Ips emarginatus, 302
Ips grandicollis, 302
Ips integer, 302
Ips pini, 302
Iris borer, 206, 207, 246
  soft rot, 206–207, 246

J

Jacquemontia pentantha, leaf curl, 48
Jacquemontia tamnifolia, yellow mosaic of, 48, 49
Jasmine, chlorotic ring spot of, 69
Jasminum, 67
Jasminum multiflorum, 155
Jasminum samba, 40, 69
Jatropha curcas, leaf curl of, 60–61
Jatropha gossypifolia, yellow mosaic of, 48–49
Jatropha multifida, 49
Jatropha podagrica, 49
Javesella discolor, 10
Javesella dubia, 10
Javesella obscurella, 10
Javesella pellucida, 10
Juglans, 266
Jute, yellow mosaic of, 46

K

Kaincope disease, 195
Kakothrips robustus, 151
Kladothrips acaciae, 155
Kladothrips rugosus, 155
Koptothrips zelus, 155
Krauselkrankheit, 69–70
Kribi, 195
Kurtomathrips morrieli, 158

L

Lactuca sativa, 66, 169, 186, 246, 272
Lactuca serriola, 66
Laemophloeus biguttatus, 300

Lagenocystis radicicola , 430
Laodelphax striatellus, 10
Largus succinctus, 230
Lasuis alienus, 240
Lasius niger, 240
Laspeyresia pomonella, 242, 267
Latania, 185, 194
Latente Rosettenkrankheit, 111
Lathridius minutus, 265
Launia asphemifolia, 63
Lavatera arborea, 67
Lavatera cachemiriana, 67
Lavatera cretica, 67
Leaf curl, 40
  virus disease, 57–65
  crumple virus, 60
  mottle, 89
Leafhopper, 8, 9–11, 116, 167, 172, 182, 186, 215, 216, 308, 330, 345, 350, 370, 392
  aster, 233, 251
  beet, 168
  cicadellid, 180
  potato, 238, 251, 255
  rose, 238
  white apple, 239
Leafminer, vegetable, 166–174
Lecanium, 273
Legendra mollisima, yellow vein mosaic of, 56
Lens esculenta, 62
Lentil, 67
Leonurus sibiricus, 51
Leperisinus fraxini, 262
Lepidocyrtinus, 306
Leptinotarsa decemlineata, 225, 232, 251, 263
Leptocera, 250
Leptocorisa acuta, 268
Leptoglossus appositus, 228
Leptoglossus phyllopus, 228
Leptographium, 303
Leptosphaeria coniothyrium, 300
Leptothrips mali, 158, 275
Lethal yellowing (*LY*) disease, 178–200
Lettuce, 169, 186, 389
  big vein disease, 376, 418, 424, 432
  big-vein virus, 419, 422, 424, 427
  mosaic virus, 169
  neurotic yellows, 8
  ringspot, 381
  rot, 272
  soft rot, 246
Leucas, yellow vein mosaic of, 56
Leucas cilléata, 56
Leucas stelfigera, 56
Lima bean, yellow mosaic of, 51
Lime dieback, 16
Limnobaris deplanata, 274
Limothrips cerealium, 151
Limothrips denticornis, 151
Linum grandiflorum, 66
Linum usitatissimum, 66
Liothrips, 156
Liothrips associatus, 155
Liothrips ilex, 275
Liothrips karnyi, 155
Liothrips pistachiae, 158
Liothrips ramakrishnae, 155
Liposcelis divinatorius, 265
Liriomyza, 168, 170–172
Liriomyza langei, 168
Liromyza sativae, 166, 170
Litargus balteatus, 299
Lithophane antennata, 242
Little cherry disease (*LCD*), 349–350, 351
Livistona chinensis, 185, 186, 194
Lobesia botrana, 311
Lolium multiflorum, 329, 334, 335
Lolium perenne, 332, 334, 335, 379
Lonchaea pendula, 247, 267
Lonchacea aristella, 310
Longidorus, 367, 377, 381–384, 386–391, 393, 395, 397, 402
  life cycle, 391–392
Longidorus africanus, 389, 391
Longidorus apalus, 377, 382, 386, 395
Longidorus attenuatus, 377, 381, 382, 386, 389, 397, 403
Longidorus caespiticola, 401
Longidorus elongatus, 377, 378, 381, 386–389, 391, 394–404
Longidorus leptocephalus, 401
Longidorus macrosoma, 377, 381, 382, 386, 387, 388, 394, 395, 397, 398, 400, 401
Longidorus martini, 377, 382, 386
Longidorus protae, 386
Longidorus vineacola, 390
Looper, cabbage, *266*
Lucilia, 235
Lucilia seriata, 235, 269
Luffa acutangulata, 55
Lupin, white, 67
Lupin hartwegii, leaf curl of, 61
Luteovirus, 3, 8
Lycopersicon, 231, 259, 274
  virus, 7, 160
Lycopersicon esculentum, 40, 49, 50, 58, 63, 67, 68, 94, 169
  leaf curl of, 61–63
Lycopersicon exculentum, 187

Lycopersicon glandulosum, 61
Lycopersicon hirsutum, 61, 62
Lycopersicon peruvianum, 61, 62
Lycopersion pimpinellifolium, 61, 62, 63
Lycopersicon pississi, 61
Lygaeus turcicus, 228
Lygidea mendax, 236
Lygocoris pabulinus, 236
Lygothrips jambwvasi, 155
Lygus, 217, 218
Lygus elisus, 236
Lygus hesperus, 312
Lygus invitus, 237
Lygus lineolaris, 208, 217, 218, 231, 232, 237, 246, 254, 272, 312
Lygus pratensis, 231, 237, 246, 254
Lygus rugulipensis, 264

M

MLO, *see* mycoplasma-like organisms
Maconellicoccus ugandae, 18
Macronoctua onusta, 206, 246
Macroplilium lathyroides, 51, 53
Macrosiphum euphorbiae, 2, 255
Macrosteles fascifrons, 107, 182, 186, 187, 233, 251
Magdalis, 135
Maggot
  apple, 223
  cabbage, 204, 227, 245, 264, 272, 274
  onion, 206, 234, 247
  rasberry cane, 243
  seed corn, 204, 246–248, 254
  squash root, 235
Maise, 332
Maize, *see also* corn
  chlorotic virus, 9
  chlorotic mottle virus, 120, 121, 127
  kernel red streak, 329
  mosaic, 10
  rough dwarf, 10
  stalk rot, 252
Makhorka tip chlorosis virus, 160
Malachra adeifolia, 67
Malachra capitata, 53
Malachra crispa, 67
Malacosoma neustria, 241
Malathion, 51
Malope trifida, 67
Malus, 233
Malva borealis, 67
Malva hispanica, 67
Malva nicaensis, 62
Malva parviflora, 52, 66–67
Malva rotundifolia, 67
Malva sylvestris, 45
Malvastrum coromandelianum, 56, 67, 71
Malvastrum tricuspidatum, 54, 63, 64
Malvaviscus arboreus, 59
  leaf curl of, 63
Manduca sexta, 263
Mango, malformation disease, 301, 348, 351
  wilt, 305
Manihot, 268
Manihot utilissima, 69
Marigold, 187
Mayetiola destructor, 274
Maytenus senegalensis, 155, 156
Mealworm, 314
Mealybug, 15–38, 180, 239, 259, 299, 306, 350
  citris, 307
  virus-cocoa complex, 17–34
  virus transmission, 16–34
Medicago sativa, 231
Megacoelum modestum, 270
Megadelphax sordidulus, 10
Megalurothrips distalis, 161
Megalurothrips usitatus, 260
Megamelus gracilis, 182
Melanaphis sacchari, 270
Melanoplus bivittatus, 233
Melanoplus differentialis, 233, 255
Melanoplus mexicanus, 233
Melanotus oregonensis, 234
Melanoplus sanguinipes, 233
Melica uniflora, 387
Melichares Keegani, 264
Melitara junctolineella, 244
Melitara prodenialis, 244
Melittia satyriniformis, 258
Melittomma insulare, 272
Melon aphid, 258
Membracid, 93–96
Memecylon, 152, 154, 155
Memecylon lushingtoni, 154, 155
Merremia aegyptia, 50
Merremia quinquefolia, yellow mosaic of, 49–50
Mesothrips jordani, 155
Metascarta histrio, 181
Methomyl, 173
Methoxychlor, 141
Micrococcus amylovorous, 233
Microgyne macrophylla, 155
Micrutalis, 184

Micrutalis malleifera, 94
Midge, 306, 307
  raspberry, 300
Mildew, powdery, 306, 433
Mimorista flavidissimalis, 244
Mimusops elengi, 155, 156
Minettia, 235
Mirperus scutellaris, 270
Mite, 296, 297, 307, 308, 310, 312, 315
  black currant gall, 336, 337
  cereal rust, 329
  disease vector, 328–351
  dry bulb, 328
  eriophid, 181, 326–356
  tetranychid, 365–369
  tetranynchus, 188
  virus vector interaction, 357–373
  warty grain, 264
  wheat curl, 327, 328, 329
Miticide, 333
Mixothrips craigheadi, 155
Mold, broad bean, 315
  corn, 314
  fig, 310
  gray grape, 311
  sooty
Monarthrum fasciatum, 303
Moneilema, 243
Monihot aipi, 70
Monihot glaziovii, 70
Monihot palmata, 70
Monilia, 302
Monilinia, 316
Monilinia (Sclerotinia) fructicola, 310
Monilinia fructigena, 310, 311
Monilinia laxa, 310
Monochirus multispinosus, 269
Monolepta, 122
Monolepta flaveola, 120
Monolepta harmatura, 120
Morus indica, 56, 340
Mosquito, 168
Mosaic diseases, 65–71
Moth, 212
  Argoumois grain, 316
  cactus, 276
  clearwing, 301
  codling, 218, 242, 267
  leopard, 274
  meal, 265
  oriental fruit, 310, 311, 316
  tussock, 241
Mucuna pruriens, 66, 67
Muellerianella fairmairei, 10
Mulberry, 340
  ringspot virus, 377, 382, 386
  yellow net vein virus, 56
Mung bean, 314, 426
  yellow mosaic, 46, 50–51
Musa, 224, 262, 271
Musa balbiniana, anthrocnose disease, 309
Musca domestica, 235, 250, 316
Muscina assimilis, 235
Muscina stabulans, 235
Mushroom, 272
  brown blotch, 263
Muskmelon, wilt, 208
Mycoplasma-like organisms (*MLO*) 9, 11, 180, 182, 184, 189, 191, 194–196
Mycetobia pallipes, 273
Mycodiplosis hemileiae, 306
Mycophila barnesi, 272
Mycophila speyeri, 272
Myelophilus (Blastophagus) minor, 302
Myelophilus (Blastophagus) piniperda, 296
Myrcia, 155
Myrobalan latent ringspot virus, 377
Myzus persieae, 8, 168, 232, 238, 265, 432

N

Nabis capsiformis, 270
Nabis ferus, 254
Nannorrhops ritchieana, 194
Narcissus, 380
  smoulder, 307
Nasturtium, 259
Nectria coccinea var. *faginata,* 298
Nectria cucurbitula, 299
Nectria galligena, 299
Negrospora oryzae, 313, 315
Nematicide, 403, 404
Nematocera, 165, 167
Nematocerus acerbus, 117
Nematode, 179, 180, 375–416
  control, 401–404
  hop cyst, 375
  vector, 383–392
  virus transmission, 392–401
Nematospora corylii, 313, 314
Nematospora gossypii, 312, 313, 315
Nemopoda nitidula, 250
Neoascia, 236
Neomegamelus elongatus, 182
Neotephritis finalis, 264

Nepovirus, 377–382, 385, 393, 401
  nematode vectors, 377
Nerium, 272
Nezara, 229
Nezara hilaris, 229
Nezara viridula, 229, 268
Nicandra physalvides, 52, 63, 67
Nicotiana, 259, 263
Nicotiana acuminata, 53
Nicotiana alata, 53
Nicotiana bonariensis, 53
Nicotiana clauca, 63
Nicotiana clevelandii, 53, 66, 430
Nicotiana glauca, 53
Nicotiana glutinosa, 49, 53, 58, 61, 62, 63, 66, 68, 94, 339
Nicotiana knightiana, 53
Nicotiana maritinia, 53
Nicotina paniculata, 53
Nicotiana rustica, 63, 339
Nicotiana sylvestris, 61
Nicotiana tabacum, 45, 48, 50, 53, 63, 65, 66, 67, 68, 152
  leaf curl, 48
  var. Harrison's special, 57
  var. White Burley, 56–57, 58
Nigrospora oryzae, 312
Nilaparvata lugens, 10
Nilaparvata wolcotti, 184
Nippaecoccus nipae, 180
Notostrix jamaicae, 181
Nudobius corticalis, 300
Nupedia infirma, 248
Nysius vinitor, 270

O

Oak, 433
  decline, 304
  powdery mildew, 433
  wilt, 303–305
Oat, 362
  blade blight, 203, 260
  golden stripe virus, 422, 431
  mosaic virus, 419, 422, 426
  pseudorosette, 10
  sterile dwarf, 10
Ochrovena hibiscae, 54
Odoiporus logicollis, 245
Odoiporus paganus, 262
Odontoglossum ringspot virus, *431*
Odontonema nitidum, yellow net virus of, 71
Oebalus pugnax, 314
Oecanthus angustipennis, 300
Oecanthus niveus, 300
Oil, virus transmission inhibition, 7
Okra mosaic virus, 118
Oleander, canker, 272
  scare, 273
Oleas, 261
Olive knot, 203, 220–222, 261
Olpidium, 420, 428, 429, 430, 434
Olpidium brassicae, 376, 399, 418–422, 424–434
Olpidium cucurbitacearum, 418, 419, 420, 422
Olpidium radicale, 418, 419, 420, 422, 430
Omalicna, 180, 181, 184
Onciders cingulatus, 305
Oncometopia nigricans, 182, 186
Oncopeltus fasciatus, 228
Onion, 331, 390
  maggot, 234
  mite, 328
  mosaic, 329, 334, 350, 359
  pathogen, 259
  soft rot, 202, 205–206, 247
  thrip, 232, 260, 275
  yellow dwarf virus, 367
Onychothrips tepperia, 155
Ootheca, 122
Ootheca mutabilis, 117
Ophiostoma, 302, 303
Ophiostoma Kubanicum, 304
Ophiostoma ulmi, 134
Opius dimidiatus, 166
Opuntia, 243, 276
Orangeworm, naval, 267
Orchestes canus, 233
Orchestes pallicornis, 233
Ormenaria rufifascia, 184
Orthotomicus caelatus, 302
Orthotylus flavosparus, 237
Orthotylus marginalise, 237
Oryza sativa, 252, 268
Oscinella frit, 264
Oscinella pusilla, 264
Oscinisoma, 269
Ostrinia nubilalis, 247, 255, 272
Oulema, 122, 124
Oulema dunbrodiensis, 120
Oulema lichenis, 120, 125
Oulema melanopa, 119, 120, 121, 125, 127
Ovulina azaliae, 158, 309
Oxalis barrelieri, 53
Oxycarenus gossypinus, 228
Oxydemetonmethyl, 55

P

P. madeiriensis, 29
P. tuberosa, 50
Pachodynerus erynnis, 189
Pachyrrhizus erosus, 50, 53
Paecilomyces farinosus, 51
Palm, African oil, 193
  aphid, 189
  Arikury, 194
  bud rot, 195
  Buri, 194
  Cabada, 194
  christmas, 194
  cluster fishtail, 194
  coconut, 349
    lethal yellowing, 177–200
    red ring disease, 179, 195
  date, 194
    Canary Island, 194
    Senegal, 194
    wild, 194
  fan, Chinese, 194
    Fiji Island, 194
    Thurston, 194
  hurricane, 194
  hybrid Maypan, 195
  Jamaican tall, 191, 192, 195
  Kentia, 194
  Kona, 194
  Latan, 194
  Livistona, 188
  Malayan dwarf, 189, 191, 192, 194, 195
  Mazari, 194
  oil, 296
  oil leaf spot, 307
  Palmyra, 194
  Pritchardia, 187, 189
  Puerto Rican Gaussia, 194
  seashore, 194
  spindle, 194
  sugar, 194
  Veitchia, 185, 186, 187, 188, 189, 190
  windmill, Chinese, 194
Pandanus utilis, 185
Pangola stunt, 10
Panicum, 332
Papaya, leaf curl of, 58
Parabolocratus flavidus, 182
Paracoccus, 18
Paraleyrodes, 181
Paralongidorus maximus, 401
Paraluperodes, 122
Paraluperodes quaternus, 117
Paramyelois transitella, 267
Paranthrene tabaniformis var. *intermedia*, 274
Paraphlepsius continus, 182
Paraputo anomalus, 18
Parascaptomyza pallida, 249
Parathion, 51, 55, 109, 166
Paratrichodorus, 376, 382, 384, 386, 395, 398
Paratrichodorus allius, 378, 392, 393, 394
Paratrichodorus anemones, 378
Paratrichodorus minor, 386, 389, 390, 391, 392
Paratrichodorus nanus, 378
Paratrichodorus pachydermus, 378, 386, 388, 394
Paratrichodorus porosus, 378, 386, 392
Paratrichodorus teres, 378, 388
Paratrioza cockerelli, 87
Parthenothrips dracaenae, 158
Paspalum notatum, 187
Passiflora foetida, yellow vein mosaic of, 71
Pavetta hispidula, 155, 156
Pea, early browning virus, 378, 382, 383, 394, 397
  enation mosaic, 8
  false leaf roll virus, 432, 433
  pigeon sterility, 345–346, 351
  stem necrosis virus, 422, 425
Peach, brown rot, 311
  mosaic, 342–344, 350
  phony, 9
  rosette mosaic virus, 377, 380, 381
  yellow bud mosaic virus, 380
Peanut, 297
  clump virus, 422, 423, 431
Pear, 215, 301
  Bartlett, 88, 89
  blight, 233
  decline, 88–89
  fire blight, 214
  Magness, 89
  prickley, 243
  prickly rot, 276
  psylla, 216, 239
Pectinophora gossypiella, 312
Pediculopsis, 315
Pegomya, 234
Pegomya fusciceps, 248
Pegomya rubivora, 218, 243
Pegomya rumicifoliae, 234
Pegomyia calyptera, 234
Pelargonium hortorum, 268
Pelomyia cruciata, 250

Pencillium, 157, 297, 313, 316
Pentapleura, 251
Pentatoma ligata, 229
Pentatoma sayi, 229
Pentopleurs, 251
Pepper, 187, 383
  bell, 160
  soft rot, 246
Peppermint, "die-back," 389
Peregrinus maidis, 10
Periwinkle, 186, 189, 190, 389
  leaf curl of, 71
  Madagascar, 88
Perkinsiella saccharicida, 10, 270
Perkinsiella vastatrix, 10
Permethrin, 173
Persimmon wilt, 304–305
Pestalotiopsis gladicola, 307
Pestalotiopsis palmarum, 307
Pestalozzia, 157
Petunia, 168
Petunia, 59, 68
Petunia hybrida, 58, 63, 168
Phaedon, 122
Phaedon cochleariae, 118, 121
Phaenicia sericata, 235, 269
Phalacrus, 254
Phaonia variegata, 236
Phaseolus, 260, 261, 268
Phaseolus aborigineus, 50, 51, 53
  leaf curl, 48
Phaseolus aconitifolius, 46, 47, 50, 53
  leaf curl, 48
Phaseolus atropurpureus, 46, 47
Phaseolus aureus, 46, 47, 50, 52
Phaseolus coccineus, 51, 53
  leaf curl, 48
Phaseolus lathyroides, 46, 47, 50, 52, 125, 157
Phaseolus longependunculatus, 52
  yellow mosaic of, 51
Phaseolus lunatus, 40, 46, 47, 50, 51, 52, 53
  yellow mosaic, 51–52
Phaseolus mungo, 46, 47, 50, 52
Phaseolus trichocarpus, 53
Phaseolus vulgaris, 46, 47, 48, 50, 52, 53, 62, 67, 70, 94
  leaf curl, 48
Pheidole megacephala, 311
Phenax sonneratii, 67
Philaenus spumarius, 238
Phlegethontius sexta, 259
Phlepsius irroratus, 255
Phleum green stripe, 10
  mottle virus, 120, 121, 124, 127
Phoenix canariensis, 194
Phoenix dactylifera, 194
Phoenix reclinata, 185, 194
Phoenix sylvestris, 194
Pheromone, 142
  aggregating, 137, 140
  feeding, 140
Phomopsis, 301
Phorate, 55
Pharbia fusciceps, 248
Phorbia brassicae, 274
Phorbia rubivora, 243
Phormia, 248
Phrasterothrips conducans, 155
Phrydiuchus topiarius, 276
Phyllanthus niruri, 59
Phyllocnistis citrella, 266
Phyllocnistis saligna, 266
Phyllocoptes, 326
Phyllocoptes abaenus, 350
Phyllocoptes fructiphilus, 346–347
Phyllophaga, 212, 254
Phyllotreta, 118, 122
Phyllotreta atra, 120
Phyllotreta nigripes, 120
Phyllotreta striolata, 117
Phyllotreta undulata, 120
Physalis, 94
Physalis ixocarpa, 66
Physalis peruviana, 45, 63, 68
Physalis wrightii, 66
Physalospora zeae, 313
Physothrips basicornis, 152
Phytocoris ulmi, 237
Phytomonas campestris, 265
Phytomonas juglandis, 266
Phytomonas manihotis, 268
Phytomonas melophthora, 261
Phytomonas michiganensis, 231
Phytomonas rubrilineans, 269
Phytomonas sepedonica, 232
Phytomonas solanacearum, 262, 263
Phytomonas stewarti, 252
Phytomonas tabaci, 263
Phytomyza minuscula, 171
Phytophaga destructor, 274, 295
Phytophora, 54
Phytophthora, 179
Phytophthora palmivora, 311
Phytoptus, 326, 336
Phytoptus inaequalis, 344–345
Phytoptus insidiosus, 342–344
Pierce's disease agent, (PDA), 10, 11
Pieris rapae, 266

Piesma, 97
Piesma audaratum, 101
Piesma capitatum, 98
Piesma cinereum, 97, 98, 110
Piesma maculatum, 98
Piesma quadrata, 97
Piesma quadratum, 97–102, 104–112
Piesmid, 8
Piezodorus purus, 229
Pigeon pea, yellow mosaic of, 46
Pigweed bug, 110
Pine, blister rust canker, 302
  burn blight, 299
  canker, 302
  lodgepale blister rust canker, 302
  red, 299
  red needle blight, 307
  Scots, 299
  zirbelkiefer, 231
Pineapple, mealybug wilt, 16
Pintalia, 182
Pinus cembra, 231
Piper, 154
Piper nigrum, 155
Pissonotus piceus, 182
Pistachio virus, 158
Pithyophthorus cariniceps, 302
Pithyophthorus pulchellus, 302
Plagiognathus politus, 237
Plane tree, stem canker, 300
Planicephalus flavicosta, 182
Planococcoides njalensis, 17–31
Planococcus citri, 16, 17, 18, 20, 23, 24, 28, 30, 31
Planococcus kenyae, 18
Planococcus lilacinus, 18
Plant lice, jumping, 87–91
Plant pathogen, bacterial, 201–226, 227–276
  fungal, 293–324
Planthopper, 8, 11, 167, 172, 182
Plantain, wilt, 262
Plasmodiophorid, 420–421, 425, 429, 431, 434
Platanus, 300
Platypedia areolata, 227
Platyedra gossypiella, 230
Pleopeltis pteropus, 152
Pleotrachelus, 420
Plum curculio, 310
Plusia brassicae, 266
Plutella maculipennis, 246, 266
Plutella xylostella, 246, 266
Poa pratensis, 213
Podagrica, 122
Podagrica sjostedti, 118
Podagrica uniforma, 118
Poeciloscytus basalis, 237
Polanisia viscosa, 64
Polistes fuscatus, 188
Pollenia rudis, 235
Polybia, 225, 262
Polybia occidentalis, 309
Polymis obtecta, 182
Polymyxa, 421, 431
Polymyxa betae, 419, 420, 421, 422
Polymyxa graminis, 376, 419–422, 425, 426, 429, 431, 434
Polyporus, 297
Poplar, bacterial canker, 273
  borer, 252
Populus, 273
Populus tremoloides, 299
Poria, 296
Portulaca, 160
Potato, 160, 247, 383, 385, 432
  Andean latent virus, 118
  Andean mottle virus, 117
  aphid, 255
  blackleg, 202, 203–205, 247
  black ringspot virus, 382
  bouquet, 381
  brown rot, 224
  flea beetle, 258
  leaf curl of, 65
  leafhopper, 238, 251, 255
  leafroll, 8
  mop-top virus, 419, 422, 423, 431
  necrosis of, 70
  psyllid yellows, 87
  ring rot, 203, 232
  seed piece decay, 247
  spraing disease, 383
  virus, (PVY), 339, 348, 350, 365
  virus X, 367, 370, 432
  wilt, 263
  yellow vein mosaic of, 71
Potyvirus, 3
Prenolepis imparis imparis, 240
Pritchardia, 186
Pritchardia affinis, 185, 188, 194
Pritchardia pacifica, 194
Pritchardia remota, 185
Pritchardia thurstonii, 185, 194
Proarna hilaris, 181
Prosapia biceneta, 184
Protalebrella brasiliensis, 182
Protoparce sexta, 263
Protophormia terraenovae, 249

Prunus, 342, 348
latent virus, 347–348, 351
necrotic ringspot virus, 377, 380, 401
Prunus americana, 343
Prunus angustifolia, 343
Prunus avium, 343, 349
Prunus besseyi, 343
Prunus cerasus, 343, 349
Prunus emarginata, 344, 345, 349
Prunus japonica, 343
Prunus munsoniana, 343
Prunus serrulata, 349
Prunus tomentosa, 343
Psallus delicatus, 229
Psenoflata brevis, 180, 181
Pseudanaphothrips achaetus, 260
Pseudococcus, 18, 180
Pseudococcus adonidum, 180
Pseudococcus citri, 297, 307
Pseudococcus comstocki, 18, 29, 297
Pseudococcus concavocerarii, 18
Pseudococcus deceptor, 297
Pseudococcus hargreavesi, 18, 30
Pseudococcus longispinus, 17, 18, 30, 34
Pseudococcus maritimus, 16
Pseudococcus nipae, 180
Pseudococcus njalensis, 299
Pseudocurly top disease, (PCT), 93–96
Pseudomonas, 221, 264
Pseudomonas angulatum, 259
Pseudomonas aptata, 259
Pseudomonas avenae, 203, 228, 260
Pseudomonas campestris, 265
Pseudomonas cerasi, 260
Pseudomonas citri, 266
Pseudomonas flectans, 157, 203, 260
Pseudomonas juglandis, 266
Pseudomonas lachrymans, 260
Pseudomonas maculicola, 203, 261
Pseudomonas marginata, 261
Pseudomonas medicaginis var. *Phaseolicola*, 157
Pseudomonas melophthora, 203, 222, 223, 224, 261
Pseudomonas mutabilis, 221
Pseudomonas phaseolicola, 261
Pseudomonas saliciperda, 252, 261
Pseudomonas savastanoi, 168, 203, 220–222, 261
Pseudomonas savastanoi var. *fraxini*, 262
Pseudomonas solanacearum, 224, 225, 262, 263
Pseudomonas tabaci, 263
Pseudomonas tolaasii, 263
Pseudomonas tumefaciens, 227
Pseudopityophthorus minutis-simus, 303
Pseudopityophthrous pruinosus, 303
Pseudoxiphydria belulae, 301
Psila rosae, 247
Psophocarpus tetragonolobus, 46
Psorosticha zizyphi, 266
Psylla pyri, 87, 88
Psylla pyricola, 87, 216, 239
Psylla pyrisuga, 87, 88
Psylla simulans, 239
Psyllid, 167
yellows, potatoe, 87
Psyllidae, 87–91
carrot proliferation disease, 90
citrus greening disease, 89
pear decline, 88–89
Psylliodes, 118, 122
Pterocomma populeum, 273
Puccinia graminis, 158
Puccinia graminis tritici, 157
Puccinia puncti, 307
Puccinia rubigo-vera tritici, 157
Pullularia pullulans, 307, 315
Pumpkin, wilt, 208
yellow vein mosaic, 55
Pycnostyanus (Sporocybe) azaleae, 308
Pygospina spintat, 184
Pyralis farinalis, 265
Pyrausta nubilalis, 247, 255, 272, 297, 313, 316
Pyroderces simplex, 230
Pyrus, 233
Pyrus communis, 89
Pyrus ussuriensis, 88, 89
Pythium, 54, 295, 296, 432, 434
Pythium graminicola, 295
Pythium myriotylum, 295
Pythium ultimum, 432, 433

Q

Quail pea mosaic virus, 117
Quedius raevigatus, 300

R

RNA virus, 116
Rabbit, myxoma virus, 168
Radish mosaic virus, 117
Raspberry, 218, 300, 379, 380, 385, 388, 389
fire blight, 243

gall, 227
leaf-curl disease, 381
ringspot virus, 377, 378, 381, 394, 395, 397–402, 404
yellow dwarf disease, 380
Red clover mottle virus, 117
Reticulitermes flavipes, 301
Rhabdophaga, 252
Rhabdophaga saliciperda, 252
Rhabdoscelus obscurus, 269
Rhabdovirus, 3, 8
Rhagoletis completa, 267
Rhagoletis pomonella, 222, 223, 224, 261, 316
Rhaphanus sativus, 63
Rhipiphorothrips cruentatus, 151
Rhizoglyphus, 261
Rhizoglyphus echinopus, 261, 296
Rhizoglyphus solani, 261
Rhizophagus brunneus, 300
Rhizophagus ferrugineus, 296
Rhizopus, 157, 316
Rhizopus stolonifer, 312, 313
Rhododendron, bud-blast disease, 308
Rhopalosiphum fitchii, 238
Rhopalosiphum prunifolium, 238
Rhubarb, 381
Rhynchaenus canus, 233
Rhynchaenus pallicornis, 233
Rhynchosia minima, 48, 67
mosaic of, 46
yellow mosaic of, 52–53
Rhynchosia reticulata, 53
Ribes, 301, 340, 348
Ribes grossularia, 64
Ribes nigrum, 337, 340
Rice, black rot, 252
black-streaked dwarf, 10
blight, 268
grassy stunt, 10, 11
hoja blanca, 10
necrosis mosaic virus, 422, 431
nematospora infection, 314
stinkbug, 314
stripe, 10
tungro virus, 9
weevil, 264, 275
yellow mottle virus, 120, 124
Ricinus communis, 269
Root, tip galling, 389, 390
Rootworm, corn, 253
Rosa, 53
Rosa bourbiana, yellow vein mosaic of, 57
Rosa eglantera, 346, 347
Rosa multiflora, 346, 347
Rose, 380, 385, 389
leafhopper, 238
rosette, 346–347, 351
yellow mosaic of, 53
Rot
apple, 261
celery heart, 207–208
cereal, 274
cocoa pod, 311
coconut bud, 259
coffee bean, 313
cotton ball, 227, 231, 259, 268, 312–313
fig, 309–310
grapefruit, 276
grass, 274
soft iris, 206–207
soft onion, 205–206
soft, potato, 203–205
softwood white, 302
stone fruit brown, 310–311
Rotylenchulus, 180
Rubber tree, *Alternaria* blight, 307
Rubiaceae, 152
Rubus idaeus, 243
Rust, 433
coffee, 306
thistle, 307
Ryegrass, 334–335
mosaic virus, (RMV), 329, 334–335, 350, 359, 362

S

Sabal, 182
Sabal palmetto, 182
Saccharomyces pini, 302
Saccharum officinarum, 269
Sage, 276
Sahlbergella singularis, 298
Saint Paul wilt, 195
Salat, 110
Salix, 252
Salmonella montevideo, 264
Salsola, 110
Salvia sclarea, 276
Salvia splendens, 53
yellow vein mosaic of, 57
Sancassania, 297
Sanctanua fasiatus, 182
Sap, sour, 233
Saperda carcharias, 273
Saperda populnea, 273

Saperda tridentata, 135
Saponaria vaccaria, 65
Sapromyza bispina, 235
Sarcophaga aurifrons, 270
Sarcophaga carnaria, 220, 236
Satellite virus, 419, 421, 422, 423
Sathrobrota simplex, 230
Sawfly, European apple, 241
Scab, *Gladiolus*, 261
Scabbia, 296
Scale insect, 298, 306
Scaptomyza graminum, 168, 207, 208, 246
Scaptomyza pallida, 249
Scatophaga furcata, 248
Scatophaga squalida, 248
Scatophaga stercoraria, 234, 248, 299
Scatopse, 250
Scatopse fuscipes, 250
Schedothrips Orientalis, 155
Schefflera, 152, 154
Schefflera divaricata, 152
Schefflera elliptica, 152
Schefflera lucescens, 152
Schefflera polybotrya, 152
Schefflera racemosa, 153, 155, 156
Schefflera scandens, 152
Schizanthus, 63, 64
Schizoneura lanigera, 238
Schizophora, 166
Sciapteron tabaniforme, 274
Sciaria tribentata, 250
Sciaria tridentata, 250
Sciaridae, 296
Scirtothrips aurantii, 151
Scirtothrips bispinosus, 151
Scirtothrips cardamomi, 150
Scirtothrips dorsalis, 151, 161
  leaf curl vector, 57
Scirtothrips manihoti, 158
Scleroderis lagerbergii, 302
Sclerotinia, 316
Sclerotinia narcissicola, 307
Sclerotium rolfsii, 296
Scoliophthalmus, 269
Scolocenus, 181
Scolytus laevis, 135
Scolytus multistriatus, 135, 136, 138
Scolytus pijri, 234
Scolytus rugulosus, 218, 234, 299
Scolytus scolytus, 135, 136
Scoparia dulcis, 63
Scopeuma stericorarium, 234
Scrophularia mottle virus, 118
Scyophophorus interstitialis, 298
Sejus pomi, 275
Selenothrips rubrocinctus, 151
Senecio vulgaris, 66
Sephena cinerea, 216, 238
Sepsis, 250
Septoria tritici, 425
Sesamum indicum, 63
Sesamum orientale, leaf curl of, 65
Sesselia, 122
Sesselia pusilla, 120
Setaria, 332
Shope's rabbit papillomavirus, 168
Sida acuminata, 53
Sida acuta var. *carpinifolia*, 67
Sida aggregata, 53
Sida carpinifolia, 63, 64, 67
  mosaic, 51
  yellow mosaic of, 53
Sida cordifolia, 53, 59, 63, 67
Sida humilis, 53
Sida glabra, 53
Sida glomerata, 53, 67
Sida linifolia, 67
Sida micrantha, 52, 67
Sida napaea, 67
Sida procumbens, 53
Sida rhombifolia, 52, 53, 63, 67
Sida urens, 53, 59, 67
Sida veronicaefolia, 63
Silene vulgaris, 152
Siphocoryne avenae, 238
Sirex cyaneus, 302
Sirex juvencus, 302
Sirex noctilio, 302
Sisal, bole rot, 298
Siteroptes graminum, 308
Siteroptes reniformis, 312
Sitobion avenae, 216, 238
Sitodiplosis mosellana, 275
Sitona, 122, 296
Sitona hispidula, 231, 296
Sitona lineatus, 117, 124
Sitophilus granarius, 265, 316
Sitophilus oryzae, 264, 275, 316
Sitophilus zeamais, 314
Sitotroga cerealella, 316
Slictocephala bubalus, 239
Smut, carnation anther, 308
  fig, 310
Soapwart, leaf curl of, 65
Softwood, white rot, 302
Sogatella furcifera, 10

Sogatella kolophon, 10
Sogatella vibix, 10
Sogatodes cubanus, 10, 182
Sogatodes kolophon meridiana, 182
Sogatodes molinus, 182
Sogatodes orizicola, 10
Soil, fungal plant pathogen, 295–297
Soil-borne virus, 375–416
  control, 401–404
Solanaceae, wilt, 263
Solanum demissum, 339
Solanum dulcamara, 66
Solanum esculentum, 52
Solanum gracile, 94
Solanum melongena, 63, 94
Solanum nigrum, 63, 385
Solanum tuberosum, 65, 67, 70, 71, 232, 247
Solenopsis, 251, 263
Sonchus yellow net virus, 8
Sonchus oleraceus, 66
Sorghum arundinaceumn, 431
Sowbane, 168
  virus, (SMV), 168, 173
Sowthistle yellow vein, 8
Soybean, 123, 296
  crinkle mosaic, 58
  yeast spot, 314–315
  yellow mosaic of, 48
Spergula arvensis, 66
Sphaerotheca pannosa, 306
Sphaerotheca phytoptophila, 301
Spangbergiella vulnerata, 181, 182, 186, 187
Sphenophorus parvulus, 274
Spilostethus (Lygaeus) pandurus, 314
Spinach, 111, 160
  soft rot, 246
Spinacea oleracea, 66, 102, 111, 246
Spinacia tetranda, 102
Spinacia turkestanica, 102
Spiroplasma, 9, 191
Spissistilus fistinus, 296
Spittle bug, 11
Spodoptera frugiperda, 271
Spongospora subterranea, 420, 421, 422, 428, 434
Spruce, canker, 302
Squash, 169
  butternut, 121
  mosaic virus, 117, 123, 126
  vine borer, 258
  wilt, 208
  zucchini, 173
Stachytarpheta jamaicensis, 63
Stellaria media, 94, 152
Stenotaphrum secundatum, 184
Steneotarsonemus laticeps, 307
Sterculia tragacantha, 19
Stereum sanguinolentum, 302
Stilbus apicalis, 254
Stilbus viduus, 254
Stinkbug, 314–315
Stirellus bicolor, 255
Stizolobium, 67
Storey's leaf curl, 16
Stragania robusta, 182
Strawberry, 385, 389
  crinkle, 8
  latent ringspot virus, 377, 380, 381, 394, 395, 399
  yellow crinkle, 380
Streptomyces, 264, 275
Streptomycin, 222
Strobaera, 184
Stromatinia, 296
Sufelula sunidesalis, 296
Sugar beet, 101, 102, 169, 274, 380, 389, *see also* Beet
  bacterial disease, 259
  Docking disorder, 388, 403
  leaf curl virus, 8
  pseudocurly top, 94
  savoy virus, 97, 98, 110, 112
Sugarcane, 295
  chlorotic streak, 345, 351
  delphacid, 270
  Fiji disease, 10
  gumming disease, 269
  mosaic virus, 422
  red rot, 298
  top rot, 269
  weevil, 269
Sunflower, capitulum decay, 264
  mosaic virus, 158
Sweet potato
  mosaic, 45, 68
  vein clearing virus, 69
  yellow dwarf, 45, 68
Syagrus, 122
Syagrus calcaratus, 118
Sylvicola fenestralis, 273
Synanthedon myopaeformis, 301
Synanthedon tipuliformis, 301
Synchtrium endobioticum, 432
Synedrella nodiflora, 63, 64, 68
Synoeca surinama, 309
Syrphus, 236

Systena, 117, 122
Systena blanda, 254
Systena frontalis, 120
Systena taeniata blanda, 254

T

Taeniothrips atratus, 152
Taeniothrips firmus, 152
Taeniothrips inconsequens, 151, 157
Taemiothrips meridionalis, 151
Taeniothrips nicipes, 152
Taeniothrips nigricornis, 157, 260
Taeniothrips pini, 152
Taeniothrips xanthoceros, 158
Tagetes patula, 187
Taraxacum officienale, 66
Tarophagus proserpina, 10
Tarsonemus, 264, 299
Teramnus uncinatus, 46, 52
Terminalia chebula, 153, 155, 156
Termite, 301
Terthron albovittatus, 10
Tetragonia tetragonoides, 102
Tetranychus telarius, 272
Tetranynchus tumidus, 188
Tetranychus urticae, 365, 367, 370, 371
Teuchothrips longus, 155, 156
Texananus, 182
Thamnotettix nigrifrons, 255
Thaumatothrips froggatti, 155
Themira minor, 250
Themira putris, 250
Theobroma cacao, 298
Thevetia nerrifolia, leaf curl of, 71
Thilakothrips babuli, 155
Thistle rust, 307
Thizopus nigricans, 312
Thomasianna theobaldi, 300
Thrip, 149–164, 167, 180, 215, 218, 309, 310, 315, 330, 345
  banded greenhouse, 255, 261
  bean, 232
  black flower, 260
  flower, 275
  gall, 150–156
  grass, 255
  infestation effect, 151
  onion, 157, 232
  pear, 242
Thrips angusticeps, 151
Thrips bremneri, 275
Thrips imaginis, 151
Thrips madroni, 151
Thrips nigripilosus, 158
Thrips tabaci, 151, 152, 157–161, 232, 260, 275
Thunbergia fragrens, 152
Thyanta custator, 229, 254, 314
Thysanoptera, 149, 315
Tmetothrips subapterus, 152
Tobacco, 48, 49
  angular leaf spot, 259
  etch virus, 169
  Granville wilt, 224–225, 263
  hornworm, 263
  leaf curl, 49, 55, 61, 63–64, 70
    vector, 41–45
  mosaic virus, (TMV), 16, 127, 168, 169, 367, 370, 422, 423, 431–434
  powdery mildew, 306
  rattle virus, 378, 382, 383, 393, 394, 395, 397, 398, 402, 403
  ringspot virus, 120, 127, 159, 367, 377, 378, 380–382, 394–398
  stunt disease, 418, 424, 425, 429
  wildfire, 263
  yellow net virus, 56–57
Tobravirus, 376, 378, 382–383, 386, 393, 397, 401, 403
Toeniothrips inconsequens, 242
Tolmetothrips granti, 153
Tomaspie inca, 307
Tomato, 49, 57, 58, 67, 160, 169, 187, 274, 385
  black ring virus, 377, 381, 382, 394, 395, 397, 398, 401, 402
  bronzing virus, 160
  brown rot, 224
  bushy stunt virus, 422, 429, 430, 433
  bushy virus, 367
  canker, 231
  enation leaf curl, 62–63
  fruitworm, 230
  leaf curl of, 61–62
  pathogen, 259
  pseudocurly top, 94
  ringspot virus, 377, 378, 380, 381, 393, 397
  spotted leaf virus, (TSWV), 158, 159, 160, 161
  wilt, 267
  yellow leaf curl of, 62
  yellow mosaic, 49
  yellow top virus of, 49
Toxaphene, 166
Trachycarpus fortunei, 194

Tree canker disease, 298–303
Treehopper, 93, 94
  Buffalo, 239
Trialeurodes abutilonea, 68
Trialeurodes abutilonia, 41, 45, 60
Trialeurodes lubia, 68
Trileurodes natalensis, 41, 45
Trialeurodes ricini, 68
Trialeurodes vaporariorum, 65, 268
Trialeurodes varporalorum, 41, 45
Tribolium castaneum, 265
Tribolium confusum, 275, 314, 316
Trichispa, 122
Trichispa sericea, 120
Trichocera, 251
Trichoderma viride, 313
Trichodorus, 376, 382, 383, 384, 386, 388, 394, 395, 397, 398
Trichodorus anemones, 398
Trichodorus cylindricus, 378
Trichodorus primitivus, 378, 386
Trichodorus similis, 378, 391
Trichodorus sparsus, 378
Trichodorus viruliferus, 378, 383, 391
Trichoplusia ni, 266, 312
Trichosporium tingens, 302
Trichosporium tengens var. *macrosporum*, 302
Trifolium pratense, 276
Trigona, 225, 262, 309
Trigona corvina, 262
Trigona corvina, 224
Trihithrum inscripta, 271
Trioza nigricornis, 88, 90
Tripsacum dactyloides, 213
Trirhithrum coffeae, 271
Triticum, 275, 333
Triticum aestivum, 213
Tritoxa flexa, 205, 247
Triumfetta lappula, 67
Trochilium apiforme, 274
Tropaeolum, 259
Troza erytrea, 87, 89
Trypanosome, 168
Tulareamia, 168
Tulip, mite, 328
Turnip yellow mosaic virus, 118, 121
Tydeus interruptus, 265
Tylococcus westwoodi, 18
Tylozygus bifidus, 182
Tylozygus fasciatus, 180, 181
Tylozygus geometricus, 182
Typhlocyba, 180, 182
Typhlocyba pomaria, 239
Tryophagus, 297

U

Ukanodes albifascia, 10
Ukanodes sapporona, 10
Ulmus americana, 381
Ulmus thomasii, 136
Uncinula necator, 306
Urd bean, 50
Urena lobata, 59
Urocerus gigas, 302
Urtica californica, 66
Ustilago maydis, 158
Ustilago violacea, 308

V

Vasates dubius, 334
Veitchia merrillii, 185, 194
Ventilago maderaspantana, 155
Veronica chamaedrys, 152
Vernonia, 55
Vernonia cinerea, 63, 64
Vespa, 251
Vespula, 241
Vicia cracca, 152
Vigna, 268
Vigna aconitifolia, 53
Vigna acontifolius leaf curl, 48
Vigna angularis, 53
  leaf curl, 48
Vigna radiata, 51
Vigna sinensis, 47
  yellow mosaic, 53
Vigna sinensis var. *sesquipedalis*, 63
Vinca rosea, 88
  leaf curl of, 71
Vine panachure, 348, 351
Virus, 149, 158–161, 167, 168, 179
  aphid-transmitted, 424
  beetle transmitted, 116–121
  bromovirus, 118–119, 121
  circulative, 3, 6, 8, 33–34
  comovirus, 116–118
  filamentous, 424, 425, 434
  fungus-transmitted, 417–438
  isometric, 421–423, 425, 427–430, 434
  leafhopper-borne, 10
  mealy bug vector, 16–34
  mite transmitted, 326–356, 370–372, 424
  nematode transmitted, 376–383, 392–404

noncirculative, 3–8
potyvirus, 424, 432, 433, 434
reovirus type 3, 168
soil borne, 375–416, 417–433
straight tubular, 423, 425, 429, 431
stylet-borne, 33–34
thrip born, 158–161
tobamovirus group, 423, 432, 433, 434
tombusvirus Group, 434
tymovirus, 118
Vitis, 376, 385
Vitis labrusca, 380
Volucella, 244

W

Walnut blight, 203, 266
Wasp, 215, 224, 225, 241
fig, 310
Watermelon mosaic virus, (WMV-1,2), 169, 173
Weevil, 122, 135, 307
apple flea, 233
boll, 312
granary, 265, 316
rice, 264, 275, 314, 316
sisal, 298
sugarcane, 269, 298
Wheat, 213, 264, 275, 335, 336, 362
curl mite, 327, 328, 329
European, striate mosaic, 10
midge, 275
mosaic virus, 419, 422, 423, 424, 425, 426, 431
Nigrospora disease, 315
rosette virus, 425
rust, 157
spindle streak mosaic virus, 419, 422, 431
spot chlorrsis, 333
spot mosaic, (WSpM), 333–334, 350
streak mosaic, (WSMV), 326, 329, 336, 350, 359 360, 362 363, 370, 371, 424
striate mosaic virus, 330
wireworm, 212, 254
yellow mosaic virus, 419, 422, 431
Whitefly, 39–85, 167, 180, 181, 202
control, 51, 55
greenhouse, 268
mosaic disease vector, 65–71
sweet potato, 268, 269
yellow mosaic vector, 45–57
Wild cucumber mosaic virus, 118
Willow, borer, 252
canker, 252, 261
gall, 227
watermark disease, 252, 261
Wilt
banana, 262
cucurbit, 202, 208–211, 257
mango, 305
oak, 303–304
persimmon, 304–305
solanaceae, 263
Southern bacterial, 224–225
Stewart's corn, 211, 252
vascular, 303–305
Wireworm, 227
wheat, 212, 254
Wissadula amplissima, mosaic virus of, 70–71
Witch hazel, aphid gall, 203, 252
Withania somnifera, 63, 64
Woodwasp, 302
Worm, cabbage, 266
canker, 241
carpenter, 135
corn root, 295

X

Xanthium strumarium, 50
Xanthomonas, 203, 271
Xanthomonas campestris, 202, 203, 265
Xanthomonas citri, 203, 266
Xanthomonas juglandis, 203, 266
Xanthomonas malvacearum, 229, 267, 268
Xanthomonas manihotis, 268
Xanthomonas oryzae, 268
Xanthomonas pelargonii, 268
Xanthomonas phaseoli, 202, 268
Xanthomonas phaseoli var. *fuscans*, 268
Xanthomonas ricini, 269
Xanthomonas rubrilineans, 269
Xanthomonas translucens, 269
Xanthomonas vasculorum, 269
Xerophloea, 181
Xerophloea viridis, 182
Xestocephalus, 180
Xestocephalus tessellatus, 182
Xiphinema, 376, 382, 383, 384, 386, 387, 388, 389, 390, 391, 393, 394, 395, 397, 402
life cycle, 391–392
Xiphinema americanum, 377, 378, 380, 384, 385, 391, 393, 395, 397
Xiphinema americanum sensu lato, 378

Xiphinema bakiri, 377, 386, 389
Xiphinema basiri, 377, 386
Xiphinema brevicolle, 377, 386
Xiphinema californicum, 378
Xiphinema coxi, 377, 381, 385, 387, 398
Xiphinema diversicaudatum, 375, 377, 379, 380, 381, 382, 385, 387, 388, 389, 390, 393, 394, 395, 397, 398, 399, 400, 401, 403
Xiphinema index, 376, 377, 378, 381, 385, 387, 389, 390, 391, 393, 394, 395, 398, 403
Xiphinema italae, 377, 385
Xiphinema mediterranean, 386
Xiphinema pachtaicum, 385, 386, 387
Xiphinema vuittenezi, 377, 381, 386, 387
Xyleborus affinis, 305
Xylobiops basilaris, 304

Y

Yellow bud mosaic virus, 380
Yellow mosaic disease, 40, 45–57
Yellow vein mosaic, 40

Z

Zakuklivanie ovsa, 10
Zea mays, 110, 188, 211, 252, 271, 272
Zelkova, 139
Zelkova serrata, 139
Zeugera pyrina, 135, 274
Zinnia, 56, 58
Zinnia elegans, 54, 63, 66
  leaf curl of, 65
  yellow net disease of, 57